The Provinces of the Roman Empire

P9-EEA-953

Pannonia and Upper Moesia

Already published

BRITANNIA
Sheppard Frere

DALMATIA
J. J. Wilkes

NORICUM
Géza Alföldy

THE PROVINCES OF THE ROMAN EMPIRE

Edited by
Glen Bowersock
Professor of Greek and Latin, Harvard University

Sheppard Frere
Professor of the Archaeology of the Roman Empire
in the University of Oxford

A. L. F. Rivet
Reader in Romano-British Studies, University of Keele

John Ward-Perkins
Director, British School at Rome

Graham Webster
Reader in Archaeology, University of Birmingham

J. J. Wilkes
Lecturer in Roman History, University of Birmingham

FRONTISPIECE Late Roman officer's gilt parade-helmet (no. 1) from Berkasovo
(p. 337)

Pannonia and Upper Moesia

A History of the Middle Danube
Provinces of the Roman Empire

András Mócsy

Professor of Archaeology
University of Budapest

Translation edited by
Sheppard Frere

LONDON AND BOSTON
ROUTLEDGE & KEGAN PAUL

First published in 1974
by Routledge & Kegan Paul Ltd
Broadway House, 68–74 Carter Lane
London EC4V 5EL and
9 Park Street
Boston, Mass. 02108, USA
Printed in Great Britain by
Richard Clay (The Chaucer Press) Ltd
Bungay, Suffolk

© András Mócsy 1974
No part of this book may be reproduced in
any form without permission from the
publisher, except for the quotation of brief
passages in criticism

ISBN 0 7100 7714 9

DG
59
.D43
M613
1974

39,665

MATRI CONIVGI PATRI
CARISSIMIS

CAMROSE LUTHERAN COLLEGE
LIBRARY

Contents

Plates

ix

Plates

x

Plates

Figures

Figures

Preface

There are various ways in which the history of a Roman province may be written. In preparing this book I was not much attracted to a routine description, subject by subject, of the history of events or wars, administration, economic life, of town and country, arts, religion and so on. One of the chief problems in the study of the history of the Roman empire is that research relating to the empire as a whole is not organically interconnected with that relating to its parts. A daunting gap separates the study of central Roman imperial history from local, often highly developed, archaeological research. This gap may be bridged only by the use of a method which explores every aspect period by period and in accordance with historical principles. This method I have chosen despite the fact that the results of local research at present available are not as uniformly useful as might be desired.

The reader will certainly notice that I have not always been able to apply my principles equally in both the provinces assigned to me. It would have been ideal for my purpose had research been conducted in Pannonia and in Upper Moesia with identical aims and with equal completeness. But such studies are still decisively influenced by the history of the last hundred years, and even in Pannonia itself the desired uniformity has not yet been achieved. Even now, the estimable efforts of several generations of Austrian, Hungarian, Jugoslav and Czechoslovak scholars have scarcely achieved the adoption of generally agreed methods and principles of study and field-work.

Nevertheless, many attempts at individual synthesis are available: I was able to base my studies on much previous work of great value. If in my footnotes and in the select bibliography I have failed to give them the prominence they deserve, the omission is not due to ingratitude but to the universal fate of scholarly studies, which this book in its turn will not escape.

In my treatment of Upper Moesia I have had to face fundamental difficulties and lack of information. In spite of the encouraging growth of Serbian research today, there is still so much work to be undertaken that I have been compelled to generalize from all too few data, work how I would. Nor must it be overlooked that, owing to the very fact that Upper Moesia is only half the size of Pannonia, less space could here be devoted to the former; moreover, for much of the period its Danube frontier was of minor political and military importance.

A further decision of method had to be made: were the two provinces to be treated separately in two parts under one cover, or parallel and together in a comparative study as a subdivision of Illyricum? I have chosen the second method, for it seemed to me that the history of the two provinces is complementary. Even if, on the larger stage of the Roman empire, they fail to play the part of an intelligible unit of history, in their own area they enjoy many close connections.

I had also to solve the problem where to sever the threads connecting the history of our provinces with that of the empire as a whole. Some of my readers will no doubt feel that I have cut them off too short, while others will miss the amplification of local detail which might be expected to form the strength of a good provincial monograph. But it would be pretentious to claim that the present volume is a reference book: its more modest aim is to offer a synthesis of what I myself have learnt—and in part developed. It must be left to the reader to decide whether the synthesis offered is premature. I believe, however, that any serious student of Roman provincial archaeology must sooner or later write his monograph.

This work, then, is no authoritative source-book, but the product of an archaeologist's urge to synthesize. It is necessarily subjective, since aspects which I believe to be important are emphasized, while others are left in the background or neglected. I have endeavoured to give due coverage to political, social and economic conditions, and to the history of civilization and religion. In the footnotes I have given primary sources priority over modern literature; where sources are numerous I refer to authors who have listed and evaluated them. Neither the footnotes nor the select bibliography contain references to standard works on Roman imperial history; these are readily available in more general works. In compiling the select bibliography I have kept in mind the needs of scholars by including publications in which references will be found to the older but still useful literature.

My original text was written in German in 1970. Later information if published by the summer of 1972 is included in the bibliography; it appears in the text only if I have been persuaded to modify my original views. But some important recent discoveries could not be given the extended treatment they deserved: for instance, the reconstruction of the Scarbantia street-system (K. Sz. Póczy), the newly excavated Roman building in barbarian territory in Slovakia (T. Kolnik), the excavations of late Roman fortified settlements in the interior of Pannonia (K. Sági, S. Soproni, E. Tóth), the large-scale excavations at Sirmium, the recently published discoveries at municipium Dardanorum (the late E. Čerškov), and especially the vast programme of Jugoslav excavations in the Iron Gates (the results of which, however, are not yet fully available).

I must thank T. Nagy, who allowed me to make use of his important discoveries concerning the topography of the legionary fortress of Aquincum (Figs 14, 22 and 23). It is a particular pleasure to have been able to refer in more than one context to work of my own pupils, and to have been compelled to reconsider my own opinions in consequence. Indeed the views expressed in this book are the outcome of twelve years of lecturing in the University of Budapest and of the stimulating contact with my pupils which has resulted from it.

I am indebted to several friends for reading the whole or parts of my manuscript, and am particularly grateful for help given by my colleagues L. Balla and J. Fitz. Above all I owe much to the friendly criticisms of J. Gy. Szilágyi, which enabled me to clarify various passages in the text and to avoid some errors. Any mistakes and obscurities that remain are my own responsibility.

I wish to express my particular thanks to Professor Sheppard Frere for his careful supervision and correction of the English translation of my text and for much editorial work in the preparation of this volume.

Budapest A. MÓCSY

Chapter 1
Thracians, Illyrians and Celts

At the time when the Greeks were already familiar to some extent with the geography and ethnography of central Europe the upper Danube was firmly in the hands of the Celts, whereas the lower reaches, from the Iron Gates to the Black Sea, had from time immemorial belonged to the settlement area of the Thracians and the Getae. While, of course, some details concerning the races who inhabited the valleys of the western and southern tributaries of the Danube between Vienna and the Iron Gates were available to the Greeks, they were scarcely adequate as a basis for an authoritative ethnic appraisal. The earliest information is to be found in the *Iliad*: 'Zeus . . . turned his shining eyes away into the distance, where he saw the lands of the horse-rearing Thracians and the Mysoi who fight hand to hand. . . .'[1] Hellenic philologists queried this passage in Homer, as the Greeks in that era knew only of the existence of the Mysoi in Mysia, part of Asia Minor. It was Posidonius, the last great scholar of Hellenism, who provided the right answer by pointing out that a tribe called the Mysoi lived north of the Thracians and could be identified as the Mysoi mentioned by Homer.[2]

This old controversy is to some extent characteristic of the whole of the Greeks' knowledge of their northern neighbours. In archaic times they had a pretty clear picture of the Danube river-system and of the inhabitants of the Adriatic coast. This information, gathered by Greek traders, was in part forgotten in the classical period, along with the decline of Greek trade with the north, and in part became intermingled with mythology and hence gradually distorted.[3] It became, for example, a self-evident truth that one arm of the Danube flowed into the Adriatic and hence that the Balkans were an island.[4] It

was not until Roman expansion to the north that a fresh breeze bearing new
knowledge put an end to what was in the main idle speculation and information
derived from books. Even as far as the Mysoi–Moisoi were concerned, Posidonius' information was only scanty, whereas a Roman army under the command of C. Scribonius Curio, advancing from Macedonia as far as the Danube
between 76 and 72 B.C., brought back more up-to-date details of the inhabitants
of the Danube valley.[5] Greek scholars were interested in the history and ethnography of central Europe only in so far as the barbarians were involved in
events in the Greek homeland. This episodic information which has been
handed down is, however, adequate for a sketch to be made of the main features
of political development on the middle Danube in the last centuries B.C.

Apart from the Mysoi referred to by Homer, the earliest peoples on the middle
Danube mentioned by name are those listed by Herodotus and Hecataeus, e.g.
the Sindoi, Sigynnoi, Kaulikoi, etc., none of whom played any part in the subsequent history of this area.[6] Later they were not included among the peoples
who lived in the Danube valley.[7] They were, however, tribes who were still in
contact with Greek trade, whereas later tribes for the most part only came within
the Greek and Roman range of vision if they had become the enemies or allies
of Macedonia or Rome. It has recently been assumed that the Sigynnoi, who
probably lived in the great Hungarian plain, traded with the Veneti in Istria,
who in turn were involved in early Greek trade on the Adriatic coast.[8] Traces
of these trading contacts are attested by a few late archaic finds in Hungary and
the northern Balkans.[9] In the fifth century B.C. the ethnographic picture at least
of present-day Serbia becomes somewhat clearer (Fig. 1). According to Herodotus the valley of the Morava (which he calls the Brongos) below Niš belonged
to the Triballi, a tribe centred north of the Balkan mountains on the Danube
around Oescus (Bulgarian: Gigen), and these people later often proved troublesome to the Macedonians. The western neighbours of the Triballi were the
Illyrians, for the Angros, a tributary of the Brongos which cannot be identified
with certainty (Ibar? Zapadna Morava? Toplica?), rises in Illyrian country and
flows into the Brongos in the area of the Triballi. So much for Herodotus.[10]
Later sources[11] report the advance of the Illyrian Autariatae into the territory
of the Triballi about the end of the fifth century B.C. and not much later there
are reports of Celtic conquests in the Carpathian region and on the lower
Danube. After the Autariatae had driven the Triballi out of the Morava valley
they themselves were subdued at the beginning of the fourth century by the
Celts,[12] who were emerging as a new influence in the Carpathian region and in
the Balkans. The known ethnic pattern of the original inhabitants in the Roman
period first began to take shape as the result of the Celticization of many areas

2

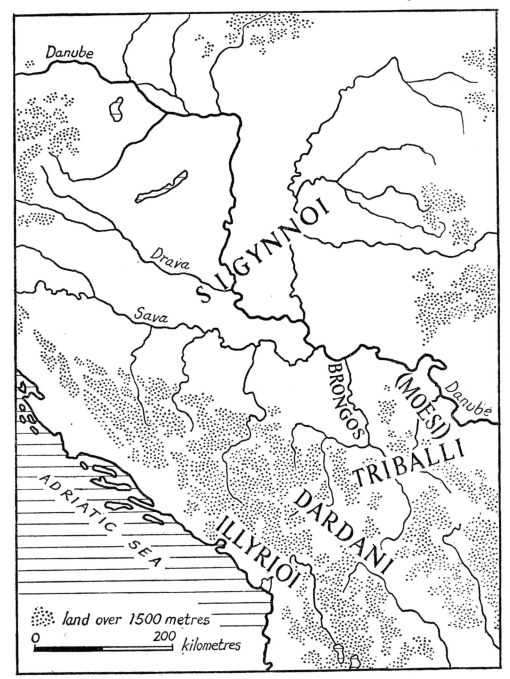

Figure 1 The area of the middle Danube in the fifth century B.C.

in south-east Europe, and it is only after the arrival of the Celts that it is possible to follow the political changes on the middle Danube more precisely.

The Triballi belonged to the Thraco-Getic ethnic group which inhabited the eastern half of the Balkan peninsula. The Autariatae were an Illyrian tribe which had settled in the western half. The third ethnic and linguistic component, the Celts, probably came in part from northern Italy and in part along the Danube from the west.[13] The Thracians, Illyrians and Celts were the three most important ethnic and linguistic groups in south-east Europe, and in Roman times they made up the native inhabitants of Pannonia and Moesia Superior. The linguistic boundaries ran through these two provinces: the Celtic–Illyrian through Pannonia and the Illyrian–Thracian through Moesia Superior.[14] This is one reason why sizeable political units were only rarely established. Neither Pannonia nor Moesia Superior were known as geographic or political concepts in pre-Roman times; both regions belonged for the most part to political structures which had their centres outside the country.

As far as the Thracians and Illyrians are concerned, the problems surrounding their origin and linguistic classification are today much in a state of flux. There was a tendency in the last decade to reject the theory of linguistic uniformity among the Thracians as well as among the Illyrians, and to reserve the terms 'Thracian' and 'Illyrian' for a smaller and more readily definable tribal group. After the abandonment of suspect Pan-Illyrism in the 1930s, philologists came to recognize that the Veneti and Liburni on the northern Adriatic coast spoke a language related to, though different from, Illyrian.[15] Analysis of the earliest information about the Illyrians on the Adriatic indicates that the name Illyrioi applies only to a small area in the south of what was later to become the province of Dalmatia;[16] a critical classification of Illyrian names[17] has established that there were two or three distinct areas in the provinces of Pannonia and Dalmatia. Finally, doubts were also raised by archaeologists who attributed the late Bronze Age and early Iron Age culture of northern Dalmatia to a people who differed from the Illyrians.[18] The Dalmatians and Pannonians, therefore, were either not Illyrians or at best were only related linguistically; but they did not speak the same language as their neighbours to the south, who alone are regarded in the sources as Illyrians. In the case of the Thracians the situation is even more confused. Classical scholars were convinced that the Getae and Dacians spoke the same language[19] and that the Thracians and Getae were in fact one and the same people.[20] The mapping of the place-names in Dacia, Moesia and Thrace recently produced the hypothesis that the Thracians in the wider sense belonged to two different linguistic groups: partly to the Thraco-Getic and partly to the Dacian–Moesian.[21] This division is based principally on

4

the suffixes (-dava, -para, -sara, etc.), the geographical distribution of which is to some extent restricted. Historically there would probably have been better justification for a division into a Dacian–Getic group and a Thraco–Mysic group; argument based on and restricted to place-names is too narrow a basis for such sweeping hypotheses. In any case Pannonia and Moesia Superior were on the periphery of both the Illyrians and the Thracians in the wider sense of those terms. For the moment it will perhaps not be misleading if, following tradition, these fringe races are regarded as Illyrians and Thracians, though we must not overlook the fact that dialects—probably more so then than now—affected linguistic uniformity. A number of tribes are known to have existed along the linguistic boundaries, right inside Pannonia and Moesia Superior, though the linguistic and ethnic groups to which they belonged have not been conclusively established. Of the tribes whose history will be described later, the Dardani are variously taken to be of Illyrian and of Thracian stock;[22] the Scordisci were, of course, a group established by the Celts; in imperial times, however, their names were Illyrian (Pannonian),[23] and some sources include them among the Thracians;[24] it is only recently that the Eravisci have been conclusively identified as a Celtic race.[25] The contact zone between Illyrians, Thracians and Celts was obviously a very broad one, and their relations with one another were subject to constant fluctuation, which ceased only with the Roman conquest.

Celtic expansion reached the Carpathian area roughly at the same time as the Celtic invasion of Italy, that is at the beginning of the fourth century B.C. According to Celtic legend, 300,000 people migrated to Italy and Illyria: Livy's account mentions Bellovesus and Sigovesus, nephews of the Celtic king Ambigatus, who sent them against Italy and against the inhabitants of the *Hercynia silva* respectively.[26] The Celtic legend in Pompeius Trogus describes wars against the native inhabitants which lasted for years and led to the gradual subjugation of the Pannonians.[27] Early La Tène finds in Pannonia suggest that the Celts advanced along the Danube and conquered only the north-west part of the Carpathian region in the fourth century B.C.[28] (Fig. 2).

Another arm of Celtic migration to the east probably started from northern Italy, for, while fighting on the Danube, Alexander the Great received an embassy from the 'Adriatic Celts'.[29] This incident coincided with the violent collapse of the Autariatae who, according to some sources, were forced to yield to the Celtic advance, and after a long and turbulent wandering here and there were finally wiped out.[30] Towards the end of the fourth century the Celts, who had already established themselves in Pannonia, renewed their raids on the Balkan peninsula; these soon brought them into conflict with the king of

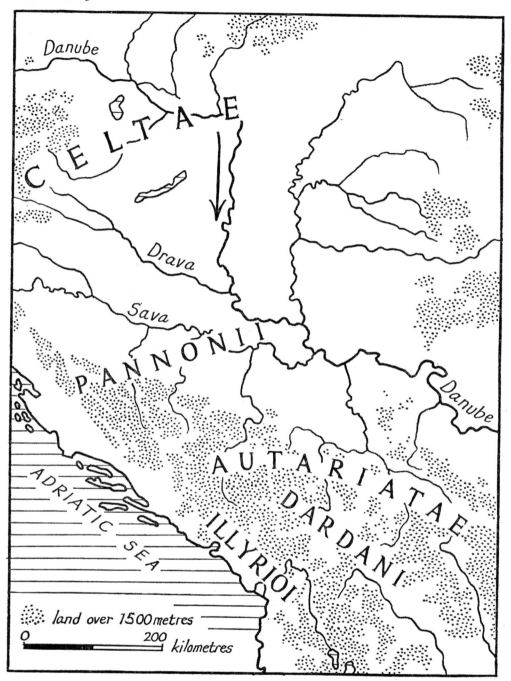

Figure 2 The area of the middle Danube in the fourth century B.C.

Macedonia on the northern border of Thrace.[31] These raids were the prelude to the great Celtic invasion of the Balkans in 279 B.C., which, in many respects, resembled the migration of more than a century earlier. On this occasion, too, the numbers involved seem to have been very large; Justin in his epitome of Pompeius Trogus mentions a charge by 150,000 infantrymen under Brennus' command,[32] and even if this figure is grossly exaggerated, since Diodorus mentions only 50,000,[33] this much is certain: the enterprise was concerned from the outset with finding new areas in which to settle. Again Celtic legend indicates that there were two leaders, Belgius (or Bolgius) and Brennus, whereas a variant states that there was a third group under the command of Cerethrius.[34] A section of the people and of the troops stayed behind to protect the homeland (*oikeia*),[35] which can probably be identified as the area in Pannonia already consolidated by the Celts. From the dating of a Greek bronze vessel found in a grave in the La Tène cemetery at Szob on the Danube bend in Hungary (Pl. 1a) the only conclusion is that it was booty from Greece.[36]

The years 280–278 were marked by fluctuating battles which are of no concern here.[37] Brennus' army plundered Delphi; one of his bands which had deserted crossed into Asia Minor, while a third group was able to establish an independent Celtic state south of the Balkan range which persisted for several decades, 'the empire of Tylis'. Brennus' army, split into several bands, withdrew to the north in 278. It is probable that these great migrations, which created more or less permanent Celtic territories in the Balkan peninsula, brought about an intensive Celticization of the areas surrounding those from which they had started, namely in the Danube valley from below Vienna to the Iron Gates, where in fact there had been earlier Celticization (Fig. 3). Tradition has it that the establishment of a very strong political structure in the north of present-day Serbia can be traced back to these migrations: the Celts on their return from Delphi settled on the mouth of the Save and called themselves Scordisci.[38] Moreover, it is probably true that those parts of Pannonia which had not been conquered in the first Celtic invasion were already Celticized by the beginning of the third century B.C. La Tène finds of the C period are distributed throughout the whole of what was later Pannonia, although it is impossible to decide whether those in certain areas, e.g. in the mid-Save valley, indicate Celtic settlement or merely the distribution of artefacts made by Celts. The area under the control of the Scordisci has also yielded fairly rich finds of La Tène type.[39]

The races north of Macedonia did not immediately recover from the upheavals of the Celtic invasion. Sources mention various rearguard battles in the Balkans;[40] heavy battles probably also resulted from the consolidation of Celtic control in the Danube valley. From the middle of the third century onwards a

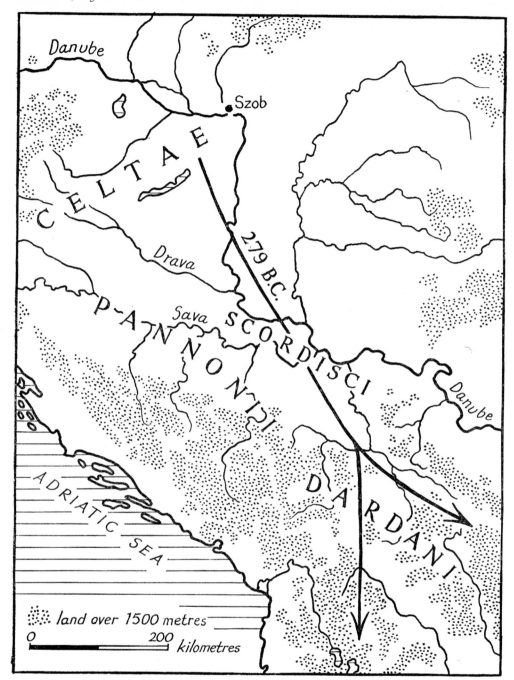

Figure 3   The area of the middle Danube in the third century B.C.

very active power was emerging in the immediate neighbourhood of the Macedonians; and being probably the first to free itself from the Celts, it was expanding to the east, south and west. This power consisted of the Dardanian race
which had settled on the upper reaches of Axius (Vardar) and in the Kosovo
polje and Metohija basins north of the huge Scardus Mons (Šar Planina). It
belonged to the Thraco-Mysic Balkan group[41] which was racially connected
with the peoples of western Asia Minor (Mysoi in Asia Minor, Moisoi on the
Danube, Dardanoi on the Axius and near Troy, Phrygians in Asia Minor and
Brigoi in Thrace, etc.).[42]

In the fifth century the Dardanians were probably subject to the Triballi.
Whether they were reached by the expansion of the Autariatae is unknown, but
it is not improbable. In the time of Philip II they fought against the Macedonians[43] and they probably suffered severely at the hands of the Celtic invaders
at the beginning of the third century. In 279 B.C. Brennus' tribe was in Dardania,[44] and it was there that Leonnorius and Lutarius (who transferred 20,000
Celts to Asia Minor) deserted him. Celtic rule in Dardania did not, however, last
long. About the middle of the third century the Dardanians were already
revealing themselves as very dangerous enemies of the Illyrians, and by the last
third of that century they had become the proverbial enemies of Macedonia.[45]
The Macedonian kings, Demetrius, Antigonus Doson and Philip V had to
wage constant war against them. The battle-grounds were in the Vardar gap
around Stobi and south of the Scardus Mons, and the wars were basically
defensive against a mountain people which was constantly engaged in raiding.[46]
Philip V wanted to apply tough measures in order to put an end to this raiding
once and for all, but before he could do so the Dardanian raids suddenly ceased
in 197. This is all the more surprising since in the same year the Macedonians
were defeated at Cynoscephalae, so that the cessation of the Dardanian attacks
cannot be attributed to an increase in strength of the Macedonian forces. Moreover the Dardanians were increasingly being regarded as the natural allies of
Rome, and thus would have been in a strong position for launching successful
attacks against Macedonia.

It is very probable that from the beginning of the second century the Dardanians were being threatened from the north and were concentrating their
forces there. There is also probably a connection with the fact that antagonism
between Rome and Macedonia had drawn the Balkan peoples into strategic
alliances. For some time after 278 B.C. nothing is heard of the Scordisci who had
established Celtic control to the north of the Dardanians. Then, under Philip
V, they suddenly emerge as allies of the Macedonian king and as the enemies
of Rome and the Dardanians. Whereas nothing is known of what was happening

9

the western edge of Pannonia that the pre-Celtic language completely disappeared. It was less effective south of the river Drave—as in the case of the Scordisci themselves whose names in Roman times are Illyrian (Pannonian).[49] This is obviously due in part to the numerical ratio of Celtic conquerors to the non-Celtic native population and in part to the length of time Celtic control lasted. The Celtic group, which after its retreat from the Balkans in 278 B.C. was able to establish itself under the name Scordisci on the mouth of the Save, formed only a thin Celtic upper class, which was gradually absorbed by the subjugated but nevertheless numerically stronger native population. It is therefore not surprising that this people is described in the sources sometimes as Celts, and sometimes as Thracians, while by the end of the second century B.C. they are actually given an Illyrian suffix and are called Scordistae.[50] Accordingly the Scordisci, who were frequently the only, and always the strongest power in the central Balkan area in the second century, are to be regarded as merely one of the Celts' political creations and not as a Celtic tribe.

After Rome conquered Macedonia the Dardanians were heard of once again when they laid claim to Paeonia.[51] Thus by 168 the Scordisci had not yet extended their control to include Dardania. Around the middle of the second century it is probable that their interest lay rather to the west, for in 156 they were defeated by the Romans,[52] possibly in the great Dalmatian war in which they fought on the side of Rome's enemies. The first battle on the Macedonian border between the Romans and the Scordisci is not attested until 141;[53] there are no records of earlier battles against Macedonia's northern neighbours. In 156, probably during the Dalmatian war, the Romans laid siege to the Pannonian town of Siscia (Segestikè) at the mouth of the Kulpa in the Save valley [54] and this advance to the east probably also involved the Scordisci. It is about this time that the first historical mention of the Pannonians occurs in a fragment of Polybius.[55]

It is possible that it was this defeat in the west, which probably occurred in the upper Save valley or in northern Bosnia, that caused the Scordisci to turn their expansion southwards. In the middle of the second century they must have subdued the Dardanians, since from 141 onwards only the Scordisci together with west Thracian tribes are constantly mentioned as being the enemies on Macedonia's northern boundary (Fig. 4); some of the battles took place where the Macedonian kings had earlier engaged the Dardanians. For a long time after the middle of the second century there is no further mention of the Dardanians. The permanent allies of the Scordisci in their many wars between 141 and 109 were the west Thracian tribes, among whom the Maidoi are repeatedly mentioned; the theatre of operations included the valleys of the Axius (Vardar),

Astibus (Bregalnica), Strymon (Struma) and the upper reaches of the Hebrus (Marica).[56] It must therefore be assumed that the area controlled by the Scordisci in the second half of the second century covered the whole of what was later to become Moesia Superior. According to Strabo, this control extended as far as Paeonia, Illyria and Thrace, and involved the subjugation of many tribes in the central Balkans.[57] Strabo also mentions that the Scordisci living west of the Morava were known as 'the big ones' and those to the east as 'the little ones'.[58]

While their hegemony lasted, the Scordisci did not give up their claim to the Save valley. When in 119 B.C. the Romans besieged Siscia for the second time,[59] a Dalmatian war was in progress in which the Scordisci were again involved. And when some years later the Cimbri were migrating through the western half of the Carpathian region, once again it is only the Scordisci who are mentioned as living in the Save valley.

The migration of the Cimbri did not bring about any substantial changes in the history of the Danube lands, but it is nevertheless of particular interest since Posidonius' description contains many important details, for instance that the Cimbri first came up against the Boii and, having been repulsed by them, met with the Scordisci, the Taurisci and then the Helvetii.[60] This list of names provides a clear picture of the power structure in the Carpathian region at the end of the second century. The Cimbri migrated from the north through the western half of the Carpathian region, the northern part of which was controlled by the Celtic Boii, the southern part by the Celtic Scordisci and the south-western part by the Celtic Taurisci. It is not known where the boundary between the Boii and the Scordisci ran, but it is very probable that the border between the Taurisci and the Scordisci lay west of Siscia. This can be deduced not only from the fact that the Scordisci took part in the battles each time the Romans besieged Siscia, in 156 and 119 B.C., but also from a passage in Strabo which states that the upper reaches of the Save belonged to the Taurisci, but Siscia to the Pannonians.[61]

As for the Pannonians, Posidonius does not even mention them. Thus at the end of the second century they must have been controlled by the Scordisci; Polybius, however, writing of an event which took place not much earlier, indicates that he was aware of their existence.[62] The Pannonians belonged to the Illyrian or to the pre-Celtic native population which was related linguistically to the Illyrians and which inhabited the north-western Balkan area; according to detailed descriptions by Strabo and Appian, they broke up into a number of tribes which settled in a fairly big area extending from the Drave as far as the Dardanians and the Ardiaei in southern Dalmatia.[63] Only a few of these Pannonian tribes belonged to the region which later became the province of Pannonia; of the tribes in the province of Pannonia, Strabo classes only the Breuci and the Andizetes as Pannonian. In Roman imperial times the Andizetes were living at the mouth of the Drave and had the Breuci as their southern neighbours on the Save. In addition to these two tribes, others living between the two rivers must also be regarded as Pannonian, namely those called Colapiani, Iasi, Oseriates, Amantini, Cornacates and Scordisci, all within Roman Pannonia. Names appearing on inscriptions from these tribal areas form the northern section of the so-called central Dalmatian–Pannonian group of names which has its distribution in the area occupied by the Pannonians, i.e. in southern Pannonia and eastern Dalmatia.[64] The original inhabitants of the north-western part of what was later Moesia Superior were possibly Pannonian too, but at an early stage they had come under the control of the Scordisci, whose centre happened to be in that very area.

In Posidonius' description of the migration of the Cimbri there occurs the first mention of the control of north Pannonia by the Boii. As already indicated, this area was under Celtic occupation from the beginning of the fourth century, and not only the wealth of La Tène material but also the obviously Celtic names of the native inhabitants of western and north-western Pannonia in imperial times point to an intensive Celticization. Names of tribes in the fourth, third and second centuries have, however, not been handed down. Posidonius was the first to reveal that control was in the hands of a tribe which had the same name, Boii, as a once powerful Celtic tribe in northern Italy. Tribes with the same name frequently occur in different countries, either because they are racially connected (subdivision occurring only later as the result of migration), or because the tribal name in the language concerned was a common ethnic term, or because during its migrations the tribe appeared at different places at different stages of history. The easiest explanation is, of course, to assume a migration, the conclusion which Posidonius reached when he suggested a racial link between the north Italian Boii and the Boii in the Carpathian area. According to this hypothesis, the Boii were driven out of northern Italy at the beginning of the second century and went to the Danube,[65] or, according to another ancient version, to Bohemia, the ancient name of which, Boiohaemum, provided welcome support for this theory.[66] Whether there is any historical basis for this idea is open to question. Thorough investigation of La Tène material does not indicate so close a relationship to finds in Bohemia—part of the original Celtic homeland—that a Celtic migration from Bohemia into Slovakia may be deduced.[67] There is even less evidence of any link with Celtic finds in northern Italy.[68] All the more striking is the close relation between the Slovakian and

Hungarian La Tène material, from which we may infer a gradual Celticization of the whole of the northern Carpathian region between the fourth and the second centuries.[69] This process—perhaps without further Celtic immigration from the west—started with the first Celtic invasion at the beginning of the fourth century; if the Celts in the north-west of the Carpathian region were in a position at the beginning of the third century to embark on further large-scale migrations into the Balkans and even as far as Galatia, then it must be assumed that gradual Celticization within the Carpathian area also occurred.

Power-relationships, as indicated by Posidonius and Strabo, soon underwent a radical change (Fig. 5). In the first place it may be assumed that there was a loosening of Scordiscan control in Dardania. In 97 B.C. the Dardanians, of whom nothing had been heard for a long time, appeared on the scene as the allies of the west Thracian Maidoi, and were defeated by Macedonia's Roman army.[70] Not much later the Dardanians and Scordisci were allied in the struggle against Sulla on Macedonia's northern boundary,[71] and shortly afterwards the Scordisci were defeated by L. Cornelius Scipio Asiagenus, a defeat from which they never recovered. It is impossible to date Scipio's war precisely within the years 88–81 B.C., as the relevant sources give rise to much uncertainty; for instance the war is said to have been started in revenge for a second plundering of Delphi by the Celts. There is also a great lack of clarity in the chronological order of events.[72] Nevertheless, one thing may be assumed as certain, namely that Scipio gained a truly decisive victory. The Scordisci never again appeared as dangerous enemies on Macedonia's northern boundary, and between 81 and 15 B.C. there is only one further mention of them. According to Appian they withdrew to the mouth of the Save and to the Danube islands; this is, of course, a gross exaggeration. In fact they disappeared once and for all from the area which was later known as Moesia and thenceforth are mentioned only as inhabiting the south-east corner of Pannonia.

It would not be in keeping with available evidence if the decline of Scordiscan power were attributed solely to Scipio's military victories. Macedonian and Roman campaigns against barbarians on the northern Macedonian boundary always had a limited aim, the establishment of peace, and there was never any decisive interference in the power-relationships of the tribes in question. The Scordisci were probably under pressure from other directions, and events which occurred soon afterwards make this a very reasonable assumption. When, for example, in 64 B.C. Mithridates set out to attack Italy via the northern Balkans— a plan which the last Macedonian kings had also seriously considered—it is the Pannonians and not the Scordisci who were said to control that area.[73] Thus the Pannonians who, apparently, had liberated themselves from the domination of

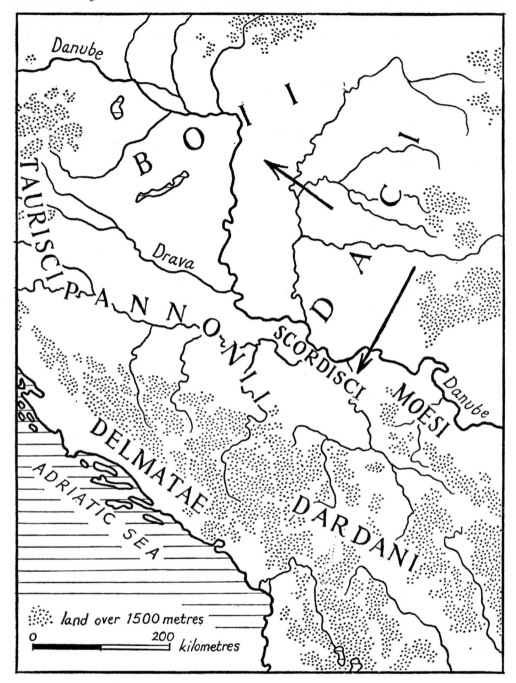

Figure 5 The area of the middle Danube in the first half of the first century B.C.

the Scordisci in the early decades of the first century were regarded as being the dominant power in the Save valley. This roughly coincided with the strengthening of the Dacian state, now united under Burebista, and he, in turn, soon attacked and defeated the Scordisci. As a result the latter became the allies of the Dacians in their struggles against the Celts in the Carpathian area.[74]

So far it has not been possible to provide conclusive evidence for all the changes in power-relationships and the ethnic picture in the first half of the first century B.C. (Fig. 5). For instance, it has only recently been accepted as probable that the migration southwards of some Celtic tribes, described by Caesar in his *Commentaries* as a Helvetian migration,[75] involved the Carpathian region. The Latobici in the upper Save valley probably arrived then, whilst the name of the Hercuniates in east Pannonia perhaps suggests that they also originated in the north. The Latobici obviously brought with them from the Saale area their strange custom of shaping their cremation vessels like houses.[76] A question still awaiting clarification is that of a western migration by a section of the Boii who, according to Caesar, besieged Noreia, the capital of Noricum.[77]

These convulsions in the central Balkans finally led to a particularism which produced many small states composed of tribes and tribal groups which acted individually (Fig. 5); the place of the Scordisci was taken by the Dardanians, Moesians and Pannonians, all of them tribes which in the second century had depended, closely or loosely, on the Scordisci.

At first the Dardanians proved troublesome to the Romans. In 76 B.C., along with Thracian tribes, they invaded Macedonia, and the war which ensued (at first under the leadership of Appius Claudius Pulcher,[78] then of C. Scribonius Curio and finally of M. Terentius Varro Lucullus) lasted from 76 to 73. This *bellum Dardanicum*,[79] which Curio waged energetically but with unprecedented cruelty, ranged over a very wide area.[80] It was at this time, too, that a Roman army reached the Danube for the first time, an event whose significance was often underlined by later Roman historians.[81] The Roman army probably advanced through the Isker (Oescus) valley and brought news of the tribe of the Moesi, which was settled in the Timok valley and on the Danube but had not been heard of before.[82] It is highly probable that it was this news which led Posidonius to interpret the passage in *Iliad* xiii, 5, correctly. Homer vouches for the fact that the Moesians were long-established inhabitants of the eastern Balkans. But for many centuries nothing was heard of them, as their more powerful neighbours, the Triballi, Autariatae, Dardanians and Scordisci had either held them in subjection or pushed them into the background. It was only after the power of the Scordisci had declined that they emerged as an independent political force.

Curio also made contact with the Dacians, but this did not lead to a head-on collision between them and the Romans. The Scordisci are not mentioned in connection with this campaign for, after their defeat by Scipio, the question of whether they were enemies or allies of Rome does not arise. At the time of the campaigns of 76–73 the Scordisci had not yet been defeated by the Dacians, and as the latter are not mentioned in 64 when Mithridates wanted to send the Bastarnae through the Save valley into Italy,[83] their defeat of the Scordisci must have occurred later. The approximate date of the Dacians' victory can be established by means of the following considerations. It resulted in their temporary supremacy in the central Balkans.[84] In the last years of Caesar's dictatorship Rome regarded a Dacian attack on Macedonia as not unlikely, and after his murder there were in fact rumours that such an invasion had taken place.[85] Thus for a time the Dardanians, too, must have been under Dacian control. They were, however, still independent when in 62 and 57 they were attacking Macedonia.[86] Therefore, the Dacian victory over the Scordisci occurred roughly between 56 and 50 B.C.

According to Strabo, Burebista carried out his conquests within a matter of a few years.[87] It is not necessary here to go into details regarding his successes in the south-east. His operations had, however, a lasting effect on the history of the Carpathian region, inasmuch as they put an end once and for all to Celticization in many areas.

As has already been pointed out, political control in the western half of the Carpathian basin and in the valley of the Save at the turn of the second century was still in the hands of the Celtic Boii, the Scordisci (who were in the last resort Celts) and the Celtic Taurisci. In the Save valley the Pannonians soon made themselves independent; in the north, however, in the first half of the first century, the Boii were still in uninterrupted control. The eastern neighbours of the Boii and the Scordisci were the Dacians; Caesar, who was familiar to some extent with the ethnic and political conditions in the Danube valley (whether or not he derived his information from Hellenistic geography and in particular from Posidonius), knew that the eastern end of the enormous undefined area of the *Hercynia silva* to which the northern Carpathians belonged was inhabited by the Dacians and the Anartii, and that these two tribes were to be found east of the Danube bend.[88] Ptolemy also mentions that the Anartii lived in northern Dacia,[89] and he is also aware of the existence of the Taurisci, the latter's neighbours. These two Celtic tribes obviously belonged to the Boian tribal federation which embraced the whole of the northern half of the Carpathian basin. The late Iron Age *oppidum* culture can be traced from the Danube bend as far as the north-eastern Carpathians (Budapest-Gellérthegy, Zemplin, Mukačevo, to

18

mention only the most important centres). Further tribes representative of Celtic control are mentioned in Tacitus' *Germania*, and these, together with later information, make it possible to draw up a list of virtually all the tribes on the northern edge of the Hungarian plain: the Osi on the Danube bend, and to the east the Cotini, Anartii and the Taurisci.[90] The zone of contact between Celts and Dacians must, therefore, have run roughly through the middle of the Carpathian region. Strabo, who gives a fairly detailed description of the battles between the Celts and Dacians, states that the boundary line was the Parisos, which is the ancient name for the river Tisza.[91] According to Strabo, Burebista's claim to certain areas led to war which ended *c.* 45 B.C. in the defeat of the Boii and their allies the Taurisci. Burebista's opponent was Critasirus, the king of what was probably a very large country. The Taurisci, over whom he ruled, were Celts in south-west Pannonia, and the Boii, who were also Celts, lived in northern Pannonia; to them the Celtic tribes on the northern edge of the plain and in Slovakia belonged. Within the Roman empire Burebista's victory made a deep impression. There was reference to a 'Boian desert' in Pannonia for which Pliny the Elder uses the term *deserta*.[92]

The Boii were no more wiped out than were the Scordisci by Scipio. But the extensive and apparently well-established Boian area of control disintegrated, while the Dacians, even if they did not found any large settlements, nevertheless established outposts in many parts of the Carpathian region. Typical small hand-made dishes with handles have been found in the later layers of late Celtic *oppida*: these Dacian dishes and various other Dacian ceramics have been unearthed in the Carpathian region, particularly in those parts into which the Boii and the Scordisci had extended their control, e.g. the Banat, the Morava valley, in east and west Slovakia and also here and there on the right bank of the Danube in the north-east of what was later to become Pannonia. The number of locations is particularly striking in Slovakia where Dacianization continued in imperial times.[93] After the Cotini had been settled in Pannonia and Moesia by the Emperor Marcus Aurelius some of them were mentioned on inscriptions; in the main they have Thraco-Dacian names.[94] Later on, Dacians who had been driven out of the Tisza area by the Iazyges also settled on the northern edge of the plain.[95]

Dacian control of the western half of the Carpathian region did not, however, last long. Burebista died about 44 B.C. and his state broke up into at least four or five kingdoms.[96] The tendency towards particularism, inherent in barbarian political structures, again put an end to the plans of a great ruler. The new Dacian kings, Burebista's successors, were, of course, involved in the struggle for power between the Roman parties after the murder of Caesar,[97] but the

Figure 6 The area of the middle Danube in the second half of the first century B.C.

great king's Dacian state gradually contracted and soon lost its leading position. In 39 the Dardanians, who had again become independent, invaded Macedonia;[98] it is possible that the Pannonians reconquered some areas belonging to the Celts, and the Dacians disappeared once and for all from what was later to become Pannonia. They were able to maintain their hold only on the plain, where later—it is not known when—they had to yield the area between the Danube and the Tisza to the Iazyges. In his description of Dacia, Ptolemy gives the Tisza as its western boundary,[99] and so it remained until Dacia was conquered by Trajan.

The only political power which was not defeated by any external enemy was that of the Pannonians. This group of tribes, in so far as its capacity to form a state was concerned, was perhaps the weakest. It is of course a commonplace to mention anarchy in ethnographic descriptions of barbarian tribes. Appian, however, in his reference to the Pannonians, does not confine himself to the mere reiteration of such platitudes: 'The Pannonians do not live in towns, but in villages and hamlets organized on the basis of clans. They do not assemble in joint councils, nor do they have joint leaders who are supreme; 100,000 of them are capable of bearing arms, but because of the prevailing anarchy they never assemble as a combined force.'[100] This primitive tribal society provides adequate explanation of why the Pannonians did not appear on the stage until after their better-organized neighbours had wiped each other out. After Caesar's murder there was no power in the Carpathian region nor in the central Balkans which could seriously have opposed Rome. Nor were the wars waged by Roman generals in these areas after *c.* 44 B.C. dictated by the necessity to intervene radically in the inter-tribal relations obtaining in the Danube area: both Octavian's campaign against the Iapodes (35–33 B.C.) and M. Licinius Crassus' war against the Dacians, Moesians and Thracians can be understood only against the background of the political situation in Rome itself.

In Caesar's last years Burebista was in undisputed control of the Carpathian region and of the north-eastern part of the Balkan peninsula. Caesar had the rumour circulated that he was formulating a large-scale plan for crushing the Dacian king,[101] and Octavian, Caesar's executor, had to fit a Dacian war into his political plans,[102] despite the collapse of Dacian hegemony in the meantime. In this context it was rumoured that certain Dacian kings would side with Octavian or with Antony; the rivals accused each other of having formed an alliance with the Dacians.[103] Even after Actium the 'Dacian problem' remained a topic of conversation in Rome.[104]

It is only against the background of this Dacian problem, inflated for propaganda purposes, that Octavian's campaign against the Iapodes becomes at all intelligible. In 35 B.C. he advanced against the Iapodes, the Alpine people to the east of Aquileia, on the pretext that they had ceased paying their taxes.[105] After hard-fought battles he succeeded in capturing their most important fortresses, including Metulum. From here he continued his advance into Pannonian country 'although they had given no cause'.[106] The final goal of this advance seems to have been the capture of Siscia, twice unsuccessfully besieged by the Romans (in 156 and again in 119 B.C.). This strong-point on the mouth of the Kulpa in the Save valley was the natural spring-board for an advance to the east, and it was in this context that Octavian's propaganda put the capture of Siscia. It was said to be the most important base from which to launch an attack against the Dacians. Appian, in his account of the war against the Iapodes, which is indirectly based on Octavian's commentaries, reports a quarrel between the upper classes and the common people of Siscia. The former were in favour of yielding to the Roman terms—the handing over of 100 hostages and accepting a Roman garrison: the people, however, resisted and in the end the town had to be taken in a battle which lasted a month.[107] This account by Appian is at variance with his above-mentioned description of the primitive social conditions among the Pannonians. The suspicion is justified that Octavian, to defend his attack, was reverting to the trick repeatedly used by Caesar in his Gallic war, the suggestion that a section in the enemy camp—always the aristocracy—was friendly to Rome, and that, therefore, the Roman general came not as a conqueror but as the supporter of the Romanophile aristocracy. Such reasoning in support of the capture of Siscia was all the more necessary because the Pannonians were not the traditional and proverbial enemies of Rome. Even Appian rightly admits that over a long period the Romans had taken no notice of the peoples living on the other side of the Eastern Alps.[108]

After capturing Siscia, Octavian divided the town into two by means of a wall, and occupied it with a force consisting of twenty-five cohorts under the command of Fufius Geminus. He made the Pannonians make submission to him without continuing his advance eastwards, and then returned to Rome for the winter. In the following year there is again no mention of an advance against the Dacians: in 34–33 Octavian pacified the Dalmatian tribes living to the south of the Iapodes.

It is now clear that Octavian's plans did not include a Dacian war after the capture of Siscia. The three-year war undertaken on this pretext[109] had ultimately resulted in the pacification of an area of great importance. By securing the Eastern Alps and the coastal strip along the northern Adriatic, a link was established which might be of importance not only in an advance against

Antony but also for the future occupation of Illyricum, which had been awarded to Octavian at Brundisium in 40 B.C.

There are no reports of Siscia's fate in the following decades, apart from the fact that a rising by its inhabitants in the winter of 35–34 B.C. was put down. It is however probable that in practical terms it remained in Roman hands. In the wars under Augustus it became the Roman army's most important stronghold. Had not its fortress been firmly held by the Romans, Tiberius' campaign against the Scordisci in 15 B.C. would have been impossible.

The Roman advance from Macedonia was likewise influenced by the Dacian problem. The general who set about putting Caesar's plan for a war against the Dacians into effect came into conflict with Octavian for that very reason, and his victories were eliminated from the official versions of the history of that war.[110] M. Licinius Crassus, a former supporter of Antony, who went over to the side of Octavian shortly before Actium, received the consulship, along with Octavian, in 30 B.C. (without having previously held the praetorship: he probably insisted on the consulship as a reward for changing sides). In 29 he became proconsul of Macedonia and in this capacity he launched a large-scale war on his own account.

This war, in which Crassus was victorious against the Thracians and Getae,[111] was officially known as the Thraco-Getic war. However, it may be inferred from a brief note by Cassius Dio that the first peoples to be defeated, and the real enemies, were the Dacians and their allies, the Bastarnae.[112] Horace extols Crassus' war in the words *occidit Daci Cotisonis agmen*;[113] the later account is, however, influenced by the official version, according to which Crassus defeated only the Moesians and various west Thracian tribes. The Dacians, defeated by Crassus, were the subjects of King Cotiso, whose rule also extended south of the Danube.[114] He had taken over the territory in the central Balkans which had been conquered by the Dacians. In view of the way history was distorted for political reasons, it is difficult to reconstruct the first part of this Dacian war, even in outline; nevertheless, the additional uncensored details provided by Cassius Dio[115] make it possible to come to some conclusions about it. After mentioning briefly the victory over the Dacians he prefaces his description of it by the remark that the Dardanians, Triballi and the Dentheletae were defeated by the Bastarnae, and that Crassus came to the aid of the Dentheletae. Now it is known that the Bastarnae were a tribe of mercenaries living at the mouth of the Danube, who frequently put their troops at the disposal of the Macedonians, Mithridates, the Dacians, etc. It is probable that at this time they were taking part in the war on the side of the Dacians, and it is readily understandable why they in particular should attack these tribes. The latter were the neighbours and natural enemies of the Dacians who were in control south of the Danube. Cotiso, when attacked by Crassus, probably appealed to the Bastarnae for help and they then attacked those tribes which wanted to get rid of Dacian control in the neighbouring area, and were consequently potential allies of Rome. It was the Dentheletae who asked Crassus for help, and after defeating the Dacians he advanced into Moesian territory. In the following year (28 B.C.) he continued his war in western Thrace.

The war conducted by Crassus not only broke Dacian control south of the Danube but also led to the pacification of the Dardanians, Triballi, Moesians and some of the west Thracian tribes. From 28 B.C. onwards there was peace in the central Balkans. Then in the year 16 there were reports of an invasion of Macedonia by the Scordisci and the Dentheletae.[116] In the same year the Pannonians invaded Istria. Cassius Dio, who records both these invasions, says that the Pannonians' allies were the Noricans; these were probably Celts living in the upper valley of the Save, that is to say they were the Pannonians' western neighbours. Cassius Dio goes on to say that the repulse of the Pannonians by P. Silius Nerva led to their renewed subjection; the indication that there had been an earlier one probably refers to Octavian's campaign against the Iapodes.

It is very probable that it was these invasions which made Augustus send the young Tiberius, probably as early as 15 B.C., against the Scordisci and some Thracian tribes, neighbours of Macedonia. This campaign might also be regarded as a step towards the pacification of the border regions of north-eastern Italy and northern Macedonia.[117] A consequence of Crassus' Dacian war was that the Scordisci had again achieved independence and, occupying as they did a key position at the mouth of the Save, could represent a danger for both Italy and Macedonia. However, when, a few years later, Tiberius had to fight the Pannonians, he was backed by allies in the Save valley who were none other than the Scordisci.[118]

Further events in the western Carpathian region and in the central Balkans took place as a result of the Roman advance towards the Danube frontier, and belong, therefore, to a new chapter in the history of that area (p. 34).

Literary sources have made it possible to trace the history of the last four or five centuries B.C., at least in its main features. As for the social structure and the basis on which state structures often developed rapidly to embrace wide areas, the sources are by no means so helpful. Nevertheless, from the previous account certain inferences may be drawn which can easily be reconciled with the very sparse direct information contained in ancient literature.[119]

One of the most striking characteristics of all the political structures was the dominant position of one tribe within each of them usually extending over a wide area. At the beginning of the period for which there are literary sources the Triballi exercised supreme authority over a very considerable region of the central Balkans. Their own settlement area and base was a not particularly large territory lying between the Balkan mountains (Haemus) and the Danube, where their existence is attested right into imperial times.[120] Their centre at that time was at Oescus (Gigen). They were driven out of their western and southern possessions by the Illyrian tribe of the Autariatae, who superseded the Triballi in the fourth century until they were themselves driven out by the Celtic onslaught; after succumbing to this they disappeared almost entirely from the historical scene. The Celts had established a similar, though more permanent, political structure in the western part of the Carpathians. In the course of the third century they were able to extend the area under their control with astonishing speed. At the same time the Dardanians established supremacy in the south, although it did not extend over so wide an area as that of the Triballi and Autariatae. In the second century B.C. it was known that there were Celtic tribes (Boii, Taurisci and Scordisci) exercising independent control within the area dominated by the Celts. The Dacians then made their appearance under King Burebista and established brief control over a variety of non-Dacian tribes.

These political structures were each centred upon a tribe which had come to the fore as conqueror and organizer and upon various local tribes, not necessarily related to it linguistically; the local population was either subdued and exploited in the harshest manner possible, or was forced into an alliance. For each of these extreme forms of treatment there is clear evidence. At the height of their power, according to Theopompus, the Autariatae had 300,000 subjects 'in the condition of helots';[121] about the middle of the first century, on the other hand, the Scordisci were forced to take part in Burebista's wars as the allies of the Dacians.[122] The military expeditions conducted by the Triballi, Autariatae and the Celts were, therefore, probably started by a martial section of the tribal society. This enterprising and mobile warrior class was able from time to time to conquer large areas and to exploit their inhabitants. There is probably an historical basis for the legend surrounding the Celtic princes Bellovesus and Sigovesus,[123] not to mention Brennus, Bolgius, Leonnorius, Lutarius, Cerethrius, Akichorius and others who set out with their wives and children around the year 279 and subsequently succeeded in establishing Celtic states in the Balkans and even in Asia Minor.

It is less easy to give a clear definition of the circumstances and conditions which determined the duration and effect of these political formations. At the beginning of the fourth century, Celtic bands had conquered the north-western part of the Carpathian region and set about its Celticization with determination. During the empire non-Celtic name-elements in north-west Pannonia are almost completely lacking. Celtic bands also established control through the Scordisci, though later sources refer to the latter as a mixed tribe composed of Celtic, Illyrian and Thracian elements, while the names found in imperial times in that part of south-east Pannonia occupied by the Scordisci clearly belong to the so-called central Dalmatian–Pannonian group and reveal no Celtic influence worth mentioning. Celtic bands also set up the 'empire of Tylis' in Thrace, but all trace of them disappeared in the third century.[124] In the absence of archaeological and linguistic evidence it has not been possible so far to establish the location of this Celtic state. At the same time as it came into being, the Celtic kingdom of Galatia was also being founded; this became an all too dangerous rival of Pergamum and was able to survive.

The impact of Celtic influence on Pannonian material culture was successful even in areas where there was evidence in Roman times that the indigenous population were not Celts. But there was no such fundamental change in the latter's burial rites. Inhumation, as practised by the Celts, never gained the upper hand, although such burials here and there in early imperial times go back to Celtic elements in the original population.[125] Because the late Iron Age in Pannonia has not been adequately investigated, and in Moesia Superior even less so, it is impossible at present to decide whether there were considerable differences in the spread of La Tène culture which could be attributed to variations in the degree of Celticization. In any case La Tène types are generally characteristic of local production in pre-Roman Pannonia. Recent excavations have produced La Tène finds in the northern part of Moesia Superior, in the territory of the Scordisci;[126] in the southern part of this province, where the Scordisci were only temporarily in control, the influence of La Tène culture is not attested. It was this area which, from the middle of the fourth century at the latest, was able to preserve its independence on a permanent basis. The Dardanians did, of course, suffer severely during the Celtic invasion of 279, during that of the Bastarnae in 179 and probably ultimately at the hands of the Scordisci but, retaining their old name, they maintained their identity until imperial times. It is also possible to say something about the social structure of this tribe. Greek ethnographers had a certain amount of information about the Dardanians who, as the neighbours of the Macedonians, had come into contact with the Greeks at an early date. Hence, there are frequent reports of underground huts covered with dung, of their proverbial dirtiness, of musical instruments and last but not least of products such as cheese and woollen goods.[127] All this suggests a race of mountain shepherds. Agatharchides supplies the further information that there was a rich Dardanian social class whose members kept thousands of serfs who had to till the land and do military service for their masters.[128] It is probable that here there was a symbiosis of warrior mountain shepherds and peasants kept in a state of subjection. This social structure was doubtless similar in many respects to that in which one tribe became dominant, and hence was in a position to repulse attacks by external tribes.

The Dardanians belonged to the older group of Balkan peoples who were racially related to those of north-west Asia Minor. This relationship had already struck the Greeks: 'There are many similar names among the Trojans and Thracians,' writes Strabo.[129] The Trojan Dardanians, attested in the *Iliad*, prompted writers in late antiquity to represent the emperors who came from Moesia Superior, Constantine the Great and Justinian in particular, as being the descendants of a Trojan, i.e. an ancient Roman, race. The Moesi (Mysoi), who first appear in the first century B.C. after the defeat of the Triballi, Autariatae, Scordisci and Dacians who followed one another in the control of Moesian country, also belonged to the Thraco-Phrygian group of races. What the pre-Celtic native inhabitants of Pannonia and the north-western part of Moesia Superior were called is not known, the reason being that Celtic control over these areas led to fundamental ethnic and social changes. The inhabitants of the areas over which the Scordisci established control were later known as Scordisci, although their language was not Celticized—on the contrary, they absorbed the Celtic ruling class. The only pre-Celtic section of the population in Pannonia whose name is known is the Pannonians, but it is not known how far their original territory extended northwards. Like the Moesians they did not appear on the scene until after the break-up of the hegemonies established by individual tribes. At that time, in the first half of the first century, they lived in the northern part of what was later Dalmatia, in the Save valley down-stream from Siscia and on the lower reaches of the Drave.[130] It is very possible that they were the original inhabitants of that part of Pannonia lying between the Drave and the Danube,[131] but there their language underwent complete Celticization.

What information we have concerning Pannonian society we owe entirely to Appian, whose brief description has already been quoted. In contrast to the Celts and Dacians the Pannonians were unable, even temporarily, to establish any kind of political unity. Their state of anarchy, underlined by Appian, showed itself even in their resistance to Augustus, when each tribe fought as a separate force under its own leaders against the Romans. Hence they were not capable of submitting even temporarily to a central power on the lines of that established by Critasirus the Boian, or Burebista the Dacian. Their primitive social institutions possibly explain why nothing was heard of them until a late date in the Danube and Balkan areas. The better-organized and technically superior Celts subdued them without difficulty; it was not until the Celtic Scordisci had been defeated that they appeared as an independent tribe in the Save valley.

The degree to which the Celts were superior to the native population can best be understood by considering the circulation of the so-called barbarian coinage[132] (Fig. 7). Money as an economic adjunct was not used in the central Balkans or in the western half of the Carpathian area until after the consolidation of Boian and Scordiscan power. All the coins are copies of various Greek tetradrachms, in particular the Philippus, though these did not themselves circulate to any marked extent either in Moesia Superior or in Pannonia.[133] Thus the copies were not a replacement for money already in circulation, but represented the first coinage to serve a function in local economic life (Pl. 2). The introduction of money originated with the Celtic tribes and even later was generally restricted to those areas under Celtic control. Mapping of coin-finds shows concentrations, in the region of Sirmium, attributable to the Scordisci (Fig. 7); in the Danube valley from Vienna to the Sirmium region, where Celtic control was already firmly established in the fourth to third centuries; and in a strip stretching from the Danube as far as the Carpathians, in the north of present-day Hungary and in Slovakia, which can be traced back to the Celtic advance in the third to second centuries. That south-west Pannonia has not produced very many coin-finds to date is perhaps due to the fact that Roman money began to circulate there at an early stage (Fig. 7).

It is impossible to state precisely just when these primitive coins first began to be minted. The standard authority on Celtic coins in the Carpathian area dates the earliest mintings to the end of the second century;[134] more recent Slovakian specialists favour a century earlier.[135] In actual fact the period of the Celtic invasion of the Balkans, when the Celts first came into contact with Greek money, must be taken as the earliest possible date. Moreover, Celtic coins were first found in excavations at late Celtic *oppida* along with moulds for dies and occasionally tools for striking the coins.[136] It was precisely in the late Celtic period that trade flourished; certain types of bronze wares, weapons and jewelry spread over a remarkably large area, indicating a very uniform *oppidum* culture. Both Polybius and Posidonius knew that there was vigorous trade along the Save. The entrepôt was Aquileia where, according to Strabo, the barbarians from the Danube area exchanged their goods, slaves, cattle and skins against wine, oil and products of the sea.[137] These commodities cannot, of course, be attested archaeologically. There is, however, evidence that Italian families in the latter part of the republican period owned slaves with Illyrian names[138] and

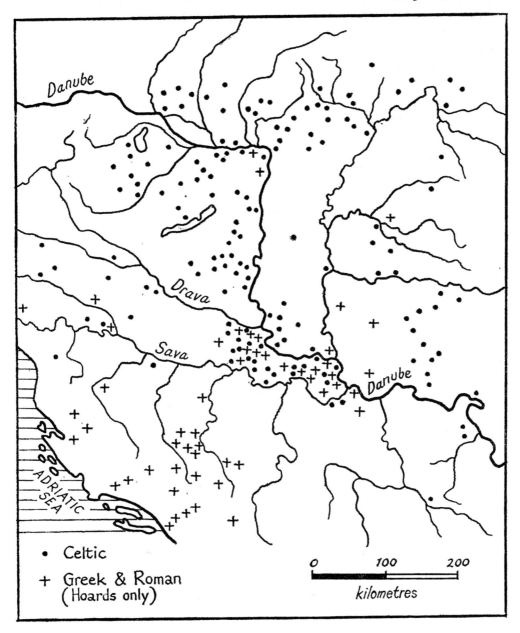

Figure 7 The distribution of Greek, Roman republican and native coins

also that Italian merchants in the same period established a trans-Alpine branch in Nauportus not far from Ljubljana.[139] Roman money did not, however, reach Pannonia and Moesia Superior until after the conquest of the Save valley by Octavian[140] (Fig. 7). Money obviously served only the needs of local trade, and consequently the area in which the individual types of Celtic coins circulated was fairly small. Most coin-finds represent a narrowly restricted series of related types. In contrast to those of other barbarian countries, the coins do not bear chieftains' names, probably because they were minted not by the chieftains themselves but by traders who then had them circulated within the radius of their activity.[141]

The only exception is the region of Sirmium, where republican *denarii*, Greek tetradrachms and coins of Apollonia and Dyrrhachium are frequently found,[142] and where, according to K. Pink, the minting of coins by the eastern Celts first started (Fig. 7). The important part played by the Scordisci in the history of the Balkan peninsula makes the circulation of coins readily understandable, or, more correctly, the important role of the Scordisci may be attributed largely to the fact that they continued to occupy an area which was the meeting-point of the most important routes in south-east Europe. Roman coins and those of the Adriatic coastal towns reached the Sirmium region along a route which, starting from the Adriatic, crossed Bosnia in a north-easterly direction. This link explains not only why the Scordisci took part in the Dalmatian wars but also why under Tiberius this very route was developed as one of the first roads in the Balkans. The reason why the Scordisci were crushed by Tiberius in 15 B.C. also becomes clear at last: their territory was of the utmost importance for the control of both Pannonia and Moesia.

Chapter 2
Conquest of the Danube region and Augustan frontier policy

Rome's interest in the region east of the Alps was first aroused when minerals, primarily silver and iron, began to be mined in the areas occupied by the Celts. Rome's first and most important approach to these areas was the founding of Aquileia, at the mouth of the Natiso, in 181 B.C., after fluctuating battles with the Celtic tribes of the north-eastern Alps. The establishment of this harbour was of vital importance in the subsequent history of the Danube provinces. There then followed a variety of attempts to advance northwards, culminating ultimately in the successful exploitation of Norican iron on Rome's behalf. During the second century Aquileia had become an extremely important Mediterranean entrepôt; here, iron from Noricum was loaded on to ships; and from here the trade of north-east Italy and that of the north-eastern Alps was organized. The extent to which the merchant houses in Aquileia were involved in the exploitation of Norican minerals is attested by recent Austrian excavations on the Magdalensberg near Klagenfurt.

East of the Julian Alps there was no subsidiary depot from Aquileia similar to that on the Magdalensberg. Whereas Appian's *Illyrike* contains a great deal of information concerning the early history of Illyrian–Roman relations, it particularly emphasizes that the Illyrian races beyond the Alps were long ignored by the Romans.[1] The manner in which the Romans gradually obtained control of Norican iron, without at the same time formally conquering and occupying Noricum, indicates that they were not concerned with the annexation of Noricum as a province but solely with acquiring control over its minerals and trade. Iron and silver meant more to the Romans than the slaves, cattle and skins which the Illyrians sold in the market at Aquileia. Roman traders were concerned to keep control over Norican iron on the mining sites north of the Alps, whereas the barbarians in the Save valley were interested in transporting their goods to Aquileia in order to sell them there.[2] Rome consequently made no diplomatic efforts to involve the races of the north-west Balkans and of the Save valley in Mediterranean trade; on the other hand she frequently had to wage war against the Alpine tribes, particularly the Taurisci and the Iapodes, who, as neighbours of Aquileia, either threatened the town itself or were even more of a threat to the latter's expanding vineyards and other important agricultural estates. Furthermore, the inhabitants of the Julian Alps turned the transport of goods over the Alps between the Save and Aquileia to their advantage by acting as guides, delaying or hijacking it (in the ancient literature they are described as tribes of brigands),[3] so that from time to time the Romans were forced to take rigorous action against them. As long as the Macedonian kingdom lasted, an advance by that power into the Save valley was feared in Rome, as it could open up the route to Italy. An advance on a larger scale beyond the Alps into the Save valley took place only when the Liburnian and Dalmatian pirates threatened the interests of Aquileia, that is to say its shipping on the Adriatic. In 156 and again in 119 B.C., on both occasions in the context of Dalmatian wars, the Romans laid siege to Siscia. However, military operations having the express purpose of occupying the upper Save valley were first carried out by Octavian. The campaign conducted by the consul C. Cassius Longinus in 171 seems only to have been an attempt to establish a link between northern Italy and the Balkans. On the pretext of attacking Macedonia he advanced over the Julian Alps into the country of the Istri, Iapodes and Carni, but returned after plundering their lands.[4] These tribes, allies of Rome, complained before the Senate in 170. It is possible that Cassius Longinus only intended to intimidate Aquileia's neighbours. Otherwise there is no evidence that Rome was bent on making strategic use of the anti-Macedonian attitude of some Balkan races.

The Romans were even less interested in advancing northwards from Macedonia. The bold campaigns of Scipio Asiagenus, Scribonius Curio and Licinius Crassus had only the limited aim of pacifying the northern part of the province of Macedonia, which, as in the days of the Macedonian kingdom, was subject to invasion by various Balkan tribes.

As already mentioned in Chapter 1, accounts of the campaigns of Octavian (35–33 B.C.) and Licinius Crassus (29–28 B.C.) were distorted for propaganda purposes to suggest that both were preventive actions against danger from the Dacians. In fact, after the death of Burebista, who ruled over a centralized state and controlled large areas acquired by conquest, the Dacian kings ceased to be dangerous enemies. Octavian's aim in advancing towards the Save valley must consequently be traced to the need to establish a land-link between Italy and the Balkans and at the same time to pacify troublesome neighbours of the north-eastern Italian colonies, primarily Aquileia and Tergeste. That these Alpine tribes were successfully pacified is attested by two late Republican inscriptions from Nauportus (Vrhnika, south of Ljubljana), which indicate that there was a settlement of representatives of Aquileian trading houses on the other side of the Alps.[5] In any case it was only under Augustus that Roman *denarii* first began to circulate in what was later Pannonia, though they did so before 11 B.C., which is regarded as the official date of the annexation of Pannonia.[6] In 35 B.C. the Pannonians in the Save valley made submission to the Romans.[7] This prompted Octavian in his speech to the army before the battle of Actium to point out that Roman soldiers had advanced as far as the Danube.[8]

Not much later Crassus also advanced from the south as far as the Danube. This campaign was the last Roman military operation that we know of in the area which later became Upper Moesia. There is just as little evidence that formal annexation of this country was envisaged as there was in connection with Octavian's activities in Pannonia. Whereas the latter was formally incorporated into the empire in 11 B.C. and the imperial frontier pushed forward to the Danube, the province of Moesia was not established until much later and then, so to speak, unobtrusively. The problem of Moesia's beginnings is still un-resolved; it has, however, much in common with that of the annexation of Pannonia, whose larger northern area was only occupied gradually and in the same unobtrusive manner.

The question when and how Moesia was occupied and became a province has puzzled modern historians, since from Crassus (29/28 B.C.) to Poppaeus Sabinus,[9] the first governor, there was an interval of roughly half a century during which there were neither wars of conquest nor rebellions. For a long time it remained an open question whether Pannonia's northern boundary in Augustan times was the Drave or whether that emperor advanced to the Danube, as claimed in the *Monumentum Ancyranum*,[10] for the Pannonian wars were waged in the Save valley and there is scarcely any evidence that the Danube frontier was occupied in Julio-Claudian times. Thus, here too, there is a gap of half a century which neither archaeological material nor historical data can fill.

Events under Augustus in the northern Balkans and in the western Carpathian regions shed light on these related questions and at least make it possible to formulate the problem itself more precisely and hence to see it in perspective. It will become clear that wars of conquest and the reduction of rebellions were no more preconditions for the establishment of a province than was a permanent system of garrison-posts and fortresses throughout the new province.

In the previous chapter the purpose of the campaign waged by the young Tiberius against the Scordisci (15 B.C.?) was interpreted as being to pacify a people holding a key position and to turn them into allies of Rome. The pattern for this operation probably derived from Roman experience subsequent to the capture of Siscia. To begin with, it proved by no means easy to pacify the Celtic and Pannonian (Illyrian) inhabitants of the Save valley. In fact, in the winter of 35/34, immediately after Siscia was taken, they rebelled, and in 16 B.C. they invaded Istria, regardless of the fact that there was a very strong occupation force in north-east Italy, and even regardless of the Roman garrison in Siscia. Local disturbances which involved only the latter were probably a frequent occurrence. The Scordisci, on the other hand, were on the decline. This led them to form an alliance with any neighbour which happened to be stronger and then to play the part of mercenaries. They were, in fact, Burebista's mercenaries, and in 16 B.C., as the allies of the Dentheletae, they plundered in Macedonia. The Romans obviously soon recognized that, as neighbours of the Pannonians, Dardanians and other tribes in western Thrace, the Scordisci were a potential danger both to Macedonia and to northern Italy; but since their services were for sale, they were not difficult to win over to the Roman side.

In the years following Tiberius' campaign against the Scordisci there are reports of renewed revolts by the Pannonians.[11] In 13 B.C. Agrippa himself had to take the field against them, and after his death Tiberius was given command.[12] The presence of Augustus in Aquileia in 12 B.C.[13] is a measure of the importance he attached to these operations. Tiberius with his allies, the Scordisci, attacked ruthlessly on this occasion. There are only extremely scanty and summary reports of the very tough fighting—*magnum atroxque bellum*[14]—which ensued. From these only the following facts emerge, that one of the Pannonian chieftains was called Bato—a very common Pannonian name—that the main enemies were the Breuci and the Amantini on the lower Save, and that the battle-ground was *inter Savum et Dravum*. After the victory Tiberius had the adolescents sold as slaves, an inhuman method of pacification which was seldom applied by Augustus. Tiberius received the triumphal insignia.[15]

In the following years the Pannonians again resorted to arms. Up to 8 B.C. Roman generals, in particular Tiberius, had to go repeatedly to Pannonia to attempt somehow to subdue the rebels who were waging guerrilla warfare.[16] Meanwhile in 11 B.C. an imperial province under the name Illyricum was constituted to include the enormous area of what was later to become Dalmatia and Pannonia; the Danube was declared its northern boundary.[17]

In his personal account it is probably no accident that Augustus mentions the Danube nor that he states that it has become the imperial frontier. He considered the Elbe and later the Rhine and the Danube as the only suitable northern frontiers of the empire. It was not that he had a preference for rivers as clear dividing lines, or that as a southerner he held rivers in special regard, or considerations of that nature. Nor was it because in his day these big rivers would have been regarded as impossible to cross—in descriptions of the Celtic–Germanic north there is abundant evidence of the Rhine and Danube being crossed in both directions: Augustus clearly recognized that not only the security of the imperial frontiers but also the consolidation of Roman rule north of the Alps could be achieved only if Rome held Europe's two great rivers. Moreover he was fully aware that transport of goods and army supplies, and safe communications in general, could only be achieved along waterways. The trans-Alpine provinces could not, however, be reached easily from Italy, and certainly not by water. If Rome was interested in trans-Alpine areas it was self-evident that the rivers linking these areas must be in Roman control.

Within this context it was only of secondary importance that the Danube and Rhine should also become imperial frontiers. The campaigns which soon followed the constitution of the province of Illyricum are evidence that Augustus was intent on ensuring Rome's influence on both banks of the Danube and thereby primarily on turning the river into an imperial traffic route. Shortly after Illyricum was established as a province the governor Domitius Ahenobarbus advanced to the Elbe and also effected changes in the balance of power among the Germans by granting the Hermunduri new settlement areas[18] vacated by the Marcomanni, whose migration to the east probably occurred at that time. The Marcomanni had driven out the Boii who lived north of the Danube, and as a consequence they remained for centuries the neighbours of Noricum and Pannonia.[19] The Romans had come to some sort of agreement with the Marcomanni which, amongst other results, made trade relations possible.[20] In the following period the Hermunduri, as is well known, were on good terms with Rome.[21] Further to the east the Quadi had been granted settlement areas, probably also during the last decade b.c. Their eastern neighbours, the Celtic Cotini, Osi, Taurisci and Anartii also came into contact with Rome at the time when Illyricum was being constituted. At this period they were probably allies or subjects of the Dacians after their defeat by Burebista half a century earlier. In 10 b.c. there was a report of an incursion by the Dacians over the frozen Danube; its repulse is actually mentioned by Augustus in his *Monumentum Ancyranum*.[22] The Roman general on this occasion was probably M. Vinicius; a fragment of the inscription honouring his exploits, which was set up in Tusculum, contains details of a campaign in which the Dacians and the Bastarnae, their mercenaries, were defeated. Vinicius then advanced as far as the

Celtic races on the northern edge of the Hungarian plain and forced them into an agreement with Rome.[23] According to Florus and Velleius he was engaged as early as 13 B.C. in suppressing the revolt of the Pannonians and so was one of Augustus' specialists in Danubian affairs.[24] Another general, Lentulus, was entrusted with the task of bringing order to the southern edge of the Carpathian area. His campaign, for which a precise date cannot be given, secured the Danube for Rome against the Dacians around the Iron Gates. Moreover, according to Florus, he was also successful in keeping not only the Dacians but also the Sarmatians 'at a distance from the Danube'. Lentulus established Roman guard-posts on this side of the river.[25]

Although the precise years in which Lentulus' campaigns took place are unknown (the dates given for them lie between 10 B.C. and A.D. 11), their context within the framework of Augustus' policy for securing the Danube is clear from the brief description given by Florus. In emphasizing that the enemy (the Dacians and Sarmatians) were 'kept at a distance' he adds that this satisfied Augustus. This is the only reference to the establishment of guard-posts (*praesidia*) on the section of the Danube opposite the Dacians. There are reports of several Dacian wars under Augustus, but neither their number nor their dates are certain.[26] Whereas the German and Celtic neighbours of the newly established Danube frontier were easily pacified either by military or by diplomatic means, it was not so easy to establish peaceful or friendly relations between the Dacian kingdom and Rome. That was probably the reason for placing garrisons on the frontier, particularly that section of it opposite the Dacians. In fact in the second half of Augustus' reign there is frequent reference to a Moesian army, although at that time Moesia was not yet a province.[27] The commander of this army was subordinate to the governor of Macedonia.[28]

Within the Augustan system, which was followed by the Julio-Claudian emperors, a very prudent balance was held between demonstrations of strength and diplomacy; wherever peace could be achieved by diplomacy, that is by alliances and treaties, military measures were deemed superfluous and avoided as much as possible. Military occupation of the Danube frontier at the time of Augustus was a last resort, and Lentulus was in fact the only general to carry out such a measure. 'The watch on the Danube' even under Tiberius merely meant that legions were stationed in a province which had the Danube as a boundary, not that they were actually stationed on the bank of the river.[29]

A further method of consolidation, often successfully adopted under Augustus, was the resettlement of large ethnic groups. We have already seen reason to suspect that the migration of the Hermunduri and the Marcomanni was brought about either at Rome's instigation or at least with her connivance. In the first decade A.D. Aelius Catus, commander of the Moesian army and possibly governor of Macedonia, had 50,000 Getae settled south of the Danube in what was later Moesia;[30] the most significant movement of this kind, however was the transfer of the Sarmatian tribe of the Iazyges to the great Hungarian plain. This settlement, so important for the subsequent history of the Carpathian area, can be understood only within the context of Dacian–Roman disputes. After Burebista's victory over the Celts the Dacians were in virtual possession of the whole of the Hungarian plain. The Dacian wars under Augustus were a clear indication that the usual methods successfully applied elsewhere were useless against the Dacians. The obvious thing to do was to interpose a buffer state in the plain, which would keep the dangerous enemy at bay. The first general to come into contact with the Sarmatians was Lentulus, who forced them to respect the Danube frontier.[31] It is possible that Lentulus was in fact the general who organized the transfer of the Iazyges to the Hungarian plain. At the time of Vinicius' campaign, at any rate, they were not yet settled in the Carpathian region,[32] but by the middle of the first century they are mentioned as being old neighbours of the Quadi.[33] According to Pliny they drove the Dacians out of the plain into the mountains on its northern edge;[34] their earliest settlement area was between the Danube and the Tisza[35] and thus they were in a position at least to keep the Dacians away from Pannonia.

The great campaign to secure the Danube was launched by Augustus in A.D. 6, when he dispatched Tiberius with a very strong force against Maroboduus, king of the Marcomanni.[36] This campaign was intended to pacify the strongest German power north of the Danube. Tiberius was already deep in enemy country when the Pannonians on the initiative of Bato, chieftain of the Pannonian tribe of the Daesidiates, revolted. Tiberius returned immediately to Pannonia after he had concluded a peace-treaty with Maroboduus, which the latter respected until he was overthrown.[37]

There now followed three difficult years for Roman rule, so recently established in Pannonia.[38] As the cause of the revolt Cassius Dio cites the mass levy of the Pannonian tribes imposed by the governor, Valerius Messalla. In fact, up to A.D. 6 Pannonians had not been enlisted in the Roman army, not even after the great war of 12–9 B.C.: at that time, it will be remembered, Tiberius had had the young men sold as slaves. In the years between, boys who had escaped slavery by reason of their youth were growing up and, according to Velleius, they gave the Pannonians confidence to throw off the unaccustomed foreign yoke.

The course of this war is better documented than that of Tiberius' war of conquest, particularly as Velleius, an eye-witness, and Cassius Dio both supply equally detailed, though in many respects differing, versions of it. Neither source, however, gives any useful information concerning the critical years A.D. 7–8. What took place in Pannonia after A.D. 6 and before the surrender on 3 August A.D. 8 can only be understood if we visualize the rebels' guerrilla tactics and Tiberius' clever and cautious generalship.

All the Pannonian tribes both north and south of the Save joined the rebel-lious Daesidiates. In the north the Breuci, under their chieftain, who was also called Bato, and in the south the Daesidiates were the leading tribes. The Pannonians who at the outset were still a well-organized and unified group attacked Salona and Sirmium simultaneously: then Bato of the Daesidiates, after being defeated by Messalla, joined forces with the Breucan Bato. Soon Caecina Severus, commander of the Moesian army, came to the rescue of the Roman garrison in beleaguered Sirmium and drove the rebels into the mountains north of the city (Alma Mons—Fruška Gora). From this point onwards reports become obscure. Leaving aside events in the area later to become Dalmatia, mention should first be made of an attack on Macedonia, led perhaps by Das-menus, which was probably repulsed by the governor Sextus Pompeius.[39] Caecina Severus with his army, now reinforced by Thracian auxiliaries, had to return to Moesia because of an invasion by the Dacians and Sarmatians. Tiberius withdrew to the south-west corner of Pannonia and established his headquarters at Siscia, on the Save, a base which had proved its usefulness.

In the following year (A.D. 7) there are reports of advances under Caecina Severus and Plautius Silvanus, the latter having meanwhile been sent to the aid of Tiberius; on the other hand, we hear reports that Augustus was dis-satisfied with Tiberius' strategy of delay and as a result that, in addition to Plautius Silvanus and others, Germanicus was ordered to the seat of war. Tiberius had, however, clearly recognized the situation and adapted his tactics accordingly. Already, while troops under Caecina Severus and Silvanus were being assembled for deployment, it became clear that the rebels were not to be lured into open encounter. A surprise attack by them west of the Alma Mons in the marshes of the *Hiulca palus* (also known as the *Volcae paludes*) almost led to the defeat of the Roman troops. The result was that Tiberius withdrew his men to a mountain ridge between the Drave and the Save—known in his circle as Mons Claudius[40]—applied a scorched earth policy and awaited the collapse of the rebellion which was showing increasing lack of cohesion. Meanwhile negotiations with some of the Pannonian leaders were probably being started, for when on 3 August in the following year the Breuci were forced by hunger and disease to capitulate on the river Bathinus (Bosna),[41] their chieftain, Bato, was rewarded with the rank of king. This same Bato also captured Pinnes, one of the rebel leaders, and handed him over to the Romans.

But Bato's kingdom did not last long. The other Bato, chief of the Daesidiates, attacked him, and the Breuci who had meanwhile risen in revolt handed over their king. The revolt was crushed by Plautius Silvanus.[42] Probably it was in this same year (A.D. 8) that Illyricum was divided into two.[43] The northern part, probably not officially called Pannonia until much later,[44] had already been pacified; in the southern half, Dalmatia, it was not until the following year that Bato, that doughty and consistent enemy, was defeated.

After the rebellion was crushed, such Pannonian youths as were capable of bearing arms were recruited as auxiliaries and withdrawn from the province. Some *alae* and *cohortes Pannoniorum* were raised at this time and in addition no fewer than eight cohorts from among the Breuci.[45] The *cohortes Latobicorum* and *Varcianorum* likewise probably now came into being. In the Julio-Claudian period all Pannonian auxiliaries attested either on inscriptions or in military diplomas come from southern Pannonia.[46] Recruitment of auxiliaries from among the northern Pannonian tribes did not begin until later, roughly about the middle of the first century.[47] It would seem that only the tribes south of the Drave took part in the rebellion: Suetonius suggests that north of the Save only the Breuci were involved.[48] The next chapter will show that after A.D. 8 Tiberius used very harsh measures of pacification, which brought about the disintegration of the formerly powerful Breuci. Thus it is very probable that before that date the Breuci, as the only rebels in Pannonia, were in control of a fairly big area in the Save valley. In any case, not all the tribes even in southern Pannonia rebelled; the Scordisci seem to have remained steadfast in their support of Rome.[49]

In the last years of Augustus' reign there was peace on the middle Danube. Lentulus' war against the Dacians and Sarmatians may have occurred at the latest around A.D. 10. There is much to be said in favour of this date, since according to Florus, our main source, Lentulus settled both the Dacian and Sarmatian problems for quite a long time to come. Yet in A.D. 6, these two tribes had attacked Moesia.[50] Thus the settlement of the Iazyges in the Hungarian plain might well have taken place around A.D. 10. The date given by most scholars, however, is 17–20, when Drusus was active in Illyricum.[51]

The years of peace which followed A.D. 10 or 11 brought no illusion at Rome that the military and diplomatic measures undertaken across the Danube had consolidated the situation in final form. Apparently Augustus wished first to embark on further steps south of the river with the aim of consolidating his rule. In 14 he sent Tiberius to Illyricum *ad firmanda pace quae bello subegerat*.[52] Details

39

of this commission are lacking as the death of the aged emperor shortly afterwards forced his return. One of his tasks may be inferred from what followed.
On the news of the death of Augustus the legions in Pannonia rose up in
revolt.[53] They had fairly substantial reasons: according to Tacitus, the soldiers—
in the main old and about to be discharged—were opposed to grants of 'un-
cultivated mountainous land and swamps' for veteran allotment. Now it is
known that the *deductio* of colonia Julia Emona (Ljubljana) was actually started
in that year. By the end of 14 the new town's wall was finished.[54] The soil
round Emona was, however, far from suitable for agriculture; the Laibach
marshes reached right up to the town-wall, and from the further edge of the
marshland there rose the spurs of the Julian Alps. At the same time or even
earlier, legio XV Apollinaris was transferred to Carnuntum.[55] From this it is
clear that Augustus wished among other things to start the colonization of
Pannonia and simultaneously to transfer a legion close to the Marcomanni.

Both tasks were carried out by Tiberius. His other work on the middle
Danube was no more than a logical, though less energetic, continuation of
Augustus' programme. When in 17 he sent his son, Drusus, with special
plenary powers to Illyricum,[56] the latter's tasks were probably not basically
different from those which his father should have carried out in 14. The subsequent period throws light on the functioning of Augustus' Danube policy.

In 17, when a conflict was about to break out between Maroboduus and
Arminius, the former, relying on a peace-treaty concluded with Tiberius,
appealed to Rome for help. The plea remained unanswered. Instead Tiberius
had Maroboduus overthrown by Drusus. The German king fled to the empire,
where he was granted asylum and interned in Ravenna. Shortly afterwards the
new king, Catualda, who had been installed by Drusus, was also overthrown,
and he, too, was given asylum in the empire. A new political structure was then
formed north of Noricum and Pannonia under Vannius, king of the Quadi,
whose rule lasted for thirty years. Drusus, who knew how to exploit these
upheavals in Rome's interest, celebrated an ovation in Rome in 20.[57]

The policy initiated by Augustus also stood the test when Vannius was overthrown in 50.[58] Rome was uninterested in the internal politics of Vannius' kingdom so long as he remained a reliable vassal and his position was sufficiently
strong. Meanwhile he had considerably extended his empire eastwards and had
made some neighbouring tribes, including possibly the Cotini and Osi, liable to
tribute.[59] He also made some kind of alliance with the Iazyges, for the royal
cavalry was recruited from them. This extension of power by a Romanophile
king was welcomed by Rome; but when Vannius' control began to wane
because of a palace revolution, the Emperor Claudius was concerned not with

supporting him but with winning over the leaders of the revolt to the Roman cause. Palpellius Hister, governor of Pannonia, was ordered not to intervene in the civil war raging in Vannius' kingdom, but to recognize the latter's rebellious nephews, Sido and Italicus. Along with his followers Vannius was settled in Pannonia and his nephews entered into their inheritance as vassals of Rome.[60]

The combination of methods pursued by Augustus in his Danube policy is seen particularly clearly in the activities of a Moesian governor. In the second half of Nero's reign Plautius Silvanus Aelianus had more than 100,000 barbarians settled in the province, intervened militarily in a Sarmatian movement which was under way, forced princes of the Dacians, Bastarnae and Roxolani to make submission before Roman standards, took hostages and confirmed certain kings in power. For his services he later received the triumphal insignia from Vespasian.[61]

This successful Danube policy lasted until the Year of the Four Emperors. So long as there was no doubt that Rome was in a position at any moment to intervene with force or diplomacy in the affairs of the peoples beyond the Danube, it was not difficult to maintain a balance. In the confusion of the years 68–9 the best troops were, however, withdrawn from the Danube provinces and in various places a power vacuum ensued between the rival Roman parties. The governor of Pannonia, Tampius Flavianus, the legionary commanders who met in Poetovio, and the procurator of the province, Cornelius Fuscus, were still able to apply Augustus' methods until, along with the Pannonian army, they became involved in the civil war. But in Moesia, which had been denuded of troops and which, of course, confronted the Dacians (among whom little sign of submissiveness was to be found), the old methods completely collapsed.[62]

Once again the example of Pannonia reveals how these methods functioned in time of crisis. After it had been decided in Poetovio to march with the Pannonian legions to Italy against Vitellius, various measures were taken to deal with the situation beyond the frontier. Sido and Italicus, princes of the Quadi, and the princes of the Iazyges not mentioned by name, were summoned to a conference to discuss the furnishing of military aid in accordance with the existing client treaty. When the 'discussion' ended, the princes of the Iazyges were detained as hostages and the troops they had brought with them were sent home; Sido and Italicus, together with their troops, were taken to Italy. The difference in the treatment of the two client-states is evidence both of familiarity with the situation and of mature judgment of it. Sido and Italicus were loyal to Rome, but were hated by their own people. If the troops which it was their bounden duty to provide had been taken to Rome without them, the

Internal conditions improved to such an extent in Pannonia during the decade and a half after the crushing of the rebellion that legio IX Hispana could be temporarily transferred to Africa against Tacfarinas. At the time, Tiberius reported to the Senate that the Danube was being guarded by two legions in Pannonia and by two in Moesia.[75] It was about this time that the latter was constituted a province. While no date can be given for this event, Appian states explicitly that Tiberius annexed it to the empire.[76] The first governor of Moesia, Poppaeus Sabinus, was at the same time governor of Achaea and Macedonia, and when he died after twenty-four years' service in the Balkans the two un-armed Greek provinces and Moesia, which was occupied in strength, were again entrusted to a single governor, Memmius Regulus. Constituting Moesia a province was thus probably only a formality and was not preceded by a war of conquest. There was no change in the command of the forces stationed in the province: their commander was responsible as *legatus pro praetore* to Poppaeus Sabinus.[77]

It would seem that Tiberius made no fundamental alteration in the deployment of troops which had been arranged in broad outline by Augustus: Pannonia and Moesia each had a legion on the Danube, there were two in the interior of Pannonia and one in the interior of Moesia. Where legio IIII Scythica was stationed is still an open question. Most probably it was somewhere in the south, close to Macedonia; Scupi springs first to mind as there is epigraphic evidence there,[78] or Naissus where auxiliaries are attested in the first century.[79] Both places were important strategic points in the link between Macedonia and the Danube.

This route was far from easy to negotiate. The Vardar ravine on the Macedonian border initially presented an obstacle, while the narrow gorges (Serbian Klisura) in the Morava valley north and south of Naissus and in the Timok valley made the deployment of troops difficult (Pl. 7). It was in fact as difficult to reach the Danube from Macedonia as it was from Italy. There was a much easier road from Macedonia through the Struma (Strymon) valley via Serdica and then through the Isker valley. Crassus had already used it in 29 B.C. This road terminated at Oescus.

In view of the difficulties of communication Tiberius vigorously set about building roads. In the summer of A.D. 14 Pannonian legions had already been assembled for this purpose and for the construction of bridges in the upper Save valley.[80] It obviously took several years to complete the roads from Aquileia to Carnuntum on the Danube and as far as the mouth of the Save. During Drusus' administration in Illyricum further roads were built; several connected Salona to the Save; that this was a natural direction for them to take is attested by the distribution of Republican coins (Fig. 7).[81] The building of the road between Salona and Sirmium in 19–20 provided a second link between the central position of Sirmium and the Adriatic.[82]

These roads, radiating from the Mediterranean to the Danube, provided only overland links which, compared with other possibilities, were always of secondary importance. Communication by water, as has been seen, could only be achieved by linking the provinces along the Danube. This became necessary at an early stage, since it was easiest to supply the legions stationed on the river at Carnuntum and Oescus by water. Either Augustus or Tiberius developed a Danube fleet. The first mention of it is in 50;[83] however, it must certainly have been in existence by the end of Tiberius' reign as his most outstanding achievement in the Danube provinces, the building of the cliff road in the Djerdap (Iron Gates) (Fig. 8), is impossible to imagine without a fleet. Just as in the last years of the reign of Augustus Pannonian legions were used to build roads in the upper Save valley, so the two Moesian legions were concentrated in 33–4 in the ravine above the Iron Gates for the purpose of carrying out large-scale construction work.[84] Once the radial roads to the Danube were finished, their termini had to be connected. This done, work on the Danube frontier was complete.

In order to appreciate properly this colossal undertaking it is necessary first to give a brief description of this section of the Danube.[85] Where it cuts through the Carpathian and Balkan massif it flows for about 130 km through a very narrow gorge; in the Carpathian basin the river in some places is 600 m wide and occasionally even wider, but it then forces itself through a narrow valley which in the Kasan gorge is reduced in places to a width of only 150 m (Pl. 8b). In fact it is only near Donji Milanovac, where the Porečka flows into it, that there is any opening to the south; otherwise the southern bank is mostly precipitous and unapproachable. There is no natural road along this side. In the Kasan gorge, which is roughly 40 km long, the rock-face is so close to the river that ships cannot even tie up there (Pl. 9a). The rocky river-bed is an additional obstacle. An outstanding feature of the section above Donji Milanovac is the rapids, though the biggest are near the Iron Gates below Oršova at the end of the Kasan gorge. The Romans probably called the rapids in the upper Djerdap (Gornja Klisura) *scrofulae*,[86] 'little pigs', a term for rock-bound waters which is also found in other languages. In ancient literature the word 'cataracts' is also used of the Djerdap section of the Danube.[87] Finally, whirlpools and the fast current make navigation very difficult in this stretch of the river.

Construction under Tiberius was directed towards making this section generally navigable and to linking the upper and lower Danube valleys. It is no

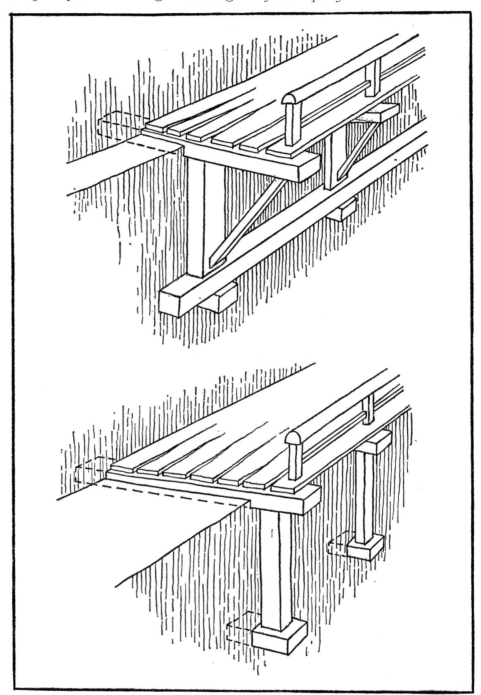

Figure 8 The tow-path in the Djerdap (reconstruction by E. Swoboda)
Top, as originally constructed.
Bottom, as reconstructed under Trajan.

longer possible to say whether and to what extent the rocks in the river-bed were removed. It is not true that all this work had to be repeated at the time of the extensive works to straighten the river-bed in the nineteenth century, for the simple reason that the work carried out then was to meet the need of steamers. In many respects the nineteenth-century works do, however, tally with the Roman, as E. Swoboda points out in his authoritative book. Particularly when travelling up-stream, ships had to be towed either by men or animals (on the new Iron Gates canal engines are used). In the Djerdap towage was, however, impossible. First of all a tow-path had to be built, a task which called for all the technical skill of antiquity. It meant either that a road had to be cut into the precipitous wall of rock, often of the hardest kind imaginable, or at least a path resting on supports had to be constructed. According to two inscriptions in the upper Djerdap, the task was completed by legiones IIII Scythica and V Macedonica in 33/4.[88] Other inscriptions indicate that work was continued under Claudius (when the same legions were involved), and under Domitian and Trajan.[89] The Domitianic inscriptions refer to the road as *iter Scorfularum* or *Scrofularum* and indicate that partly through use and partly owing to flooding the road had to be rebuilt. Trajan's *Tabula* in the Kasan gorge (Pl. 9b) encouraged scholars for a long time to regard the tow-path there as a new construction concerned with his Dacian wars. The road was, however, only useful if it made navigation possible through the whole of the Djerdap, from Golubac (Cuppae) as far as Oršova (Dierna). A new reading of the *Tabula* by E. Swoboda ultimately proved that Trajan only rebuilt the road (*viam refecit*). This rebuilding must not, however, be underestimated, since, according to the text, it was then that the road was first cut in many places into the cliff walls; as the supports and joists of the earlier road were now superfluous, they were removed. The holes hewn at regular intervals into the rock below the road and into which its supports were fixed can still be clearly seen at many places in the Djerdap. But this later road, which in places was cut into the rock to a depth of 2–3 m, also required a wooden construction. It is obvious that such a road needed constant maintenance and protection. Otherwise, on this stretch of the Danube there was hardly any danger from Dacians and Sarmatians, as the towering cliffs, and the very steep slopes on the river's southern bank, doomed any attack to failure from the start. The chief danger came from the elements, since frequently at the end of the winter ice piled up so high in the narrow gorges that the wooden supports were carried away with the melting waters. There was also the further risk that marauding barbarians would set fire to the supports and thus interrupt supplies for the fleet and troops. It was obviously to counter this danger that Tiberius had already stationed auxiliaries on this section of the frontier. But on a section 130 km long there were only a few places suitable for normal-sized forts. It can be assumed that even in later times there were auxiliary forts only at Golubac at the entrance to the Djerdap, at Čezava (Novae) and at Donji Milanovac (Taliata).[90] The remaining guard-posts were fortlets and watch-towers.[91]

The disposition of troops on the middle Danube was modified under Claudius to the extent that legio IX Hispana was moved from its Pannonian fortress (Siscia?) never to return, and one legion each in Pannonia and Moesia was replaced. Legio XIII Gemina took the place of VIII Augusta in Poetovio and IIII Scythica was replaced by VII Claudia, which was thereafter permanently based in Moesia. In Pannonia fortresses remained unchanged (Carnuntum and Poetovio); whether VII Claudia moved into a new base in Moesia or into the (unidentified) fortress of its predecessor is so far impossible to say. The theory is frequently entertained that it moved to its later base at Viminacium under Claudius, but in the present state of our knowledge it is almost impossible to show that a legionary fortress existed at Viminacium, Naissus, Scupi or anywhere else in Moesia in the first century.

Under Nero two legions were stationed in Pannonia and one in what was later Upper Moesia. In all the Roman period this was the time when the occupation army of Pannonia and Moesia was of least importance. It seemed as if here Rome had no far-reaching provincial plan apart from guarding and securing the Danube frontier, that even the requirements of frontier policy were restricted to a minimum. Nor did the transfer of troops in the final years of Nero's reign bring about any appreciable changes. Legio XV Apollinaris, dispatched to take part in his war in the east, was replaced by X Gemina, then by VII Gemina and also by XXII Primigenia.

The question now arises of the distribution of the not inconsiderable auxiliary forces[92] in these provinces in the Julio-Claudian period. It is only when this is appreciated that the strategy of Roman policy on the Danube and sensitive elements in the internal situation become clear. For a long time it was difficult for scholars to realize that although the Danube had been declared an imperial frontier under Augustus it is not until later that its military occupation is attested. Mommsen and subsequently many others interpreted *protulique fines imperii usque ad ripam fluminis Danuvii* as referring to the short stretch of the river from the mouth of the Drave to that of the Save[93] and assumed that the conquest of the area to the north of the former river was a gradual process. Even after it had been established beyond doubt that a legionary fortress had been set up at Carnuntum at the latest under Tiberius it was still believed that the conquest of the northern and eastern parts of Pannonia was postponed to the period between the reigns of Augustus and Vespasian. The Rhine frontier, with its strong concentration of troops and camps, can hardly be cited as comparable, for the political situation on that river was completely different. From an administrative point of view it is, however, probable that some auxiliary troops were sent forward to the Danube: there are also direct, though scanty, indications that there were Julio-Claudian auxiliary forts there.

As for the administration of the newly conquered province, it will be seen in the next chapter that the indigenous communities (*civitates*) came under the military control of a high-ranking officer, frequently a member of the Equestrian Order, from a neighbouring Roman unit. We would probably not be far wrong in assuming that in the Augustan–Tiberian period there was at least one auxiliary unit for every two civitates. The eight civitates whose territories extended to the Danube thus required at least four auxiliary regiments; it is probable that legio XV Apollinaris controlled some of the north Pannonian civitates, and that therefore the number of auxiliary troops transferred to the Danube for internal political reasons was not particularly large. The same was true in Moesia, only there it was probably not until the time of Tiberius that there was consistency in the military administration.

Neither in Pannonia nor in Moesia have archaeological excavations produced evidence of pre-Flavian auxiliary forts. Perhaps the only exception is at Adony (Vetus Salina) south of Budapest, where a short-lived fort may possibly have existed in the middle of the first century.[94] There have been fairly extensive excavations of the forts of the later *limes* in Hungary, particularly between Arrabona and Intercisa, but nowhere, apart from Adony, is there any evidence that can possibly be dated to the pre-Flavian period. Either pre-Flavian forts were not sited on the Danube or later forts were not built on pre-Flavian sites. At Vetus Salina fort II, which is possibly pre-Flavian, was discovered south of its successor. More recently, excavations in Százhalombatta (Matrica) and at Intercisa have revealed traces of fort ditches outside the area of the later sites, but as yet they have not been dated.[95]

Small finds which can be dated, such as pre-Flavian samian or coins, yield even less conclusive evidence of auxiliary forts on the Danube. More reliable evidence for such sites at some points along the river comes from tombstones of pre-Flavian auxiliary soldiers and veterans. Such tombstones are unknown on the Moesian frontier but are numerous in Pannonia. The fact that these auxiliaries were almost without exception troopers belonging to various *alae* is probably the result of a tactical principle of the early occupation period. On the evidence of tombstones the following places along Pannonia's Danube frontier can be regarded as sites of pre-Flavian auxiliary forts: Arrabona,[96] Brigetio,[97]

Aquincum,[98] Lussonium,[99] Teutoburgium[100] and Malata[101]—a chain of pretty evenly distributed sites. The question now arises as to whether they all existed under Augustus and Tiberius. In view of the complete lack of supporting archaeological evidence, the view here suggested is that most of the fort-sites listed and possibly others at present unknown were not established until about the middle of the first century. Proof of this is, of course, impossible. At Arrabona, where ala I Augusta Ityraeorum and ala I Pannoniorum left behind many pre-Flavian tombstones, a recent study of early Roman finds[102] suggests that the site was not established until after the middle of the century; at Brigetio, where Nero's praetorian prefect C. Nymphidius Sabinus as *praefectus alae* set up an altar,[103] the evidence of samian likewise begins after the middle of the century;[104] at Aquincum in the fort of ala II Asturum and ala I Hispanorum early Roman archaeological material goes back to the same period.[105]

Whether the establishment of these forts was an innovation on the part of Claudius it is impossible to say. Be that as it may, he was the emperor who founded the first *colonia* north of the Drave; moreover the first definite indications that auxiliaries were being recruited in northern Pannonia also date to the Claudian period.[106]

The main body of the occupation force in pre-Flavian times was concentrated either with the legions or in the interior of the province. Here the problem is even more difficult than it is with the occupation forces on the frontier. At least we know that a significant number of the *auxilia* were accommodated in or around the legionary fortresses; they are, however, epigraphically attested only at Carnuntum, and even there only in small numbers.[107] In the early period it was exceptional for auxiliary soldiers to set up tombstones: they probably did so only where there was an active stonemason's yard or where one within easy reach could deliver the finished product. Stonemasons who could chisel a tombstone, and especially those who were specialists in that field, were to be found in pre-Flavian times only at a few places in western Pannonia, mainly near legionary fortresses and at Emona, Savaria, possibly at Scarbantia, and perhaps also at Aquincum in eastern Pannonia. Except at Carnuntum tombstones of auxiliary soldiers have been found only at Aquincum, on some fortsites on the Danube and in the vicinity of Scarbantia. Tombstones of soldiers belonging to ala I Hispanorum and ala I Pannoniorum[108] have been found at Mattersdorf, north of Scarbantia, and at Gyalóka to the south of it respectively. It can be deduced from these finds that cavalry units were stationed on the road from Savaria to Carnuntum in the middle of the first century. This stationing of cavalry units on roads leading to the imperial frontier is regarded as the most important characteristic of the Julio-Claudian system of occupation. Whether the recently discovered earthwork fort at Gorsium (Tác) was also part of it has not yet been established. It is possible that it was garrisoned by the ala Scubulorum.[109] Apart from that at Gorsium, mention should also be made of a fort discovered by excavation at Nadleški Hrib, south-east of Emona, which has, however, not yet been firmly dated. Its site is close to that of a hill-fort (Ulaka) belonging to the native inhabitants and must have served to keep the latter under observation.

In Moesia pre-Flavian tombstones of auxiliary soldiers have been found only at Naissus and possibly at Ratiaria. That at Naissus commemorates a Cretan soldier who had probably been discharged under either Claudius or Nero.[110] The stone at Ratiaria might alternatively be of Flavian date.[111] In any case the former allows us to infer that auxiliary units were stationed on the road through the Morava valley in pre-Flavian times. North of Naissus, near Ravna (Timacum minus) in the Timok valley there is epigraphic evidence of an early Flavian auxiliary fort;[112] it is highly probable that it had a predecessor on the same site. Place-names in the Morava valley, such as Horreum Margi and perhaps Praesidium Dasmeni and Praesidium Pompeii, suggest that there was a road there with guard-posts and supply depots. If legio IIII Scythica and also perhaps temporarily VII Claudia were stationed in southern Moesia (at Scupi? or Naissus?) it is reasonable to suppose that auxiliary units held a road through the valleys of the Morava and the Timok.

The *auxilia* attached to the army of occupation in Pannonia were already being augmented in the Julio-Claudian period by local recruits. Tacitus makes special reference to this with regard to the year 50.[113] Some tombstones referred to above are those of south Pannonian cavalrymen. Tombstones and military diplomas clearly indicate that the civitates on the Save and Drave provided the army with a very large number of auxiliaries. The Breuci, Colapiani, Cornacates, Sisciani, Varciani, Iasi and Latobici, all of them south Pannonian tribes,[114] are mentioned by name. Only the first four of these are attested as contributing to the Pannonian army;[115] the young men from the others were used outside the province. This pattern may be due to fortuitous transmission of information; nevertheless it is striking that the Colapiani lived in the Kulpa valley in the vicinity of Siscia and had therefore been under Roman control since 35 B.C.; Siscia had also been Tiberius' headquarters not only in 11–9 B.C. but again in A.D. 6–9, so it may be assumed that the Colapiani, who were among the earliest natives to be pacified, were also among the most reliable. The same was true of the Varciani, western neighbours of the former. The gradual extension of local recruitment of the *auxilia* is easy to trace. By the middle of the first century not only the Iasi but now north Pannonians, too, were being recruited into the

Pannonian army.[116] It is possible that more Pannonians were being drafted into the navy, since in the Year of the Four Emperors the fleets of Ravenna and Misenum were both largely manned by Pannonians and Dalmatians.[117] Tribes which were not yet thoroughly pacified were probably considered suitable material for the navy. That the gradual extension of recruitment areas is a palpable symptom of the consolidation of Roman rule is proved by the fact that until late in the second century Upper Moesians were not only not called up for service within the province but were not used outside it. It should be borne in mind that it was the governor's order for recruitment which was the immediate cause of the Pannonian–Dalmatian revolt in A.D. 6.[118] In Moesia, which was never acquired by conquest, drastic intervention was avoided; in contrast to the cohortes and alae *Breucorum*, *Latobicorum*, *Varcianorum* and *Pannoniorum* there was only a single auxiliary unit bearing the name of a Moesian tribe: this was the ala *Vespasiana Dardanorum*, but even so it was a creation of the Flavian period.

Chapter 3
Native population and settlement

The principal peoples and tribes of pre-Roman times are all mentioned also in the Roman period as inhabitants of Pannonia and Moesia. Our most important source is Pliny the Elder, who in Book 3 of his *Natural History* describes the provinces and lists their inhabitants. These lists, arranged alphabetically, are based on a description of the empire, begun by Agrippa and continued under Augustus; there is every reason to believe that such lists were the official registers of the *civitates peregrinae*. A later listing of indigenous tribes, arranged geographically and probably also based on official sources, is the work of Ptolemy; however, the probability is that it is neither complete nor administratively accurate. As far as Pannonia is concerned, there is substantial agreement between Pliny's list and that of Ptolemy; for Moesia, however, there is a marked discrepancy which is not without a certain significance.

Apart from the peoples who were of importance before the province of Illyricum was constituted (the Boii, Taurisci, Breuci, Andizetes, Amantini, Scordisci and Latobici) (Fig. 9), Pliny's list[1] contains a whole series of peoples who are mentioned first in imperial times, some of them only by him. It is conceivable that such peoples were not included in descriptions of conditions and events in pre-Roman times because, being of little importance, they were under the control of their more powerful neighbours. This may well have been true of some, but is not generally applicable. A considerable number of peoples in Pliny's list have Latin names, which suggests that they were in fact new groups organized by the Romans. Some of these 'tribes' were called after places: Cornacates after Cornacum (Sotin, on the Danube, east of Mursa), Oseriates (place-name not attested), Varciani after Varceia (a place which, although

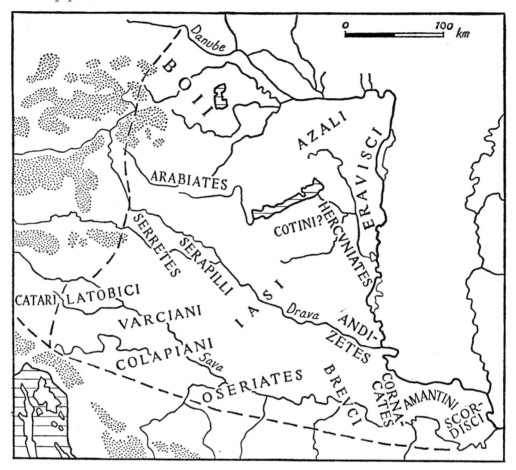

Figure 9 The native peoples of Pannonia

attested,[2] cannot be located); others were called after rivers: Colapiani (Colapis, Kulpa), Arabiates (Arabo, Rába, German Raab). To these may perhaps be added the Hercuniates, called after the *Hercynia silva* (this probably preserves the legend that they came from the north). Apart from the Hercuniates and the Arabiates the peoples with Latin names lived in the Save valley (Fig. 11). Normal tribal names not attested in pre-Roman times are to be found in the Drave valley (the Serretes, Serapilli, Iasi) and in northern Pannonia (the Azali and Eravisci).[3]

It is well known that the tribes who put up the strongest resistance to Tiberius, and who also led the Pannonian–Dalmatian revolt, were located in the Save valley; our sources mention the Breuci and the Amantini. But between them in Roman times the Cornacates appeared, a group which had without doubt been organized by the Romans. In the Save valley, west of the Breuci, were the

Oseriates, Colapiani and Varciani, all of them with Latin names. It would seem that after the subjection of the Pannonians, the extensive area formerly controlled by the Breuci and Amantini was dismembered, and some of the new administrative units created by the Romans were either given the names of the old tribes, or, in most cases, were called after places or rivers. This drastic interference in the balance of power is readily understandable after a *magnum atroxque bellum*.⁴ Tiberius also adopted a measure which was otherwise seldom used: in 11 B.C. he had Pannonian youths sold as slaves.⁵ North of the Save valley such ruthless measures were unnecessary. In the Drave valley and in north Pannonia potentially dangerous tribal alliances and concentrations of power were probably dissolved; as a result small tribes were able to gain their independence under Roman control. Only in the case of the Arabiates and Hercuniates were completely new groups formed; both 'tribes', however, lived in boundary zones; the Arabiates probably between the spheres of influence of the Boii and the Taurisci, and the Hercuniates between those of the (Boian) Celts and Pannonians. As a result the Boian empire, already weakened by the Dacians, was dissolved; living in its former territory were to be found Boii, Azali, Eravisci, Arabiates and Hercuniates.

There is additional evidence in respect of the Azali. In Chapter 1 it was seen that the Celticization of north Pannonia had already begun in the fourth century B.C.; substantial settlement of Celts in the Danube valley is attested by numerous finds of the La Tène B and C periods. In the area occupied by the Azali the personal names of the indigenous population in the imperial period are very closely related to those of southern Pannonia and northern Dalmatia. Azalian names, for which there are general analogies, occur in the same form in the Save valley.⁶ The assumption that it was in Roman times that the Azali were first transferred from south Pannonia to the Danube is supported by a recent epigraphic find in Bulgaria. A tombstone set up at the latest under Claudius for an Azalus states that he had served in a cohort of freedmen formed by Augustus.⁷ But when and how was a north Pannonian enslaved under Augustus? The only possible occasion seems to be the suppression of the Pannonians in the Save valley in 11 B.C., when Tiberius sold their youths in Italy. It is therefore quite possible that the Azali were still living in the Save valley in 11 B.C. and that it was only by Tiberius that they were transferred to the Danube, where they were given an area in the *deserta Boiorum*. It is probably not too bold to assert that by this move Tiberius wished to smash Boian power. Thus in the middle of the area dominated by the north Pannonian Celts a non-Celtic tribe was settled.

It is not surprising that Tiberius adopted such radical measures when reorganizing local balances of power. Roman initiative was behind the eastern migration of the Marcomanni, the settling of the Iazyges on the plain and the transfer of enormous masses of people to Moesia under Augustus. The probability of even more ruthless action is obvious in an area which was to be constituted a province, particularly one *perquam vicinum* to Italy.[8]

Drastic measures were necessary only in south Pannonia. There is no indication that Tiberius either had or wished to take military action north of the Drave. The military occupation of the north was perhaps based on treaties, since Carnuntum, for example, which had become a legionary fortress by A.D. 15 at the latest, and possibly a little earlier, is attested at the time of the outbreak of the large-scale revolt in A.D. 6 as 'a place in the kingdom of Noricum'[9] with which Rome had an alliance. There is no information concerning radical intervention or military action by the Romans east of Carnuntum. The settlement of the Azali was probably made possible by the fact that the land allotted to them was deserted after the Boian–Dacian war. In north-eastern Pannonia the Eravisci actually minted their own coins under Augustus on the Roman *denarius* standard. Without some kind of agreement by the Romans this would have been unthinkable. In the main the silver coins of the Eravisci were an imitation of republican *denarii* of the period 90–70 B.C., which had enjoyed very wide circulation in Pannonia since 35 B.C. The legends on these imitations are sometimes abbreviated personal names (DOMISA, ANSA, DVTEVTA, etc.) but most often the tribal name in the form RAVIZ or IRAVSCI (Pl. 2). They were probably minted in Aquincum, since their distribution occurs within a fairly narrow radius of Budapest. Only a few reached Gaul and free Dacia. It is possible to date the coins, as it was not until the end of the first century B.C. that Roman money began to circulate in north-east Pannonia. In the hoards of RAVIZ coins Augustan *denarii* appear, and one hoard even contains a Caligulan *denarius*.[10]

No doubt it can only have been on the basis of a special treaty that a people within the imperial boundaries established by Augustus was allowed to mint coins according to the Roman standard, though with the imprint of their own names. After the province of Illyricum was established, groups among the indigenous population were not all treated alike. Even after Bato had surrendered on the river Bathinus in A.D. 8, Tiberius appointed him king under Roman overlordship, and Bato set about establishing his political power in his new role in a rather self-willed and independent manner. He was, of course, soon deposed, but not at Tiberius' instigation. The peoples in north Pannonia against whom military action was unnecessary had probably acknowledged Roman control in a variety of ways; in some cases the requirements exacted from them probably did not differ from the diplomatic arrangements made, for example, by Vinicius with the Cotini, Osi and Taurisci outside the empire. It is not known which tribes formed *civitates liberae, foederatae* or *stipendiariae*; it is, however, probable that the treatment meted out to the Pannonian tribes in the Save valley was the most unfavourable, whereas those in the north were dealt with more leniently.[11] In north Pannonia sporadic instances of citizenship granted to natives are attested as early as the first half of the first century. For example, C(aius) I(ulius) Magimarus from near Aquincum was granted citizenship either by Augustus or Caligula.[12] The fact that the north Pannonians were not obliged to furnish auxiliaries until the middle of the first century is perhaps also attributable to their more lenient treatment enshrined in a treaty. It was probably only in the course of the first century that the differences between *civitates foederatae* and *stipendiariae* faded into insignificance.

North Pannonia was also used in the Julio-Claudian period for the transfer of peoples from barbarian areas. There is literary evidence of this only in the case of Vannius, king of the Quadi, who, together with his followers, was settled in Pannonia in 50. The early Roman material excavated in Pannonia, mainly in the area between Lake Pelso (Balaton) and the Danube, consists of graves and scattered finds of a Dacian and early Germanic character; this suggests that there had been transfers of peoples from the area north of the Danube.[13] We should probably not regard all these finds as the property of Vannius' people. The Dacian finds in imperial Pannonia go back to the Dacianization of the northern Carpathian region, which began in the middle of the first century B.C. and so may also be evidence of non-Dacian settlements. Particularly relevant in this respect are the Cotini who were given land in Pannonia before the beginning of the second century. In Ptolemy's lists they appear as Kytnoi in the north-eastern part of Pannonia Superior.[14] The place-name Osonibus in north-east Pannonia could well suggest the inference that the Osi, neighbours of the Cotini, were settled there.[15] Most of the Cotini and Osi, however, remained beyond the frontier until the time of Marcus Aurelius. By the time he incorporated the Cotini into the empire in the course of his great war, the majority bore Dacian names.[16] Even after the time of Marcus the Osi continued to live outside the empire.[17] The Cotini were basically a Celtic people who in the first century A.D. were not yet completely Dacianized. In his *Germania* Tacitus says they spoke Celtic.[18] Thus the Cotini, when settled in Pannonia, were linguistically related to the local population. It is possible that this was also true of Vannius' followers, as the name Vannius is frequently attested in Noricum and west Pannonia among the original Celtic inhabitants and must, therefore, be a Celtic (Boian) name.[19] One group of the Boii is also mentioned by Tacitus[20] as living in south-west Slovakia, that is, beyond the frontier. It is not improbable that Vannius, king of the Quadi, was descended from a noble Boian family and that his followers consisted of the Boians outside the empire who had come under German rule. A significant proportion of the groups which were settled in northern Pannonia in the course of the first century were thus probably of Celtic origin.

Among them was a woman belonging to the Celtic tribe of the Anartii, who had settled at Aquincum in the first half of the first century and had been given Roman citizenship.[21] The settlements of Celts and Germans from outside the empire were dispersed over a fairly extensive area and their communities attached to the existing administrative units (*civitates*). There is no indication that they formed their own *civitates peregrinae*. In Ptolemy's lists the Cotini alone are mentioned as having a continuous settlement area; however, his map of Pannonia is so distorted that it is impossible to locate them accurately. The most likely area would be south of Lake Pelso, which in Roman times was heavily wooded and in part marshy. According to Ptolemy the Kytnoi must have inhabited the area between the Azali and the Iasi.

On the basis of indigenous family names, inscriptions, local costumes depicted on tombstones and the jewelry which went with these costumes, it is possible to give a fairly accurate outline of the ethnic scene in Pannonia in Roman times. The sources—stone monuments and grave-goods—although they begin only towards the end of the first century, permit of inferences which are valid for its first half. It will probably be easier to understand the later periods if the information derived from the relevant source-material is summarized at this point.

Largish closed groups of indigenous (i.e. non-Latin) names[22] are found only in some areas of the province. In other parts they occur only sporadically in inscriptions. There are also large areas in which no inscriptions have been found, but in many cases we can supplement their evidence with that of military diplomas which were issued to members of Pannonian tribes, and which bore the name of the auxiliary soldier and of the tribe to which he belonged. Since names on military diplomas can be classed in the same onomastic group as those on inscriptions in the area in question, the diplomas can also be used as sources for the classification of family names. Non-Latin *cognomina* in the three-name system are taken to be indigenous names, provided it can be proved that the bearer of the Roman name was not a foreigner, as also are the single names of peregrini in so far as it can be proved that they were neither slaves nor foreign auxiliaries.

Peregrini are attested in closed groups in the upper Save valley, in north-west Pannonia (particularly in the region of the Neusiedler See, Lake Fertö) and also in the north-east of the province. A well-defined group was found in the upper Save valley, south-east of Emona. Within this group there are names which otherwise do not occur in Pannonia but in contrast are all the more frequent in Istria and north-east Italy, where they form the northern Adriatic group of Venetic–Liburnian family names. Among the tribes in Pannonia, though without indicating their settlement area, Pliny lists the Catari; they were probably related to the Catali who dwelt around Tergeste (Trieste). The Venetic–Liburnian group of names in the Emona region can therefore be regarded in the same light as those of the Catari. Elsewhere in the upper Save valley the names are clearly Celtic; there are analogies mainly in neighbouring Noricum. Names in north-west Pannonia are also Celtic, again with their closest analogies in Noricum. Those of the Leitha area, however, are somewhat influenced by German names, such as Strubilo, Tudrus, etc., which are often thought to reflect the settlement of people from outside the empire and, more specifically, that of the followers of Vannius. Those with indisputably German names were, however, freedmen of the original inhabitants with Celtic names; they must, therefore, have come into the province via the slave trade and not as the result of the settlement policy described above. In north-east Pannonia indigenous names fall into two distinct groups. In the western part of the region where they occur some have either no analogies or are attested in southern Pannonia, particularly in the region of Sirmium and among the Breuci (e.g. Bato, Breucus, Dases, Dasmenus, Licco, Liccaius, etc.). Although there are Celtic names (e.g. Aturo, Busturo, Ciliunus and Saco) on the eastern edge of this area, these are unknown among the rich variety of their eastern neighbours' names. The western group of names, which can certainly be identified with the Azali and the eastern group of the Eravisci, is contiguous with a north-east to south-west line from the bend on the Danube to Lake Balaton, which formed the boundary between Pannonia Superior and Inferior. According to Ptolemy the Azali dwelt in Superior and the Eravisci in Inferior. The Azalian names suggest that a south Pannonian Illyrian people was superimposed on a numerically weak Celtic population. The Eraviscan names are most closely related to the Boian in the Leitha area and to Celtic names generally in Noricum. Non-Celtic elements are found only sporadically and peripherally among the Eravisci, e.g. Bato in the south of the tribal area. Otherwise the names of the Eravisci were not only undoubtedly Celtic but equally undoubtedly belonged to the Norican–Boian stock. Two passages in Tacitus' *Germania* seem to rebut this. In describing the barbarian areas of Germany he says that the Cotini and Osi who lived between the Germans and Sarmatians were neither German nor Sarmatian; the Cotini spoke Celtic and the Osi 'Pannonian'.[23] Otherwise, there is no mention in ancient literature of a 'Pannonian language'. Earlier researchers took the *lingua*

59

CAMROSE LUTHERAN COLLEGE
LIBRARY

Native population and settlement

Pannonica to be Illyrian, and since in another chapter of the *Germania*[24] Tacitus lists the Eravisci and Osi as being linguistically related tribes living on either side of the Danube, it was inferred that the Eravisci were Illyrians. Recent researchers in the analysis of indigenous names have now proved that the names of the Eravisci were completely Celticized; the use of 'compound' and 'descriptive' names (e.g. names with the suffix –rix = king) indicates an understanding of Celtic.[25] Thus there is no doubt that the Eravisci spoke Celtic. What then was the 'Pannonian language' mentioned by Tacitus? He supplies the answer himself when explaining his hypothesis concerning migrations to and fro across the Rhine and the Danube. He assumes that these rivers are linguistic boundaries, i.e. that the Rhine separated the Germans and Celts; according to him Celtic tribes east of the Rhine and German tribes west of it were settlers who had migrated across the river. A linguistic relationship between tribes on either side of the Rhine and Danube had therefore to be explained by migrations across these rivers. Since the Osi and the Eravisci lived on the opposite banks of the Danube and spoke the same language, either the Osi must have migrated from the Eraviscan area or vice versa. Tacitus is, of course, unable to prove what happened, although it was he who propounded the whole migration hypothesis; so he admits that the direction taken by the migration is uncertain. In his main source, the Elder Pliny's history, he had only found the statement that the Osi and Eravisci spoke the same language, and as he did not regard Pannonia as belonging either to the 'Gallic' (i.e. Celtic) or to the German linguistic area he was forced to the assumption that there was a separate Pannonian language which he then ascribed to the Osi.[26]

These passages in Tacitus have called for closer examination since his theory of a migration by the Osi or the Eravisci has played a key role in the reconstruction of the history of the first century B.C.[27] The only grain of truth in Tacitus' hypothesis to survive is that the Osi, Cotini and Eravisci belonged to one and the same Celtic group of peoples in the northern Carpathian region. It was not until the first century A.D. that fate intervened to separate them; the Cotini and Osi outside the empire came under the control of the Quadi and Sarmatians and were later gradually Dacianized, particularly after the large-scale flight from Dacia in the time of Trajan, whereas the Eravisci were able to maintain their Celtic language and culture under Roman rule. At one time, however, these peoples were in the 'same state of misery' according to Tacitus.

As for the areas where indigenous names occur only sporadically or in military diplomas, it would seem that those of the Varciani were still Celtic, but those of the Colapiani already Illyrian[28] (Pannonian). Those of the Iasi, Breuci, Amantini and Scordisci are south Pannonian–Dalmatian. It is therefore highly probable that the names of the Oseriates who lived in the Save valley between the Breuci and Colapiani also belonged to the south Pannonian group. The same assumption may be made for the Andizetes, in view of the fact that Strabo regarded them as Pannonians. Celtic names with a Norican character are also met with in the Drave valley west of the Iasi, and occasional names north of the upper Drave as far as the Neusiedler See are likewise Celtic–Norican. Sporadic Celtic names are also found to the south of the Eravisci.

There are of course gaps in this distribution of names in Pannonia, as for some sizeable areas there is no available material. Nevertheless, a fairly clear picture of the linguistic groupings of the indigenous population does emerge. Celtic tribes inhabited the western and northern parts of the province, and Pannonian (Illyrian) tribes the remainder as well as a wedge-shaped area within the Celtic population in north Pannonia (Fig. 11, p. 64). We have seen that this south Pannonian linguistic island in northern Pannonia can be traced back to a settlement ordered by Tiberius. In the first century B.C. Celts were still living in the area of the Azali; it was only possible for a south Pannonian people to take it over after the Dacians under Burebista had laid waste the land of the Boii. Quite a small area around Emona belonged to the Venetic–Liburnian tribe of the Catari.

The stock of Celtic names is strikingly uniform; from this it may be inferred that the Celticization of west and north Pannonia was both protracted and intensive. Contacts with the Celtic inhabitants of Noricum were of the closest. A surprisingly large number of Celtic names in Pannonia also occur in Noricum. Celtic names attested in Gaul or in other western Celtic areas are much less frequent.[29] Among them there is, however, a small group which indicates contact with Spain (Aturo, Ciliunus, Teitia, Anbo, etc.). This detail could be ignored were it not for the fact that these names occur in the north-eastern corner of Pannonia, where a very striking type of tombstone from the early imperial period reveals the closest possible contact with Spain.[30] These tough, very simply executed tombstones were set up for indigenous peregrini; they are characterized by the so-called 'astral symbols' (Fig. 10) such as the sickle moon, sun disc (often depicted as a rosette), 'gates of heaven', stars and so-called 'keys of heaven', etc., all of them symbols which occur on tombstones in Salamanca. How these characteristically Spanish influences found their way into tombstone symbolism and affected personal names in Pannonia cannot as yet be determined. They do, however, point to the possibility that Celtic culture in Pannonia does not derive solely from the Norican–Alpine Celtic province. Research is faced with the task of separating out the different components of this Celtic culture in Pannonia. It has only recently been established that Celts from the *Hercynia*

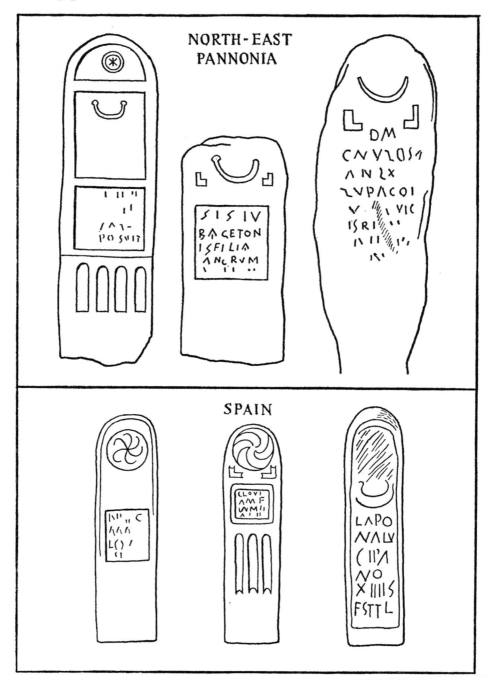

Figure 10 Tombstones with astral symbols

silva also went to Pannonia (Latobican house-urns; cf. also the tribal name Hercuniates).[31] A passage in Caesar may be quoted in support of the southern migration of the Latobici.[32] The very confused legend of the return and settlement of the Tectosages in Pannonia can perhaps be considered in connection with the Spanish contacts.[33]

The ethnographic characteristics illustrated by local costume and jewelry suggest that there was an extensive Norican–Pannonian cultural province of the Celts. This province, as far as women's clothing was concerned, was divided into several smaller costume areas (Norican hood, turban-veil, apron, etc.) which can unfortunately only be attested where women are depicted by full-length figures, or at least by busts; jewelry, however, which is the same both in gravestones and in finds is distributed over the same areas where Celtic names occur. This jewelry consists in the main of the so-called Norican–Pannonian wing-brooches (*Flügelfibeln*) (Pl. 4a), a characteristic belt-fitting frequently ornamented with open work, and a variety of other pieces of jewelry, e.g. a richly ornamented neck-charm and pair of brooches, attested on gravestones but only recently found in graves (the so-called bow-tie shaped brooches (*Maschenfibeln*)). These pieces of jewelry are restricted to the western and northern parts of Pannonia. They do not occur among the genuine Pannonians in the south.[34]

Information concerning ethnic and linguistic divisions among Moesia's indigenous population is very sparse (Fig. 11). The reason for this is not so much the lack of investigation but rather that neither indigenous names, costumes nor anything else which could lead to a grouping of the original inhabitants has come to light. There is only sporadic epigraphic evidence of names, and, to make matters more difficult, it is not always clear whether the non-Latin names on the inscriptions are those of native local inhabitants or of natives of some other Balkan area. For example, at the end of the second century A.D. in the mining area of Mount Kosmaj, south of Belgrade, there is mention of a considerable number of peregrini with Illyrian and Thracian names. It would appear, however, that they went there to work in the mines when these were started under Marcus Aurelius.[35] The only known group of local peregrini is that in the Metohija in the Dardanian area.[36] There, too, the many busts on tombstones are to be regarded as depicting local costume, but the very primitive quality of the stonemason's work permits only of very general inferences, as, for example, that brooches were probably unknown and that a little cap was worn, not unlike the felt caps popular even today in the central Balkans. Women are mainly depicted wearing a veil which hung straight down from the head and which likewise has persisted to the present day.[37] These general observations

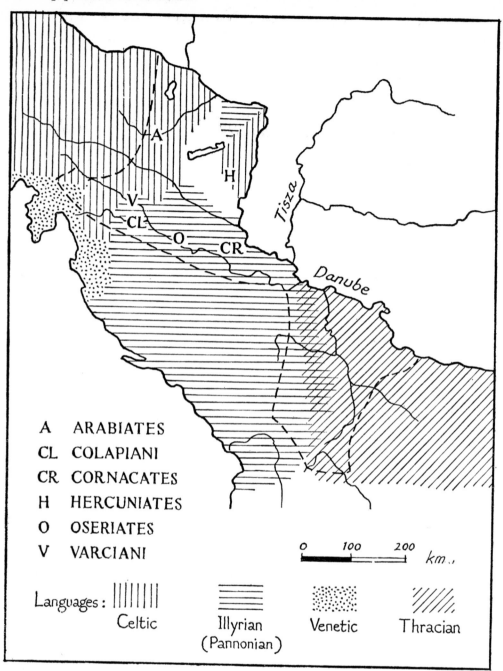

A   ARABIATES
CL COLAPIANI
CR CORNACATES
H   HERCUNIATES
O   OSERIATES
V   VARCIANI

Languages: Celtic | Illyrian (Pannonian) | Venetic | Thracian

Figure 11   Languages and tribal names with Latin suffixes

are, however, not of much help, since there is a lack of material from other parts of the province on which to base a comparison. Apart from these the analysis of names is based on those of legionaries from Scupi and Ratiaria, some of whom had indigenous names. From this scanty material the following outline emerges.

Names, where certainly local inhabitants are concerned, are clearly Thracian with a south Pannonian–Dalmatian (Illyrian) colouring which is found almost everywhere but is not completely characteristic. The list of names of the 196 discharged veterans of legio VII Claudia[38] includes, for example, soldiers from Scupi and Ratiaria called Dassius. Perhaps the name Andio, variants of which (Andinus, Andia, etc.) are attested in Dardania, should also be added. The rest of the soldiers from Scupi and Ratiaria on this list have good Thracian names, such as Bitus, Sinna, Dolens, Drigissa, Mucco, Auluzon, Mucatral and Daizo. Apart from Illyrian influence, there is also evidence of southern contact; the name Mestrius, the most frequent among the natives of Scupi, is just as common in northern Macedonia among Paionians and Pelagonians. With both Dassius and Mestrius (Mestrianus) it is a question of single names. Thus it is possible that popular names were taken over by neighbouring linguistic groups.

The Illyrian element is stronger among the Dardanians in the Metohija.[39] In addition to the commonly occurring Das(s)ius, the names Scerviaedus and Andia may both be regarded as Illyrian. Otherwise, names in this group are either Thracian (Sita, Nanea) or uniquely occurring names (Cittu, Timens, etc.). This material is too meagre to provide an answer to the old question whether the Dardanians were Thracians or Illyrians. However, a general conclusion may be permitted, that the original inhabitants of Moesia Superior were in the main Thracian, but had been exposed to Illyrian influence from the west, with the result that the Dardanian area in particular emerges as the contact zone between the Illyrian and Thracian languages. The inhabitants of Scupi probably spoke Thracian, as a Roman soldier born there in the third century considered himself a Bessus.[40] In late antiquity Bessus was the normal term applied to Thracian-speaking inhabitants of the empire; the *lingua Bessica* was Thracian. Thracian and Illyrian names occur in one and the same Dardanian family in the Metohija, e.g. Scerviaedus Sitaes (Sitae filius)[41] (Pl. 5a). Since, however, it lies on the border between Dalmatia and Moesia the Metohija cannot be regarded as the yardstick in determining linguistic classification throughout the whole of Moesia Superior. Hence it is better to assume that the indigenous inhabitants of that area spoke Thracian.

Names show hardly any evidence of the Celtic element represented in Moesia Superior by the Scordisci in particular. Some names in the western Morava valley in the Dalmatian–Moesian border area were in fact taken to show traces of

Celtic influence, but this is anything but certain.[42] Strabo and other sources mention an Illyricization and Thracianization of the Scordisci, depending upon which race had subdued the latter.[43] It has been established for Pannonia that the names of the Scordisci are south Pannonian–north Dalmatian, i.e. they have been Illyricized; the Thracianization obviously refers to those Scordisci who had managed to retain their hold on the northern part of Moesia Superior.

Evidence of the settlement of barbarians in Moesia is provided by two written sources. Aelius Catus in the reign of Augustus settled 50,000 Getae,[44] and Plautius Silvanus Aelianus in that of Nero had more than 100,000 'trans-Danubians' settled in Moesia.[45] There is much to support the assumption that part at least of these settlements took place in the area which later became Moesia Superior. In all probability those settled in that area were Getae, and were linguistically related to the indigenous population of Moesia. Finally, the Dacians in Moesia must be considered. A diploma of 71 was issued to one Dacus,[46] which suggests that there was a Dacian civitas on the right bank of the Danube. This will, however, be dealt with later.

After the constitution of the provinces the native population was assigned to administrative units known in most of the imperial provinces as *civitates peregrinae*. These civitates were either called after the local tribes and peoples (*civitas Boiorum, Azalorum*, etc.) or were given names derived from the organizational activity of the Roman government (*civitas Colapianorum, Cornacatium*, etc., see p. 53). The territories of the civitates did not necessarily correspond in size to those areas in the hands of a given tribe in pre-Roman times. As has already been pointed out, the pre-Roman tribal area of the Breuci was probably much bigger than the territory of the civitas Breucorum. Organizing the civitates probably took quite a long time. The radical measures adopted by Tiberius after the conquest of Pannonia clearly required several years, as did those taken after the rebellion had been suppressed. It is very possible that the administrative framework was not finally completed until after A.D. 8.

The *civitates peregrinae* mentioned by Pliny and Ptolemy, in largely corresponding lists, were, moving down-stream along the Danube, those of the Boii, Azali, Eravisci, Hercuniates, Andizetes, Cornacates, Amantini and Scordisci; down-stream on the Save, those of the Catari, Latobici, Varciani, Colapiani, Oseriates, Breuci, Amantini and Scordisci; down-stream on the Drave, those of the Serretes, Serapilli, Iasi, Andizetes and, finally, between the Serapilli and Boii, the Arabiates. The Belgites, mentioned only by Pliny, cannot be located. Only later in the course of the first century A.D. was the civitas Cotinorum organized—probably to the south of Lake Pelso.

There is a marked discrepancy between the lists of Moesian tribes given by

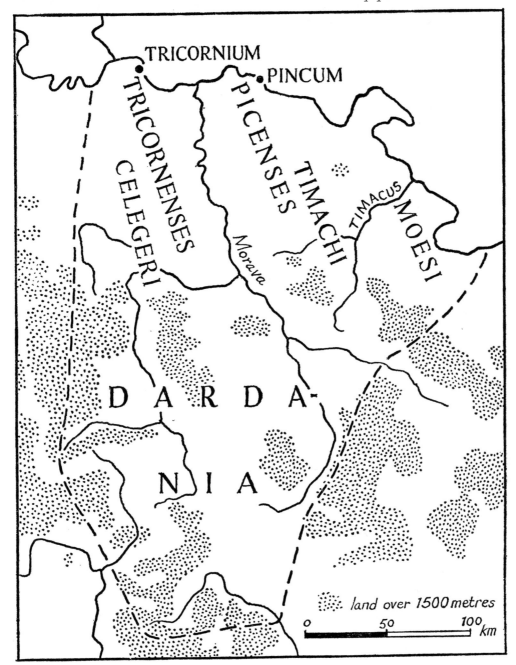

Figure 12   The native peoples of Upper Moesia

Pliny and those given by Ptolemy.[47] This suggests that there was probably a reorganization of the *civitates peregrinae* in Moesia in the first century. Pliny lists the Moesian tribes as follows: Dardanians, Celegeri, Triballi, Timachi, Moesi, *Thraces Pontoque contermini Scythae*. The sequence is geographical, not alphabetical, and hence is not based on the material collected by Agrippa for the map of the empire. Since Moesia was constituted a province only after Augustus, Pliny's list probably dates to the time of Tiberius. In support of this dating is the fact that east of the Triballi, in the area which was later Moesia Inferior, only the territory of the *regnum Thraciae* and the Scythians are mentioned in general. The Celegeri and Timachi are attested only by Pliny; they lived respectively to the north of the Dardanians on the western edge of the province and in the Timok valley. There is satisfactory proof from other sources that the Moesi lived around Ratiaria.[48] In contrast, Ptolemy lists the following tribes: the Tricornensii on the Dalmatian boundary, the Moesi on the river Ciabrus (=Cibrica), and between them the Picensii and the Dardanians in the neighbourhood of Macedonia. This list probably corresponds to the situation in the first decade of the second century. It would seem that at some time between Tiberius and Trajan a reorganization took place which was perhaps connected with the establishment of the Roman administration of what later became Moesia Inferior. This reorganization involved replacing the Celegeri and Timachi by the Tricornenses and Picenses, tribes called after place-names on the Danube. Tricornium is the ancient name of Ritopek, east of Belgrade on the Danube, while Pincum is the ancient name of Gradište at the mouth of the Pek (=Pincus). The native communities east of the Triballi listed by Pliny also took their names from places on the Danube, e.g. Dimenses, Oetenses, Obulenses and Appiarenses. This reorganization in Moesia Superior and Inferior therefore probably took place before the division of that province in 86. The actual timing was either *c*. 46, when Tullius Geminus, the governor, was ordering territorial affairs, or during the governorship of Plautius Silvanus Aelianus, who under Nero was settling large numbers of trans-Danubians in Moesia. In any case it is significant that the administrative framework in Moesia was established only gradually by the imperial government, in stages; interference in the affairs of the native population was neither so radical nor so abrupt as in Pannonia. There were probably also short-lived communities, such as that of the Dacians, which neither Pliny nor Ptolemy mention. However, according to a diploma dated 71 they also must have formed a *civitas peregrina* in the second half of the first century.[49]

The renaming and reorganization were obviously connected with the settlement of large numbers of people from the left bank of the Danube. From

Tiberius until the second century perhaps only the Dardanians, Moesi and Triballi escaped these reforms. These races are the only ones mentioned by both Pliny and Ptolemy. In the early decades of the first century they were already organized into communities under Roman control; in fact they inhabited areas which were the first to be placed under military occupation. The fortress of legio IV Scythica must have been somewhere in Dardania; that of V Macedonica was transferred at a very early date to Oescus on the Danube and a number of auxiliaries were stationed in the vicinity.[50] Ptolemy refers to Dardania as a special district in Moesia Superior.[51] It is very probable that in the last years of Augustus, even before the province was constituted, the Dardanians formed a community under Roman control. There is evidence that this was so also in the case of the Moesi, and Triballi, who were organized as civitates roughly in the second decade of the first century A.D. under the names *Moesia et Treballia*. Supreme control over these civitates was in the hands of the senior centurions belonging to the legion stationed in Oescus. In this capacity C. Baebius Atticus is epigraphically attested as being *primipilus leg. V. Mac. praefectus civitatium Moesiae et Treballiae*.[52] In Dardania the primipilus of IIII Scythica may well have been *praefectus* of the civitas Dardanorum.

The administration of the *civitates peregrinae* in Pannonia and Moesia was probably established along the lines of these early *praefecturae*. There is evidence of a *praefectura* of the civitas Colapianorum under Claudius: Antonius Naso, as primipilus of legio XIII Gemina stationed at Poetovio, was entrusted with it.[53] In the early Flavian period L. Volcacius Primus, prefect of cohors I Noricorum stationed at that time at Arrabona, was simultaneously *praefectus ripae Danuvii et civitatium duarum Boiorum et Azaliorum*.[54] This military control over the *civitates peregrinae* outlasted the Julio-Claudian period. How the *praefecturae* were allocated between officers of Equestrian rank and primipili there is no means of telling. The primipilus probably did not hold the *praefectura* over all the neighbouring civitates; even Antonius Naso was *praefectus* only of the Colapiani and not of all the other civitates in south-west Pannonia. The Boii who lived in the neighbourhood of legio XV Apollinaris had Volcacius Primus, a cohort prefect, as their *praefectus civitatis*, whereas the primipilus of that legion was probably charged with the supervision of another civitas. The civitates of eastern Pannonia, which were at a considerable distance from the legionary fortresses at Carnuntum and Poetovio, probably had as *praefectus* the commanding officer of an auxiliary unit stationed either within the area or near the civitas in question. The Julio-Claudian forts, mentioned in the previous chapter, were also probably the headquarters of military administration. It is not known whether all the civitates came under this military administration. Some of them, which had offered no resistance under Augustus and hence had perhaps become *civitates liberae* or *foederatae*, probably had no military *praefecturae*.[55] The late establishment of auxiliary forts on the Danube possibly supports this hypothesis. A definitive statement is, of course, impossible.

The officers and their soldiers could not carry out all the tasks involved in administering the communities. Even the military were dependent upon a local body or at least some representatives of the population. The probability is that the local social system, its class structure and its leading bodies were not abolished out of hand. Drastic interference in the social system may perhaps be inferred from an epitaph to a young man of the Amantini, who died while held hostage in Emona, probably under Augustus. It is emphasized that he was *Amantinus gente Undius centuria secunda*.[56] The division of the tribal clans into *centuriae* could have been based on a Roman enactment, whereby the Amantini who were engaging in active resistance were organized into *centuriae* as a means of more effectively controlling them. It is, on the other hand, just as likely that the hostages detained in Emona were organized into *centuriae*.[57] Inscriptions from the second half of the first century and from the first half of the second century contain references to native *principes*. The earliest to be attested epigraphically is perhaps *Iucundus princ. Azalus*.[58] These *principes* were members of a body on which Roman control could rely. The criteria for the selection of these leading men probably did not differ from those applied universally by the Roman government. The same formula is used by Claudius, Aelius Aristides, Antoninus Pius and others: 'the best (i.e. the most loyal) and the richest'.[59] The local aristocracy, whose members were gradually given a share in Roman control as *principes*, and adapted their interests to those of the Romans, acquired Roman citizenship in increasing numbers—some of those attested already possessed it[60]—and by the second half of the first century already formed the most important support for Roman control. A very good yardstick in assessing the consolidation of this control, as has already been seen, is the fact that there was a gradual relaxation of military occupation in the interior of the province; on the one hand the total number of troops was reduced (Claudius actually withdrew a legion) while on the other the centre of gravity was gradually shifted to the Danube frontier.

From the Julio-Claudian period there is only very sparse evidence that the local aristocracy in the civitates copied the Roman way of life and acquired Roman industrial articles. The reason was probably that this class was numerically weak at that time. So far neither cemeteries nor settlements containing imported Roman goods from the pre-Flavian period have been uncovered. The cemeteries in the legionary fortresses and colonies are, of course, the exception.[61] In graves which are clearly those of natives, Roman pottery, glasses and bronze objects have been found only on the western edge of the province along the amber road, and particularly in the *tumuli* between the Drave and Danube.[62] From this we may infer that the native population could neither afford foreign (Roman) goods nor was interested in acquiring them. Roman traders had adapted their trade to the demands of the military and foreign settlers, while Roman craftsmen had established their workshops only where there was a call for their wares. Along the amber road, from Aquileia to Carnuntum, trade was already developing in the Julio-Claudian period; it was thus possible for the native population living along this road to buy imported Roman goods. The market to which the north Italian traders directed their attention was, however, created by foreigners.

Foreigners, in the main north Italian newcomers, were already beginning to appear in Pannonia under Augustus. According to Velleius, on the outbreak of the rebellion of A.D. 6 'Roman traders were massacred'.[63] It is, however, not out of the question that his statement is no more than an indispensable ingredient of any description of an anti-Roman rebellion. Somewhat later there are reports of Roman traders settling permanently in the Marcomannic empire;[64] in fact cemeteries of early imperial date in Bohemia have produced a very rich store of imported Roman goods, particularly bronze vessels, altogether a very much richer yield than that in Pannonian graves of the same period.[65] Rome's trade with the Marcomanni—in his mention of it Tacitus refers to the *ius commercii*— probably started after the legionary fortress at Carnuntum had been established. Along the amber road, at Emona, Poetovio and Carnuntum, inscriptions referring to Italian, mainly Aquileian, civilians date back as early as the first half of the first century. Among them is a striking number of freedmen who, as representatives of north Italian trading houses, managed branches which had been established in Pannonia.[66] An important part in developing this network of trade was played by the fortresses at Carnuntum, Poetovio and possibly Siscia, where traders and craftsmen who catered for the needs of the army had settled. This group of small- and large-scale entrepreneurs, at times only tolerated and at others welcomed, led to settlements around the fortresses as early as the first half of the first century; in the oldest cemeteries at Poetovio[67] and Carnuntum[68] tombstones set up to both legionaries and civilian traders— mostly freedmen—were found. We must include here a tombstone of a man from Tergeste (Trieste) located at Siscia, where investigations have not yet begun,[69] and dating to the time of Augustus or Tiberius. There is copious epigraphic evidence of branches of foreign (north Italian) trading houses in Emona, Poetovio and Carnuntum.[70]

Archaeological material has so far failed to provide sufficiently clear information about the nature of this trade. The earliest products of Italian industry in imperial times have been discovered in only a few places and mainly in the form of loose finds; it is, however, no accident that Arretine *sigillata* (but only later forms) (Fig. 20), and bronze vessels from Campania, Aco-beakers and north Italian glass-ware dating from the Julio-Claudian period have been found only in the settlements along the Amber Road and in a few places in the valleys of the Drave and Save.[71] There is much to be said for the view that these goods were bought only by the military and by foreigners who had settled in the province. The indigenous population was still living under conditions which had existed since time immemorial, and used only local products which cannot accurately be described, since there is still insufficient information about the material culture of the later pre-Roman Iron Age. Once again it may be possible to draw an inference from a later period: in the second century there were still cemeteries and settlements in which only local goods were found, and imported goods are either not present at all or only on a very minor scale. In general it is right to assume that the indigenous population in the Julio-Claudian period maintained their late Iron Age mode of life intact.

At present it is impossible to decide whether the only thoroughly investigated late La Tène settlement in Pannonia, that on Mount Gellért at Budapest,[72] can be used as a basis for assessing the transitional period. This settlement on the steep rocky hill in the centre of the Hungarian capital acquired importance because the Danube below is very narrow at this point and routes from the Carpathian basin have to cross here. As a consequence Budapest was always the most important crossing-place in the area. An *oppidum* was established on Mount Gellért in the first century B.C., and at its foot, near the ancient ferry in the part of the town known as Tabán (Fig. 14), an open settlement with a marked industrial and commercial character developed. This settlement was protected by the *oppidum*. The traditional red-and-white ceramics of the late La Tène period were produced in the industrial centre at Tabán. The reinvestigation by É. Bónis of the rich archaeological material has revealed that there was a wide ramification of cultural and commercial contacts. There is no evidence of Roman influences, although part of the material must be dated to the early decades of Roman control. Only the imprint of an intaglio, depicting a seated Victoria with shield, with which the red and white painted vessels were stamped, shows any Roman influence (Pl. 4b).

The Mount Gellért hill-fort was abandoned around the turn of the first century B.C.; the settlement of Tabán probably lasted somewhat longer. After the Roman conquest native fortified hill-settlements were either evacuated or gradually abandoned throughout the whole of Pannonia. Excavations on a smaller scale have proved this to be so in the Braunsberg near Carnuntum,[73] the Leopoldsberg near Vindobona[74] and at others, too. Nor did the big *oppidum* of Gomolava near Hrtkovci,[75] not far from Sirmium, produce any Roman material worth mentioning. Most earlier investigators were of the opinion that these fortified *oppida* represented a danger to Roman security and hence had to be evacuated. It is, however, not unlikely that these uncomfortable, overcrowded sites, which were apparently only used as defences in an emergency, simply ceased to have a purpose after the Roman conquest. It is probable that after the abandonment of the Mount Gellért *oppidum* a settlement developed to the north of it on the site of the later legionary fortress (Fig. 14), whence archaeological material mainly of the late La Tène type has been excavated. It is assumed that this settlement sprang up around the auxiliary fort at Aquincum. Such a fort is in fact assumed to have existed there from the middle of the first century.[76] It is possible that the industrial quarter at Tabán was also transferred to the settlement, where the military provided a market for its products. It was at Aquincum, too, that the gradual transformation of industry and the native way of life received its first impulse from the army. At sites where no troops were stationed life went on as before.

What has just been said applies primarily to Pannonia. Investigation of Julio-Claudian forts and native culture in Moesia Superior has still to be carried out. Such loose finds of imported Roman goods as have so far come to light are, however, of a remarkably late date and belong to the Flavian period. Whereas the settlement of Italian traders in Pannonia can also be attested epigraphically from the first half of the first century onwards, in Moesia Superior there are no inscriptions either to soldiers or to civilians of the Julio-Claudian period. This absence of inscriptions is all the more important for judging conditions in the country, since from the time of Augustus, as in Pannonia, two legions had been stationed in Moesia. In the fortress of Viminacium inscriptions do not begin until very much later, while owing to this absence of inscriptions the other fortress has so far not even been located. We have already mentioned that, as early as the beginning of the first century, civilian settlements were established around Pannonian legionary fortresses, and tombstones were set up by their richer inhabitants. That such stones are unknown in Moesia is possibly proof that Italian traders did not arrive until much later.

The vicinities of fortresses were not the only places where foreign settlers established themselves. A small group of auxiliary veterans settled, for example, north-west of Scarbantia, near Walbersdorf.[77] Their tombstones suggest that they were set up to several interrelated families who kept slaves and freedmen.

Pliny calls Scarbantia *oppidum Scarbantia Iulia*,[78] which points to some autonomous group of Roman citizens in the neighbourhood of Scarbantia. Such groups of citizens, occasionally enjoying a kind of autonomy, settled mainly, of course, near legionary fortresses. In Scarbantia the name Iulia suggests that autonomy had been granted by Tiberius. Similar groups were to be found in west Pannonia, in the upper Save valley and probably elsewhere, though of the latter so far there is no proof. In Scarbantia one of the earliest tombstones was set up to a *dec(urio) Scarb(antiae)* who probably was not yet a decurion of the Flavian *municipium* (otherwise the inscription would have read *dec. mun. Fl. Scarb.*), but a decurion of the *oppidum*.[79] It is also not impossible that the citizens of Emona who dedicated an inscription in Savaria to Aequorna, a goddess from their homeland, were already settled in this area in pre-Claudian times. On this beautifully executed plaque they refer to themselves as *consistentes finibus Savar(iae)* and not *finibus coloniae Claudiae Savariae*.[80] From the time of Claudius this latter title was consistently used on inscriptions in the town.[81]

In the Julio-Claudian period not a single town with Roman municipal status was established in Upper Moesia, and there were only two in Pannonia. Both were *coloniae* which were founded by settling legionaries who had completed their military service, and both were sited on the amber road (Fig. 37, p. 220).

Emona (Ljubljana) (Fig. 13) was founded by Tiberius, but settlement of veterans began in the last years of Augustus.[82] The connection between the mutiny of the legionaries in A.D. 14 and the beginnings of the *deductio* have already been touched upon in the previous chapter (p. 40). It is possible that the uprising was responsible for the rapid building of the town; its wall was erected either at the end of 14 or at the beginning of 15. It encloses a rectangle measuring 522 by 435 m. The ditch, gates and posterns of the town have been excavated. Excavations carried out by W. Schmid before the First World War revealed some *insulae* and the main features of the street-grid. This rectangular network incorporates all the principles applied in the founding of veteran settlements. The excavations, which have recently been begun, will probably supply much information concerning the earliest periods which did not emerge from previous work. The extensively excavated cremation cemeteries, which have revealed a wealth of material, extended along the road from the north gate, and have provided evidence that there was a considerable settlement from the beginning of the first century A.D. The citizens settled in Emona on the basis of the *deductio* were veterans of the legions stationed in Pannonia. It is mainly those from XV Apollinaris that are attested, though there is an inscription to a veteran from VIII Augusta. Most of the persons attested by the earliest epigraphic evidence, were, however, civilians; from this it may perhaps be assumed that

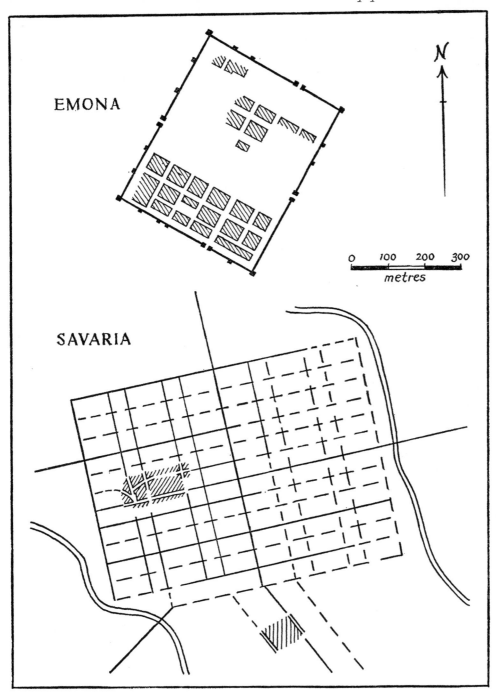

EMONA

SAVARIA

0   100   200   300
metres

N

Figure 13   Town plans 1

civilians were also involved in the *deductio*. The civilians concerned were mainly north Italians, a few from other parts of Italy, and some southern Gauls. There were also many civilians from Aquileia; this is in itself not surprising in view of the important part this neighbouring town played in the commercial life of the Save valley and the amber road. Citizens involved in the *deductio* were allocated plots of land in the upper Save valley north-west of Emona, whereas the nearby marshland, south-east of the town, was left to the native inhabitants. Evidence of the former is provided by numerous inscriptions found round the present-day village of Ig. By reason of its position Emona was destined to become an entrepôt and the most important centre in Pannonia of the commercial network based on Aquileia. This may well explain why the group of citizens from Emona so soon appeared in Savaria. The traders' settlement and entrepôt at Nauportus, established under the late republic, soon lost its importance and became a *vicus* within the territory of Emona, despite the fact that in A.D. 14 it was a settlement *municipii instar*.

Savaria (Szombathely) lies in a fertile plain on the western edge of the province. Both soil and position made it much more suitable for a *deductio* than was Emona, which had the sole advantage over Savaria that it was a communication centre. In the choice of site for founding a colony the main criterion was probably the need to establish a strong-point on the long stretch of the amber road between the legionary fortresses at Poetovio and Carnuntum. It is probably not fortuitous that the Tiberian colony of Emona was roughly midway between Aquileia and Poetovio, and the Claudian *colonia* roughly midway between the latter and Carnuntum. The *deductio* of veterans under Claudius at Savaria thus fits into the plan which originated at the time of the initial occupation of the country.

Before the time of Claudius the territory of the colonia Claudia Savaria, along with many other parts of Pannonia, belonged to Noricum. Velleius emphasizes that in A.D. 6 Carnuntum was 'a place within the *regnum Noricum*', while Pliny introduces his description of Pannonia with a brief statement about the *deserta Boiorum*, adding 'but now in the *deserta* there are the settlements of the oppidum Scarbantia Iulia and of the Claudian Colonia Savaria'.[83] The Amber Road on the western edge of Pannonia and on the eastern edge of Noricum was thus probably treated as a special area (*Noricis iunguntur deserta Boiorum*); its native inhabitants shared in the Julio-Claudian rights of citizenship granted to Noricum, indeed even the *deductio* of the colonia Savaria could easily have fitted into the Claudian municipalization programme devised for Noricum. Hence Mommsen still held the view that it was only after Claudius that Savaria was attached to Pannonia. It was in fact only gradually during the first half of the first century that places along the Amber Road became part of Pannonia: the exception was Celeia (Celje = Zilli) which remained in Noricum. Even in later times the stretch of the Amber Road between there and Poetovio ran through Noricum.

Veterans of legio XV Apollinaris were involved in the *deductio* of Savaria, a fact underlined on tomb-inscriptions.[84] There is no evidence of a similar involvement of the other Pannonian legion, XIII Gemina, which had recently been stationed at Poetovio. The veterans were allotted plots of land in the town's fairly extensive territory, as inscribed tombstones set up by Roman citizens with Italian names to the west, north and south-west of Savaria tell us.[85] They were probably mainly descendants of veterans. The earliest veterans must have had a considerable number of slaves, since, frequently, their tombstones mention several freedmen. The veteran Sextus Uttiedius Celer had at least six freedmen, which presupposes a minimum of twelve slaves.[86] Some of the veterans, however, did not remain in Savaria but returned to Carnuntum where they settled close to their old garrison. This can be deduced from the fact that on tombstones from the camp cemetery along the street starting from the Porta Principalis Sinistra some veterans of legio XV Apollinaris are mentioned as being from Savaria or belonging to the Tribus Claudia of colonia Savaria.[87] It is not known whether these veterans had sold their plots of land or whether they had been discharged with a *missio nummaria*. It is, however, not improbable that they had left Savaria because where the Amber Road crossed the Danube at Carnuntum there was a better chance of work. The veterans in Savaria bought their slaves in the slave market at Carnuntum; a freedwoman belonging to the veteran Uttiedius was, for example, called Carnuntina. In later times, too, Savaria had close contacts with Carnuntum.

Another component of the citizen-body of Savaria, and one not to be underestimated, was a number of north Italian civilians. Among its leading families there was a preponderance of Aquileian traders, principally the Caesernii, Canii, Caesii and Petronii; a number of other citizens belonged to the tribe Velina, which was the *tribus* of Aquileia. The most famous Aquileian family was that of the Barbii, who by the second half of the first century were already decurions of Savaria,[88] a fact which calls for special emphasis, since this family otherwise did not take much part in the commercial life of Pannonia. The opposite was true of the other important family from Aquileia, the Caesernii,[89] who, starting from Savaria, spread throughout the northern part of the province. Epitaphs to citizens of Savaria have also been found in Aquileia,[90] and a freedwoman of the Petronii in Aquileia was called Savarina.[91]

These interrelationships within the urban aristocracy are attested from all the towns on the Amber Road between Aquileia and Carnuntum. There is always something attractive about discovering personal and family connections from the study of family and individual names (*gentilicia* and *cognomina*). To this end there is abundant epigraphic material from the first two centuries. For example, a certain L. Atilius Saturninus had his son, a soldier in legio XIIII Gemina, buried on the southern boundary of the territory of Scarbantia. Now there is a freedman with the same name buried in Aquileia whose tombstone states that he was a citizen of Scarbantia and had been murdered by *latrones*—probably on a business trip to Aquileia.[92] Traders (*negotiatores*) are mentioned on a number of tombstones in western Pannonia (Pl. 5b); these were mainly freedmen of Aquileian trading families and bore the latter's names.[93]

As for Savaria, a group from Emona arrived there within the framework of these contacts, possibly before the town was founded. Shortly afterwards a man from Emona who was not even a Roman citizen (*Lucius Maximi filius*) was buried in Savaria.[94] Most of the evidence, however, belongs to a later date. Both in Savaria and Scarbantia inscriptions did not become customary until after the middle of the first century; and trading contacts are also first attested by finds in the second half of the century. It was, however, under Tiberius, Caligula and Claudius that trade was developed, when the group of Roman citizens was emerging at the *oppidum* of Scarbantia and colonia Savaria was being founded.

As far as can be established, the town of Savaria consisted of a large quadrilateral some 700 by 550–600 m with a rectangular street-system (Fig. 13). The town-walls were probably built at a later date. It is likely that they enclosed a bigger area than the quadrilateral colony of the *deductio*, but so far definite traces of them have not been established. Excavations to date have also failed to produce evidence of the earliest building period under Claudius. It is possible that the town acquired its urban character only gradually. The *capitolium*, probably a very imposing structure, was not built until about the middle of the second century; on the southern perimeter of the town, in the first century, there were still primitive huts built of sun-dried tiles. The sewage-system, water supply and street paving probably date back to the founding of the town by Claudius. The streets were paved with very large basalt slabs, while a subterranean aqueduct supplied water from a considerable distance (*c.* 50 km); in order to ensure the necessary gradient this conduit was adapted to the terrain, and so to begin with it ran from north to south and turned east only when it reached the altitude of Savaria.[95]

At the latest when the veteran settlement was founded under Claudius, and possibly even earlier, the land of western Pannonia was surveyed by centuriation. Until recently traces of the field boundaries fixed in accordance with all the rules of the Roman *gromatici* could be detected on modern maps.[96] The *centuriatio*

78

of the *ager Savariensis* was uniform with this system, having a deviation of 22–23 degrees from north to west. The same orientation was applied to the streets of the town. The land thus centuriated was not all allotted to veterans. The native inhabitants, particularly in the northern and western parts of the territory, were left undisturbed on their old properties; in many places in these areas there are hundreds of *tumuli*, doubtless native burial places. This is confirmed by inscriptions to peregrini with Celtic names and also by inscriptions found in the *tumuli*.[97] Indigenous Celtic names persisted in these parts into the second century; instances of marriages between natives and citizens from elsewhere are also attested.[98] Thus natives living in the open country were not simply expropriated in order to make way for the settlement of veterans. Some of them took part in the life of the town and were actually granted citizenship under Claudius; they were not completely relegated to the background.[99] In the pre-Flavian period a tombstone (Pl. 5b) was set up to a certain *Atta Bataionis filius* who was described as a *negotiator*.[100] Attention has already been called to the relatively rich store of objects found in *tumuli* in west Pannonia; bronze and glass vessels and *sigillata* pottery are not particular rarities there and indicate that there was a tendency among the native upper class to adapt its way of life to the material culture of the foreign settlers. This process obviously took a fairly long time. Even in Savaria only the framework of its urban character was established under Claudius, and the use of imported Roman goods was but the first step in the transformation of the native way of life. Nevertheless, the cemeteries so far excavated in Savaria indicate that the local production of objects of a Roman type was sufficiently considerable in the second half of the first century to oust their own native products.[101] This is in striking contrast to what is found in the northern and eastern parts of Pannonia, where, in the same period, native industry largely dominated the market and was also able to satisfy the demands of the auxiliaries.

Chapter 4
The Danube frontier from Vespasian to Marcus Aurelius

Experiences on the Danube frontier in the Year of the Four Emperors, and not least the defeat of Fonteius Agrippa, prompted the new Flavian government to revise the Augustan scheme. This does not mean that troops stationed in the interior of the province were immediately transferred to the frontiers; on the contrary, this process occupied the period from Vespasian to Trajan. Nevertheless, the occupation of the frontier areas by stronger forces and the building of garrison-posts were regarded as urgent tasks. It is reported that the governor of Moesia, Rubrius Gallus, ordered construction work to be undertaken on the Danube in the earliest years of Vespasian's reign:[1] C. Calpetanus Rantius Quirinalis, governor of Pannonia, had similar work carried out in that province. He is mentioned on four building-inscriptions, with almost identical texts. Three of these, which were found at Carnuntum and are dated to the year 73,[2] refer to a building of legio XV Apollinaris, whereas the one from Aquincum[3] was set up by an *ala*, possibly ala Hispanorum I; it is probably of the same date as the Carnuntum stones. These inscriptions attest military buildings, both in the fortress at Carnuntum and in the auxiliary fort at Aquincum; there is, however, no means of establishing whether these stations were being rebuilt in stone. The inscription from Aquincum is particularly big (roughly 3 m in length); hence the building which carried it must have been of some importance within the fort. The older reading of this inscription led to a variety of theories: for example that a stone fort had been built at Aquincum as early as A.D. 17–20 and that, therefore, the legionary fortress of Carnuntum must likewise have been of stone.[4] Since, however, it has recently been established that it is not [*T*]*i*(*berius*) *Caesar* but *T*(*itus*) *Caesar* who is mentioned on the Aquincum inscription, the date of the

stone fort is again an open question. The inscriptions of the governorship of
C. Calpetanus Rantius Quirinalis, however, do not prove the erection of stone
forts then at Carnuntum and Aquincum, as a stone building-inscription is
known from Carnuntum dated to the years 53–4.[5] However this may be, these
inscriptions of Calpetanus fit into the framework of the new Flavian frontier
policy. This can be illustrated with some further details.

First of all we have proof of the existence of the fort at Taliata (Donji Milanovac) in the Djerdap in the early years of Vespasian. Recently a military diploma
issued on 27 April 75 [6] to a cohort soldier from Antioch was discovered there.
He had presumably settled near his old garrison. Apart from this, we have no
definite evidence that auxiliary forts were established on the Danube. Flavian
military diplomas prove that it was under Domitian that the auxiliary forces in
Pannonia and Moesia were first considerably reinforced. Despite this there is
evidence that fresh auxiliary troops were arriving under Vespasian: for instance,
cohors I Noricorum equitata, which took over the fort at Arrabona,[7] cohors I
Thracum Syriaca, which was now stationed at Timacum minus (Ravna in the
Timok valley),[8] and perhaps also cohors I Raetorum, which may have been
transferred to the Danube from Syria along with cohors I Thracum Syriaca in
the expedition of Licinius Mucianus.[9] The stationing of cohorts of infantry on
the Danube frontier perhaps indicates a change in tactics, as previously it had
always been held by *alae*. In the absence of further excavation it is impossible to
say which Danube forts were constructed under Vespasian. Generally speaking,
however, it is very probable that some at least were built for the newly arrived
cohorts. It is possible that there was some connection between them and the
reorganization of the Danube fleet. There is, of course, evidence that a *classis*
was already in existence in the first half of the first century, yet both the Pannonian and the Moesian fleets were termed *Flavia*.

The new plan for guarding the Danube frontier took the form not so much of
a sudden increase in the forces in the provinces as of an alteration in their deployment. Vespasian transferred a considerable number of *auxilia* from Pannonia to the Rhine, amongst them ala Asturum II, ala Scubulorum and cohors
XXXII Voluntariorum c. R.; there is no evidence that such large numbers of
troops arrived under him as under Domitian. The first indications of the new
frontier policy are the increasing role of the cohorts and the strengthening of the
occupation forces in the frontier zone. In Moesia the catastrophic events of
69–70 made radical intervention necessary. The only legion (VII Claudia) in
the area of what was later Moesia Superior proved inadequate in face of the
Dacian danger. Accordingly, Vespasian transferred legio V Alaudae there.
Where it was based is not known; that it was somewhere in Moesia in the early

Flavian period is attested by an inscription on a tombstone set up to a veteran at Scupi.[10]

The military precautions taken by Vespasian ensured peace on the Danube for a decade and a half. There is no evidence that hostilities broke out either under Titus or in the early years of Domitian. It was, however, during this period that the reunified Dacian state came into being. The Dacians, who had risen in force in 69–70 against Rome—an indication that this people was recovering its strength—had been united, or rather reunited, into a single state by King Diurpaneus, the predecessor of Decebalus. The centre of his state, the royal seat at Sarmizegethusa, lay dangerously near the Danube frontier of Moesia.

It would seem that the outbreak of the Dacian war took the Romans by surprise. The first attack on Moesia at the end of 85 or the beginning of 86 was so unexpected that the governor, Oppius Sabinus, together with a legion, probably V Alaudae, was defeated.[11] Domitian, accompanied by Cornelius Fuscus, the praetorian prefect, went in person to Moesia. In succeeding years all the necessary conclusions were drawn: the provincial armies on the Danube were increased to several times their original strength; most of the legions were transferred to the bank of the river; command of the long Danube frontier in Moesia was subdivided. These measures initiated a very troubled period of reforms which lasted until Hadrian. Research has by no means reached the point where it is possible to outline the history of troop movements in these three decades. While there is no dearth of hypotheses, lack of cogent evidence and of reliable clues is equally impressive.

Domitian's initial measures affected the threatened province of Moesia. To replace the annihilated legio V Alaudae, IIII Flavia was transferred from Dalmatia, and from the west two more legions, I and II Adiutrix, were moved into the province. These greatly strengthened forces were divided between two commands: from 86 the Danube frontier as far as a point east of Ratiaria was assigned to the newly created province of Moesia Superior, while the lower Danube down to the Black Sea belonged to Moesia Inferior. From the start the governors of the two new provinces were consulars. The first governor of Moesia Superior, L. Funisulanus Vettonianus, had been governor of Pannonia and probably arrived in the threatened province immediately after the defeat of Oppius Sabinus. In order to unite the whole of the Dacian section of the Danube frontier under one command the region of Sirmium was attached to Moesia Superior in 86.[12] It is usually assumed that, of the newly arrived legions, I and II Adiutrix were stationed in the neighbourhood of Sirmium and IIII Flavia was in the same fortress as VII Claudia at Viminacium. This was probably the headquarters. Ratiaria and Singidunum are also possibilities. Moreover, we cannot exclude the possibility that not all the legions were stationed on the Danube frontier of Moesia Superior; one might well have been assigned to Moesia Inferior. From a geographical point of view it is difficult to understand why Ratiaria, separated as it is from Moesia Superior by the Stara Planina mountain chain, was not attached to the lower province. If, however, we can assume that one legion was stationed at Ratiaria, and the two legions already in Moesia Inferior (I Italica and V Macedonica) were augmented by a third, we can then provide an explanation why Ratiaria was assigned to Moesia Superior. For in this way each of the Moesian provinces was allotted three legions. If this is so, then there could have been only one legion at Viminacium. In any case, double fortresses were at this time increasingly coming to be regarded as unsatisfactory because of the danger they offered to internal security.

In the same year, 86, Domitian handed over command to the praetorian prefect, Cornelius Fuscus, and returned to Rome. The province was being cleared of Dacians, but in his subsequent expedition Fuscus was disastrously defeated,[13] losing both his life and his army. Domitian once again hurried to the seat of war. The command was entrusted to Tettius Iulianus, who finally won the battle of Tapae in 88. Decebalus, the Dacian king, sued for peace.[14]

At first sight it seems surprising that Domitian did not follow up his victory, or at least exercise stricter control over an enemy responsible for the defeat of one legion and the death of three Roman generals (Fonteius Agrippa, Oppius Sabinus and Cornelius Fuscus). He remained content, however, with one Dacian victory and immediately moved against another enemy, the Germans on the Danube. There is much to support the view that in the early 80s he had already been mobilizing against them; at that time he was already concentrating troops on the German Danube front,[15] and it is not out of the question that the sudden appearance of three legions in Moesia in 86 was made possible by previous preparations for an assault on the Germans. The Dacian attack came as such a surprise to the Romans simply because the government was concentrating all its attention on the Germans. This had probably not escaped the notice of Decebalus.

The causes of the threat of war on the Danube frontier of Pannonia and Noricum are not known. There is no further mention of Sido and Italicus, the kings of the Quadi who had proved loyal vassals of Rome during the Year of the Four Emperors, beyond the fact that they took part in battles in Italy on Vespasian's side and were involved in the plundering of Cremona; nor do sources mention any change of dynasty. According to Tacitus both Sido and Italicus were hated by their people, and he adds that both remained loyal to Rome.[16] By the 80s both were probably already dead, as they had, of course, been reigning since 50. By the time of Domitian they must have been advanced in years. According to Cassius Dio the immediate cause of war was the refusal of the Germans to supply Domitian with military aid for his Dacian war.[17] Such a demand was by no means unusual: Sido and Italicus had been required to take up arms on the side of the Flavians in the civil war of 69. Thus the Roman–German alliance involved mutual aid in the event of war. When the Marcomanni and Quadi refused to supply troops, Domitian attacked them and actually had their peace-emissaries put to death. These events must have taken place in the months following the victory at Tapae, at a time when Domitian's hands were somewhat freer. In order to engage the Germans the Roman troops had to take the shortest route from the Dacian theatre of hostilities; one inscription[18] indicates that some of them had to march through Dacia, i.e. northwards through the Hungarian plain. Episodic accounts available permit no satisfactory chronological reconstruction, nor do they provide any indication of results. Domitian soon left Pannonia for the Rhine, but had to return immediately (in 89) as Roman troops had suffered a defeat. It was at this point that peace was formally concluded with the Dacians; Decebalus, who did not wish to be present in person, sent his brother Diegis to the negotiations as his plenary representative. He received the diadem from Domitian and appointment as a client-king. Considering the tense situation in Pannonia, the peace terms were very mild.[19]

In Rome in the same year Domitian celebrated a triumph over the Dacians and the Chatti, but not over the Germans on the Danube. This suggests that the war against them in Pannonia was still in progress. There is, however, no information concerning events in that province immediately after 89. It is probable that Domitian succeeded in winning over northern neighbours of the Marcomanni and Quadi to the Roman cause, thus enabling him to attack the Suebi later from the north. Not long afterwards the Lugii appealed for Roman help against the Suebi; in reply Domitian sent them formal help in the shape of a hundred cavalrymen. The result was that the Suebi then involved the Iazyges on their side, and hostilities, which—probably by diplomatic means—Domitian had managed to postpone after 89, now broke out as the Suebian–Sarmatian war.[20]

There is very little information about this war. In the work of later historians there is, however, frequent mention of a catastrophic defeat, which attracted attention not only because a whole legion was destroyed, but also because this defeat could be laid at the door of Domitian.[21] The legion, possibly XXI Rapax which had recently been transferred from the west, was defeated in Pannonia by the Sarmatians and this caused Domitian to return once again to the province, where he remained for eight months in either 92 or 93.[22] On this occasion he did not return to Rome unsuccessful, for he celebrated an *ovatio* over the Sarmatians.

The modest ovation, instead of a triumph, suggests that Domitian still had further plans in view in Pannonia. It would seem that the Germans were not finally defeated. The fact that the war had no definite name in contemporary terminology implies that it was not regarded as finished.[23] Indeed there are reports of a Suebian war undertaken by Nerva. The end of the war coincided with the adoption of Trajan.[24]

The Suebian–Sarmatian wars of Domitian were of decisive importance for the military situation in Pannonia. As has been seen, the auxiliary forces had already been reinforced before the outbreak of the war, a fact which is attested by military diplomas from the years 80, 84 and 85. Once the war had started it was necessary to bring in a legion. This was XXI Rapax, which was soon destroyed and was immediately replaced by XIIII Gemina. There are indications that these two legions were based in the south-east of the province, near Mursa.[25] As a result of the transference of XXI Rapax, or rather of XIIII Gemina, to Pannonia, this became once again a three-legion province. A little later the army of occupation was reinforced by a further legion, and so the position remained until late antiquity. The Dacian and Suebian–Sarmatian wars under Domitian resulted in Pannonia becoming once and for all one of the most important military provinces of the empire.

At the same time the difficulties of tracing the distribution of the Danube legions under Domitian and Trajan must not be glossed over. It is known, of course, that the occupation army in Pannonia and Moesia was almost doubled at the outbreak of the wars, and that as a corollary the Rhine army, from which the troops transferred to the Danube had in the main been drawn, lost its primacy. It is also possible to list the legions and a considerable number of newly arrived auxiliary units by name; what is, however, completely uncertain is where they were based. Legionary fortresses, including double fortresses, in addition to the old attested ones at Carnuntum and Poetovio, may well have existed at the time of Domitian and Trajan at Brigetio, Aquincum and Mursella in Pannonia, with another in the neighbourhood of Sirmium; and in Moesia Superior, as well as the known Viminacium fortress, at Singidunum, Ratiaria and possibly also at Naissus. The only late Flavian to Trajanic fortresses to be partly revealed by excavation are those at Vindobona, Brigetio and Aquincum. But even with these it has not been possible to determine when they were built. This is all the more regrettable, since it is clear from the start that the legions marshalled against the Dacians also took part in the Suebian–Sarmatian war without exposing the

Dacian front. At this point a factor which is often ignored must be underlined: as we have seen, after the outbreak of the Dacian war the region of Sirmium was attached to Moesia and only after the defeat of the Dacians in Trajan's reign was it returned to Pannonia. It is also definitely known that the western boundary of Decebalus' Dacian empire was the river Tisza; the greater part of the Sirmium region was, therefore, opposite Dacia. It is, then, very probable that part of the Danube army was stationed near Sirmium; evidence of this is provided by the great number of late Flavian and Trajanic inscriptions discovered in Sirmium which refer to Danube legions (IIII Flavia, I and II Adiutrix).[26] Somewhere in the Sirmium region, which was now playing its accustomed leading role, a fortress must certainly have been placed, but its precise site is unknown. The colonia Sirmium, founded under Vespasian, is the least likely place.

It is fairly certain that legio IIII Flavia was transferred to Aquincum immediately after the end of the Dacian war, that is in 89 or shortly afterwards, and began building the fortress there. Its camp prefect, L. Naevius Campanus, had altars erected, and it is probable that a big Domitianic building-inscription[27] was cut in the same stonemason's yard. The fortress at Aquincum was built north of the late La Tène *oppidum* on Mount Gellért at the point where a north-west to south-east valley reaches the river-bank (Fig. 14). This valley provided a good connection with the west–east stretch of the Danube further upstream; a diagonal road by-passing the river-bend north of Aquincum led to the *porta decumana* of the fortress. The chronological relationship here between legiones IIII Flavia and II Adiutrix is still uncertain. There are certainly traces of the latter at Aquincum dating to the late Flavian period.[28] Perhaps it was represented only by a vexillation or perhaps it was transferred there for a time before the outbreak of the German war and was only later moved to the Sirmium region. The problem is made even more difficult by the fact that soldiers belonging to II Adiutrix were buried south of the fortress and at a considerable distance from it, at a spot where tombstones of auxiliaries have also been found. On the basis of these tombstones and the large number of samian sherds of the end of the first century found in the area, it is assumed, perhaps rightly, that there was another Flavian fort there (Fig. 14) which, because it was sited on a narrow part of the river-bank, could not have been large enough to accommodate a whole legion.[29] In view of the difficulties with which Domitian had to contend on the Danube frontier of Pannonia between 89 and 93 (particularly in connection with the formerly peaceable Iazyges) it is by no means improbable that for a time Aquincum had many troops. Within the Carpathian basin it did, of course, provide the best crossing-point of the Danube; and it should not be forgotten that up to the destruction of the independent Dacian state by Trajan the main settlement area

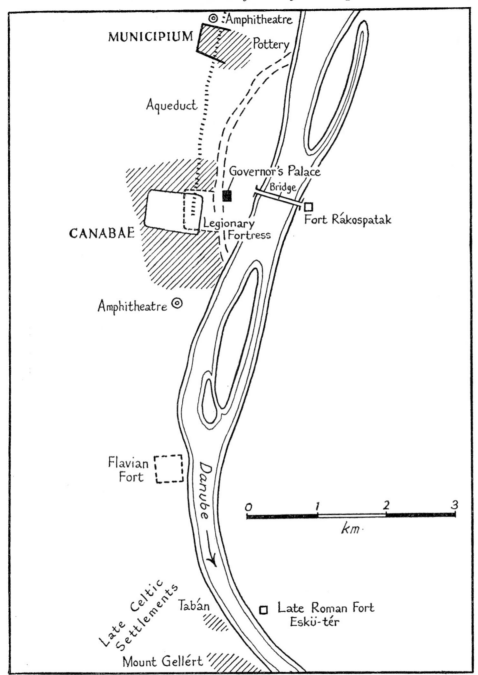

Figure 14   The environs of Aquincum

of the Iazyges was in the north-western part of the Hungarian plain, exactly opposite Aquincum.[30]

The legionary fortresses at Vindobona and Brigetio were built at the latest around the turn of the century. A building-inscription from Vindobona is dated no later than 103,[31] but it is probable that legio XIII Gemina, transferred from Poetovio, was already stationed there before this. It left behind a large number of stamped tiles and building blocks[32] which can be precisely dated, particularly as this legion was attached to Trajan's expeditionary army in the Dacian war and remained in Dacia thereafter. Legio XIII Gemina also took part in the building of the fortress at Brigetio. The tiles stamped VEXIL III refer to vexillations of three legions, which, as other stamps found in Brigetio tell us, were XIII Gemina, XIIII Gemina and XV Apollinaris. After the departure of the first the work was continued by the other two legions; thus, a very early date is unlikely for the start of construction.[33]

As a result of the transference of XIII Gemina from Poetovio to the Danube and the abandonment of its former fortress, the army in Pannonia was concentrated on the river, and the interior of the province remained demilitarized for a long time. As far as present knowledge goes, all the later auxiliary forts on the Danube were built between 89 and 102, with the result that there was a dense chain of them along the river. From the interior of the province, on the other hand, we have no evidence for auxiliary or other stations during the second or third centuries. Later Roman historians place the establishment of the Danube *limes* in Pannonia precisely in this period and connect it with the Suebian–Sarmatian war.[34]

At present, Flavian auxiliary forts are certainly attested only at Vetus Salina (Adony), where L. Barkóczi discovered four earth-and-timber forts with round towers.[35] He is also responsible for the very reasonable theory that Flavian auxiliary forts should be sought where so-called Po valley *sigillata* pottery has been found. Further clearly interpretable evidence makes it possible to draw up a list of Flavian auxiliary forts, though it cannot, of course, be considered complete. Auxiliary forts either built under the Flavians or already in existence may be assumed at the following places: Cannabiaca (Klosterneuburg), Ad Flexum (Magyaróvár), Arrabona, Brigetio, Albertfalva, Vetus Salina, Lussonium (Dunakömlöd), Ad Militare (Kisköszeg-Batina), Teutoburgium (Dalj) and Malata (Banoštor).[36]

It is very probable that the *limes* of Moesia Superior was also consolidated in the late Flavian and Trajanic period. The important rescue operations which are being carried out in the Djerdap will doubtless supply the evidence; at present only interim reports are available, which indicate that the earliest layers of the forts there date from the end of the first century. A fragmentary Trajanic building-inscription from the fort at Prahovo (Aquae?)[37] must be dated to the early years of this emperor, since it is unlikely that after the conquest of Dacia there was any need for new construction-work on the stretch of the Danube which ceased to be an imperial frontier.

The extensive building activity on the Danube line in Pannonia is not attributable to any danger from the Dacians. The decision to double the army of Pannonia from two to four legions, and to concentrate troops on the Danube, probably arose from the recognition that Augustus' old plan of defending the frontier by diplomatic means was no longer viable. The new Flavian government reached this conclusion at the outset and took steps which would not have suited the Augustan scheme. A basically different frontier policy *vis-à-vis* the Danube Germans and Sarmatians probably formed part of Domitian's plans, for even before the outbreak of the Suebian–Sarmatian war he had concentrated troops in Pannonia. The killing of the Suebian peace-emissaries revealed a radical break with the old methods, and the bitter experiences of the war which ensued went further to convince the emperor that security for the imperial frontiers was only possible from a position of strength, that is to say from a constant demonstration of strength.

We do not know what kind of peace-treaty was concluded with the Suebi and the Sarmatians under Nerva. In any case Trajan did not revert to the old scheme but had construction work on the *limes* continued. The legionary fortresses at Vindobona and Brigetio were apparently built in the very earliest years of his reign.[38] He was, nevertheless, concerned not to aggravate the tense situation in the Carpathian basin. In the *Germania*, written in the first year of the reign, Tacitus relates that the Marcomanni and the Quadi had kings by the grace of Rome. The same source states that the old dynasty of Tudrus had been replaced by 'foreign' kings. These kings were installed not later than 98, when Trajan, who had just been appointed emperor, made a detour to the Danube before returning to Rome.[39]

There is a good deal in favour of the view that the much debated Roman buildings in Slovakia and Moravia (Fig. 15) are likewise connected with Trajan's treaties with the Germans. As for their date, we can at least be certain that the beginning of the second century, when legio XV Apollinaris was still stationed in Pannonia, is the latest possible context, for this legion provided tiles.[40] The tiles, coming as they do from military units, mainly legions, have prompted scholars to regard these buildings as Roman forts, or at least as the houses of commanders of Roman troops. It is known that, in the last years of Marcus, Roman units were stationed in Marcomannia, and from the time of Valentinian there are also reports of Roman forts in the territory of the Quadi. But apart from

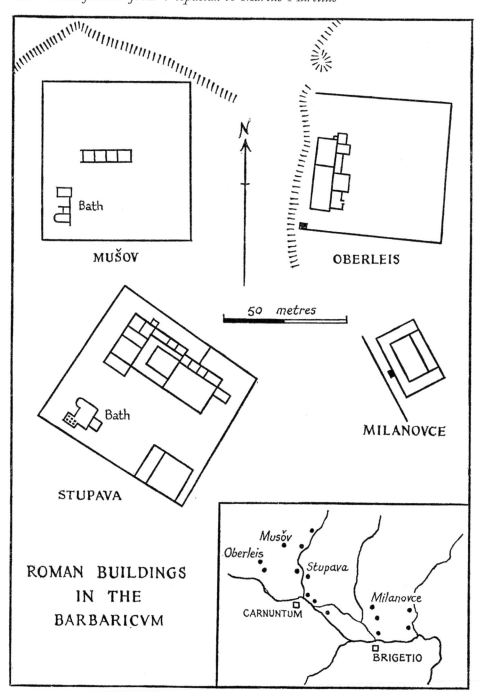

Figure 15 Roman buildings in the Barbaricum

the tiles from military tile-works, these buildings have nothing military about them. They were simple dwelling-houses, furnished in the Roman manner with a certain degree of comfort, and stood in a courtyard surrounded by a wall. The walls were some 50–60 cm thick, but of a rampart or ditch there is no trace. Had they been inside the province, these buildings could easily have been taken for small villas; it is impossible to regard them as the fortified houses of military commanders, still less as strong-points or forts for troops. Some years ago E. Swoboda came up with the idea that they were built by Roman troops for barbarian chieftains; there is nothing odd about this as, after all, according to Tacitus, the power of these chieftains rested on the *auctoritas Romana*, which supported them financially—*raro armis saepius pecunia*. Why should not the Romans also have built residences for these client-princes?[41]

At the beginning of Trajan's reign, if not before, a new treaty was concluded with the Sarmatians (Iazyges), for during Trajan's Dacian war they provided the Romans with troops, and immediately after it was over invoked certain agreements.[42] Trajan's attention was, however, primarily directed to the Dacians. Even after the peace of 89 tension on the Moesian front was not fully relaxed. Dio Chrysostom, who hurried there from his partly self-imposed exile in Dacia after hearing the news of Domitian's death, found himself in the middle of preparations for war. The description in his *Olympic Discourse*[43] refers to a big camp in Moesia, probably Viminacium: 'There one could see everywhere swords, everywhere corselets, everywhere spears, and the whole place was crowded with horses, with arms, and with armed men. Quite alone I appeared in the midst of this mighty host, perfectly undisturbed and a most peaceful observer of war . . . desiring to see strong men contending for empire and power and their opponents for freedom and their native land.' Around the year 100 large numbers of military units were concentrated in Moesia Superior. On a diploma of 93 three *alae* and nine cohorts are mentioned, whereas a diploma of 100 refers to no fewer than twenty-one cohorts.[44] Preparations for war, of course, involved troop movements. In the apparent confusion of moving whole legions here and there one principle emerges to which Trajan's supreme command adhered, and probably not without reason. Legions newly moved in from the west were stationed in fortresses in Pannonia, whereas those which had already been there for a fairly long time were sent into battle against the Dacians. This in turn brought about a change in the distribution of legions in Pannonia, and to a lesser degree in Moesia Superior: in the latter province legio VII Claudia, already at Viminacium, remained there, and IIII Flavia moved to its final base at Singidunum a few years after the end of the war. During Trajan's Dacian wars Moesia Superior was, of course, the launching area for all military operations, and it is probable that the emperor established his headquarters at Viminacium. Of the Pannonian legions, IIII Flavia, XIII and XIIII Gemina and XV Apollinaris joined Trajan's expeditionary army, though it is more likely that the last two were represented only by vexillations. Legiones I and II Adiutrix were also involved in the Dacian war. Both X Gemina and XI Claudia were moved from the Rhine as replacements for the legions sent to Dacia, but this probably did not happen before the end of the first Dacian war. Legio XI Claudia was quartered in the recently built fortress at Brigetio,[45] and X Gemina at Aquincum, where it left behind many traces of its building activities, and where some of its soldiers are attested on tombstones.[46] Evidence that this legion was stationed at Aquincum over quite a long period is provided by the fact that its veterans settled near the fortress.[47] After Dacia was conquered legio XI Claudia was transferred to Oescus in Moesia Inferior and XXX Ulpia Victrix, a legion recently raised by Trajan, took over the camp at Brigetio. Legiones XIIII Gemina and XV Apollinaris returned to Vindobona and Carnuntum respectively.

It is very possible that the division of Pannonia into two provinces under Trajan (Fig. 16) was directly connected with the distribution of the legions. In 102 the province was still undivided.[48] In 107 Hadrian, the later emperor, is attested as praetorian legate of Pannonia Inferior;[49] the division must then have occurred between 103 and 106. However, it is not certain that he was the first governor of Pannonia Inferior; it is not impossible that he was the first praetorian governor but followed either one or two consular governors of the province, and that therefore the province might have been divided as early as 103. If so, the immediate cause of the division may have been the presence of a fifth legion in Pannonia, the recently raised XXX Ulpia. The first governor of Lower Pannonia would accordingly have been a consular, because of the presence of two legions in his province. All this, however, is at present pure speculation. Nor is it necessary to suppose a purely military reason. After the division of the two other big frontier provinces—Moesia in 86 and Germany in 90—Pannonia remained the only province with a disproportionately long external frontier and a disproportionately large army. In addition, the Sirmian section of the frontier, between the mouths of the Drave and Save, which had been added to Moesia Superior in 86, was returned to Pannonia in 106 at the latest.[50] The Pannonian section of the frontier, 700 km long, faced two barbarian tribes which not only belonged to different linguistic groups with fundamentally different ways of life but also required basically different tactical approaches in both war and peace. The experience of a century of treaty-relationships had amply shown this to be so. The division of Pannonia meant that the Upper province had the Germans as

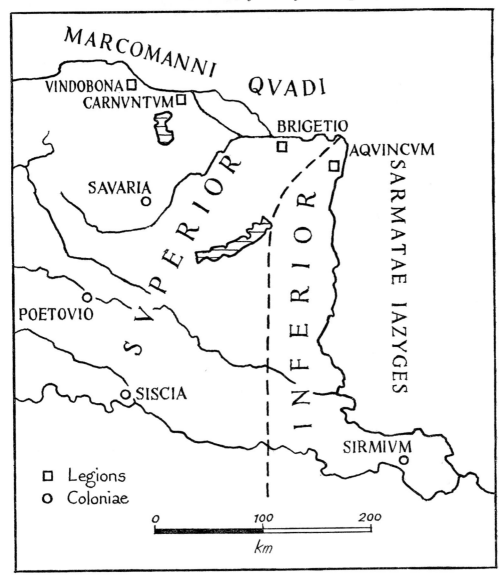

Figure 16 The division of Pannonia under Trajan

neighbours, while the Sarmatians, always known as the Iazyges, bordered the Lower. Moreover, the new inter-provincial boundary was drawn in such a way that the whole of the largely urbanized western part was assigned to Pannonia Superior, and this province, in relation to its area, had a fairly short stretch of frontier. Pannonia Inferior, on the other hand, possessed only a single urban community at the time it was established, colonia Flavia Sirmium, and its territory never achieved a width of more than 50 km; yet it carried a much longer frontier zone. The legions were assigned in proportion to the relative size of the provinces. Pannonia Inferior was allotted one, whereas the Upper province, with three times the area, was given three. At the time of the division the legions received recruits from the provinces in which they were stationed, and it is therefore not unlikely that in fixing the provincial boundaries the question of recruitment was considered. Finally there was another overriding principle which was consistently applied under Trajan in all the European frontier provinces: the number of *coloniae* corresponded to the number of legions in the province. In this connection it should be pointed out that at the end of the first century Emona no longer belonged to Pannonia, but to Italy.[51] Pannonia Superior had three *coloniae* (Siscia, Poetovio and Savaria) and three legions (XIIII Gemina, XV Apollinaris and XXX Ulpia), Inferior one *colonia* (Sirmium) and one legion (X Gemina). Here, too, recruiting policy was involved.

The rank of the governors of the new Pannonian provinces varied according to the number of legions under their command; the governor of Superior with his headquarters at Carnuntum was of consular rank and was given the province immediately after tenure of this office, while the governor of Inferior, based at Aquincum, was a praetorian and, generally speaking, became consul designate while still *legatus Pannoniae Inferioris*. Hence it frequently happened that the governor of Inferior soon became governor of Superior, although it is not possible to lay down hard-and-fast rules about it. The governors of Inferior were appointed fairly regularly from among the legionary legates in the Upper province. That they were primarily *viri militares* occasions no surprise.

The establishment of the province of Dacia in 106 had an importance for the future of the Danube frontier which must not be underestimated. The most active and the least docile people within the European barbarian area, and the only barbarian race in Europe that had managed from time to time to establish a stable state, was now finally eliminated. Its role was taken by Sarmatian tribes of horsemen, the Iazyges and the Roxolani, who, until the time of Trajan, had generally played only a secondary role in Roman–barbarian relations on the Danube. Down to Trajan it was Roman relations with the Dacians which were of supreme importance, and it was only in this context that the Sarmatian tribes acquired significance as possible allies or buffer states. Iazygan horsemen fought on Trajan's side against the Dacians, and those of the Roxolani on the Dacian side against the Romans.[52] In the first century A.D. the Iazyges were wedged between the Romans and the Dacians in the northern part of the Hungarian plain—it is not unlikely that the Romans had established them there as a buffer state—but a section of them was still in the plain of Oltenia in the time of Trajan.[53] When Decebalus conquered this area the Iazyges inevitably became hostile. They therefore fought on the Roman side, though fifteen years earlier they had been responsible for the disaster of legio XXI Rapax. After the defeat of the Dacians in 106 the Iazyges naturally demanded the return of Oltenia.[54] But as Trajan was determined to constitute Dacia a province, it was out of the question to cede Oltenia to a barbarian people. The result was war on the Danube frontier of Pannonia, waged successfully by Hadrian during his governorship.[55] Details of the peace-treaty are lacking, but it is very probable that Rome was inclined to accede to some of the demands of the Iazyges, and that the latter were easily satisfied. The distribution of early Iazygan archaeological material indicates that in the first century this tribe of horsemen had occupied only the northern, or more precisely the north-western, part of the Danube–Tisza plain.[56] Under Decebalus the Tisza was the western boundary of the Dacian state.[57] Once this state had disappeared in 106 a power vacuum arose in the area east of the Tisza. Trajan did not incorporate the plain between the river and the Transylvanian mountains into Dacia but left it to the Iazyges. Archaeological finds from the second century indicate that during that period they conquered the whole of the plain on both banks of the Tisza; vessels found in some of the so-called Iazygan cemeteries reveal fairly strong Dacian influence.[58] It is not improbable that Rome gave the Iazyges the Tisza plain in return for Oltenia. A westerly migration of the Iazyges from Oltenia through the newly established province of Dacia is also not unlikely; half a century later they were allowed an unimpeded passage through Dacia.

Here we have the main problem which faced the Romans after the establishment of a province on the left bank of the Danube. The Dacian state had, of course, been eliminated, but the mounted Sarmatian tribesmen in the Carpathian basin and on the lower Danube were not people prepared to enter into a system of treaties such as had been made with the Germans. The new province, wedged between the Iazyges and Roxolani, was to some extent a bastion which held them in check on two fronts, but at the same time it was an obstacle to the westerly migration of these peoples and hence a cause of constant unrest on the *limes*. Nor was the solution to the Iazygan question, which was reached in 107, a long-term solution, for ten years later the Romans faced a frontal attack by the Sarmatians both west and east of Dacia.

The military occupation of Dacia required of the imperial army some legions and a large number of *auxilia*. The earliest army of occupation naturally consisted of parts of the expeditionary force. Those Pannonian and Moesian legions which had taken part in the expedition soon returned home and only

XIII Gemina and IIII Flavia [59] are later attested in Dacia. It was in 110 that the occupation forces were first reduced; in that year a fairly large number of auxiliary regiments were sent back to the provinces from which they had come. [60] A further section of the expeditionary army was ordered to the east when Trajan started his Parthian war in 114. The auxiliary units of the new army of Dacia were largely drawn from the expeditionary army and from the standing army of Moesia Superior. At the same time the number of auxiliary units in the latter province was considerably reduced, for the establishment of a province on the other side of the Danube had a second consequence which must not be over-looked, in that the river ceased to be the frontier of the empire in the stretch between Viminacium and the mouth of the Alutus (Olt)—in other words along the extremely important section of the Djerdap. The Danube forts, built earlier mainly under the Flavians, were presumably evacuated for the most part, and that stretch of the river, which until 106 had been strongly held by troops, was demilitarized. Thereafter it was only on the relatively short stretch of the river between Singidunum and Viminacium that Moesia Superior was contiguous with barbarian territory; yet this sector of the frontier had two legions, IIII Flavia in Singidunum and VII Claudia in Viminacium, and some auxiliary units. Down-stream from Viminacium the chain of forts was thinned out to such an extent that one can no longer speak of a military frontier. The strong-points still held on the border between Upper Moesia and Dacia were probably the follow-ing: Pincum (Veliko Gradište), Cuppae (Golubac), Novae (Čezava), Taliata (Donji Milanovac) and Egeta (Brza Palanka) where tile-stamps and inscriptions relating to auxiliary units suggest that there were garrisons in the second and third centuries. [61] This list of forts, mainly those of auxiliary units, will probably be supplemented when excavations in the Djerdap are completed, but in the meantime a few facts can be established. First of all, the extremely small number of stations is striking. They are mainly to be found east of Viminacium in the upper Djerdap (Gornja Klisura) and extending as far as the mouth of the Porečka, near Taliata. Its valley led into the mining district of the Pek and Mlava valleys and was the only place in the Djerdap from which the interior of the province could be reached by road. The Gornja Klisura could easily be reached from the west by barbarians and was the reason why this stretch of the tow-path had to be strictly guarded. The stretch below Taliata was not so strongly held: Egeta had to have a unit (cohors I Cretum), [62] as the road which by-passed the great Danube bend of the Kasan defile and the Iron Gates met the Danube road at this point. Down-stream from Egeta there are no definite traces of second- or third-century auxiliary forts. As for the auxiliary regiments which in those centuries were attached to the army in Moesia Superior, the most striking feature is their small number;[63] their strength was just over half that of the two legions (some 6,500 auxiliaries against somewhat more than 10,000 legionaries), extremely weak compared with auxiliary forces in Pannonia, Dacia and Moesia Inferior. There was only a single *cohors milliaria*, and the forts listed above were correspondingly small, as a rule not more than two ha (roughly five acres), though mostly not more than one ha.[64] The demilitarization of the *limes* in Moesia Superior did not involve the interior of the province. It was in fact the only province on the right bank of the Danube where the hinterland continued to be occupied in the second and third centuries; a post-Trajanic auxiliary fort is attested at Timacum minus (Ravna) in the Timok valley on the road between Naissus and Ratiaria, which was garrisoned at least from the reign of Vespasian, notably by cohors I Thracum Syriaca in the period from Vespasian to Marcus Aurelius.[65] Auxiliary units were stationed not only on the Danube frontier of Moesia Superior (where there were some seven auxiliary forts in the second century) but also in the interior, as is shown by the fact that the provincial army included a total of some twelve auxiliary units. The discovery of further forts in the interior would not be surprising. The military occupation of the interior was necessary for internal reasons. While problems relating to the defence of the frontier were not forgotten, they were of secondary importance. Of more account was the fact that Timacum minus was situated at the point where the road from the Danube to Naissus, and ultimately to the Adriatic, joined the Timok valley. At this point the shortest route from the lower Danube to that sea could be blocked.

This east–west connecting route through the Balkan peninsula began to acquire significance immediately after the establishment of the province of Dacia and because of it; the stretch of the Danube which had been evacuated was also important as providing a link between Dacia and the Mediterranean. The new province was cut off from the south by the very high range of the southern Carpathians, nor was it readily accessible from the west. The best routes linking it with the Balkans ran along its boundaries—through the Banat from Viminacium and through the Olt valley. From the Danube a valley also opened to the north near Dierna (Oršova), but even this short route had no good connections on the south bank. Transdierna (Tekija) could only be reached from the west along the cliff road in the Kasan defile and from the east only along the Danube valley. The *Tabula Peutingeriana*, which fortunately includes the road network in Dacia, shows three roads entering from Moesia with starting-points at Viminacium, Taliata and Egeta, none of which could have been actual crossing-points. From Viminacium the road followed the southern bank until it finally crossed the river near Lederata (Banatska Palanka); from Taliata it utilized the cliff road through the Kasan defile as far as Transdierna, the first point at which the river could be crossed, while from Egeta the road followed the left bank to cross at Pontes (Kladovo), opposite Drobeta (Turnu Severin). In short, all the crossing-points could be reached only by using stretches of road along the valley, and this led to brisk traffic along the river-bank.

The most important crossing-point was, of course, Drobeta, where Trajan had his famous bridge built by Apollodorus of Damascus. The work was carried out by units from the first occupation army of Dacia, which left behind large quantities of stamped tiles.[66] Tradition, hostile to Hadrian, has it that he had the bridge's woodwork removed;[67] without casting doubts on the statement, one may say that it is highly improbable that it was not replaced. Be that as it may, Drobeta was the most important crossing-point, simply because the parts of Dacia lying south or north of the Carpathians could equally well be reached from it.

The establishment of Dacia as a province contributed to the unification of the interior of the Balkans, especially Moesia Superior, with the rest of the empire. Communication between Drobeta or Viminacium and the Mediterranean was available along the old waterways of either the Danube or the Save, but a direct route which terminated at Brundisium ran along the north-east to south-west diagonal through the Balkan peninsula; this acquired importance only after the conquest of Dacia. It started on the Adriatic at Lissus (Leshja, Alessio), and ran north of the Scardus Mons via the Kosovo Polje,[68] following the Timok valley until it reached the Danube near Ratiaria. From there it followed the river up-stream as far as Drobeta. It was perhaps no accident that it was at Ratiaria that Trajan established his only *colonia* in Moesia Superior, for it was along this road from Lissus to Ratiaria that the earliest urban centres in the new province were founded, at Ulpianum, Naissus, Timacum minus and Ratiaria itself.

Trajan's reign was the first important turning-point in the history of the Danube area since Augustus. After the transitional period under the Flavians it was now that the deployment of both legionary and auxiliary garrisons, the direction of the main lines of communication and many other things too were settled for a long time to come. After Trajan there were no changes of critical importance within either the military or the trading spheres, although it was not until Hadrian that troop movements ceased. Some changes did occur in the distribution of the legions in the first years of the latter's reign, but these were the result of Trajan's Parthian war. Unfortunately we do not know when Pannonia acquired the four legions which until late antiquity formed the back-bone of its army; but it is probable that these legions arrived in time to take part in the battles which were fought on the Sarmatian frontier in 117.

The legions which returned from the Parthian war were I and II Adiutrix, which took over from XXX Ulpia at Brigetio and X Gemina at Aquincum respectively. Of the legions stationed in Pannonia under Trajan, XV Apollinaris and XXX Ulpia left the Danube area for good; legio X Gemina moved into the fortress at Vindobona and XIIII Gemina into that at Carnuntum. Legio IIII Flavia, which in the early years after the conquest of Dacia was attached to the army of the new province, probably did not become part of the army of Upper Moesia until the beginning of Hadrian's reign. The distribution of legions in Pannonia and Moesia Superior from the time of Hadrian to late antiquity, familiar to geographers and historians of imperial times,[69] was as follows:

Vindobona	legio X Gemina
Carnuntum	legio XIIII Gemina
Brigetio	legio I Adiutrix
Aquincum	legio II Adiutrix
Singidunum	legio IIII Flavia
Viminacium	legio VII Claudia

Thus three legions faced the Suebi in the Carpathian and Czech regions, and three faced the Sarmatians of the Carpathian basin. The Upper Moesian legions were stationed, so to speak, as reserves in the middle of a long frontier which enclosed the Sarmati (Iazyges) on three sides. The greatest pressure from the Sarmatians was felt on the long sector of the Pannonian frontier from the Danube bend to the mouth of the Save; it was chiefly legio II Adiutrix and the many Lower Pannonian *auxilia* which fought in the battles of the second and third centuries, and it would seem that less use was made of IIII Flavia and VII Claudia. It is quite possible that the land which was wedged between Pannonia and Dacia north of the Upper Moesian section of the frontier had to some extent become a military zone under the command of the governor of Dacia and, therefore, that the Upper Moesian legions were not in contact with the Sarmatians. Even more did legio IIII Flavia act as a permanent reserve in Pannonia Inferior. On every occasion that legio II Adiutrix joined an expeditionary army, it was replaced at Aquincum by IIII Flavia; this explains why the latter legion left a number of monuments behind in the capital of Lower Pannonia.[70]

The conclusion that the Upper Moesian frontier and army did not play a key role in fighting the Sarmatians emerges from the events in the years 117–19. As already stated, Hadrian's solution during his governorship in 107 to the problem of the Iazyges was short-lived. The latter, possibly reinforced by those who had migrated from Oltenia, conquered the plain east of the Tisza; in the course of this conquest they doubtless came into conflict with the 'free Dacians' who had remained outside the Dacian province. Information about these battles is lacking, though the result of this extension of their power became clear in the year of Trajan's death, when simultaneously the Roxolani attacked the Lower Moesian *limes* and the Iazyges that of Lower Pannonia. The joint action undertaken by these two tribes points to a common interest which doubtless lay in the fact that their communications with each other were hindered by having the province of Dacia wedged between them. After defeating the Dacians Trajan had entered into an alliance with the Roxolani; in 117 the reason they gave for the war was that Rome had reduced the *stipendium* paid them by treaty.[71] There are no sources which throw light on the demands of the Iazyges; they were, however, probably more considerable than those of the Roxolani, since they continued to fight after their allies had come to terms. In the attack mounted by these tribes Dacia's governor, C. Julius Quadratus Bassus, was killed and Hadrian had to go to the seat of war,[72] where he invested his trusted agent, Q. Marcius Turbo, with exceptional powers over Dacia and Pannonia Inferior.[73] The Roxolani soon settled down, and it is possible that Hadrian replaced their king and had the former ruler interned in Pola after granting him Roman citizenship.[74] Marcius Turbo held the command in Pannonia Inferior and Dacia until 119; this suggests that the union of the two armies was necessary not only on account of the joint action of the Iazyges and Roxolani but also in order to continue the fight against the Iazyges alone. In other words, by 117–19 the Iazyges were already on Dacia's western boundary, and had conquered the whole plain between the Danube and Transylvania.[75]

The war ended with the Iazyges sending an embassy to Rome, where, it seems, a client-treaty was concluded.[76] This time the solution must have been satisfactory, as for the next half century nothing is heard of the Iazyges. It was during this period of peace that the road through their country was finally built. A direct connection between Pannonia and Dacia obviously became necessary after the latter was conquered, and Trajan had probably attempted to build a road running from west to east across the plain. Aurelius Victor credits him with having had a road from Gaul to the Pontus built by 'wild peoples'.[77] However, the wars of conquest waged by the Iazyges between 107 and 117 are hardly likely to have encouraged its use. After 119 and before the outbreak of the Marcomannic wars under Marcus Aurelius it was a busy road, particularly as it was possible for a long distance to go by boat along the river Maros (Marisia, Marisos). The entrepôt was at the mouth of the Maros, at Szeged, where in the course of demolishing the medieval fort in the last century Roman stone sculptures, tiles and an inscription were uncovered.[78] The inscription was to a certain Mercator (Fig. 17), overseer of the station (*vilicus*), who was superintended by

Figure 17 Inscription of Mercator from Szeged

the *praefectus vehiculorum*.[79] So far the course of the road has not been investigated; since, however, Szeged is a fairly well-defined site its direction will have been west–east, starting from the auxiliary fort of Lugio (Dunaszekcsö), for the importance of which there is other evidence, and running along the Maros as far as Dacia. These peaceful conditions probably led to the abandonment of the greater part of the Banat by the Romans. It has recently been established that legio IIII Flavia was based after 106 at Berzovia in the eastern part of the Banat.[80] However, it was soon transferred to its later base at Singidunum. Its withdrawal was perhaps the result of the evacuation of the greater part of the Banat, which was given to the Iazyges. This may have been the part of Dacia which, according to some sources, was evacuated by Hadrian shortly after he came to the throne,[81] an event that was, perhaps, brought about by the embassy of the Iazyges to Rome.

Scholars have recently connected some layers of ash found in the auxiliary forts of Campona, Vetus Salina and Intercisa with the destruction caused by the Iazyges between 117 and 119.[82] Coin-hoards and evidence of fire in the legionary fortress at Aquincum have also been dated to this period[83]—whether rightly so, it is difficult to say. It is equally uncertain whether the Marcomanni and the Quadi took part in the war of 117–19. The only evidence for such an assumption is one mention in the *Historia Augusta*[84] and one undated reference to an *expeditio Suebica et Sarmatica.* The reference in the *Historia Augusta* states only that Hadrian had installed a German king, but this may not necessarily have had any connection with a war, and indeed may equally well refer to west German tribes. The *expeditio* mentioned in the *cursus* of a certain senator would be better dated to the time of Marcus.[85]

Definite evidence of a crisis on the German frontier of Pannonia comes only in Hadrian's last years. If the periods of peace on the Upper Pannonian Danube frontier are compared, it emerges that between every two wars or diplomatic crises there was a period of peace which lasted for a generation: (i) from the time when Vannius became king in A.D. 19 to his overthrow in 50; (ii) between the latter date and Domitian's Suebian war in 88–93, or more correctly until Nerva's victory and the establishment of the new order by Trajan in 98; (iii) between 98 and the outbreak of a Suebian war in 136 or just before, and (iv) between the conclusion of peace *c.* 140/4 and the outbreak of war in 167; each of these periods lasted three to four decades, i.e. a generation. This regular repetition of crises in the client states probably had fairly profound reasons. In the first place there were the dynastic changes among the Germans, which led to fresh negotiations with Rome and to new hopes of a more favourable outcome on the part of the Germans. In Roman–Sarmatian relations there were also periods of peace, the only difference being, it would appear, that they lasted longer. From the immigration of the Iazyges at the beginning of the first century A.D. there were no fewer than eight or nine decades until the war under Domitian, and from the peace of 119 until 167 almost five decades. Solid evidence of joint action by the Germans and Iazyges is available only for the time of Domitian. The system of client-states, then, stood the test even after the Danube frontier had been occupied by troops in the Flavian–Trajanic period. Every war ended with the appointment of a German king by the grace of Rome, and he was able to guarantee the peace as long as his life lasted.

The war which broke out in Hadrian's last years also ended in an investiture. The exact date of its outbreak is unknown. Coin-hoards discovered on the Upper Pannonian *limes* include terminal coins from the years 130/1, 132/4 and 137, and layers of ash in Aquincum are also regarded as being connected with this war.[86] As will be seen later, these terminal coins give no indication of the date at which the hoards were buried. One thing is certain, however: L. Aelius Caesar, who had been singled out as Hadrian's successor, was entrusted in 136 with the joint governorship of the two Pannonian provinces. According to his biography in the *Historia Augusta* he had had a few military successes and hence acquired 'the reputation of being a moderate general'.[87] When he died in 138 the joint governorship was not repeated; this suggests that perhaps the greatest danger was over. The war was continued by the new governor of Pannonia Superior, Ti. Haterius Nepos, who was awarded the *ornamenta triumphalia*, probably also in 138—the last general in Roman history to be so distinguished.[88] His victory was doubtless followed by the usual negotiations which resulted in the investiture of a king of the Quadi. The reverse of a coin of the Emperor Antoninus Pius belonging to the years 140/4 depicts the investiture: the emperor is shown handing the diadem to the king and the inscription runs REX QVADIS DATVS.[89] As such representations on coins are comparatively rare, this one is not without importance; it is possible that the negotiations were protracted and the coin depicted long-awaited news.

Peace on the Danube under Hadrian and Antoninus Pius was only interrupted by each having to wage war against the Sarmatians and the Quadi. Recently the possibility has been frequently considered that a period of unrest began in the 150s. The *Historia Augusta* agrees that the war of 167–80 did not break out suddenly but had been postponed for years by the governors of the provinces concerned.[90] Now there is a quite large number of coin-hoards whose latest coins date to the 150s. At first sight this suggests that there was a long period of unrest under Pius and in the early years of Marcus, in which people had buried their valuables; and that as the owners of much of the buried treasure perished in the confusion, it remained in the ground. If we pursue this idea to its logical conclusion, we must assume from the distribution of these coin-hoards that there were several invasions by the barbarians, who penetrated deep into Pannonia and Noricum, and even as far as central Italy. This was in fact the conclusion reached; the decade and a half preceding the outbreak of the Marcomannic war were interpreted as a period of increasing insecurity and regarded as the 'prelude to the Marcomannic wars'.[91]

Few will suppose that the war broke out suddenly, still less that it was unexpected. Yet repeated sorties by the barbarians into the Danube provinces in a period of general peace under Pius would have left an echo. The invasion of Italy by the Danubian Germans in the early years of the Marcomannic wars made so deep an impression that not merely contemporaries, such as Lucian, but even Ammianus Marcellinus considered it worth mentioning. Therefore we ought rather to connect the coin-hoards ending in the middle of the second century with the later wars. If a person hoards money over a period of time, his collection will not inevitably contain coins of the last few years. The circulation of money, particularly in remote parts of the provinces or among the lower class of agricultural workers, was probably a fairly slow process, and even if a person came into possession of newly minted coins it is quite possible that he did not hoard them. Recently it was pointed out that two hoards of coins, unearthed south of Carnuntum and containing terminal coins of different years, must unquestionably be connected with the same barbarian attack: the difference was caused by the circumstances surrounding the hoarding and the income of the owners.[92] The coin-hoards of the middle of the second century were, therefore, probably not buried until the time of Marcus, during the catastrophic years following 167. It should also be remembered that money had not yet depreciated to the same extent as it did in the third and fourth centuries. As will be seen, even in the third century coin-hoards containing terminal coins of varied dates can be traced to the same war; this applies even more to hoards containing good second-century money. People in later centuries collected these coins by preference, for instance Sarmatian graves of the third and fourth centuries mostly contain second-century silver *denarii*.

In the history of the Danube frontier the years of peace under Hadrian and Antoninus Pius were a period of final consolidation but also one in which inflexibility set in. As already indicated, the *limes*-forts on the Danube, with very few exceptions, were built between 89 and 102, and at the same time the garrisons in the interior of the province were withdrawn. At the present time it is not possible to describe the system of occupation within the interior of Pannonia in the first century, as the number of suspected or definite forts is still very small (Gyalóka, Mattersdorf, Ulaka, Tác—only the last two are attested by excavations, though they still have not been dated).[93] It is, therefore, very significant that after the beginning of the second century there is neither archaeological nor epigraphic evidence that Roman troops were stationed behind the Pannonian *limes*. It follows that in contrast with the situation in Upper Moesia, all the known auxiliary units of the *exercitus Pannoniae Superioris* and *Inferioris*, whether attested on inscriptions or diplomas, could be accommodated in the *limes*-forts. There were more than thirty such forts (Fig. 18), and second-century diplomas record on average 10–12 units in Pannonia Superior and 13–18 in Inferior. Those *limes*-forts so far excavated were manned continuously in the second and third centuries, and the considerable number of inscriptions referring to auxiliary soldiers and veterans or their units which have been found at other places on the *limes* lead to the same conclusion—that the rest of the forts were also continuously manned.[94]

The forts on the Pannonian *limes* from the boundary of Noricum to the mouth of the Save have in the main been located and identified, but we cannot exclude the possibility that further forts will be discovered. For a long time it was tacitly assumed that all of them were listed in the *Antonine Itinerary* or marked on the Peutinger map; field-work and excavations both suggested that those places, and only those, were sites of forts. Recently, however, an auxiliary fort of the second or third century was discovered and investigated at Albertfalva (Budapest, XI District); it was probably the first *limes*-fort south of the fortress at Aquincum, though Campona, the thoroughly investigated fort at Nagytétény (Budapest, XXII District) is listed in the *Itinerary* in this position. The *Itinerary* does not mention the fort of Albertfalva, probably because it was destroyed in the third century under Gallienus.[95]

Down-stream from the mouth of the Drave the sites of the forts have been more or less pin-pointed with the help of the *Itinerary* and the discovery of inscriptions, but owing to lack of field-work and even more so of excavation it is by no means certain which sites are to be regarded as *limes*-forts of the second and third centuries. On this stretch of the Danube a great deal of building was carried out in late antiquity; hence those places mentioned in the *Itinerary* were not necessarily fort-sites in the earlier period.

The distribution of auxiliary forts on the Pannonian *limes* reflects the working of the military mind, inasmuch as a basically mechanical plan, worked out at a desk, was adapted to the terrain; the essence of the plan was even distribution. The forts were 20 km or 15 Roman miles apart, even in marshy areas subject to flooding where a less compact chain would have been sufficient (for instance east of Carnuntum, or in the modern county of Tolna south of Intercisa). At the same time the sites were so chosen that the forts could control as long a stretch of the Danube as possible, or block the entrance to a lateral valley. The principles involved are still somewhat obscure. In many places the loess plateau on the *limes* of Pannonia Inferior ends on the river-bank in a high, very steep escarpment or cliff which is being undermined by the scour of the river; the steep edge of the plateau is constantly giving way. Despite this, some of the most important auxiliary forts were built on the edge of the plateau. Though they faced the river, these forts were quite inaccessible from it; moreover the *limes*-road along the bank which connected them had to overcome considerable changes of level. Because of the collapse of the cliff edge the forts had sometimes to be moved further back. For instance, under Hadrian the fort of Intercisa was moved 20 m to the west, obviously because the front wall was undermined or had actually fallen into the river. Of those forts sited on the plateau only parts of the *retentura* have survived (e.g. at Annamatia or Lugio). Neither for offensive nor for defensive strategy did these forts offer any advantage; they were, in fact, nothing more than favourably situated observation-posts, and watch-towers would equally well have served the same purpose. Forts were only built on the loess plateau where there was a cliff edge along the river. Wherever the cliffs did not rise abruptly from the river and there was flat ground between the slope of the plateau and the water the forts were built down below on the bank. In such cases they were mostly sited at the entrance to a valley or to a cutting in the plateau, and generally at one end of the flat land rather than in the middle of it. In marshy areas subject to flooding, small island-like hills were chosen; if it happened to be impossible to build the *limes*-road through the marshland, then a link road to the fort was built. For instance, according to the *Itinerary* the fort at Alisca (Őcsény) lay[96] *ad latus*—off the *limes*-road. Where the river-bank was not steep and the hinterland was mainly flat, the forts were sited fairly close to the bank, with the result that they were sometimes washed away. Only the rear face of the auxiliary fort at Vetus Salina has survived, and in some cases the whole fort has disappeared; it should also be remembered that Poigen, where the station *Villa Gai* of the Peutinger map was situated, vanished without trace in the sixteenth century through erosion.[97]

The earliest periods of the forts so far investigated have so much in common that a general description is appropriate. In the past, all the forts so far located down-stream from Brigetio as far as Intercisa have been excavated to a greater or lesser extent:[98] more recent excavations have produced many details of some above Brigetio in present-day Hungary.[99] In all the forts excavated by modern techniques—and these fortunately form the majority—traces of earth-and-timber forts were found, consisting of a ditch, rampart and post-holes and, what is rarer, sun-dried clay bricks and the foundations of wooden barracks. Except at Intercisa, where the superimposed stone fort does not coincide with the palisaded one, the remains of the timber fort were destroyed in erecting the buildings of the stone fort, since the latter corresponded precisely in both shape and size with its predecessor. At Intercisa only a single palisaded fort could be identified and in other forts there are traces of only a single timber period below the wall and ditch of the stone structure. This can hardly be accidental. The earth-and-timber forts built in the Flavian period could hardly have survived without re-building for three, four or even five decades until their replacement in stone under Hadrian or even later. The timber forts below the latter probably date only to Trajan, and the Flavian forts (and even earlier ones here and there) were not on the same sites as the later stone structures. This assumption is supported by the fact that exceptionally at Vetus Salina several forts of earlier periods have been identified underneath the stone fort. However, on other well-excavated sites rescue work has revealed traces of earthworks south of the stone fort, as for instance at Matrica and Intercisa.[100]

This means that the sites of the *limes*-forts were not finally fixed until Trajan. Previously, troops and forts were much less permanently sited. The building activity undertaken in Trajan's reign marked the beginning of a new policy with regard to the forts on the *limes*; Hadrian merely continued it when he had them rebuilt in stone, or at least when he started large-scale building operations which in some places were possibly not completed even under Pius.

Though information concerning types of earth-and-timber forts in use is still lacking, there is a substantial amount available about the plan and construction methods of the stone ones. Since they show no basic difference from the well-known type found in Britain and on the Rhine, they must be associated with the building activity started by Hadrian (Fig. 18). Stratigraphic observations confirm this dating.

The stone forts built under Hadrian or in the mid second century have rounded corners, rectangular or, often, square gate-towers and internal corner-towers built to a trapezoidal plan. The thickness of the walls is generally no more than 1·20 m, though 0·80 m was not unusual. So far hardly any buildings within the forts have been uncovered, but one thing is certain: the defensive walls were strengthened inside with a rampart embankment. Construction on the *limes* was probably started after Hadrian's visit in 124; at the same time the *limes*-road connecting the forts was paved with stone. North of the fort of Matrica a fairly worn coin of Hadrian inscribed COS III (i.e. not minted before 119) was found underneath the thick metalling.[101] It is probable that the *limes*-road was metalled before the forts were built; hence it would not be wrong to date the forts to the middle of the second century. In the stone fort at Quadrata in Pannonia Superior, which has recently been investigated, the rampart was thrown up, in places at least, a good deal later.[102]

In addition to forts, smaller fortifications, mostly watch-towers, were built during these first extensive operations on the *limes*. Trajan's Column, with its reliefs of *burgi* on the Danube, provides indisputable evidence of their existence. These *burgi* were made of wood, but so far very few have been uncovered—there are, for example one near Fischamend and others near Dunaalmás and near Nyergesujfalu, all on the Upper Pannonian *limes*.[103] The one near Nyergesujfalu was stone-built. The number of such *burgi* from the early period which have so far been investigated is so small that no general deductions may be drawn. It is probable that a number of the towers and fortlets identified by surface inspection on the bank of the Danube in the Djerdap by the older generation of archaeologists, particularly by E. Swoboda, date to the earlier imperial period, since forts could be built in only a few places in the Djerdap (Pl. 8a), but the cliff road had to be constantly guarded.[104]

It is probably no accident that the earliest milestones in Pannonia date from Hadrian and Antoninus Pius (some of the stones, with their unusual way of indicating distance—e.g. *a Malata Cusum*—are memorial stones to Nerva which

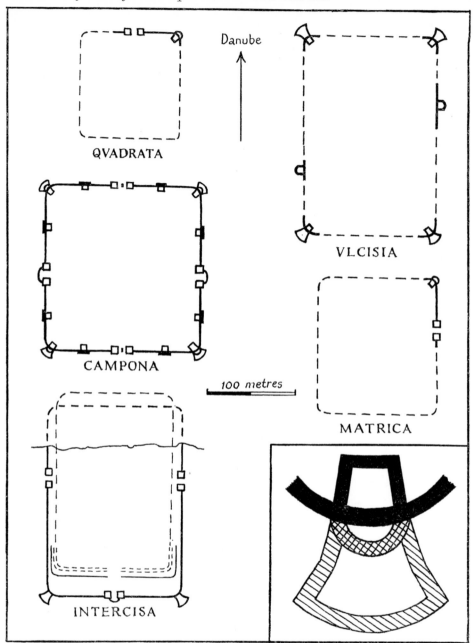

Figure 18   Danube forts in Pannonia

had been renewed).[105] The road network in the province was obviously not finished until the mid second century, though certainly no later. An enormous amount of labour and material was involved in the building of the *limes*-forts; the stone had frequently to be transported from quarries which were more than a day's journey from the building-site. Stone, tiles, lime and wood were probably carried, where possible, by water. Heavy demands were made on the *classis Flavia Pannonica* as well as on privately owned ships, but without roads construction work was impossible. Moreover, this work, which probably took a decade to complete, would have been unimaginable without some central planning and guiding inspiration. Contemporary and modern historians both credit Hadrian with this inspiration, and his negative attitude towards his predecessor's conquests seems to suggest that they are right. What is striking, however, is the fact that the plan for defending the frontier was apparently drawn up under Trajan; Hadrian only drew from it the inevitable conclusion and had the forts built of stone as permanent structures. If Appian of Alexandria, writing not much later, describes the empire as being surrounded by a closely woven chain of fortifications which protect it 'as if it were an estate',[106] and Aelius Aristides refers explicitly to a wall round the empire,[107] then this new conception, formulated by Trajan and put into effect by Hadrian, found significant expression. It would perhaps be idle to attempt to distinguish the various stages of this plan ranging from Vespasian to Hadrian. It was probably the state of foreign policy and experience over many years that led Trajan, despite or perhaps precisely because of his policy of conquest, to furnish the empire with plans for a different system based on the best possible frontiers, and Hadrian to put these plans, with certain modifications, into effect.

It might reasonably be assumed that the same kind of building work was carried out on the short sector of the *limes* in Moesia Superior. There was, however, an important difference here in that the interior of the province was not demilitarized. In fact the occupation army had to be reinforced under Marcus Aurelius.

As a result of the concentration of the Danube army on the *limes*, a process started by the Flavians and completed under Trajan, the river became one of the most important routes in the empire. It goes without saying that Trajan paid particular attention to shipping. The famous Trajanic *tabula* of the year 100 in the Kasan gorge confirms that the cliff road had been improved by widening (Pl. 9b); it had been hewn much deeper into the rock-face, and the wooden structure which had jutted out over the river was replaced by something simpler. A recently discovered inscription of Trajan attests the construction of a canal in the Iron Gates: *ob periculum cataractarum derivato flumine tutam Danuvi navigationem fecit*. It was not until the end of the nineteenth century that a later canal with the same purpose was made in the Iron Gates.[108]

No useful details are available about the harbour installations on the river.[109] The earlier landing-places close to the water had probably already been washed away. From certain landing-places which have survived from late antiquity we may draw the general inference that the earlier ones—mainly Flavian or Trajanic—were floating jetties. The army needed them, not only because of the considerable amount of transport by water but also because, since at least the time of Trajan, Roman strong-points had been built on the left bank of the river which could only rarely be reached from the opposite bank by a pontoon bridge. The posts of a bridge which connected the legionary fortress at Aquincum with the bridge-head fort on the opposite bank near the mouth of the Rákos (Budapest, XIII District) were discovered in the river-bed in the last century.[110] They came either from a wooden bridge built for some specific purpose and not intended to last longer than one summer, or from a pontoon bridge supported on extremely strong piles. It is hardly likely that a wooden bridge would have survived a winter in view of the very severe ice drifting. Most of the Danube crossings were probably made by ferry.

Many of the forts on the left bank of the Danube have disappeared without trace. The only one to be thoroughly excavated was an auxiliary fort covering an area of 175 by 176 m, discovered opposite the legionary fortress of Brigetio;[111] opposite Carnuntum the only discovery was the remains of a building (the Ödes Kloster) which may not originally have been sited on that bank,[112] and opposite Aquincum the fort on the Rákos was a walled quadrilateral with an inner portico on three sides.[113] The fort across the river at Brigetio, Iža-Leányvár, can be identified with Celamantia, mentioned by Ptolemy; it is not directly opposite the fortress but a little down-stream; at Aquincum, too, we may assume that there was a fort near the ferry down-stream from the fortress and opposite the late La Tène *oppidum* on Mount Gellért. It was here that Roman hypocausts were found outside the late Roman fort.[114] Further fortifications have been discovered opposite the auxiliary forts of Lussonium (Dunakömlöd) and Lugio (Dunaszekcsö):[115] their date is, however, not at all certain. There are inscriptions from the time of the Severi built into the walls of the fort opposite Lugio[116] which suggest a late date; but as Lugio was the terminal of the road through Sarmatian territory (attested on an inscription from Szeged), it was necessary for the crossing to be guarded on the left bank. It must, therefore, be assumed that some auxiliary forts had counter-forts across the river as early as the second century. It is probable that fully developed forts were built only at certain places, mainly opposite legionary fortresses. The others seem to have been simple rectangular open areas of about 0·5 ha enclosed by walls.[117] *Sigillata* pottery from the fort at Iža-Leányvár begins with the so-called Po valley products of the Flavian period,[118] and suggests that it was built at the time of the Suebian war before a legion was stationed at Brigetio. After the end of the Danubian wars these forts on the left bank probably assumed a quite different role. They were an organic part of the *limes* system; they did not serve as the spring-board for advances planned as part of an expansionist policy, which in any case came to an end with Trajan. It goes without saying that Rome realized the advisability of continuing to demonstrate its 'military presence', not least as a means of supporting its client-princes. But this could have been achieved equally well by the chain of forts on the right bank. The counter-forts served chiefly to safeguard peaceful contacts with barbarian territory. This did not mean that transport always had to have a military escort, but that a prudent system of treaties had to be guaranteed. The protection of communications under such conditions was, therefore, probably more of a policing task.

Nothing throws into sharper relief the self-confidence of the Romans, and the peaceful conditions which were seldom disturbed, than the palace of the governor of Pannonia Inferior which was built at the latest in the mid second century.[119] It stood on the bank of the Danube in front of the fortress at Aquincum; the ceremonial apartments, a suite of imposing rooms decorated with mosaics, stucco and murals, coincided, so to speak, exactly with the outer line of the imperial frontier, and the main front of this imposing building faced the barbarian lands. It is impossible to imagine that it could have been conceived in a fit of aberration or forgetfulness on the part of the architect. It was a conscious demonstration, to those within as well as those without, of that contemporary feeling to which, for example, Aelius Aristides gave literary expression. Such a striking architectural style, particularly in the case of a governor's palace, must be attributed not to a whim but to carefully considered political strategy: of this the governor had to be the prime embodiment.

Chapter 5
The first age of prosperity

The interval between the Year of the Four Emperors and the outbreak of the Marcomannic wars, lasting almost a century, can be divided into two periods of more or less equal length, which, despite the exigencies of foreign policy, emerge as logically connected stages of a systematic plan. The latter was, of course, only the consequence of applying in the peaceful post-Trajanic period conclusions drawn from a policy which had developed gradually under the Flavians and Trajan. In Pannonia this policy was crowned with success; in Moesia Superior, however, it came up against difficulties which caused development to be diverted to an entirely different direction. In general the first period, lasting from Vespasian to the moment when troop deployment on the Danubian *limes* was complete, can be regarded as a warlike situation; in the second period, roughly from the last years of Trajan to the outbreak of the Marcomannic wars, the framework thus brought into being was consolidated.

The spread of cities

An indication of the Flavians' new policy was the foundation of two colonies soon after Vespasian came to the throne. The Elder Pliny calls Siscia a *colonia*,[1] though he does not as yet apply that term to Sirmium, the other colonia Flavia in Pannonia; both towns were, however, given as the *origo* of Roman citizens who were witnesses to a military diploma[2] dated 73. Thus a *deductio* must have taken place in Sirmium as well as Siscia in the period 70–3.[3] A military diploma was issued in 71 to a man serving with the Ravenna fleet,[4] which, according to the formula *et sunt deducti in Pannoniam*, was treated as an exceptional case and

awarded the *missio honesta* through a *deductio*. The sailors attached to the fleets at
Ravenna and Misenum were probably treated in this privileged manner because
they had supported Vespasian in the Year of the Four Emperors.[5] At present
there is no evidence that there were legionary veterans at either Siscia or Sirmium; on the other hand, the tombstone of a discharged sailor at Siscia[6] suggests that the foundation of the colony at Siscia, and probably also of that at
Sirmium, resulted from the grant of land to discharged members of the fleet.
The phrase *sunt deducti* implies that land was granted in a colony. The sailors who
had distinguished themselves on the side of the Flavians were mainly natives of
Pannonia and Dalmatia, who on settling in the Save valley were probably not in
a position to make any significant contribution to the Romanization of the two
new colonies. At Siscia and Sirmium the number of tombstones of newly
settled citizens and their descendants is smaller than in Savaria and suggests that
the former were less intensively Romanized. The deductions at Siscia and Sirmium, which were exceptional in concerning members of the navy, were not
simply due to desire on the part of Vespasian to reward the *classiarii* who had
fought on his side and to settle them for political reasons in their home province.
Part of them were in fact settled in Paestum.[7] The choice of Siscia and Sirmium
was therefore not accidental. There is epigraphic evidence that the spread of
Italian traders into the Save valley had been restricted to the upper reaches.[8]
Hence, by settling the sailors in the two most important centres in the valley,
Vespasian probably also wanted to encourage shipping on the river. These
former sailors, now settled in the harbour towns of the Save, were probably
given special privileges with respect to traffic on the river. In this context it is not
without significance that the Pannonian fleet, though of earlier origin, from the
time of the Flavians bore the title 'Flavia'. The possibility has already been
considered that the reorganization of the Pannonian fleet dates back to
Vespasian.

The founding of two colonies on the Save is in some way connected with the
transfer of legio V Alaudae to Moesia and with the early stages of the construction of the fort-system on the Danube frontier below the mouth of the Save.
This stretch of the Danube could most easily be reached from Italy by way of the
latter river; in developing the supply network, the Save, and therefore also the
discharged marines at Siscia and Sirmium, assumed first-class importance.

There is still insufficient information concerning the topography of Siscia and
Sirmium to make any reconstruction of the original town-plans possible. Although extensive excavations in Sirmium in the last decade have established in
the main the extent of the late Roman imperial city as well as the main streets,
many public buildings, the line of the town-wall and the site of the early Chris-

tian cemeteries (Figs 29, 52), the pattern emerging is relevant only to late antiquity, by which time the town had probably increased several times over.[9] Excavations have so far failed to discover the site of the earlier town-centre, that of the Flavian colony. The samian pottery found then or brought to light in earlier excavations and now awaiting examination in the museum at Sremska Mitrovica is, however, in such quantity that a vigorous commercial life and a wealthy population in the first and second centuries must be assumed. The town probably stood a little distance from the Save, since the harbour, which was enclosed by the town-walls in late antiquity, was linked then to the town by a double wall and was clearly therefore not adjacent to it. Siscia stood at the confluence of the Colapis (Kulpa) where both rivers bend, so that the town, whether on the right or left bank of the Kulpa, had rivers on three sides. It is still not certain where the Flavian colony, or even where the Pannonian fortified town of Segestiké, captured by Octavian in 35 B.C., was situated. If we follow Count Marsigli (1658–1730), an able military engineer who made a plan of Siscia, we should seek the town on the left bank of the Kulpa where, according to him, the town-walls and the gates were sited between the Kulpa and the Save at Stari Sisak. On a hill on the right bank of the Kulpa and opposite Stari Sisak is Novi Sisak, which the surface configuration would suggest might be the site of Segestiké, were it not for the fact that there are still dykes round Stari Sisak which could be those of Tiberius.[10] The topography becomes even more obscure when we consider that the search for Segestiké, for the headquarters of Tiberius, for the assumed legionary camp and finally for the Flavian colony need not necessarily all involve the same place.[11]

The territories of Siscia and Sirmium were inhabited by the Pannonian tribes of the Colapiani and the Amantini respectively. It is probable that their *civitates peregrinae* ceased to exist in 71 and that their tribal aristocracy was gradually merged in the city councils (*ordo*). Roman citizens who had acquired that status under the Flavians are attested from both towns.[12] Part of the land was, of course, divided among the sailors. An interesting monument, a boundary stone[13]—the only one of its kind in Pannonia—was discovered at Čerević north of Sirmium. This stone was on the boundary of the *vicus Iosista*, all of which had been assigned to an officer of the Equestrian Order. This may be deduced from the phrase *adsignatus Tib. Claudio Prisco pr(aefecto) al(ae) I c(ivium) R(omanorum)*. If Priscus as a *iudex* or *arbiter* had assigned the *vicus* lands to someone else (to whom?), then the inscription would read *per Tib. Cl. Priscum*. Plots of land were therefore granted not only to sailors but also to high-ranking officers. From the officer's name it follows that it was only under Claudius or Nero that he was granted citizenship; hence it is highly improbable that as early as 71 he was already in a position to function as an *ala*-prefect of equestrian rank. It was probably under the Flavians that he reached this command—the third stage of the *militia equestris*—from which it may be deduced that the allotment of land in Sirmium did not end with a single act of *deductio* in 71. It is obvious that the land allotted to a common sailor was not the same size as that granted to an officer. The land attached to a *vicus* was, relatively speaking, extensive, and this explains why the big estates in the neighbourhood of Sirmium, of which more later, had their origin in the first century.

The other towns created by the Flavian policy of urbanization in Pannonia were municipia, which will be considered in another context in this chapter. Here it need only be emphasized that the three municipia—Neviodunum, Andautonia and Scarbantia—lay on roads which in the Flavian period were the most important in the province (Fig. 37). The founding of Neviodunum and Andautonia resulted in a chain of urban communities on the upper Save which made the founding of further towns unnecessary there. This chain obviously formed part of the plan, mentioned above, for encouraging traffic on and along the Save. Scarbantia, which as the *oppidum* of Scarbantia Julia had enjoyed a certain corporate organization since the time of Tiberius, was situated on the Amber Road between the Claudian colonia Savaria and the legionary fortress at Carnuntum. The municipalization of this existing group of citizens was probably not such an important turning-point in the history of this settlement as a *deductio* would have been. Of greater importance was the new policy which led to the founding of the first municipia, three of which now made their appearance in Pannonia. The grant of municipal status to Scarbantia itself could just as easily have been made under Nero, or even Claudius. That it was given this status under the Flavians means that this step suited their new policy, one which led to the creation of five cities in Pannonia, more than twice as many as under the Julio-Claudian emperors.

In Upper Moesia the founding of the first city is also attributable to the Flavians. It is, however, significant that in this province it was not possible at that time to create municipia, as the necessary conditions did not exist; there were neither groups of citizens made up of foreign settlers nor any well-established, more or less Romanized, *civitates peregrinae*. The Flavians also proceeded very cautiously with their policy of urbanization based on the *deductio*. The only Flavian colony was in the Axius (Vardar) valley, not far from the Macedonian–Moesian boundary and in an area which was largely isolated from the rest of the province, and which even later on lacked much cultural connection with the northern parts. From the point of view of ease of communications and land suitable for agriculture some places in the Kosovo Polje or in the lower

Morava valley would have proved to be better choices for a *deductio* than the neighbourhood of Scupi (Skoplje). The upper Vardar was probably chosen because at that time it was regarded as being the most completely pacified area, and the north–south line of communication running through the Vardar valley, the Kosovo Polje and the Morava valley was regarded, no doubt with reason, as having a certain importance. South of Scupi in the Vardar valley a tombstone was discovered of a veteran who had previously served in IIII Macedonica, a legion disbanded by Vespasian: its owner must have settled there in the pre-Flavian period.[14] This implies a settlement near Scupi – and one which exercised a certain attraction – in pre-Flavian times. This settlement had perhaps been established close to a legionary fortress. If this assumption is correct, then the colony was founded on the site of the abandoned fortress, a frequent occurrence.

From its original title D(omitiana), which was soon given up, it can be concluded that the colony came into being at Scupi as the result of a *deductio* in the early years of Domitian's reign.[15] The name colonia Flavia D (. . .) has been found on only two tombstones, which also happen to be amongst the earliest set up by veterans. The early dropping of the epithet Domitiana can only be explained by the *damnatio memoriae* which befell the founder of the colony. Moreover, it must have been founded before the outbreak of Domitian's Dacian wars, since veterans who were settled there had formerly belonged to legio V Alaudae, which was destroyed in the early days of the fighting.[16] Fortunately it is possible to list virtually all the legions which provided veterans for the new colony. Most of them came from VII Claudia, and some from V Macedonica, I Italica and V Alaudae, in other words from legions which were stationed at the time in Moesia. That VII Claudia provided most of the veterans and also, according to tombstones from Scupi, the earliest decurions, is possibly due to the fact that it was veterans from this legion who were the first to be involved in the *deductio* and were joined by those from other legions at a later date under Domitian, or even under Trajan. The term *deductus* or *deducticius* occurs only on tombstones set up to veterans from VII Claudia.[17]

The colony was not within the area of modern Skoplje but on a hill a little to the west at the confluence of the Vardar and Lepenac within the district of Zlokućane and Bardovce. Apart from a few houses only the theatre has been investigated,[18] and there are still no details of the different periods of the street-grid. It was only after the disastrous earthquake of 518 that a new town was built and fortified under Justinian on the site of present-day Skoplje. Many tombstones were discovered in the cemeteries at Bardovce and Zlokućane, but two-thirds of those set up to veterans were found north and south of the town; and it is from this distribution that certain inferences may be drawn. The first significant detail is that only one of the veterans buried at the colony was not a decurion, and of those who were, only two were buried outside the town. The earliest *missiones agrariae* were probably established north of Skoplje and east of Lepenac where, with one exception, all the tombstones are those of veterans of legio VII Claudia, who were either decurions or described as *deducticii*. This latter term is found only on tombstones in this area. Later allocations of land to other legions are attested west of Lepenac and south of the Vardar; discharged praetorians and auxiliary veterans settled in the southern part of the area. In the eastern half of the territory, north of the Vardar, there is no evidence of veterans, although there are just as many tombstones there. The veterans' plots were round the town within a radius of about 20 km, and their owners were buried on their own land (*in praedio suo* according to one tombstone).[19] This custom was also later continued by the urban aristocracy. This class was descended from the veterans; the decurions and the wealthy citizens who were buried on their estates round Scupi were the descendants of the veterans who had been settled at the end of the first century. It is significant that east of Scupi, where there is no evidence of veterans, decurions of a later period also left no inscriptions. On the other hand there is evidence of some Augustales,[20] who probably were able to acquire land only there.

The veterans were an extremely mixed group. They probably included more citizens of the provinces than Italians. Apart from one from Placentia, only Dalmatians (Salona), Macedonians (Beroia, Stobi), southern Gauls (Lucus Vocontiorum) and Syrians (Berytos) are attested.[21] In view of the method of recruitment to the legions concerned, it may also be assumed that Noricans, men from Asia Minor and Spaniards were represented. This is of particular importance, since in a province such as Moesia Superior, where the Italians were not so numerous and had arrived there later than they had on the Adriatic coast or in western Pannonia, the heterogeneous nature of the population was such that it was possible for Latin to become the mother tongue of certain of the towns. Syrians, Gauls, Macedonians and Italians could only communicate with each other in Latin, which they had all learnt during their long service with the legions. This was the reason why the new colonia Flavia on the southern border of Upper Moesia, and very close to Greek Macedonia, could become a Roman town.

All that was possible in the circumstances was probably achieved under the Flavians in the way of new municipia and *coloniae*. The founding of towns in Pannonia along the two main arteries of trade and troop-movement, the amber road and the Save valley route, was virtually complete. Such towns as were established later on these roads could hardly be regarded as towns in the strictest sense of the word. In Moesia conditions conducive to the founding of towns were much less favourable. Under Trajan the situation hardly changed, or, more accurately, the process started by the Flavians continued; Hadrian was the first to grasp its administrative implications. Trajan's policy for *coloniae* in Pannonia and Moesia was in part a continuation of that initiated by the Flavians and in part was a reflection of military policy, and hence was only indirectly connected with local conditions. If it is assumed that under Trajan the number of legions in each provincial army corresponded to that of the colonies in the province, and if it is borne in mind that the permanent number of legions in each province was to all intents and purposes fixed by him, then the conclusion follows that he created the new colonies (Poetovio in Pannonia and Ratiaria in Moesia) as new recruitment areas for the provincial armies.

Poetovio was a legionary fortress throughout the whole of the first century, and was only evacuated shortly before the establishment of the colony took place. It is not impossible that up to the time of Trajan there was also a legionary fortress at Ratiaria. In Upper Moesia, which had been pacified to a very limited extent, the choice of site for the new colonia Ulpia possibly rested on the fact that the vicinity, as a former military district, was regarded as pacified. An additional consideration was possibly their strategic commercial positions: Poetovio, at the point where the Amber Road crossed the Drave, was also connected with the Danube by a route down the Drave valley and with Aquincum by a road striking off to the north-east. Ratiaria lay at the spot where the shortest route connecting the Adriatic with Dacia reached the Danube.

In Poetovio there are inscriptions to veterans of legiones I and II Adiutrix, and possibly also of IIII Flavia, who became citizens of the new town on the basis of several *missiones agrariae* and *missiones nummariae*.[22] The town also yields positive proof that the deduction of a colony was not necessarily the result of a single mass settlement of veterans since, according to the inscription on his tombstone, a veteran was granted land in Poetovio on the basis of a second *missio agraria*.[23] As well as *missiones agrariae* there is also specific mention of a *missio nummaria*.[24] The colony was probably founded on the site of the abandoned legionary fortress which lay on the south bank of the Drave in the area of present-day Zgornja and Spodnja Hajdina near Ptuj. But a bridge was also built over the river and a new part of the town soon sprang up on the north bank.[25]

Ratiaria's veterans came from legio VII Claudia, but only a very few can be identified.[26] This is also true of the veterans at Poetovio. The majority of those for whom there is epigraphic evidence at Ratiaria were also decurions. Their inscriptions are concentrated in the town. If this situation is compared with that revealed by the epigraphic material at the Claudian colony at Savaria and at the

Flavian colony at Scupi, where legionaries were likewise settled and where many more inscriptions have been found—not confined to the town—we are forced to the conclusion that there had been a change in the policy of *deductio*. Perhaps urban development which had already begun in the vicinity of the abandoned fortress formed the basis of a city. Epigraphic and archaeological material establish pretty clearly that this was so at Poetovio.[27] If Ratiaria was a Flavian legionary fortress, a civilian settlement near by must also be postulated. Thus the nucleus of a town already existed, and the few veterans who were granted land in the new colony there were only the aristocrats among a town population already in being.

Trajan's veteran settlements were the last examples of city foundation by *deductio* in the history of the empire. Throughout his reign the founding of colonies was still linked to the allotment of land to veterans. Poetovio is referred to in a note on Trajanic land-allotment in Pannonia by the gromatist Hyginus.[28] On the other hand it is uncertain whether the Hadrianic colony at Mursa was based on a *deductio*.[29] This colony was the last to be established in the Danube area, possibly in the empire, as a fresh autonomous city. Later colonies started as municipia and were given the title of colony under the Severi or occasionally later. But the colonia Aelia Mursa was not based on an earlier municipium. Hadrian's well-known remark, quoted by Aulus Gellius,[30] to the effect that he was surprised that a city which could enjoy its own customs and laws as a municipium should wish to change its status to that of a *colonia*, would suggest that this was unlikely. Moreover, our sources are unanimous that Mursa was founded by Hadrian. It is not known whether veterans were settled there, nor does the still obscure topography of the town provide any information on this point.[31]

Development of trade

The founding of towns was thus based either on the settlement of veterans or on existing settlements of Roman citizens who had come in from outside the province. The Flavian–Trajanic period was one of massive immigrations, but it was not until Hadrian that they formed the nucleus of an urbanization policy. Without exception all the towns founded both by the Flavians and by Trajan were sited on the two roads which at just that time were losing their importance to the Danube road. By contrast, the immigration of the Flavian–Trajanic period was affecting the frontier zone where urbanization had not yet begun.

In fact one should speak of a second immigration into Pannonia. The first, already touched upon in Chapter 3, involved Italian traders from northern Italy, mostly from Aquileia, who took over the newly established markets on the Amber Road and in the upper Save valley and probably also the external trade and its chief entrepôt at Carnuntum. Although this network of trade was in its heyday at the beginning of the Flavian period, it was still by no means equal to the new demands made upon it. The very moderate needs of the partially Romanized soldiers in the few auxiliary units stationed on the Danube could be met either from local native production or from army supplies. It would seem characteristic of the pre-Flavian Danube army that it was supplied with imported goods only to a very limited extent. Whereas the Julio-Claudian forts on the Rhine can be dated on the basis of Arretine pottery and other finds of luxury goods, these are completely absent in the Danube forts, and even in areas in west Pannonia more exposed to Italian trade they have not been found in appreciable quantity. Then, under Domitian, units suddenly appeared, whose members either came from provinces in which trade was more advanced, or were so numerous that they were able to create a new situation on the Danube. Instead of a few cohorts or *alae*, each composed of some 500 or 1,000 half-barbarian soldiers, in a very short space of time several legions and an auxiliary army many times its original size had to be reckoned with, and all this in an area which had not previously been faced with a supply problem of this magnitude.

It is obvious that the first to operate in the new markets which had come into being as the result of the establishment of new forts were the Italians, that is to say the representatives of the Aquileian trading houses which already possessed branches in west Pannonia. Analysis of names on tombstones from the end of the first and the first half of the second centuries occasionally provides detailed information about how this network of trade arose.[32] The families of the Canii and Caesernii may be cited as examples of those involved.[33] Both were originally from Aquileia but had settled during the first century in Pannonia, the Caesernii at the beginning of the century in Emona and the Canii later on in Savaria. From the Flavian period onwards we find further establishments along the diagonal road leading into the interior of the province and ultimately to the Danube; the Caesernii finally settled in Brigetio, Aquincum and Sopianae and at the mouth of the Sala (Zala) not far from Lake Pelso; the Canii also settled near this lake and at Aquincum (Fig. 19). At the southern corner of the lake the only good point at which to cross the Sala's marshy river-mouth was near Zalavár and Fenékpuszta, where tombstones and a rich cremation cemetery indicate that there was a very early Roman settlement. These simple tombstones record not only Tiberii Iulii but also the Opponii, another family from Aquileia.[34] The diagonal road from Savaria to Mursa passed through Fenékpuszta, and its most important trading post before Mursa was Sopianae. Evidence that the Caesernii there came from

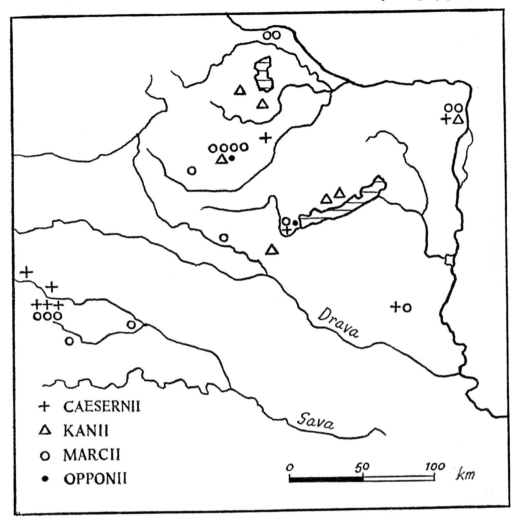

Figure 19 The distribution of some Italian families in Pannonia

the Savaria branch of the family is provided by their relatives, the Marcii, who are frequently recorded on inscriptions in Savaria and on tombstones both at Zalavár, not far from Fenékpuszta, and at Sopianae, as being related to the Caesernii.[35] Around the turn of the century there is epigraphic evidence at Aquincum of one *C(a)esern(i)us Zosimus natione Cilix* who obviously at one time had been a Cilician slave of the Caesernii.[36] They reached Brigetio by way of the short diagonal road along the river Arabo (Rába) and were not the only Italian family who found their way via Savaria to this fortress. In addition to *Q. Antonius Q. f. Cl. Firmus Savaria*[37] it is probable that other civilians left Savaria to

121

settle in Brigetio in the first half of the second century.[38] The Canii also came from Savaria. There is evidence of them north and south-west of Lake Pelso and at Aquincum, which suggests that they took the other important diagonal road which ran from Poetovio north of Pelso as far as Aquincum and crossed the Savaria–Mursa road near Fenékpuszta.

It is impossible to list here the names of the numerous Italian families who settled at Brigetio, Aquincum and other places of lesser importance on the Danube at the end of the first century. It is, however, probably no coincidence that the distribution of late Italian *sigillata* pottery (Fig. 20) and the so-called Po valley wares coincided with the migration of the Caesernii, Marcii and Canii, to cite only them as examples. Po valley pottery (of forms Drag. 1, 2 and 25) is extremely rare in the interior of the province but examples have been found in many places near Lake Pelso, as for instance in a cemetery not far from Fenék-puszta (at Keszthely-Ujmajor) and at Ságvár, a settlement at the junction of the road south of the lake with a minor one from Mursa to Brigetio. The same is true of the distribution of late Italian *sigillata* bowls with barbotine decoration. The most recent distribution map clearly indicates the importance of the Amber Road and the roads from Savaria to Brigetio and from Poetovio to Aquincum.[39]

In a variety of ways archaeological material reveals the direct links connecting the west Pannonian towns with Aquincum or Brigetio. The tombstone reliefs at Brigetio are probably derived from those of Savaria,[40] with the possibility that stonemasons moved from there to Brigetio at the beginning of the second century. It is typical that in producing such simple objects as earthenware lamps the craftsmen at Aquincum copied the products of Poetovio.[41] Under Domitian the demand for industrial goods probably increased by leaps and bounds in the Danube forts, with the result that a brisk traffic developed on the roads which linked them with the industrially more advanced towns in west Pannonia. In the second stage west Pannonian–Italian entrepreneurs settled on the Danube; this meant that it was no longer necessary to transport goods over long distances as they were made on the spot. Excavated material from the Danube area of the period from Domitian to roughly the reign of Hadrian contains a wealth of imported goods found neither before nor afterwards.

Only part of these imported goods came from Italian or west Pannonian workshops. As soon as there was a good number of garrisons along the whole length of the Danube, commercial shipping began to flourish and long-distance river-borne trade soon proved superior to that along the roads. Industrial centres in Gaul and Germany soon came to recognize that their goods could be carried safely and cheaply on the Danube to Pannonia and even to Moesia. South Gaulish manufacturers had tried earlier on to dispatch their goods to

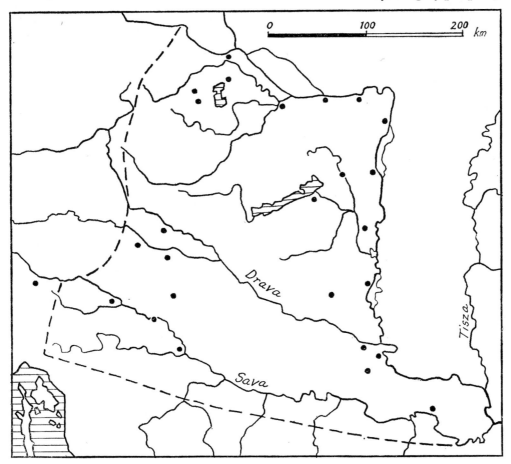

Figure 20　Italian *terra sigillata* in Pannonia

Pannonia via Aquileia, so that at a fairly early date their samian began to appear in the south and west of the province. The Danube, however, provided a unique opportunity to extend their markets. Around the turn of the century there was a flourishing Gallic–German trade along the river, and soon it succeeded in ousting that of Italy wherever the great distances overland resulted in goods from the west becoming competitive.

It is impossible to overrate the importance of the results of this change in the pattern of trade. To begin with, the reverse side of the picture should be considered. The area around Lake Pelso at the junction of the two most important diagonal roads (Savaria–Mursa and Poetovio–Aquincum) was excluded from the increase in traffic, as it was, of course, no longer necessary to transport imported goods overland over long distances, from west Pannonia to the

Danube. The decline both in economic life and in cultural standards is fairly clear in the area around Lake Pelso. The tombstones found north of the lake, which were the work after all of craftsmen in Poetovio and Savaria, date without exception to the first half of the second century.[42] The cemeteries show signs of becoming poorer, and in place of the general prosperity which was beginning on the Danube at this time we find that only large villas, including here and there some luxurious ones, were being built; these were no doubt put up either for rich owners of estates or for their bailiffs. The best known is that at Baláca, north-east of the lake, which is decorated with murals of very high quality in the earliest post-Pompeian style.[43] On the other hand, there is no evidence at all that the native population shared in this opulence, although this heyday of villa-building north of Pelso occurred at the very time when the natives of north and east Pannonia were beginning to set up tombstones and the aristocracy there was introducing the expensive custom of wagon-burials. This decline in the centre of the province is all the more significant if one considers that the native population particularly in this area had begun at a fairly early date to acquire some of the trappings of Roman life. In a native cemetery at Cserszegtomaj north of Fenék-puszta a two-handled vessel of late La Tène type (Pl. 1b) was found bearing the burnished inscription *da bibere*.[44]

The owners of these villas in the Pelso area were probably Italians, who had settled there and acquired estates during the Flavian–Trajanic period when economic conditions were favourable. It is possible that these were managed by servile bailiffs (*vilici*). *L. Petronius L. lib. Licco*, who married a native woman, *Galla Cnodavi f.*, may have been a freedman who acted in this capacity.[45] Whether he was identical with a freedman of the same name who appears on a list of *liberti* at Aquileia[46] is not certain, but by no means improbable. Another component of the wealthy upper class in the area was the veterans, who at the time when economic conditions were favourable had shown a preference for settling along the diagonal roads.[47]

Once western trade dominated the Danube it was not only Gallic–German goods which found their way to Pannonia but their manufacturers too. Although they secured a strong foothold in towns such as Carnuntum, Savaria and Brigetio,[48] which traditionally belonged to the Italian sphere of influence, it was primarily in Aquincum that they outnumbered the Italians. Here the contributory factor was that the legions transferred to Pannonia under Domitian and Trajan came in the main from the Rhine army; and in particular X Gemina, which was stationed in Aquincum for a longish time, came from the lower Rhine. At an even earlier date funerary sculptures at Aquincum reveal evidence of Rhenish influences[49] (such as military tombstones with full-length figures, or

Figure 21   Inscription set up by the burial-club of the *cives Agrippinenses*

eagles with outspread wings (Pl. 10b)). Rhenish ornament then appears on the tombstones of the Rhenish immigrants and beautifully illustrates the powerful influence suddenly introduced from the Rhineland.[50]

As early as the time of Trajan the Rhinelanders at Aquincum had formed an organization called *cives Agrippinenses Transalpini*, which carried out the functions (Fig. 21) of a *collegium funeraticium*.[51] The members of the many *collegia* which had been established at Aquincum at the beginning of the second century were foreigners who had settled there when trade was flourishing. The *collegium fabrum et centonariorum* saw to the burial of Caesernius Zosimus,[52] and also made itself responsible for the funerals of the Gaul C. Secconius Paternus[53] and the Thracian freedman *L. Valerius Seutes domo Bessus*.[54] A *collegium*, significantly called *collegium negotiantium*, is mentioned on an altar to Juppiter Optimus Maximus, Neptune and Mars, thereby indicating that the prosperity of these traders was attributable, in addition of course to the chief god of the empire, to shipping and the military.[55]

Apart from the *collegia*, the foreigners who had congregated at Brigetio and Aquincum also had an officially regulated organization. That in Aquincum was called *veterani et cives Romani consistentes ad legionem II adiutricem*,[56] whereas in Brigetio only the shortened form *cives Romani* is attested.[57] It is odd that despite the fact that there are many inscriptions from the early period in Carnuntum, where there was already a civilian settlement in the first half of the first century, so far no similar organization is attested. The corporate organization of *cives Romani* also had a council consisting of decurions and headed by two magistrates. Initially at Brigetio the over-all control was in the hands of a *curator civium Romanorum*. From the legal point of view these corporations were not unlike the numerous *cives Romani consistentes* of the late republican and early imperial periods, which had been founded by the Roman traders permanently domiciled abroad with a view to establishing autonomous control over their interests. Corporations of this type, too, probably existed in Pannonia; their late emergence, however, and the fact that they persisted until the time of the Severi is attributable to quite different causes. Civil settlements which are often surprisingly large developed near the legionary fortresses, surrounding the camp mostly on two sides though occasionally on three. Traders, shopkeepers and craftsmen—and, of course, the many veterans who planned to invest their money from the *missio nummaria* in a business near the fortress—settled in areas which were outside the jurisdiction of a self-governing city or of a *civitas peregrina*, but were under military control. Thus the Roman citizens living round the camps were neither *incolae* of a Roman community nor privileged foreigners in a *civitas peregrina*. However, as soon as such a settlement had reached a certain size which could no longer be administered without some kind of organization, a corporate form had to be devised. The original name *cives Romani consistentes ad legionem*, frequently attested in other provinces also, is a clear and legally accurate description of the situation.

There is evidence of the expansion of these civil settlements round all four legionary fortresses in Pannonia (Fig. 22), but only those at Carnuntum and Aquincum have been investigated to any extent (Fig. 23).[58] While there are clear indications that they had sewers and a water supply and many other adjuncts indispensable to urban life in antiquity, rectangular street-grids are not typical and town-walls were not a feature of the period of the principate. Behind the fortress at Carnuntum and not far from the *porta decumana* there was a rectangular market-place enclosed by a portico.[59] Inside it several wells were discovered which suggest that its primary purpose was to serve as a cattle-market. Trade in the Danube region is known to have consisted largely of the import of cattle and other foodstuffs from the barbarian areas. It is highly probable that

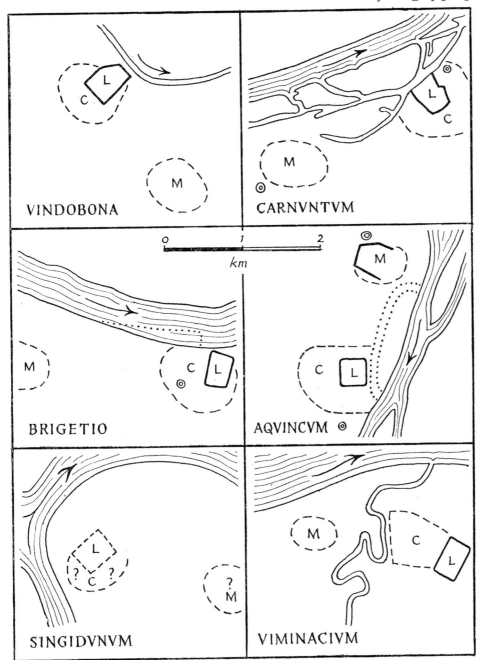

Figure 22 Legionary fortresses: *canabae* and *municipium*

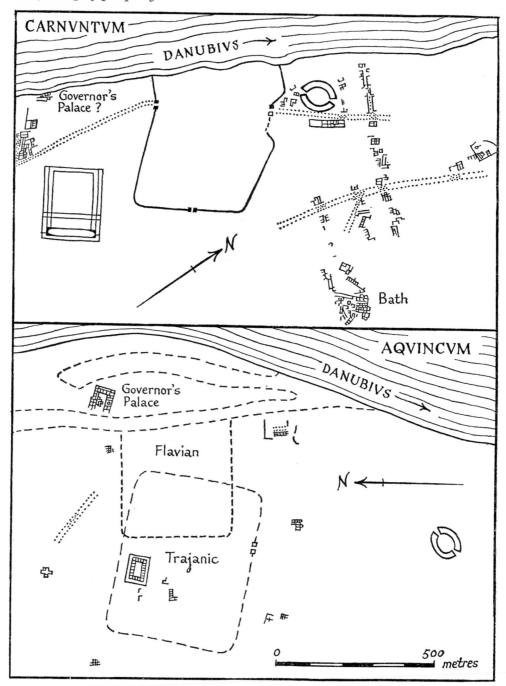

Figure 23   Canabae legionis: Carnuntum and Aquincum

this external trade was organized to take place in the vicinity of the fortress and under the control of the legion. In this way the fortresses became increasingly centres of trade. They were a source of attraction for all kinds of manufacturers, and hence made a considerable contribution to the development of many of the settlements on the Danube.

One of the most important elements in this external trade was slaves; dealings in them may well have been encouraged by the wars under Domitian and Trajan. The considerable number of slaves in Pannonia who came from beyond the frontier are mainly recorded in inscriptions dating to the Flavian–Hadrianic period and mostly found near Carnuntum and Aquincum; as examples we may quote *Peregrinus sutor caligarius natione Dacus, Naevius Primigenius Naristus, Vibius Logus Hermundurus, Tudrus, Strubilo*, etc. near Carnuntum, and *Scorilo Ressati lib.* from Aquincum.[60] It is strange that so far there is no evidence of Sarmatian slaves, but then there was hardly any Roman export trade to that area until the time of Marcus.

It was only indirectly and belatedly that Upper Moesia began to share in the economic prosperity of the Flavian–Trajanic period. The reasons for this are more fundamental than the simple explanation that this province was further away from Italy and the Rhineland than Pannonia was. The large-scale troop concentrations during the Dacian wars of Domitian and Trajan probably attracted army contractors and the usual hangers-on to Moesia, but war conditions, not to mention the two serious defeats under Domitian, were not conducive to the establishment of a permanent outlet for goods. It is also very questionable whether any long-term trade on the frontier which faced the hostile Dacians could be depended upon. In the early Flavian period there had of course been traffic in goods on the Danube below the confluence of the Save, and this was responsible for the sporadic appearance of Po valley *sigillata* pottery;[61] but everything which attests the settlement of foreign manufacturers and the establishment of civilian settlements near the legionary fortresses dates to the time after the Dacian wars, and this is probably no accident. If the great wealth of epigraphic material from Carnuntum from throughout the whole of the first century is compared with the complete lack of inscriptions from Viminacium, where a legion was stationed from at least the middle of the first century, then it must be admitted that conditions of military life in pre-Trajanic Upper Moesia were completely different from those in Pannonia. The earliest tombstones at Viminacium date to the beginning of the second century. Before that probably no one thought of setting up tombstones in the legionary cemetery, simply because there are likely to have been very few relatives permanently settled near the camp; hence it was not possible to establish a civilian settlement, particularly as there was little scope for traders. The founding of colonies at Siscia and Sirmium, clearly connected with Vespasian's attempts to encourage shipping on the Save, is an indication that even in the relatively peaceful period before the Dacian war artificial methods were necessary to give a fillip to traffic on the Danube below the mouth of the Save.

There was a sudden and fundamental change in the situation once Dacia became a province. Tombstones began to be set up—a definite sign that community life was being consolidated along Roman lines [62]—and demand soon led to the opening of a stonemason's yard at Viminacium. The earliest and still very simple stones were erected to high-ranking legionary officers; but before long they were becoming elaborate, occasionally by no means inartistic (Pl. 6b), and both in composition and style reflected the work of the south Pannonian stonemasons. [63] Both at Singidunum and Viminacium there are tombstones to families which originally came from Aquileia, for example that of *L. Barbius l. lib. Nymphodotus.* [64] The stretch of the Danube from Singidunum to Viminacium was of importance in west–east trade, particularly in trade between Italy and Dacia, which was carried along the Save and Danube as far as Dierna and Drobeta. Traders from Aquileia were as much involved in this as were those from the Rhineland. A married couple, in Aquileia *consistens*, had an altar erected in Singidunum [65] and a *negotiator Daciscus* from Cologne was buried in Aquileia. [66] The question arises why the Cologne traders found it necessary to make a detour via Aquileia in order to trade with Dacia, instead of going via Pannonia. Aquileia's central role was perhaps the deciding factor: Gallic and German goods were also on sale in that town.

The part played by north Italians and Rhinelanders in the economic development of the Upper Moesian section of the Danube area was neither substantial nor exclusive. At the same time as they were making their appearance at Singidunum and Viminacium there was an appreciable influx from the eastern provinces, and the considerable number of oriental colonists were putting their stamp on urban life in newly conquered Dacia. These easterners first appeared at Singidunum, Viminacium and Ratiaria in the first half of the second century and both in number and in influence were in no way inferior to those from the west. As in Pannonia it was the military from the Rhineland who brought the Rhenish civilians in their wake, so in Moesia the appearance of those from the east cannot be separated from the military. Soldiers from the eastern provinces and from the Greek provinces of the Balkans regularly served in the Upper Moesian army, both in the legions and in auxiliary units. [67]

The settlements in Upper Moesia developed along the same lines as those in Pannonia. At Viminacium a large civil settlement developed round the fortress [68] and a smaller but definitely attested one round that at Singidunum (Fig. 22).[69] Inscriptions have so far failed to throw light on how the communities were organized. *Cives Romani* with a *curator* at their head are attested from Margum at the mouth of the Morava.[70] This place, which had an importance for the internal trade of the Balkans, probably underwent the same development in the second century as the settlements round the legionary fortresses.

The mines

From an imperial viewpoint the economic development on the Danube was probably less important than the activity begun by Trajan in Upper Moesia and lasting until Marcus, which aimed at exploiting the very rich mineral resources of the province. The economic development of many Danube settlements was also encouraged by the exploitation of the mines, as the products of these mainly remote workings had to be transported to the Danube along its tributaries (Ibar, Toplica, Morava, Pek, Mlava, Timok). The *cives Romani* at Margum were probably engaged in the transport of raw metals along the Morava.

The opening up of the mines in Upper Moesia also led to an influx of population, though this did not have the same far-reaching effects as that of the north Italians and Rhinelanders into Pannonia and into the Danube towns of Upper Moesia. The social status of the miners, it would seem, was not as high as that of the manufacturers who had established their business in the prosperity of Flavian–Trajanic times. Also the management of the mines was in the hands of imperial authorities (procurators and freedmen) whose official position gave them little interest in the establishment of community life along urban lines.

The main source of information concerning the opening of the Upper Moesian mines comes from the so-called mine coins, small bronze coins of the second century with the names of Danubian mines on the reverse (Pl. 13b).[71] These mines lay in Noricum, Pannonia, Dalmatia and Upper Moesia, and practically all of them have been located. The Upper Moesian mines mentioned on the coins were *metalla Ulpiana*, obviously those east of Priština in the present-day Kosovo Polje (Pl. 14a); *metalla Dardanica*, probably those in the Kopaonik mountains east of the river Ibar and to the north of the Kosovo Polje; *metalla Aeliana Pincensia*, called after the river Pincus (Pek) and hence identical with the north-east Serbian Erzgebirge; and the *metalla Aureliana*, probably connected with the place-name Auriliana mentioned by Procopius,[72] and hence to be identified with the mines round Bor in north-east Serbia. Identifying these mines is not as simple as it looks; a little thought, however, will perhaps show that to identify them in any other way is not possible. Moreover, all of them must be regarded

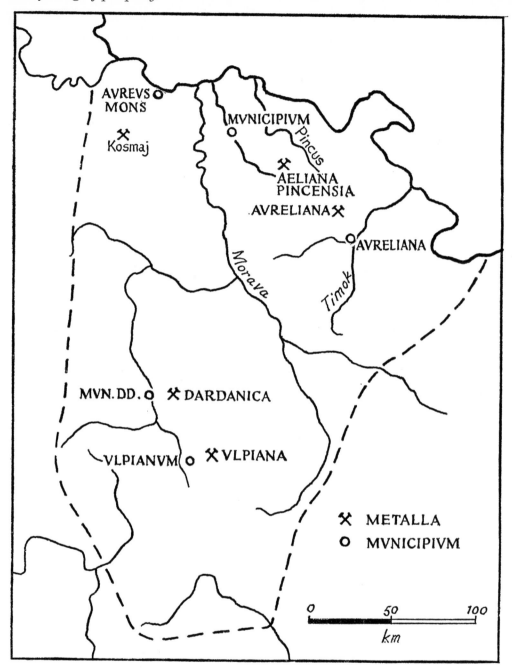

Figure 24 Mines and municipia in Upper Moesia

as identical with the best-known mines still operating today; and it is obviously not fortuitous that a *municipium Ulpianum* and a *municipium Dard* (. . .) are known to have existed not far off. Ulpianum was south of Priština and to the west of Mount Žegovac and Dard (. . .) in the Ibar valley, west of Kopaonik. At a later date the name Ulpianum became Ulpiana; this suggests that the Aureliana listed by Procopius in the χώρα (chora) of Aquae among the place-names in the Timok valley was perhaps originally Aurelianum, and hence that there was a municipium there. This is a justifiable inference, since it is known that there was a place called Municipium,[73] probably municipium Aelianum, not far from the mines which must be identified with the *metalla Aeliana Pincensia*. It will be seen later that in the course of the second and third centuries municipia were established near all these mines (Fig. 24), and that each took the name of its respective *metalla* (pp. 223 ff).

The Upper Moesian mines produced primarily lead and from it silver; but probably copper and here and there other minerals were also extracted.[74] The output was very high, and in some areas mining was very intensive. Near the mines on Mount Kosmaj south of Singidunum, which first went into production under Marcus Aurelius and for that reason are not mentioned on the earlier mine coins, it was estimated last century that the tips contained a million tons of waste material.[75] It was further established that the ores had been extracted right down to ground-water level.

The mine coins start with Trajan and mention *metalla Ulpiana* in several Danube provinces. Hence the name Ulpianum originated under him, a fact which is otherwise established; it was also under Trajan that there was the first mention of the Dardanica (Pl. 13b). Thus he was responsible for the opening and organization of the south Moesian mines. Somewhat later under Hadrian the Aeliana and possibly also the Aureliana came into production. The former, which derive their name from him, are mentioned on coins minted under Hadrian but reference to the latter occurs only on undated mintings. The name Aureliana may perhaps be connected with the young Caesar Marcus Aurelius and may therefore also be dated to the year previous to Hadrian's death or possibly a little later. Whether it was for security reasons that the opening of the mines in the south preceded those in the north, which were closer to the frontier, can only be a matter of conjecture. It was Marcus Aurelius who had the mines on Mount Kosmaj nearest the frontier started, and he had to protect them with auxiliary units.

Unfortunately there is very little information about the way these mines were managed. The one certain fact is that originally they were in the hands of pro-curators (*procuratores metalli*) each of whom was probably responsible only for one *territorium metalli* and as a rule was an imperial freedman.[76] Freedmen of the emperor also provided the technical experts as, for example, *P. Aelius P. Lib. Menander centurio officinarum*, or *P. Aelius Plato mensor*.[77] The mines were, however, leased to coloni who also had a kind of autonomy. On an inscription from the *metalla Dardanica* occurs the formula *locus datus decreto colonorum*, which is reminiscent of *locus datus decreto decurionum*,[78] the form current in municipal districts. Under Hadrian the *coloni argentariarum* erected a shrine to Antinous after his deification; this perhaps stood in the middle of the store-houses (*horrea*) (Pl. 20a) belonging to the mine administration.[79] It is probable that the coloni brought considerable numbers of miners from Thrace and Dalmatia into the province, and perhaps also some slaves. These then formed what was tantamount to an urban population in the mining settlements. The upper class in these settlements was represented by the richer coloni, the majority of whom were probably freedmen. While there is some reference to them on inscriptions from the *metalla Ulpiana* and *Dardanica*, the strongest evidence comes from the mining district on Mount Kosmaj whose greatest period of prosperity occurred under the Severi.

Civitates *and* municipia

Despite extensive colonization and large-scale immigration, foreigners were not the only ones to be involved in the new Flavian–Trajanic policy. There was also a fundamentally new approach to the native population, despite considerable differences between Pannonia and Upper Moesia. It was at this point that the development of the two provinces diverged.

The *civitates peregrinae* were granted autonomy either under the Flavians or at latest under Trajan, or to be more precise they were freed from military control and were placed under *praefecti* chosen from the ranks of the native aristocracy (*principes*). Naturally it is not certain just when this change took place, and it is even less certain whether this form of local self-government was granted simultaneously to all civitates. Possibly some north Pannonian civitates were not freed from control by military *praefecti* until the time of Trajan. It would seem that the introduction of local *praefecti* was connected with fairly widespread grants of citizenship; but grants under the Flavians are attested in only a few civitates in the south and west of Pannonia, and in some areas they began only under Trajan. For a long time analogies in Dalmatia had suggested that the military *praefecturae* were abolished towards the end of the first century and replaced by local functionaries. It was not until 1956 that positive proof was provided with the discovery of the tombstone of T. Flavius Proculus, *pr(inceps) praef(ectus)*

Scord(iscorum) who had been granted citizenship by a Flavian emperor.[80] This tombstone also revealed that, unlike the native *praepositi* in Dalmatia, the heads of the civitates were known as *praefecti* like their military forerunners. Additional confirmation that local self-government was granted under the Flavians comes from another civitas. In 1951 a tombstone set up to his wife by M. Cocceius Caupianus PR.C.B. was discovered in the enormous villa of Parndorf in the Leitha district.[81] From his name it follows that he had acquired citizenship under Nerva. The abbreviation can stand either for *pr(aefectus)* or *pr(inceps) c(ivitatis) B(oiorum)*, and either expansion could be defended were it not for the fact that in the civitas Boiorum a surprisingly large number of Flavii had acquired citizenship before Nerva's reign. They retained their traditional Celtic names (e.g. T. Fl. Cobromarus, T. Fl. Biturix, T. Fl. Samio, etc.) and were apparently very rich, to judge from the unusually large number of their slaves and freedmen referred to on inscriptions.[82] Since it is clear that the Boian aristocrats had been granted citizenship under the Flavians it is not very likely that one of their most respected members did not acquire it until the time of Nerva. The interpretation of PR as *princeps* is therefore preferable. From this wholesale granting of citizenship to the inhabitants of the civitas Boiorum it follows that it must have become self-governing at the same time. In the early years of Vespasian's reign the civitas was still under military control which at that time was exercised by Volcacius Primus in his capacity as *praefectus alae I Noricorum*.[83] His headquarters were in the *ala*'s permanent fort at Arrabona which the Boian and Azalian civitates adjoined. Volcacius was the *praefectus* of both civitates, which perhaps indicates that military control was being relaxed. It was, however, still in force under Vespasian.

It is possible that the abolition of military control over the civitates in south and west Pannonia was connected in some way with the founding of towns by the Flavians (Fig. 37, p. 220). Colonia Siscia was placed in the territory of the civitas Colapianorum which under Nero had had Antonius Naso as *praefectus*; the probable result of its foundation was that the territory of the civitas was absorbed by that of the colony. The Flavian colonia Sirmium was founded in the territory of the civitas Amantinorum, and it is probable that the *territorium* of the Flavian municipium of Scarbantia was carved out of the civitas Boiorum, as was the *territorium* of colonia Savaria at an earlier stage. Two south Pannonian civitates were raised to the status of municipia under the Flavians, and were the first indications of a new policy of urbanization which after Trajan completely superseded the old system based on colonization or settlers from outside. The municipium Flavium Latobicorum (Drnovo) was founded close to the mouth of the Korkoras (Krka); its name shows the connection with the local tribe of the

Latobici. Further examples of this type of name for municipia in both Pannonia and Upper Moesia at a later date will be quoted below (p. 223). It is probable that the new Flavian municipium was simply created out of the civitas Latobicorum, that the latter's *principes* became the decurions of the new town, and that the raising of its status was coupled with a fairly wide grant of citizenship to its inhabitants. Some Flavii are in fact mentioned on the not very numerous inscriptions from the vicinity.[84] The name municipium Latobicorum was soon disused: on later inscriptions it is called Neviodunum and as such it is listed in the *Itineraries*.[85] There are many examples from the middle Danube area of this kind of place-name change. The local Celtic name for the settlement near Drnovo was Neviodunum and continued to be used, even after the area of which Neviodunum was the centre had been given a new official title. The latter was unable to establish itself, a fate which was later shared by such names as municipium Iasorum in Pannonia, colonia Ulpia Traiana Dacica (Sarmizegethusa), colonia Flavia Felix Domitiana (Scupi), etc.

The other Flavian municipium in the upper Save valley was Andautonia, which had been founded in the territory of the civitas Varcianorum (Sčitarjevo not far from Zagreb) (Fig. 9). Although there is no evidence that it was given a new name, this is not unlikely. In any case by the time of Trajan it was again called Andautonia.[86] Both Neviodunum and Andautonia probably owed their relatively early emergence to shipping on the Save. Although there is evidence of citizenship being granted to the Celtic natives on the upper Save by the Flavians and even earlier, and although the name municipium Latobicorum is evidence for the native origin of the settlement, a section of the urban upper class was descended from the Italian entrepreneurs who had settled along the Save in the course of the first century. North Italian names (Aquileians) such as Marcii, Rustii, Capenii, Annaei, Trotedii, Firmidii, etc.,[87] are epigraphically attested at Neviodunum, and from Andautonia the Iuventii, Caesernii, Aconii, Lusii, etc.[88] Many of them were freedmen, or indeed slaves, as for example a certain . . .]*medus Trotedi negotiatioris servus*. It was significant that they set up altars to the gods of shipping—to Neptune as a general rule, but also to Savus. For instance it was to the latter that a certain M. Iuventius Primigenius made a dedication along with his business associates (*socii*). Inscriptions set up by north Italians and also by some southern Gauls (Eppii)[89] come from the near vicinity of Neviodunum and Andautonia, whereas those of the natives have been found mainly on the periphery of the territories.

Thus these Flavian municipia were based equally on the native population and on foreigners. Nevertheless the aim was clearly to create towns from the *civitates peregrinae* and, as will be seen later, this aim both under and after Hadrian became the guiding principle. The necessary conditions did not, however, obtain everywhere; these were obviously the existence of a settlement of a more or less urban nature, for which of course the foreign settlers provided the main impetus; the presence of a local upper class which not only numerically but also financially and politically was capable of forming an urban aristocracy; and finally a certain number of Roman citizens on whom the early stages of urban life would depend. Most of these conditions were to be found only on the south and western boundaries of Pannonia, the areas of the most important traffic routes in the early period. That administrative measures were necessary even here in order to open up shipping on the Save has already been mentioned (p. 113) in connection with the foundation of the colonies of Siscia and Sirmium.

The conditions just mentioned did not exist on the banks of the Danube until the late Flavian–Trajanic period and even later in the interior of the province. As we have seen, large settlements were established by foreign traders near legionary fortresses on the Danube, and we may suspect the beginnings of a similar process at various points along the diagonal roads from Poetovio to Aquincum and from Mursa to Savaria, for instance at Sopianae, although from the outset it was hindered and finally brought to a standstill by the development of trade on the Danube. The beginnings of systematic grants of citizenship both in the interior of the province and on the Danube likewise go back to the late Flavian–Trajanic period. The Flavians bestowed citizenship on the native aristocracy only in certain areas. Even in the municipium Latobicorum the majority of the natives remained peregrini; they are mentioned on numerous inscriptions from the western half of the urban territory.[90] It is only with the appearance of Ulpii of native origin that we can trace a universal, albeit gradual, granting of citizenship. These Ulpii are mainly attested in the following civitates and towns: Eravisci, Azali, Boii, Andizetes, Scordisci, Iasi, Breuci, Cornacates and Savaria, Sirmium and Poetovio. Probably the first to embark on this step was Nerva, who granted citizenship not only to the Boian *princeps* Caupianus, as already mentioned (p. 135), but also to Florus, a *princeps* of the Eravisci and son of Matumarus.[91] Often it is not possible to distinguish between civilian members of the upper class who had become citizens and veterans who had been similarly honoured. In north-eastern Pannonia there is clear evidence that citizenship was granted to civilians.[92] In the long run it is of course irrelevant whether the daily increasing number of natives with Roman citizenship were civilians or veterans. Trajanic new citizens (Ulpii) or their descendants are attested in all the areas where Hadrian founded new towns.

The situation was completely different in Upper Moesia. Neither the Flavians nor Trajan founded municipia there; moreover they showed considerable reluctance in granting citizenship. It is not in fact surprising that municipia could not be founded, since, as we have seen, settlements on a large scale only began to develop in the reign of Trajan, while native civitates had not yet reached the stage when municipalization could take place. As already pointed out in Chapter 3 the *civitates peregrinae* in Upper Moesia were reorganized in the second half of the first century, doubtless as the result of massive settlements of barbarians from the left bank of the Danube; these had probably not reached yet the stage at which any relaxation of strict military control would have been possible. Nor was it accidental that the mines were first developed under Trajan; it obviously was not as though the presence of minerals had just been discovered. Even with mines of such great importance to the empire, security was the prime consideration. There is no evidence at all before Trajan for a grant of citizenship in Upper Moesia. Even he granted it only in a single restricted area, that of the Hadrianic municipium of Ulpianum (Fig. 25, p. 146), coinciding roughly with the autonomous Jugoslavian district of Kosovo and Metohija (Kosmet). This area, contiguous on the north with the Flavian colony of Scupi, was probably well pacified, not only because it was near a colony but also because it was an extensive agricultural district, as the Greeks of the classical period knew.[93] It is precisely from this area that one of the few *latifundia* (huge estates) on the Danube or in the Balkans is attested; it belonged to the senatorial families of the Furii and the Pontii, the latter having acquired it at the latest under Hadrian.[94] A further factor was that the road from the Adriatic to Dacia passed through the Kosovo Polje. The Trajanic new citizens in the Kosmet, unlike the Boian Flavii, mostly had nondescript Latin cognomens but the names Ulpius Andinus, Ulpia Andia and M. Ulpius Timentis f. Maximus, as well as the single Illyrian or Thracian names possessed by relatives of the Ulpii, indicate their local origin.[95] There is, incidentally, no information concerning the fate of the *civitates peregrinae* in Upper Moesia.[96] It is, however, probable that the civitas Dardanorum, whose territory roughly corresponded to the Kosmet, became self-governing under Trajan at the same time as its inhabitants were granted citizenship.

Hadrian's provincial policy on the Danube, it would seem, was to continue and conclude the development started by his predecessors, particularly Trajan. Even if we confine our attention to the number of municipia Aelia in Pannonia, the importance of Hadrian in the history of the province clearly emerges. According to present evidence he created eight municipia (Fig. 37, p. 220), though it should be emphasized that this may well not be the final figure. Three out of the seven became known only a few years ago, hence the discovery of further municipia created by Hadrian or even by later emperors cannot be ruled out. He also created at least two municipia in Upper Moesia.

It is quite possible that the municipia in Pannonia and Upper Moesia were founded during Hadrian's visit there in 124. He reached the Danube from the east via Thrace, and from Pannonia continued his journey in the direction of Dalmatia. A few minor incidents which occurred during his progress are recorded on inscriptions; for instance, according to an epitaph to a favourite horse, possibly composed by himself, he took part in a boar hunt in the wilds of Pannonia.[97] He also had a trooper belonging to the cohors Batavorum report on his outstanding deeds of bravery.[98] An inscription in honour of Hadrian, incised on what was probably the plinth of a statue, suggests that Aquincum was raised to the status of a municipium during an imperial visit.[99] The other municipia on the Danube, Viminacium and Carnuntum, were probably created in like manner. These were, of course, the seats of the governors whom Hadrian had doubtless visited.

Canabae *and Hadrianic* municipia

The three municipia created by Hadrian on the Danube (Carnuntum, Aquincum and Viminacium) put the final touches to urban development near the legionary fortresses begun under the Flavians or even earlier. By the time of Trajan, at the latest, this development had produced the corporate autonomy of the *cives Romani consistentes*. While the topography of the 'camp towns' on the Rhine and Danube still remained obscure it used to be thought that the municipia founded in the vicinity of the fortresses could be identified with the settlements there which had developed into towns with the *cives Romani* or the *canabae* as their pre-municipal form of administration. But after it became clear that near almost every legionary fortress on the Rhine and Danube there was not one but two civil settlements, the problem became much more complex and many of its aspects still await clarification. The following topographical details are typical of Danube legionary bases (Fig. 22, p. 127): around each was a civilian settlement, the creation and development of which ran parallel with those of the fortress itself. The Roman citizens who had established themselves in this settlement formed the pseudo-autonomous corporate body of *cives Romani consistentes ad legionem*. This civil settlement, later called *canabae*, shared the neighbouring cemeteries with the military. At least one Roman mile from this settlement and from the fortress there was another civil settlement with its own cemetery. Between the two settlements was a space which was neither built upon nor used as a burial ground. It was this second settlement at a distance from both *canabae* and fortress which became a municipium, whereas that near the camp continued to have a semi-municipal autonomy which in legal terms was quite

139

different from that of the municipium. That the two civil settlements were on different sites has been clearly established at all the fortresses in Pannonia and Upper Moesia. At Carnuntum, Brigetio, Aquincum and Viminacium the settlements which sooner or later became municipia were above the fortress, either to the west or north, whereas at Vindobona and Singidunum they were below it, that is to the east. The reasons for these double settlements would not be difficult to identify if those which became municipia under Hadrian had also been founded under him. In that case the duality could be explained by the fact that settlements close to fortresses could not be municipalized since they occupied ground belonging to the military, where no municipal territorium could be provided. Hadrian would then have had a town founded at some distance from the camp and have granted a municipal charter; in so doing perhaps he might also have wished to check the development of the settlement *ad legionem*. The matter is not, however, as simple as that. Investigations have already proved that the settlements which became municipia under Hadrian or later were already in existence at the time of their promotion and were probably founded at the same time as the settlements *ad legionem*, or not very much later.[100]

Our starting-point must be the territorial situation. It is usually assumed that settlements *ad legionem* were established—just as the fortresses themselves—on land which was not only owned but also administered by the military. *Prata legionis* (pasture-lands belonging to the legion) are in fact attested in the first century in Spain, Dalmatia and possibly also in Lower Germany. On this land, where territorial sovereignty was in the hands of the military, municipal autonomy was impossible, and it follows that Roman citizens who had settled there could only form a corporate community, that of *cives Romani consistentes*. Unfortunately there is no information concerning the size of the *prata legionis* on the Danube. There is, however, a strong possibility that the settlements which later become municipia did not lie on military ground. The fact that two civil settlements were established near the camps may best be explained by the supposition that the circumstances under which the settlers established themselves and the communal life in the two types of settlement differed right from the start; moreover, the entrepreneurs, soldiers' relatives, veterans and other foreigners could choose to settle where the legal and administrative conditions seemed more appropriate to the kind of life they had in mind. When Hadrian brought urban development near the camp within the scope of public law he could only raise the settlement which lay at some distance from the camp to the status of a municipium, for otherwise he would have had to deprive the military of control over the area closest to the fortress. Septimius Severus was the first to take this step.[101]

140

Thus those settlements which became municipia under Hadrian were probably on territory belonging to the neighbouring *civitas peregrina*; and there is much to support the view that the latter was municipalized at the same time. This emerges, for example, from the fact that the decurions of the municipium Aelium Carnuntum have left inscriptions in the former territory of the civitas Boiorum, and some of them had acquired citizenship from the Flavians or Trajan.[102] It is probable, nevertheless, that the *ordo* of Carnuntum was largely composed of foreigners who had been living for a long time in the vicinity of the fortress. New settlers are also attested in the *ordo*: the decurion C. Domitius Zmaragdus, who was sufficiently wealthy to pay for the building of the amphitheatre at Carnuntum out of his own pocket, was an immigrant from Antioch and one of the earliest Syrian traders to settle in Pannonia.[103]

The *ordo* of municipium Viminacium likewise consisted of foreigners, exclusively so, it would seem.[104] It is from here that there is definite evidence of a decurion being the son of a freedman.[105] Freedmen were doubtless one of the most important components of the *consistentes ad legionem* in the Flavian or Trajanic periods. In towns where there was a flourishing economic life it was the freedmen who had grown rich who later provided the urban aristocracy.

The development at Aquincum was entirely different. Doubtless here too Hadrian intended to build up the life of the new municipium on the basis of the wealthiest inhabitants of the settlement *ad legionem* or, more precisely, to transfer their not inconsiderable wealth to the new municipium; but in this, it would seem, he was unsuccessful. The by no means small number of decurions of the municipium known to us from inscriptions in the period from Hadrian to Marcus were without exception natives who owed their citizenship to Trajan or to his successors, Hadrian and Pius (Fig. 25).[106] Foreigners did, of course, play their part in the life of the town as members of the numerous collegia,[107] as did the veterans of the legio II Adiutrix, but they were either not represented in the *ordo* at all, or if so only to a limited extent. Despite everything, the civitas Eraviscorum was not dissolved nor did it disappear without trace in the newly founded municipium Aelium Aquincum: it survived after Hadrian's time and was even mentioned on an altar of the third century.[108]

The rich epigraphic material from Aquincum and its environs provides certain clues enabling us to reconstruct a situation which was legally extremely complicated. In the first century the civitas Eraviscorum had *principes* who were called *decuriones* under Trajan.[109] Under Hadrian and later, however, there is evidence of only one *tabularius*[110] and one functionary *arm*(. . .),[111] the latter otherwise attested nowhere else; hence some change in the status of the civitas must definitely have taken place under Hadrian. In the same area where the *principes* or *decuriones, tabularii* and the other ordinary members of the civitas Eraviscorum are epigraphically attested, inscriptions to the decurions of the municipium have also been found. Consequently the latter owned property in an area which up to the time of Hadrian had belonged to the civitas Eraviscorum. Nevertheless it cannot be assumed that the territory belonging to the civitas was merged in that of the municipium, because the community itself, that is the civitas, although it had undergone a change, persisted even after Hadrian. Since the decurions in Aquincum came from the territory of the civitas, it must be assumed that the latter was in some sort of position of dependence on the municipium. The probability is that the administration of the civitas was taken over by the *ordo*, the members of which were, of course, well equipped to represent the civitas, since some of them came from it. The civitas which continued to exist as a territorial entity had only a few administrative officials (*tabularii*), and perhaps one or two elected honorary heads (*arm . . .?*) who were at the same time decurions. The inhabitants of the civitas, not unlike those in Latin municipia, were able to acquire citizenship on becoming members of the *ordo*, or as auxiliary veterans since recruitment was continued: diplomas made out to Eravisci in the post-Hadrianic period are not infrequent.[112] Basically, the relationship between the municipium and the civitas resembled the *attributio*, though this term is not attested. Thus it was natural for two duoviri from Aquincum, both natives who had recently acquired citizenship, to set up an altar to the well-being of the civitas Eraviscorum,[113] and since the latter was never formally abolished it could be commemorated by an augur on an altar in the third century.[114] At that time it probably existed only as a community based on a religious cult, since between the time of Hadrian and that of Caracalla all legal and administrative differences between the two kinds of community had probably ceased to be significant.

It is possible that Hadrian also reformed the administration of the settlements *ad legionem*, as it was under him that a change in their corporate organization actually took place. On inscriptions of the second century the settlement at Aquincum is called *canabae*;[115] although this term was not applied everywhere as a regular thing to these settlements in the second century, it was more often used than the old term *cives Romani consistentes*. The latter had had *curatores* or *magistri*, but the *canabae* always had decurions in addition and all the magistrates found in a municipium or colony. This pseudo-municipal form symbolized the urban character of the 'hutted settlement' (*canabae*) but at the same time it represented a compromise made necessary by the fact that the territory on which the *canabae* was sited belonged to the military. Thus during the second century near the fortresses on the Danube there were two separate self-governing communities and actually three at Aquincum, where the municipium, the civitas Eraviscorum and the military each had their own territories. That of the municipium was probably to the west and north of the town, that of the *civitas peregrina* to the south and south-west. Where the military territory lay, lack of information makes it impossible to say. It may, however, be assumed that the land round the auxiliary forts was part of it. An inscription records the *cives Romani consistentes* near the fort of Vetus Salina.[116]

Hadrian's other municipia were sited in the interior of the province and hence had their origin in purely civilian developments. Two types can be distinguished —that which developed out of a *civitas peregrina* whose decurions, it would seem, were natives who had recently acquired citizenship; and that which granted autonomy to groups who were already citizens, and in part at least of foreign origin. In making this distinction it would, of course, be wrong to insist on consistency; at the moment it is still not possible to say whether in the first type of municipium foreigners also had a say in affairs.

Included in the first type in Pannonia were municipium Iasorum, Cibalae and Bassiana, and in Upper Moesia municipium Ulpianum. Those in Pannonia were on the road from Siscia via Mursa to Singidunum, and as the result of their foundation the whole of the Sirmian region and the greater part of the Drave and Save valleys were municipalized. Municipium Iasorum was situated roughly midway between Siscia and Mursa at the spa-site of Aquae Balizae (Daruvar).[117] The official name given to it is a clear indication that the civitas Iasorum had become a municipium. Evidence that its territory was extensive is provided by an inscription giving the *origo* of an *eques singularis* at Rome, who hailed from the vicus Iovista on the Drave, which was said to belong to Aquae Balizae, otherwise called municipium Iasorum.[118] The name municipium Iasorum is also attested on a tomb-altar of a decurion, but as with the name municipium Latobicorum it failed to catch on. In the *Antonine Itinerary* the place is called Aquae Balissae. Cibalae lay south of Mursa (modern Vinkovce) at a natural intersection of routes and in an area in which practically all the important battles in Pannonia took place.[119] The area belonged to the civitas Cornacatium, the former centre of which, Cornacum (Sotin), was not far from Cibalae on the Danube. Bassiana (Petrovci), between Sirmium and Singidunum, was in the territory of the civitas Scordiscorum.[120]

Only one decurion from each of these three Hadrianic municipia is attested, namely *P. Aelius P. fil. Aelianus scriba, decurio* and *IIII vir* of municipium Iasorum,[121] M. Ulpius Fronto Aemilianus, decurion of Cibalae[122] and P. Aelius Dasius, decurion of Bassiana.[123] Despite the sparse details it is hardly accidental that the granting of citizenship to these decurions dates to Trajan or in two cases

to Hadrian. The *ordo* of these municipia was obviously formed from the aristocracy of the *civitas peregrina*. Mention should also be made at this point of the
municipium Mogentiana, since it also had Aelii among its decurions.[124] Its
territory was, however, north of Lake Pelso, where there is evidence that
foreigners had settled at an early date and owned large estates. Thus a decurion
was called Cassius and a *scriba* possibly Dubius.[125] Unfortunately it has so far
proved impossible to locate Mogentiana, though it lay somewhere on the road
between Savaria and Aquincum. Taking into account the contradictory distances given in the *Antonine Itinerary*, its site should be west or north-west of the
Bakony mountains, the most likely places being Somlyóvásárhely or Sümeg-
csehi. A further difficulty in locating its site is that, apart from one tombstone
with a doubtful reading (*scriba municipii M.*), all the epigraphic evidence comes
from areas which definitely did not belong to Mogentiana's territory, as for
example the neighbourhood of Brigetio, Aquincum and Tokod on the Danube,
east of Brigetio.[126]

The municipium Mursella, too, which was probably also founded by Hadrian,
was situated in an area in which foreigners and veterans had settled in the first
century. Its territory probably took in both banks of the Arabo (Rába) south of
Arrabona (Györ). There is epigraphic evidence of two of its officials, a *scriba* by
the name of Gallonius and a decurion called Iulius who belonged to the
Equestrian Order.[127]

Recently a new inscription came to light at Aquae Iasae,[128] attesting a
Hadrianic municipium named Salla. Salla (or Sala) is probably to be found at
Zalalövö or at any rate somewhere in the valley of the river Zala, where the
place-name Salla is recorded on the Poetovio–Savaria road. The territory of
this small insignificant city lay between those of Poetovio and Savaria.

A striking fact is that in the towns of the interior of Pannonia almost as many
scribae are attested as are decurions; in the municipium Iasorum there was a
scriba who was also a decurion and even became a member of the Quattuorvirate.
As a rule the *scribae* were paid employees and thus were not members of the
wealthy urban aristocracy. Where in these and other Pannonian towns (for
instance municipium Faustinianum) they appear to possess wealth and also
belonged to the aristocracy, there must have been cogent reasons for their
exceptional status. We have seen that Hadrian's policy of urbanization aimed at
involving the local native aristocracy in the administration of the new municipia:
in more general terms the native community was to be directed along the path of
municipal development. It is possible that the aristocracy was not equal to this
task, and in consequence the *scribae* who held the chief posts within the urban
administration acquired added importance. They may well have headed the

administration and consequently possessed a much higher social position than they would have enjoyed in larger towns, where they would have remained petty officials in an administration controlled by the aristocracy.[129] It is also worth noting that the inscriptions of decurions from Salla, Mursella and Mogentiana were found neither in the towns, nor even in their territories. This paradoxical situation is possibly attributable to the fact that the decurions were hardly involved in the public life of the town. If we remember that in both Mursella and Mogentiana land and wealth were firmly in the hands of foreigners even before they became municipia, it becomes understandable if the decurions showed little interest in participating in the life of the town, the founding of which had brought them no real advantage.

The situation was also similar in the only municipium founded by Hadrian in the interior of Moesia; Ulpianum near Gračanica, not far from Priština.[130] It must have been connected in some way with the *metalla Ulpiana*, otherwise it would not have been called after the mines. Lessees, miners and the mine administrators probably formed the nucleus of urban life, since the town was not far from the mines on Mount Žegovac. Nevertheless, the decurions so far attested were Ulpii and Aelii, whose inscriptions were significantly found not in the town but west of it in the Metohija,[131] where there is considerable epigraphic evidence of citizenship-grants to natives, both under Trajan and later. Meanwhile a peculiarity, also characteristic of Aquincum, should be noted: the Ulpii are found mostly in the territory, whereas the Aelii tended to live in the town (Fig. 25).[132] Thus Hadrian's policy of granting citizenship was harnessed to the service of urban development.

Even after his intensive policy of municipalization some *civitates peregrinae* remained untouched by it. In Pannonia Superior the Cotini (Kytnoi), Azali, Serapilli and Oseriates and in Inferior the Hercuniates and Breuci were still *civitates peregrinae*; at any rate there is so far no evidence that they became municipia under Hadrian. In Moesia Superior probably all the *civitates peregrinae* persisted as such after his day, with the possible exception of the Dardani, whose area largely coincided with the territory of Ulpianum and Scupi. The two colonies and the two municipia in Moesia Superior occupied less than half the province. Viminacium's territory was remarkably small.

Pannonia's native population left behind a great variety of late first-century and second-century archaeological material, which deserves special attention, principally because there are marked differences between certain zones which do not fit into the process of municipalization described above. Even after making allowances for the tardiness of investigation in some south Pannonian areas, it is nevertheless a striking fact that no wealth of epigraphic material, or evidence of

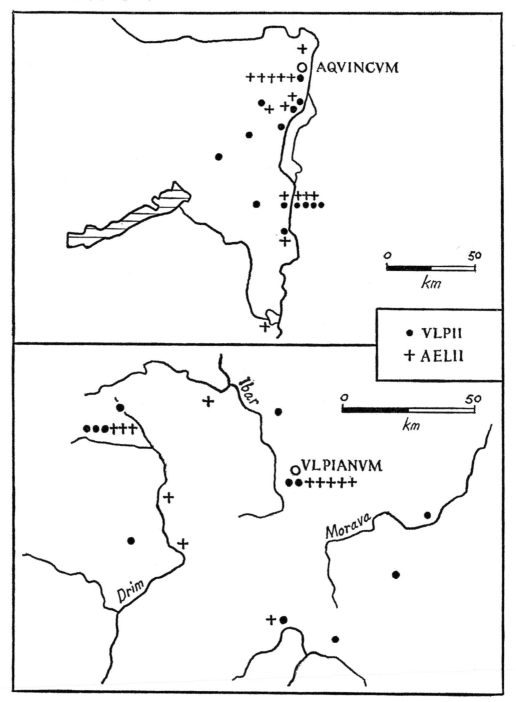

Figure 25 Ulpii and Aelii in the *territorium* of Aquincum and Ulpianum

certain special burial customs indicative of a prosperous or wealthy native class, has normally been found in the territories of the municipia founded under the Flavians and Hadrian; on the contrary, among the natives Roman customs seem to have been more deeply rooted in areas which were not municipalized until later in the second century, or even subsequently. Evidence of this sort is scarce in Flavian municipia. Before this problem can be dealt with in any detail it is necessary to consider an apparent paradox. Local (non-Latin) names, local costume and some traditional burial customs are to be found in the very areas where Roman influence on the natives was strongest. Attention has frequently been drawn to these local peculiarities simply as proof of the weakness and superficial nature of Roman influence. The stubborn persistence of native names, for example, it was urged, was the result of this influence being insufficiently powerful. The right view is, however, precisely the opposite. Celtic, Illyrian (Pannonian) or Thracian names have come down to us because they were recorded on Latin epitaphs. If the natives had not taken over the Roman custom of setting up tombstones, the identification of traditional or Latin names would have been impossible. In the same way the fact that the natives carved reliefs of their dead on tombstones provides clear evidence of persisting fashion in women's dress. Thus native dress, native names and many other aspects of local culture could only be transmitted via Roman forms; traditional elements could be expressed only through the intervention of external media. It can probably be assumed that Romanization of the local mode of life and way of thinking was least evident where there were neither Roman tombstones nor other adjuncts of Roman culture.

Burial rites, tombstones and their social significance

The same also holds good for some native burial-rites and their ideas of life after death. The so-called astral symbols (Fig. 10), already touched upon in Chapter 3, were probably of ancient origin, but are first met with as embellishments on tombstones. The most impressive example, however, is that of the cart-burials. The journey of the dead to the next world in carts, boats or on horseback is an ancient and widespread concept. Vessels shaped like carts, parts of vehicles found in graves and isolated wagon-burials are evidence for the currency of these ideas in both the earlier and later periods of the Iron Age.[133] The same concept also found expression on tombstones in Roman times; the dead man is depicted in relief in a four-wheeled vehicle drawn by a team of horses or oxen.[134] Such tombstones made their appearance before the end of the first century, were fairly numerous in the second and only ceased at the time of the Severi. Most of the dead, particularly those commemorated on the earliest tombstones, were natives, mainly peregrini with single, Celtic names. In the first half of the second century the custom of wagon-burials became fashionable.[135] The dead man, depicted in relief on the tombstone as travelling in a vehicle, was buried together with a four-wheeled wagon, horses, a set of sacrificial utensils (tripod, pitcher and plate), and with certain luxury objects such as equipment from a palaestra (aryballos and strigil), iron folding chairs, etc. (Pl. 15). Everything speaks of an extraordinary degree of luxury; the vehicle was decorated with bronze reliefs and had rich mountings, the horse's harness was definitely of a ceremonial character and the rest of the equipment was all imported, mainly from the bronze workshops of the Rhineland, but also from the eastern provinces. Pictorial representations were mostly taken from the Dionysiac cycle. That the richest natives were buried in these graves is substantiated not only by the fact that there is evidence that the idea of the journey to the next world by wagon pre-dates the actual wagon-burials (scenes depicting wagons on tombstones and pre-Roman finds) but also by the distribution of these wagon-burials in precisely those areas (including those where foreigners played a leading role) where tombstones with wagon-reliefs have not come to light.

There is no doubt that these wagon-graves originated from a very wealthy native aristocracy still very much tied to tradition, which, although its way of life as far as externals were concerned had been largely Romanized, had not abandoned its ancient beliefs about the next world. In fact it may be said that these wagon-burials only became possible as the result of Roman control, which led to the aristocracy acquiring wealth and adopting a civilized 'Roman' way of life. The man buried in the wagon-grave at Sárszentmiklós was probably one of the highest functionaries of the civitas Eraviscorum, since his grave also contained a lance-shaped badge (*hasta*) made of bronze, to which only representatives of executive power were entitled.[136] The ancient view that the dead entered the next world in vehicles was expressed in visual terms from the moment the natives began to set up tombstones, but to put it into practice by actually dispatching the dead in a wagon was possible only for the wealthiest.

Although this form of burial was probably restricted to the tribal aristocracy who owned vehicles, burial under *tumuli* was a socially less restricted custom. The indication of a grave by a *tumulus* is such an obvious idea that it was adopted everywhere and at all times. *Tumuli* were customary in Pannonia in the early part of the Iron Age, occurred sporadically in its later stages but were most frequently employed in the Roman period. These *tumuli*, some 1–3 m high and 5–12 m wide, were raised over graves containing cremated remains, and began to appear about the middle of the first century in western Pannonia and

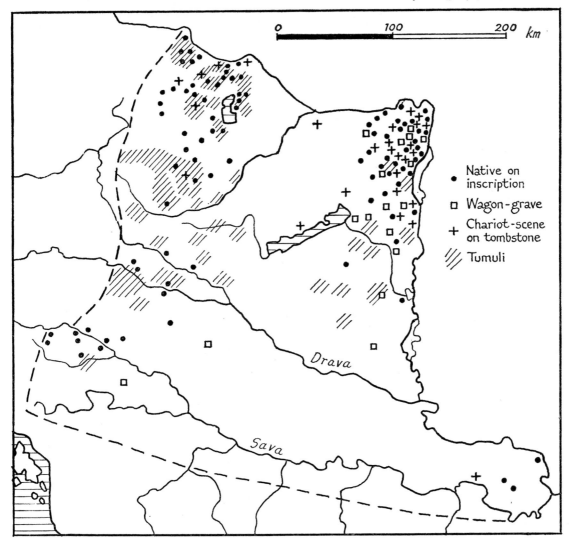

Figure 26 Monuments to natives in Pannonia

about the beginning of the second century in the eastern part of the province.[137]
This too suggests that once conditions had been consolidated old customs
flourished anew and even became more widespread than before. Of course this
applied only to areas where the native population had reached a certain level of
prosperity. Hence the ability to establish where these customs prevailed is not
unimportant.

Attention has been drawn to some special features of native culture for the
purpose of showing that they derive from traditional customs and ideas of

149

which we should have remained ignorant but for the effect of Roman control. A distinction must, however, be made. There were some local, non-Roman cultural aspects which were widespread but which required Roman forms for their expression. That native (Celtic or Illyrian) names were not abandoned in the first and second centuries follows *a fortiori* from the fact that they appear on Latin epitaphs. Similarly, the idea of conveyance into the next world in a vehicle was probably widespread, but we are only made aware of it from portrayals on tombstones or in some other practical manner. The style of native women's dress obviously persisted everywhere, but again our only information concerning it comes from reliefs on tombstones. In all these instances the habit of setting up tombstones was a symptom that Roman customs had been adopted, and at the same time it provides information concerning widespread native customs. On the other hand there were aspects of this culture which owed their dissemination to the Romans; wagon-graves and *tumuli*, while they reflect ancient attitudes of mind, were first given practical expression when the natives came under Roman control, and then only where there was no social handicap.

These native peculiarities of provincial Roman culture which had their roots in native tradition are often regarded as symptoms of the so-called Celtic, or Illyrian, Renaissance. From what has been said, however, it will be seen not as the result of any decline in or relative weakness of Roman culture, but rather that Roman influence either gave expression to or revived local characteristics.

It is highly significant that in Pannonia all these local characteristics have been found either in association or in the same area. Hence it may be assumed that in these areas the native population was either under no social handicap or, what comes to the same thing, was able to adapt its interests to those of its rulers. The demands which Roman provincial policy made on the ruling class were frequently a clear indication of the latter's wealth and loyalty which found expression in their singular customs. Epitaphs presuppose a certain wealth on the part of the families, but they also indicate the tendency to copy the Roman custom of setting up tombstones. Wagon-graves are not only a sign of wealth but also reveal that certain externals of Roman culture had been taken over (palaestra equipment, sacrificial utensils, etc.), though at the same time many traditional features of the native way of life and ideas were not abandoned.

It cannot, of course, be denied that certain differences existed in the areas where this loyal native upper class is attested (Fig. 26). Tombstones depicting the dead being ushered into the next world in wagons and actual wagon-burials are almost evenly distributed. In the Leitha district, however, only reliefs of wagons have been discovered, but no wagon-graves. The Boian aristocracy was also familiar with the notion of the dead journeying into the next world, but it was not prepared to give practical expression to it in the form of the expensive and grossly extravagant custom of wagon-burial, probably because living as they did in the hinterland of Carnuntum they had long been in contact with Roman culture. Even in the first half of the first century they had become accustomed to Roman imported goods, they kept imported slaves, built villas at least by the end of the first century, were granted citizenship and self-government under the Flavians and were ultimately made members of the *ordo* in Carnuntum under Hadrian. Thus they could look back on a long 'Roman' past, whereas the Eravisci, who did not become involved in Roman trade or play any important part in provincial politics until the end of the first century, probably clung more persistently to their traditions. The majority of wagon-burials come precisely from this civitas of the Eravisci. To the south of it in the civitas Hercuniatium, though wagon-graves occur, there are no tombstones to natives. The richest members of the tribe were probably culturally and materially in a position to allow themselves the luxury of these burials, but the upper class in the main lacked either the necessary wealth or any cultural or political inducement to set up tombstones on Roman lines.

The areas in which, on the basis of what has been described, the native aristocracy was on the point of becoming a Roman provincial aristocracy were oddly enough not so much the territories of the Flavian and Hadrianic municipia but certain zones, the boundaries of which did not always coincide with those of the self-governing communities (civitates and municipia).

A smallish group of tombstones comes from the native Celtic Latobici in the western part of the territory of municipium Latobicorum.[138] Somewhat to the north of that begins the broad zone of the west Pannonian *tumuli*, which includes the whole of the Norican–Pannonian frontier area from the Drave to the Danube (Fig. 26). The only parts where there are no *tumuli* are where there is epigraphic evidence that foreigners had settled in the first century, namely close to Poetovio (colonia Ulpia), Savaria (colonia Claudia) and within the area of the Flavian municipium Scarbantia. The *tumuli* are not indications of humble burials, since they often reveal a wealth of finds: for example walled chambers are not unusual, some of them with a dromos in the Mediterranean manner. Some chambers have inscriptions recording peregrine Celts or even Celts with Roman citizenship.[139] Although tombstones of natives occur only sporadically in the *tumulus* area their significance lies in the fact that sculpturally they do not differ from those found in the towns. A beautiful example is that of the tombstone of *Quartus Adnamati filius*, which was found not far from Savaria and depicts the she-wolf and twins.[140] *Tumuli* are less frequent in the Leitha district; in contrast there are many more tombstones of members of the Boian aristocracy, and these from the point of view of their reliefs form a class by themselves. The over-all composition was, of course, borrowed from first-century tombstone reliefs at Carnuntum, but the dead were depicted in local dress. The stonemasons seem to have been local men.

It was no accident that the whole of this west Pannonian zone coincided with the Amber Road area, where in the first century brisk traffic had developed between Italy and the barbarian north via Carnuntum. In this the natives were involved: a *negotiator* named *Atta Bataionis filius* is recorded on a tombstone (Pl. 5b), found north of Savaria, and the stone itself is a moderately successful production of a Savarian workshop.[141]

The other zone in which natives set up tombstones also coincides with a *tumulus* area and lies within the distribution of wagon-graves. In the north this zone begins at the Azalian–Eraviscan boundary area and includes the whole of north-east Pannonia from the Danube bend to roughly the Mecsek mountains. There were no *tumuli*, wagon-graves or tombstones there before the last years of the first century and they only become numerous in the first half of the second century. Thus archaeological finds suggest that compared with west Pannonia there was a time-lag of some fifty years, a result which agrees with the historical development in this area. Progress did not begin until the late Flavian period, when trade on the Danube and along the diagonal road from Poetovio to Aquincum started to flourish and when two new legionary fortresses and their numerous auxiliary units were established along the river.

Trade and the military, however, made up only one component of the Romanization process. As already seen in Chapter 3, the Celtic and Celticized natives of Pannonia were better at exploiting the advantages of Roman rule, a circumstance probably connected with the social structure of the Celtic tribes. It was no accident that the two zones in which the native population had been Romanized were Celtic, nor, probably, that in the Azalian area, where, it will be remembered, south Pannonians had been settled by Tiberius, only the narrow eastern strip closest to the Eravisci shared in this process.

Everywhere else the relevant archaeological finds which might testify to the Romanization of the native population are either extremely sporadic or completely absent. It is not surprising that in municipia where foreign settlers were present before their founding (Andautonia, Mogentiana, Mursella) these men set the tone and became the leaders in the communities, but in municipia where this was not the case the failure of natives to appear in either the epigraphic or the archaeological record must be attributed to the backwardness of native society. These municipia were, of course, in the Illyrian–Pannonian area (municipium Iasorum, Cibalae, Bassiana), and, as we have already seen, not even the members of their *ordo* were completely Romanized; the control of municipal affairs was obviously in the hands of foreigners and the *scriba* who had been specially appointed for the purpose (p. 144). Thus efforts on the part of the Romans to transform the tribal aristocracy into a municipal one met with no success.

In Upper Moesia it was only in the territory of the municipium Ulpianum that the native population left any important quantity of epigraphic material, and it is only here that there is evidence for upper-class native members of the *ordo* (Fig. 25, p. 146). In all other urban and non-urban territories there is absolutely no evidence that the native population was involved in the development of the province. Traditional religious beliefs and customs are not recorded on monuments, extremely few native Thracian or Illyrian names are to be found in the whole of the epigraphic material from the province, and native dress is not depicted on tombstones. In general it may be concluded either that, apart from the municipium Ulpianum, the native population resisted Roman influence or that it was socially handicapped.

The reasons for this are obvious. As we have seen, Rome proceeded very cautiously in occupying and constituting the province of Moesia. No war of conquest preceded its founding and for a long time the completion and occupation of the Danube road was Rome's only achievement in and on behalf of Moesia. The country was not of prime importance for Balkan commerce; the natives living in the high mountains and narrow valleys were both inaccessible and bellicose, and were far from having an appeal for entrepreneurs and colonists. Rome naturally often intervened in local affairs by establishing settlements or by reorganizing the civitates, but the exploitation of the mines, obviously of great importance to Rome, was only put in hand from the time of Trajan. On the other hand there are reports from about the middle of the second century of *latrones Dardaniae*;[142] thus even at this late date it was not possible to pacify the mountainous parts of the province.

The background of these Upper Moesian robbers is explained by the social conditions of the natives. As early as Greek times the Dardanians were known as wild mountain shepherds who kept a large number of peasants in bondage (p. 27). With the gradual development of urban centres, the opening of the mines and the appearance of the few self-governing towns, the peasants were able to free themselves from the shepherds who then gradually took to banditry. Under such circumstances stable conditions could not be established and compromise between native and Roman interests became even less possible.

Army recruitment

Even regular recruitment of Upper Moesians into auxiliary units did not begin until the reign of Marcus Aurelius.[143] Under Tiberius the Romans probably still recalled that the Pannonian–Dalmatian rebellion was sparked off by the recruiting ordered by Valerius Messalla. Even when Upper Moesia became a province levies were not raised, and later conditions remained so unstable that the government decided against having Upper Moesians in the army. The only unit in which they served was the ala Vespasiana Dardanorum raised under Vespasian, which did not, however, maintain its connection with Dardania through subsequent recruitment. Men from the colonies at Scupi and Ratiaria served in the legions in the second century; but, apart from these two instances, recruitment took place only rarely and as an exceptional measure: for example, it happened occasionally at Ulpianum[144] or with sons of soldiers serving with auxiliary units[145] in the province, though these did not recruit from Upper Moesian natives, except some Dardanians.

The Upper Moesian army of the second century was composed of legionaries, about half of whom were foreigners and the remainder natives, and of auxiliaries who were almost exclusively foreigners. The legionaries recruited in the province itself were mainly citizens of Scupi or Ratiaria, but not infrequently peregrine *incolae* of these colonies were also enlisted and in consequence became citizens.[146] The foreign legionaries came from neighbouring provinces— mainly, of course, from the unarmed provinces of the Balkans (Dalmatia, Macedonia and Thrace) and from Asia Minor. Among the auxiliaries mostly easterners are attested. There is evidence that a considerable military class was developing in this period, for sons of legionaries very often joined the same legion as their fathers.

Recruitment in Pannonia was entirely different. In the Flavian–Trajanic period many Italians, Gauls, Spaniards and Noricans were still serving with the legions; the composition of the auxiliary units was even more varied, depending on where the unit came from or, not infrequently, the area from which it was later reinforced. Syrians, Thracians, Dalmatian Illyrians, natives of the Alps, Britons, Spaniards and Germans made up a not insignificant part of the auxiliary personnel. The majority of these soldiers went with their units to Pannonia; later recruitment from the provinces or tribes after which the units were named was customary only for those with a special tactical function, as for instance the Syrians of ala I Ityraeorum[147] or some Thracian units.

The date at which the transition to local recruitment took place cannot of course be given, since obviously it was not the result of a single decree. Citizens of the Pannonian colonies were already serving in the legions in Flavian times,[148] but they never completely succeeded in ousting the foreigners. It did, of course, happen that a legion or a vexillation from it was sent as an expeditionary force to another province, from which it returned with a number of soldiers recruited there. Thus orientals joined legio XV Apollinaris during the latter's long stay in the east (from about 62 to 71),[149] while a substantial number of Macedonians, men from Asia Minor and Syrians joined legio II Adiutrix when it was engaged in putting down the Jewish rebellion of Hadrian's reign.[150] Moreover, provinces on the Danube and in the Balkans which had no legions of their own also provided recruits. The majority, certainly from the time of Hadrian, gradually came to be drawn from citizens of the Pannonian colonies, though both Pannonia Superior and Inferior produced recruits only for their own legions.

By the time of Claudius Pannonian natives were already serving in the auxiliary units stationed in the province. The personnel of the units transferred there in the Flavian–Trajanic period consisted largely of foreigners, but once their deployment was stabilized they were supplemented locally. Such evidence as there is of this is, however, extremely sparse. Auxiliaries in the second century left very few inscriptions, but as against this we possess a substantial number of military diplomas (Fig. 27, Pl. 22). The Pannonians who received them in the second century were without exception Azali or Eravisci, which cannot be regarded as a coincidence in view of the number of diplomas which have come to light from this period, particularly in Pannonia.[151] At this time both tribes still had their *civitas peregrina*, and these were also recruiting areas for auxiliaries. What is not clear, however, is why the other civitates did not provide recruits for the auxiliary units or indeed for the legions. So far no diplomas made out for Hercuniates, Breuci, Oseriates, Serapilli, etc., have been found. In the Julio-Claudian period the Breuci had provided a very large number of auxiliaries and there is also some slight evidence of first-century recruitment among the Hercuniates.[152] Around the turn of the century there were levies of auxiliaries in the Flavian municipia and actually in Siscia. T. Flavius Bonio from Andautonia was a trooper with the ala Frontoniana,[153] and the *cives Siscii et Varciani et Latobici* who were serving with the cohorts of Upper Pannonia are probably those attested under Trajan as being with a vexillatio in Syria.[154] The possibility cannot be dismissed that after Trajan men from the municipia were being recruited into the legions, though naturally this cannot be proved. Although under Hadrian there were no fewer than eleven municipia in Pannonia, inscriptions giving legionaries' birth places all show that without exception they were citizens of the colonies; it is therefore not surprising that, according to the large, almost complete list of those recruited for legio VII Claudia in 169, a very large

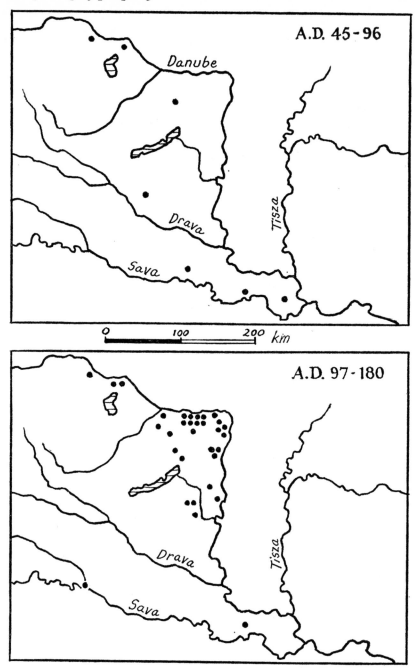

Figure 27 Auxiliary *diplomata* from Pannonia

number were from Scupi and Ratiaria but none from the Upper Moesian municipia of Viminacium and Ulpianum.[155] It has already been pointed out that under Trajan there was an equal number of colonies and legions in every province, possibly because recruitment to the legions was restricted to the colonies. What is not clear is why the citizens of the municipia and of some *civitates peregrinae* did not serve in the auxiliary units either. Possibly the reason lay simply in the transition to local recruitment. The auxiliary units in Pannonia were stationed on the *limes* sector, most of which lay on the edge of Azalian and Eraviscan territory. The civitates of these two tribes provided by far the largest number of recruits as, of course, the other civitates whose territories extended to the Danube had only one or two auxiliary forts in their vicinity. The long stretch of the river below Aquincum, which was at the same time the eastern boundary of the Eravisci, perhaps explains why their civitas was not incorporated into Hadrian's municipium of Aquincum but continued to exist in an odd state of dependence on the latter. On the *limes* sector of the civitas Eraviscorum alone there were no fewer than nine or ten auxiliary forts.[156]

What has been said in no way affects the fact that during the second century men from the *canabae* were recruited in increasing numbers into the legions. Obviously only sons of soldiers were enlisted. These were mostly both formally and legally citizens of the colony from which their parents had come or to which they legally belonged.[157] Although it cannot be proved statistically, there is nevertheless the strong possibility that the sons of these soldiers formed an increasingly large part of the legions' personnel; by the end of the second century an hereditary military class was already in existence, and even before this the majority of the legionaries were men whose origin was not recorded on their tombstones, and who bear colourless names typical of soldiers.[158]

Despite the increasing role of the hereditary military class the second-century army had a socially definable character even if it is somewhat difficult to grasp. The first point for consideration is that although until about the last quarter of the second century the auxiliaries were already Roman citizens for the most part, and therefore were being given diplomas in decreasing numbers, they originally came from the native population of peregrine status; and as far as can be judged from the places where their diplomas were discovered some of them returned there after discharge.[159] Others remained near their forts, possibly because their sons were still serving there. What the criteria were in the selection of auxiliary recuits is, of course, impossible to say, but it is very unlikely that they were descended from the tribal aristocracy. In the second century the latter were probably in the main already Roman citizens, so that their sons had no reason to join a unit in order to acquire citizenship through a long term of service. Hence it must be assumed that auxiliaries were recruited from the lower classes in the civitates.

Probably legionaries recruited from a colony were also from the lower classes. The veterans settled there were probably rich enough to keep their sons out of the army. Since we have to admit that property in the colonies tended to become concentrated in the hands of the few, this means that there were also poor families who could supply part of the legionary reserves. The other group of soldiers from the colonies only acquired citizenship on joining the legion, and in consequence they were classified as *incolae* of peregrine status; in other words, they were the native inhabitants of the territory of a colony and were socially no higher than an auxiliary from a *civitas peregrina*.

To sum up: the social composition of the army may be regarded as decidedly rustic and plebeian, with the reservation that many veterans remained in the vicinity of the camp and their sons provided part of the reserve. That the *canabae* continued in the second century as a special form of community, and indeed now with regulated organization, was probably due to the need for recruits. The earlier form of organization in these communities was called *veterani et cives Romani*; emphasis on the veterans who were also Roman citizens can be explained by the fact that they had a special role. The government was probably aware that a group of men uprooted from their birth place and closely linked to the army could more reliably meet the annual demand for recruits than levies in a community where the recruits' background, of necessity doubtful, could lead to incalculable consequences, both politically and socially.

The hereditary character of military service established in practice, though without legal regulation, produced a decidedly military class in the frontier zone, but one which was not in any way isolated from or independent of the provincial population. The richer members of this class often became decurions in the neighbouring towns, and on the other side the poorer sections of the civil population always had the option of improving their lot by joining the army. Thus the army provided the link between the two classes of the population on the frontier. If we speak of a society with a military bias or even of a military society in the frontier provinces, we have this intermediary role of the military in mind. The results of this development, which began in the second century, were clearly evident in the course of the next.

Given these conditions there was probably no 'veteran problem' such as occurred in the late republican and early imperial army. The veterans either returned to their birth places, which were probably not too far away, or they settled near the camp; the only problem is that of their role in a *civitas peregrina*. As Roman citizens they occupied a privileged position there, but were not members of the native aristocracy, who were citizens by birth and obviously of long-standing wealth and highly respected. A comparison of the wealth of the Eraviscan *principes* and decurions who could afford wagon-burials with that of auxiliary veterans leads to the conclusion that relations between them and the aristocratic citizens by birth were strained. Under such circumstances it was probably better for the veterans to settle near their former garrisons,[160] even though living conditions there could not exactly be described as Romanized.

This leads to the complex question of living conditions in the first and second centuries, to which there can only be a tentative answer in view of the dearth of reliable data. Such as exist are pointers, rather than sources upon which definite conclusions can be based. The reason for this is that some types of settlement assumed to exist have not yet been excavated, while earlier excavations which often revealed extensive residential areas were unsatisfactory from an archaeo-logical view-point. They failed to recognize the buildings of wood or clay. Nor were those who carried out such work able to distinguish the different building periods. Moreover, these earlier excavations were confined to urban settlements and to the living and bathing quarters of the larger villas, and thus provided no clue to the rural forms of settlement or to those of a definite village character. Most of the results of more recent and archaeologically better-organized excava-tions have either not yet been published, or the work has been restricted to a fairly small area, so that, as far as forms and types of settlement are concerned, we can only make assumptions. The towns will be considered first; investigation of some of them has been going on for a long time and is approaching a solution to some of the most important problems which they raise.

The towns and their archaeology

On the basis of what is so far known, and also from the point of view of town-planning, a distinction must be made between the *coloniae* and the municipia. At present information concerning the street system and the origin and growth of Flavian and Trajanic *coloniae* is lacking, but the so-called Hippodamian plan of the *coloniae* at Emona and Savaria (Fig. 13, p. 75) makes it reasonable to assume that the others were planned along the same lines. However, the ground plans of the municipia which have so far been explored differ somewhat. The eastern half of Aquincum and a slightly smaller part of Carnuntum have been excavated, and there is a certain amount of information available from Scarbantia and Bassiana. Aquincum[161] had an irregular, trapezoid residential area with a west–east length of some 650 m and a north–south width of about 440 m (Fig. 28). Somewhere in the middle ran the aqueduct from north to south: it was supported on piers and

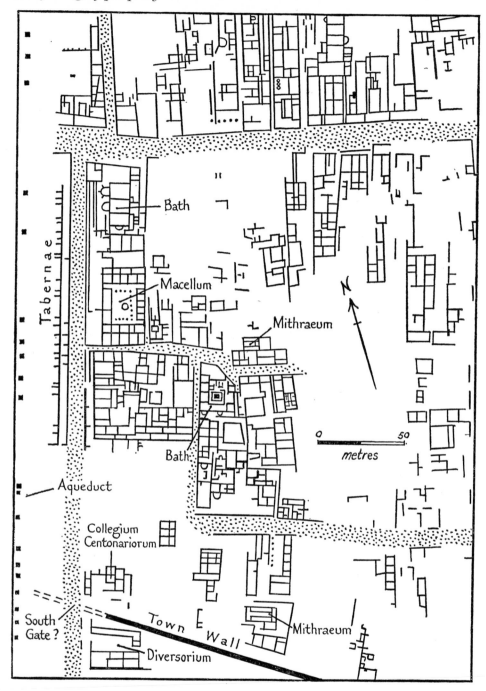

Figure 28   Detailed plan of the municipium of Aquincum

supplied the fortress from the springs of the so-called 'Roman bath', which are still in use today. Excavations west of the aqueduct were first started a few years ago, while to the east of it a section some 200 m wide has been completely exposed. Two virtually parallel streets linked by narrow alleys ran on a west–east alignment through this part of the town. So far no street running through the whole town from north to south has been discovered. Hence there is no evidence that the Hippodamian principle, according to which a grid of streets ran in a straight line across the town from one boundary to the other, was applied here, although the streets on the whole were laid out in the form of a rectangular grid. The town-centre has not yet been discovered, although there was a commercial centre, where public affairs were also conducted, roughly in the middle of the residential area east of the street which ran northwards from the south gate. Shops and businesses occupied a colonnade along the west side of this street and included shops selling Rhenish and Gallic *sigillata* pottery. On the east side was a *macellum* (food market), consisting of a square peristyle with shops and a round building in the middle (Pl. 16a). North of it was the so-called large bath-house (Pl. 17). Two public baths have been discovered, but apart from shrines there is no definite evidence of other public buildings. The *collegium centonariorum* with its famous organ was discovered near the south gate.

The private houses to the south and south-east of the *macellum* were biggish buildings with an inner courtyard round which a single house or several together formed a roughly square *insula*; in the northern part of the excavated area were narrow houses with a north–south aspect and mostly separated from one another by narrow alleys. Small workshops producing metal, leather or pottery goods were found in these houses; it is possible that similar workshops also existed here and there in those south of the *macellum*. Outside the town-walls, near the south gate, there was an inn (*diversorium*) with a big yard and stable. Which of these buildings belong to the earliest period is not yet clear. The street-system suggests that the settlement developed gradually without any pre-conceived plan. Support for this assumption is provided by its earliest traces, which date to the Flavian period, approximately half a century before the town itself was founded.

The town-walls have been located on the north, south and western sides and show every sign of having been built under Hadrian, or at least in the second century. They enclosed a fully built-up area which—as far as can be judged today—was not completed until after the walls were built. On the southern perimeter the outer walls of the houses follow the line of the town-walls, whereas the houses themselves conform to that of the town sector. The public utilities date at the latest to Hadrian. The aqueduct was not built for the town but for the legionary fortress, though the former was allowed to use its water. Among the recently exposed well-heads at the so called 'Roman bath' were some votive stones, one of which had been donated by a *decurio municipii et decurio canabarum*. The drainage system follows the twists and turns of the narrow streets, which suggests that it was added later. The amphitheatre, a small earth-bank one, with enclosure and retaining walls of stone, was completed in its final form at the latest under Pius, to judge from an altar dated 162 in its shrine to Nemesis. The excavations were carried out more than eighty years ago and did not of course establish whether this amphitheatre had been built on the site of an earlier one made of wood.

The cemetery of the municipium lay to the west of the town on a road which linked it to the diagonal Aquincum–Brigetio road. The tombstones and burials in this cremation cemetery date its beginning to the pre-Hadrianic period; for instance the *Agrippinenses Transalpini* contributed the sum of 72 *denarii* towards some of the funerals (Fig. 21, p. 125). It is also possible that a veteran of legio X Gemina was buried there before the time of Hadrian (Pl. 5c).

To the east of the town a large pottery quarter has been exposed where crafts-men were at work from the mid second century if not earlier. We shall return to this below.

Aquincum has been described in some detail since there has been little excava-tion at other municipia. While it is true that excavations in the so-called zoo-logical gardens (Tiergarten) at Carnuntum[162] have revealed a large section of the town, they have unfortunately not yet revealed its over-all plan. Just before the outbreak of World War II a start was made on the excavation of a large public building north of the gardens (the so-called palace ruins) and the work is still in progress (Fig. 29). In the fifties a large section of the town south-east of the Palastruine was expertly excavated. These three sites are a considerable distance apart. If a roughly east–west rectangle is drawn connecting the palace ruins and the part of the town to the south-east of them (which is essentially on the same orientation) with the remains in the Tiergarten, this will give an area of some 950 by 600 m (roughly 57 ha), which considerably exceeds the size of Aquincum (some 29–30 ha). This is not at all surprising as the colonia Savaria originally covered 40 ha and as early as the beginning of the first century Carnuntum was already one of the biggest towns in Pannonia. It is possible that the rectangle just mentioned may have corresponded to the area of the municipium, an assumption supported by the fact that the so-called palace ruins are centrally situated, while the very simple houses excavated in the fifties are round the edges of the rectangle (Pl. 18). Moreover, these houses resemble the free-standing houses on the northern perimeter of Aquincum, at least in so far as they both

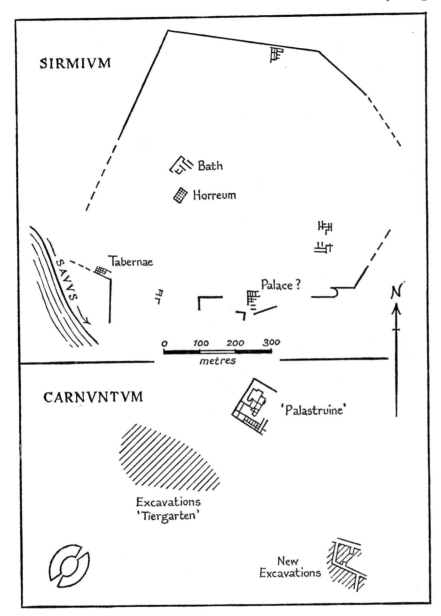

SIRMIVM

Bath

Horreum

SAVVS

Tabernae

Palace ?

0   100   200   300
metres

N

CARNVNTVM

'Palastruine'

Excavations
'Tiergarten'

New
Excavations

Figure 29   Town plans 2

stood on long narrow plots. The streets in this part of Carnuntum are not all the same width; two of them running parallel in a west–east direction are linked by two others which are not quite parallel and do not continue south. Thus this roughly rectangular street-grid does not conform to Hippodamian principles. Somewhere near the south-west corner of the rectangle was the amphitheatre (Pl. 19a), which was probably built in the mid second century if not earlier. Whether there was a town-wall has not yet been established, but it is very probable. Where it could have stood is still an open question, since certain indications suggest that not all the built-up areas of the town were in simultaneous occupation.[163]

There is no doubt that the palace ruins were those of a public building, but it was not possible to establish just what its doubtless various purposes were. The discovery of a base[164] belonging to a dedication donated by a *collegium* in the long narrow courtyard of the south precinct suggests that there may well have been an assembly room for the *collegium* within the building, but the whole lay-out with its state rooms, hypocausts, etc., makes it hardly likely that it was nothing more than such a centre, even if all the *collegia* of Carnuntum had been lodged there.

The attention of archaeologists has been drawn to some of the houses in Carnuntum, as two neighbouring dwellings there reveal a marked similarity to a villa in the Leitha district, at Winden am See, and this type was probably also to be found in Aquincum.[165] It had a colonnade or arcade on its narrow front, and a corridor running through the house and dividing it into two equal parts. There are numerous examples, even from pre-Roman times, of this type of house with its T-corridor, but whether it goes back to local houses of pre-Roman date it is impossible to say. Nor would it be right to regard it as the typical Pannonian house-form, as out of at least thirty already excavated in Aquincum only two at most are of this type, and there are only one or two examples among the villas. As for the rest of the houses at Carnuntum they are more closely related to a type found at Aquincum—and one difficult to define. These houses have hitherto resisted all attempts to classify them; their one common feature is their irregularly shaped inner courtyard—a square or rectangular one is a rarity (Fig. 30). The striking thing about these modest little houses is that they occupy very little ground. As the outer rooms were built first and the interior ones added later it was inevitable that the inner courtyard was irregular in shape. If the plot was very narrow, then the courtyard was either omitted or became an insignificant rather than a dominant feature. These houses should not be regarded as representative of one or more types.

It was the poorer section of the community which lived in these small houses,

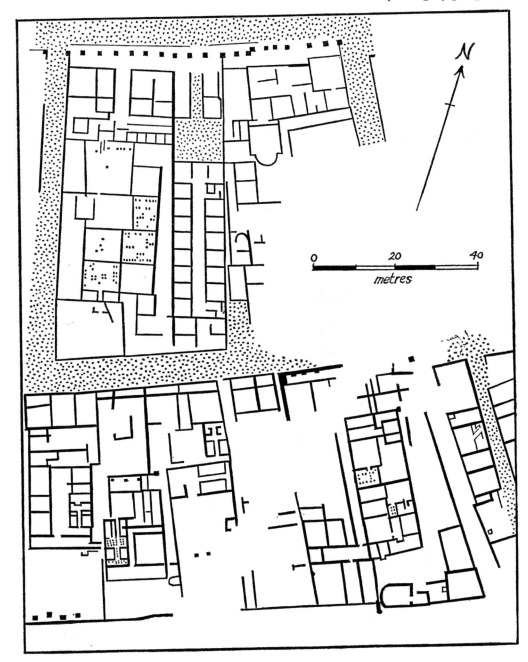

Figure 30 Detailed plan of part of the municipium of Carnuntum

which also usually served as workshops. The richer members lived in houses which were similar in style to the villas. A small number of such houses has been excavated, as for example the so-called large house at Aquincum, with its central courtyard, which, in the final analysis, had its origin in the atrium of the classical Roman house (pl. 16b).

Of the other municipia only Neviodunum is under excavation,[166] though a plan is not yet available. The part investigated, in the south of the town, lay on the banks of Korkoras and was the business and harbour area. On the other hand we have a reasonably clear picture of the lay-out of the Hadrianic municipium at Bassiana; although it has not yet been excavated, this emerges from a highly successful aerial photograph which is unique in our region.[167] There was a single west–east street running right through the town, and some running north–south but only through the northern or southern halves. Some buildings away from the town-centre were not sited in accordance with this lay-out, hence Hippo-damian principles were not observed here either. It is, of course, not known how many of the buildings and streets are late Roman or early Byzantine in date. Bassiana had a certain importance up to the end of the sixth century (civitas Bacensis), but it is very unlikely that in siting later buildings the street lay-out was disregarded.

The area enclosed by the town-walls was somewhat oval in shape and its south and east sectors were either not built up at all or the buildings were of clay or wood. Even in the town-centre it would seem there were some open spaces. It is probable that the town actually consisted of a loose mass of houses and that only the centre where the official buildings were had an urban appearance. Bassiana was, of course, one of the smallest towns (*c.* 19 ha) and, lying as it did between Sirmium and Singidunum, was an administrative rather than a communications centre. The same is probably true of the small municipia in the interior of the province. In the first place these were often such small settlements that even establishing their sites poses problems for modern investigators (Mogentiana), and secondly neither in the Middle Ages nor in more modern times were their sites regarded as suitable for the founding of towns. It is also probable that these municipia were mostly open settlements which did not acquire walls until the third or fourth century. Those of Scarbantia continued in use and were rebuilt in the Middle Ages (Pl. 20b) for the town now known as Sopron (Ödenburg) and are still standing today; recent excavations have established that they are Roman in origin[168] and were only built towards the end of the third or the beginning of the fourth century. Moreover, not all the buildings were within their circuit but only the densely built-up area in the centre of the town, the *continentia aedificia*. Since the beginnings of Scarbantia go back to Tiberius it is

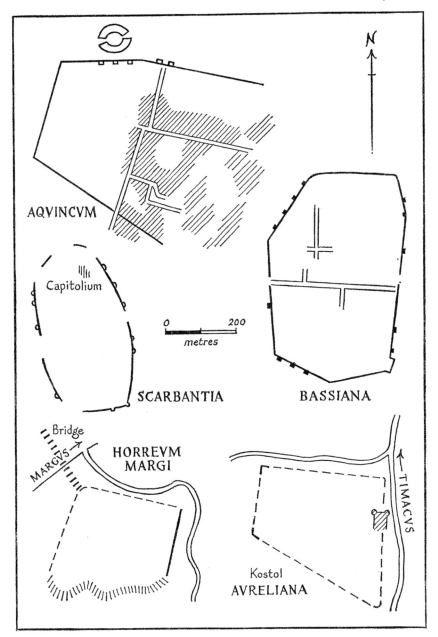

Figure 31   Town plans 3

not surprising that by the third century it had become a considerable town surrounded by scattered settlement, and obviously larger and more important than Bassiana. The walls in both towns are markedly similar in that they enclose an irregular polygon, roughly oval-shaped. Those at Bassiana probably also date to late Roman times.

If the walls in these two towns are compared with those of Emona and Aquincum, which enclose a more or less rectangular area and were either certainly erected at the time the town was founded (as at Emona), or are assumed to date from that time (as at Aquincum), then the assumption may be made that walls enclosing rectangular areas were built at the same time as the towns were founded, whereas those enclosing irregular polygonal areas are of a later date. Thus it is not impossible that the walls of Ulpianum,[169] which enclose a large rectangular area of some 500 by 500 m, were built under Hadrian.

Since settlements which received charters even under the Flavians or Hadrian did not necessarily have an urban appearance, it cannot be assumed that there were towns which did not receive this status in the first or second centuries. The only exceptions are the *canabae* of the legionary fortresses, where urban development began with the establishment of the camps; but they were in no way inferior to the municipia in the second century. As far as their area was concerned they were bigger than the corresponding municipium; on the other hand, excavations at Aquincum suggest that the population was not evenly distributed over these large settlement areas and that there was even less conformity in the type of house. Large houses which indicated no small degree of luxury stood cheek by jowl with small primitive shacks and workshops, and there were probably more buildings made of wattle and daub than in the neighbouring municipia. This does not, however, mean that the *canabae* had none of the communal services indispensable to a Roman town. Some of the streets were paved and sewered and some parts of the *canabae* were definitely urban in character. The urban part of the *canabae* at Viminacium,[170] investigated in the last century, had not only paved and sewered streets but a colonnade along one of them, while those at right angles to one another suggest some degree of town-planning. This is by no means improbable, as the *canabae*, at least since the time of Hadrian, had possessed not only decurions but also all the magistrates responsible in an urban community for local affairs. Apart from the big market-place at Carnuntum no public buildings were discovered in the *canabae*, but this may well have been fortuitous. The so-called military amphitheatres on the perimeter of these settlements were probably purely military structures. A building-inscription of the time of Antoninus Pius which mentions legio II Adiutrix is—in view of the place where it was found—no doubt rightly taken to refer to the

The first age of prosperity

military amphitheatre at Aquincum. Civilians from the *canabae* were also probably admitted to the games held in the amphitheatres. Those at Carnuntum and
Aquincum were bigger than the ones belonging to the *municipia*, and the size of
the *caveas* indicate that they were designed to accommodate a substantial
number of spectators.[171]

Nowhere is there evidence, nor is it probable, that the *canabae* were walled. It
would have violated all the principles of tactics if the fortress had been protected
by a second wall which enclosed a very large number of civilians.

Villas and rural settlements

Whereas some information is available concerning the history of urban
settlements this is far from being so with rural ones, so that even a tentative
description of them is out of the question. All that is possible at present is
to indicate some of the house types, so that by putting them into categories
and establishing their geographical distribution certain inferences may be
drawn.

Villas which can certainly be dated to the time of Marcus Aurelius are few in
number and found only in some parts of the province. 'Villa' is here taken to
mean houses or groups of buildings which either formed part of an evident
agricultural complex or which, in default of investigation, may be assumed to be
such. Seemingly detached buildings may well have been small houses in a village
settlement, parts of staging-posts or several other possibilities. The important
thing is that houses built in the technique introduced by the Romans (stone wall,
mortar and roofing tiles), but which cannot be regarded as villas, have so far
been found only in the third and fourth centuries. Thus Roman building technique seems to have been restricted in rural areas to villas and to what were
obviously military and official buildings; it was only later that it spread to
settlements of non-villa type.

In Pannonia villas dating to the early imperial period have been discovered in
those areas where foreigners, mainly north Italians, settled in the course of the
first century or else in areas where, although there were no external settlers, the
native aristocracy came to terms at an early date with Roman control—a compromise which was then sealed by the granting of citizenship. This means that so
far early-imperial villas have been found only on the western borders of Pannonia, in the Leitha district (Boians) and north of Lake Pelso. At the same time
it should be borne in mind that there has been so little excavation between the
Drave and the Save that the fact that no villas have been found there cannot be
used as an argument. The situation is entirely different on the northern and

169

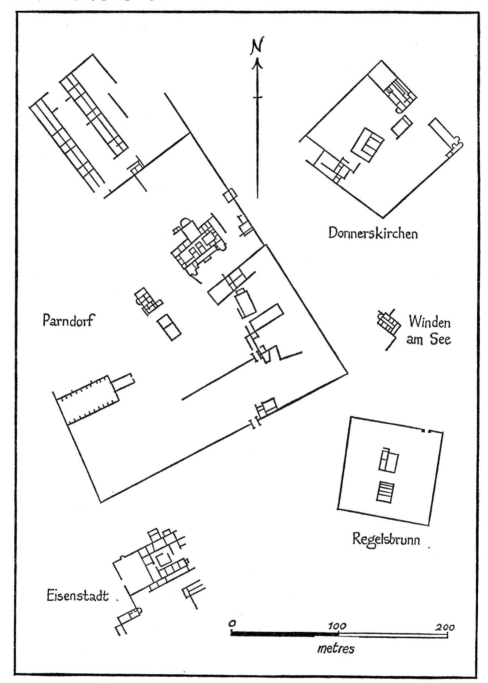

Donnerskirchen

Parndorf

Winden am See

Regelsbrunn

Eisenstadt

N

0   100   200
metres

Figure 32   Villa plans 1

eastern borders of the province where neither earlier nor more recent excavations of non-urban settlements have revealed villas of the early empire.

A clearly distinctive type is represented by villas in which the domestic quarters had a central courtyard (Fig. 32), probably unroofed.[172] In some cases this type of courtyard, which owed its design to the classical atrium, had an impluvium. It is not necessary to regard the rooms grouped round the small courtyard as alae, tablinum, vestibulum, fauces, etc. That such pedantry is out of place becomes clear from the fact that the entrance was not normally on the narrow side of the building opposite the main room with the apse. In some instances this apsidal room was missing or had been added in a haphazard fashion, as at Eisenstadt. The bathrooms were either in a separate bath-house or in the house itself. The deciding factor was the size of the whole complex.

In view of the small number of ground plans available it is inadvisable to attempt any further differentiation of types, as it is not possible to distinguish between a type and an individual case (Figs 32, 33). Among 'types' we may place big palaces with two central rooms,[173] long rectangular houses with a suite of rooms and one large interior one,[174] the house with a T-corridor, etc.[175] A more important classification is that based on the size of the domestic quarters and of the whole complex. In a few cases the farm buildings and the walls enclosing the rectangular yard have been exposed. These walls were not a defensive measure as they are mostly very narrow and there is no evidence for fortified gates or ditches. Their purpose was probably to protect the yard where corn, produce and implements were kept and in particular to keep stock from escaping. The house was cut off from the yard by an additional inner wall as a protection for the orchard, the kitchen garden or even just the flower garden, or perhaps it was to provide the residents with quiet and privacy.[176]

The biggest complexes had a yard measuring some 200 by 200 m, or possibly more (Baláca, Parndorf and Šmarje) with a house no less than 30 by 40 m; in every case the bath-house was separate (Fig. 32). The farm buildings, it has been established, were along the yard-walls and are only rarely to be found in the middle of the yard; what these buildings were used for it is difficult to say with certainty, but there are instances of granaries (*horrea*) having been found with more or less typical lay-out, and as they are of a striking size they point to the existence of very large estates. The villa at Parndorf has the biggest yard, house, bath-house and granary. It is in the territory of the civitas Boiorum and was built towards the end of the first century. It is probably correct to assume that it was the seat of a very rich member of the Boian aristocracy.

The smaller villas have a rectangular yard, some 90 by 100 m and mostly small houses (roughly 10 by 15 to 20 by 25 m) which also contain the bath-rooms

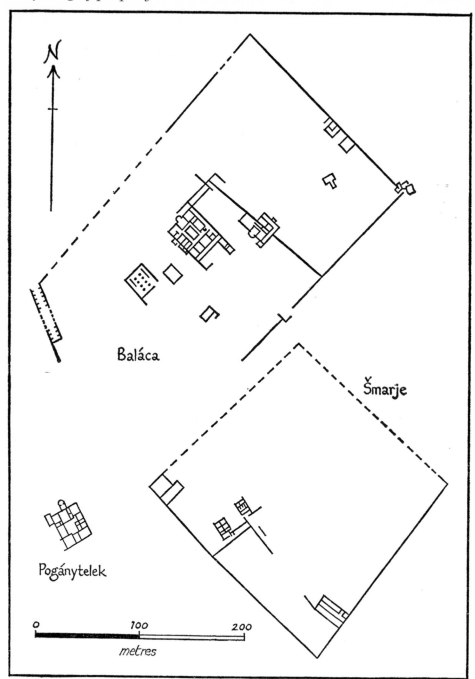

N

Baláca

Šmarje

Pogánytelek

0   100   200

metres

Figure 33   Villa plans 2

(Fig. 33). Such measurements made it possible only in rare instances to build a house based on the atrium type (the peristyle villa): a central courtyard required a larger complex.

The ethnic and social background of the owners of these early-imperial villas is fairly clear. Those in the upper Drave valley, the Pelso region and some near Scarbantia belonged to foreign settlers who had managed to acquire land and had introduced the Roman method of farm management. In the Drave valley the latter probably started with the veteran settlements made at Poetovio under Trajan, but near Scarbantia veterans and north Italian traders had already formed a corporation in the time of Tiberius and in the Flavian–Trajanic era were establishing themselves on the shores of Lake Pelso. In the Leitha region, where both large and small villas have been found, it was obviously Boian landowners who adopted the Roman way of life and its farming methods. Some Boians were very rich, others of more moderate means. It often happened that the smaller estates were merged with the bigger ones; evidence of this was found in the fact that despite the small number of villas so far excavated some were not inhabited after the end of the second century,[177] whereas the huge villa at Parndorf and other large establishments were occupied right into the late fourth century and show signs of having been considerably altered.

The villa with the T-corridor at Winden am See was in the territory of the civitas Boiorum. As far as present knowledge goes there is no justification at all for seeing a connection between this type of house and one present locally in pre-Roman times. Unfortunately it is not known how or where those who worked on the Boian estates lived. In other parts of the province, however, some late La Tène and early Roman houses have been excavated which represent a completely different type from that with a T-corridor (Fig. 34). To attempt a generalization from data now available would be premature; there is, however, a type which stands out as it occurs both in late La Tène and Roman contexts. In the *oppidum* on Mount Gellért at Budapest É. Bónis explored some roughly rectangular timber houses, with rounded corners, occupying shallow excavations in the soil; the hearth or the oven—which was made of clay and sealed when in use—was inside the house, near one of the long walls. These houses were at least 4 m long and 2–3 m wide. The roof was supported by two beams, which stood in the centre of each of the shorter sides.[178] Similar houses from the second century have been discovered at Gorsium,[179] and the type is known beyond the frontier in the Roman Iron Age of Slovakia and Moravia.[180] The civil settlement attached to the auxiliary fort at Matrica (Százhalombatta) also consisted in the second century of similar dwellings, although the ground plans have not been established in detail.[181]

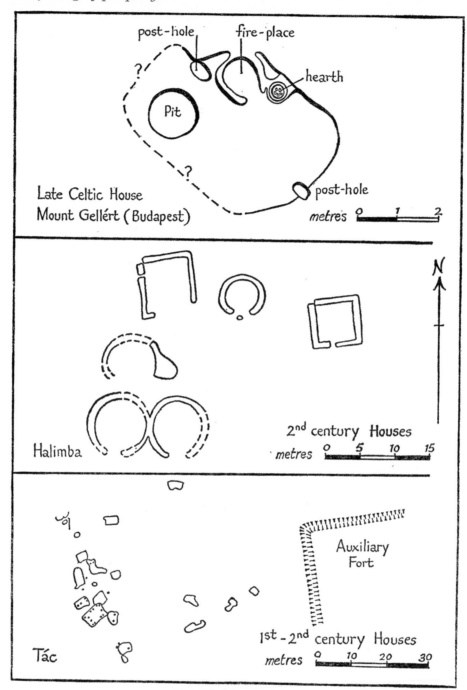

post-hole   fire-place

hearth

Pit

post-hole

Late Celtic House
Mount Gellért (Budapest)

metres   0   1   2

N

2nd century Houses

metres   0   5   10   15

Halimba

Auxiliary
Fort

Tác

1st - 2nd century Houses

metres   0   10   20   30

Figure 34 Plans of native houses

To date no houses which were not below ground level have been discovered. The probable reason for this is that they are very difficult to trace. Walls of terre pisée or timber soon disappear completely where surface conditions are the same today as they were then. It is only at Halimba that ground plans can be recognized of huts, made of wood and clay.[182] If this interpretation is correct, we have a further type of native house (Fig. 34). On the basis of the ground plans at Halimba it was small and either round or square, with a single entrance; the former had a diameter of 4–8 m, while the latter was 5–7 m across. Their dating approximately to the period from Hadrian to Marcus is attested both by loose finds from the site and also by graves in the vicinity containing cremated remains. Round and square huts are also attested for the same period outside the frontier; it is only necessary to draw attention to the reliefs on the column of Marcus Aurelius.

The discovery of these houses has, however, not brought nearer a solution to the problem how and where the native aristocracy in northern and eastern Pannonia lived. In the first place it is hardly likely that a *princeps* or a *decurio* who could afford to have an extravagant wagon-burial and indulged in gymnastics (palaestra-equipment) lived in such miserable pit-dwellings. An auxiliary veteran might perhaps be content to live in a bigger house built in this manner, since, after all, the houses in the auxiliary civil settlements of the second century were hardly better.[183] The discovery at Aszár of a hoard of some bronze vessels and simple pieces of silver jewelry (Pl. 21), belonging to an Azalian called *Atta Nivionis filius*, according to a diploma (Pl. 22) found with them, does not indicate any great wealth on the part of the owner.[184] The tribal aristocrats were much richer and more exacting in their demands. Perhaps some indication of their mode of life will be provided by the finds at Gorsium (Tác), where, according to local inscriptions, some native decurions of Aquincum lived.[185] It was in Gorsium that an elaborate tombstone of imported crystalline limestone and of first-class design was set up to a decurion. But the buildings which can be dated to the second century so far excavated there—those of stone are mainly late Roman—belong to the probably once very impressive shrine where the *Ara Augusti Provinciae Pannoniae Inferioris* stood, and where the *Concilium Provinciae* met regularly. This shrine was called *Templum Provinciae* (Fig. 41, p. 257; Pl. 23a).[186]

There is probably little hope of discovering any great number of estate centres belonging to the east Pannonian tribal aristocrats, since the latter were not very numerous. Nor is it known whether and to what extent they adapted their farming methods to Roman techniques. If there were no such change, then, while there would be houses and estate centres, we cannot expect to find villas. It was precisely these aristocrats who indulged in the traditional and extremely extravagant wagon-burials, whereas the Boian aristocrats who lived in villas had no such custom. Native dress and names persisted much longer in north-east Pannonia than among the Boians, where tombstones with Celtic names and with the dead depicted in native dress were becoming less frequent in the first half of the second century: in north-east Pannonia they persisted in large numbers right up to the Marcomannic wars. The mere fact that natives set up tombstones with Latin inscriptions by no means implies that they lived in stone houses with hypocausts and bathrooms; in north-east Pannonia this Roman standard of living can be assumed only for the richest native aristocrats who were buried in wagon-graves along with equipment from the palaestra and other Roman luxury goods.

Industrial development: the pottery industry

We cannot assume that a radical change took place in the living standards of the broad masses from the mere fact that everywhere in the second century local industry was in a position to meet the demand for goods which, both in design and craftsmanship, were becoming increasingly Romanized. In the first place the late La Tène pottery and bronzes were not technically inferior to the Roman, and, second, the spread of Roman designs was largely a matter of their being the latest fashion. 'Roman' goods have been found in settlements in north-east Pannonia alongside those based on local tradition, regardless of whether the settlement was a purely civilian one in the hinterland of the *limes* or was an auxiliary *vicus*. By the end of the first century Roman designs had gained the upper hand in west and south Pannonia, but in the north-east of the province late Iron Age designs were still being widely used in the second century, particularly in pottery in the Eraviscan area (the so-called 'grey ware of Pátka').[187] The pottery industry in Pannonia can be traced back clearly to that of the late Iron Age,[188] the influence of which in design and technique persisted into the Roman period. From that time until late antiquity potters used the same type of kiln which had already been in use during the Iron Age. The so-called Roman kiln, the square, walled type, arched with tiles, was much rarer and primarily used for firing tiles.[189]

The most characteristic products of local potters were vessels with stamped decoration (Pl. 24a); these were made and used throughout the whole of Pannonia.[190] From the point of view of technique they derive from the stamped pots of the late Iron Age. The earliest west Pannonian examples were, however, imitations of north Italian dishes which had the potter's stamp in the centre. The shape of these vessels was imitated by the local potters but given a simple leaf

176

stamp. Around the end of the first century a Celtic potter, perhaps from Aquincum, stamped his name RIISATVSFIICI (*Resatus feci*) on similarly shaped dishes (Pl. 24a); in the first half of the second century potters started imitating Gallic samian. Instead of dishes with leaf stamps, vessels of Dragendorf form 37 were produced with the relief decoration replaced by stamped decoration (Pl. 24a). These local imitations of samian were widely distributed throughout the whole of the second century and probably continued into the Severan period, particularly in those areas into which genuine *sigillata* had not yet penetrated.

As far as other kinds of pottery are concerned it is not always possible to distinguish between imported goods and those locally produced, since among the so-called Po valley and late north Italian *sigillata* there was a simpler shape with a poor glaze which may well have been made in Pannonia.[191] It is very unlikely that masses of simple glass and pottery vessels were imported into the province when they could have been produced on the spot and so have saved high transport costs. Mass imports can only be attested for brief periods, as for example during the Flavian–Trajanic period, when they were destined for the Danube garrisons. As soon as local industry could meet demands, it was also able to oust imported articles, taking over their designs and method of production. From then onwards imports were restricted to certain special lines such as samian pottery, which were, however, distributed only along the main routes, in particular the waterways. Thus in the more remote areas greater importance attached to imitations, as for example to samian copies with stamped decoration.[192]

Even locally produced samian did not reach areas off the beaten track. One of the oddest features of Danubian trade and Pannonian industry is that whereas potters were producing *sigillata* the province was hardly aware of the fact. So far two potteries producing *sigillata* have been discovered—one at Aquincum and the other the so-called Siscian pottery. On the Danube bank east of the municipium at Aquincum, excavations have revealed an extensive pottery quarter where domestic pots, tiles, terracottas and samian were produced.[193] Large quantities of punches, moulds for bowls, and other tools used in making *sigillata* came to light and made it possible to distinguish two craftsmen, the so-called 'first master' and the potter Pacatus (Pl. 25a). The typical decorations they used were vine leaves, vine tendrils with hand-incised stems, hares, dogs, wild boar, figures of Pan and a peculiar figure of Vulcan. On the basis of the well-known moulds, which have been found in such large numbers, it should be easy to identify the products among the finds in the settlements, but none has been found in Pannonia. Thus it is all the more remarkable that a bowl mould and some sherds were found at Mursa and a fragment of stucco with Pacatus' stamp on it at Tricornium (Ritopek), east of Singidunum.[194] This suggests that the

The first age of prosperity

Pacatus workshop either exported along the Danube to Moesia or was attempting to do so. The mould found at Mursa is evidence of the regular effort to move the pottery nearer to the market. The existence of the so-called 'Siscia' pottery is known only through the finding of its products in south Pannonia and Moesia. L. Nagy, who identified this pottery (Pl. 24b), suggested its location at Siscia; but it could equally well have been in some other town in the Save valley.[195]

Both potteries were in operation roughly in the decades following the middle of the second century, that is to say at the very time when Lezoux, Rheinzabern and Westerndorf wares, to mention only the most important, had practically a monopoly in Pannonia. Local potteries could not drive them from the market, but were probably able to compete on the lower Danube, where their transport costs were lower than those of the potteries of the west. It is probably also significant that at this time local demand could not be met locally, although Pacatus was turning out a fairly successful product with a good glaze. Thus the conclusion cannot be avoided that the price differences between Gallo-German and Pannonian *sigillata*, which must have operated in the Pannonian market, were not sufficiently great for the demand on the part of the masses—who could not afford western *sigillata*—to be met from local production. Local products could probably only oust imports if their quality was equally high or their prices much lower. That his pottery closed down fairly quickly is evidence that Pacatus could not produce cheaply enough. It is also probable that he was unable to win the Moesian market; under Marcus Aurelius the pottery east of the municipium of Aquincum was producing only domestic pots and terracottas.

Locally produced wares consisted of vessels derived from those of the late Iron Age but influenced by imports, or else of direct imitations of the imports themselves which gradually made the latter superfluous. Specifically Pannonian shapes of vessel are derived either from the weakening traditions of the Iron Age or from the forms of imported pottery. Especially influential among those types which did not go back to the late Iron Age were the ones that had reached the province at a critical period either by way of import or in the train of foreign craftsmen. This critical period always set in when demand suddenly increased several times over as the result of mass immigration either of soldiers or of civilians. In western Pannonia this occurred at the time of the colonial deductions and the immigration of north Italians, whereas in the eastern part of the province it coincided with the building of forts on the Danube in the late Flavian–Trajanic period.

Sculpture

What has been said above applies equally to artistic activity, if that is the right word. Just as pottery has been used to represent the whole range of industrial production, so stone sculptures, primarily grave sculptures, must serve as an example of 'art'. The only difference between the two is that the native population has no tradition in the field of sculpture, and particularly not in that of funerary sculptures. Attempts to link some of the characteristics of early stone sculptures with pre-Roman art are not convincing. The only explanation for the clumsy and inexperienced workmanship is that the native stonemason was not accustomed either to working in wood or to any kind of sculpting. The clearest example of stonemasons' work revealing training in wood, or lack of it, is to be seen in the so-called astral symbols on tombstones (Fig. 10). The whole of Pannonian stone sculpture goes back to designs introduced in the critical periods. Since the custom of setting up tombstones first came in with the Romans, there is an absence of any native influence, except perhaps for the astral symbols. The earliest stonemasons came from northern Italy and introduced their own views concerning composition and ornamentation (Pls 10a, 11a, b). When in some areas the natives began to set up tombstones they either ordered them from masons in the towns or, if these were too far away, from native stonemasons who in the early days were completely unskilled. These imitated the work of the yards already operating in Pannonia, and produced interesting and extremely characteristic groups of monuments which both in composition and ornamentation reveal a direct link with the work of neighbouring stonemasons (Pl. 6a). Such groups in the Leitha district[196] date to the Flavian–Trajanic period and in north-eastern Pannonia to the late Flavian–Hadrianic period.[197]

Until about the time of Hadrian the 'artistic' output of the stonemasons' yards was thus of two kinds: the yards which were developing demand in the big centres were working according to the traditions of northern Italy and with craftsmen who either came from there or were apprenticed to northern Italians. At the same time native stonemasons were working for the native population in areas to which the products of the 'Roman' yards could not be transported. The difference was therefore not who had ordered the tombstone but where it was to be erected. From the time of Hadrian there was a levelling out; the better craftsmen began to open yards wherever there was a demand, and the primitive products of local stonemasons disappeared. The 'specifically Pannonian characteristics' in stone sculptures were thus designs introduced by stonemasons from northern Italy and taken over by their successors. Thus the earliest influences remained for ever decisive.

Upper Moesian stone sculptures developed along the same lines, with the difference that primitive imitations by local masons are found only very sporadically because of the lack of demand. The earliest workshops were in the Flavian colony of Scupi, where simplified imitations of Macedonian sculpture were produced. This type of production, to the exclusion of the others, persisted right into the third century. The workshops which were opened somewhat later at Singidunum and Viminacium (Pls 6b, c) were influenced by those in northern Italy, or to be more precise, by northern Italian designs produced in south Pannonia, whereas the sculptures from Ratiaria were partly influenced by the latter and partly by the workshops which had already started up at Oescus in the first century. Those from Ulpianum and Timacum minus are an emanation of the traditions which had arrived there along the Danube (Pl. 26).[198]

Both in Pannonia and Upper Moesia the most important characteristic of tombstone sculptures from both town and country was that once these initial influences were accepted they proved decisive, whereas the specifically local contributions consisted overwhelmingly of quite unoriginal characteristics. Thus tombstones in south and west Pannonia are tympanum stelae (Pl. 11) on which the space for representation either played a subordinate role or was left out altogether; portraits are found only rarely. It is the design itself—the tympanum on columns or pilasters—which creates the dominant impression. In northern Pannonia, on the other hand, where tympanum stelae had also become customary to the exclusion of other styles, the stele is covered with sculpted portrait-busts (Pls 11c, 12a). The Upper Moesian stonemasons' work was even more limited in scope. The tympanum, half-columns or pilasters, the space for a wreath or portrait, the framing tendril-ornament round the border, the vase-and-tendrils motif and various iconographical designs were all taken over from south Pannonia (Pls 5a, 26a). They were, of course, not all made use of everywhere, but only a limited number of them in each town, which then came to have a few types of its own. For instance, Ratiaria had stelae with an arched top, a wreath with a rosette and the vase-and-tendrils motif, Timacum minus had these with portrait-busts in addition, while Ulpianum had only busts, the tendril-frame and the vase-and-tendrils motif (Pl. 26). The limited nature of the motifs was so pronounced that at Ratiaria and Timacum minus the vase-and-tendrils motif is cut off from the tendrils on the edges and given a space to itself; at Ulpianum, on the other hand, the tendril is always shown growing out of a pot (cf. Pl. 26a). This vase-and-tendril motif is not found at Scupi, Viminacium or Singidunum. Originality on the part of the stonemasons thus consisted solely in the way they combined certain decorative elements.

It is only rarely that the products of stonemasons' workshops can be regarded as works of art. Most of the tombstones were mass-produced, as in the case of the stelae with wreaths which were made at Aquincum[199] in the first half and middle of the second century for the *collegia*, and were then copied by native craftsmen (Pls 5c, 6a). It is only at Viminacium and Savaria that there is any evidence of artistic quality (Pl. 6b, c; 11b). If the citizens of a town wanted to commission a work of art of high quality, they sent for a foreign sculptor. About the middle of the second century statues to the Capitoline Triad were set up at Savaria and Scarbantia, and were the work of sculptors either from northern Italy or from Virunum (Pl. 27a). The stone, too, was imported.[200]

Temples and the development of religion

The building of Capitoline temples in two west Pannonian towns must be regarded not only as a symptom but also as a symbol of economic prosperity. It was in addition the time when amphitheatres were either being built or rebuilt in stone, and when the theatre at Scupi was probably constructed. It could be said that by the mid second century, in the eyes of the urban aristocracy, the province had been given its finishing touches and had become Roman. This rare moment in which there was a sense of having reached maturity and fulfilment was, of course, captured in the Roman Oration of Aelius Aristides.

The *capitolia* at Savaria and Scarbantia are also important for the history of religion. In contrast to the west European provinces where native places of worship were turned into big Gallo-Roman or Romano-British shrines, in Pannonia, and probably in Upper Moesia too, it was to official Roman gods that large shrines were erected. Consequently we have hardly any gods with native names or at least with native titles, or even possessing native characteristics.[201] The evidence of first- and second-century monuments shows that religion in Pannonia and Upper Moesia was unmistakably Roman. There were slight exceptions to this, such as the Egyptian gods worshipped by the urban aristo-cracy in west Pannonia, especially at Savaria (Pl. 27b), where the cult had been introduced via Aquileia.[202] Moreover at an early date the legions who had been fighting in the east under Trajan and Hadrian brought back with them the cult of Jupiter Dolichenus; the earliest record of him in the west is the votive inscrip-tion set up at Carnuntum by the *iuventus colens Iovem Dolichenum*.[203] The cult of Mithras was also introduced at Carnuntum at a very early date;[204] the earliest Mithraeum in Pannonia, at Poetovio, dates to the reign of Antoninus Pius, the cult being founded by customs officials.[205] Thus both the military and the administration had an impact on the development of religious belief.

Neither Roman forms nor any local ones throw light on the religious beliefs of the natives. Up to Severan times, apart from the goddess Aecorna or Aequorna at Emona, there is no evidence for any native gods or native cults. This also fits in with the characteristics of Romanization in Pannonia and Upper Moesia, even if it is not possible to formulate the reasons precisely. In the first place the conservatism of all religions must be underlined; it was responsible for the retention of traditional customs in the forms of worship. The natives did not erect altars to their own gods—if 'god' is the right word here—while even those they erected to Roman gods are extremely rare; they are primarily official dedications, such as that set up to Jupiter by the duoviri of Aquincum. Thus the native population showed little interest in the religion of their conquerors, which is the reason why the native gods were not identified with those of the classical Pantheon. Identifications were the work of foreigners, who in the first and second centuries likewise showed scant interest in the religious beliefs of the natives. Savus, Danuvius and similar gods who were worshipped by traders and officials[206] were nothing more than the *genii loci* whom outsiders honoured wherever they happened to be. Thus the reason why it was precisely the Severan period in which native religious characteristics emerged and were documented is not hard to understand. Even at that time politics and religion were closely connected, and as long as the province remained politically of no consequence its religion too showed little individuality. As soon as the situation changed, however, foreigners and the local inhabitants were equally interested in harnessing the native gods to the service of group interests. Foreigners gave native gods identifications with classical ones, while the natives suddenly showed interest in those of the foreigners.

Chapter 6
Crises on the Danube:
the rise of the Illyrican soldiery

'Dum Parthicum bellum geritur, natum est Marcomannicum, quod diu eorum, qui aderant, arte suspensum est, ut finito iam orientali bello Marcomannicum agi posset.' ('While the Parthian war was still in progress the Marcomannic war broke out but was skilfully postponed by the men on the spot so that the eastern conflict could be finished first.') This statement is found in the Life of Marcus in the *Historia Augusta*,[1] and there is no reason to doubt it. The fact that when the Parthian war began in 161 a Marcomannic war could not yet be foreseen, is proved by the fact that Marcus withdrew legio II Adiutrix from the Danubian army and employed it in the Orient.[2] It is true that this legion was replaced at Aquincum by IIII Flavia,[3] but, as we have already seen, this legion was considered a permanent reserve for Pannonia Inferior. If there had been any immediate danger of war on the Danube around 161, Marcus is unlikely to have ordered one of the most important Danubian legions to the Orient. Moreover, vexillations of practically all the Danubian legions took part in the Parthian war;[4] this is sufficient proof that the Danubian region was quiet. Danger of war on the Danube became evident only later, when Marcus even decided to recruit new legions. It was presumably to the credit of the governors of Pannonia, and perhaps also of Noricum and Dacia, that the imminent war was postponed by diplomatic means until the Parthian war could be terminated and the troops which had been sent thither from the Danube could return home.

The fact that it was possible to delay the outbreak of the war for some years by negotiations is of importance in determining the causes of the war. The *ars eorum qui aderant* presumably meant that the governors knew how to prolong negotiations with the barbarian client-princes. This, however, was possible only if the client-princes had made demands whose fulfilment they could not consider to be wholly impossible. This implies that the negotiations started within the framework of the system of *clientela*; the governors were able to raise in their opposite numbers the illusion that their demands would be considered and fulfilled at the highest level. These demands, however, were certainly novel and unprecedented; otherwise the emperor would not have felt it necessary to make serious preparations for war while the negotiations themselves were still going on. The recruitment of two legions—II and III Italicae—was in itself a measure that did not fit into the pattern of 'ordinary' frontier wars, waged from time to time in spite of, or because of, the system of client-kingdoms. The unusual demands of the barbarians called forth unusual military measures: this is a fact which requires particular emphasis. The recruitment of new legions can best be explained by an intention of the emperor to establish new provinces,[5] and it is in fact reported that Marcus intended to annex two new provinces: Marcomannia and Sarmatia.[6] If our assumption that the two new legions were intended for the two projected provinces is correct, then we must also conclude that Marcus had already planned the annexation before the outbreak of the war. Although the *Historia Augusta* at one point attributes the failure of this programme to the revolt of Avidius Cassius and at another to the death of Marcus—a chronological contradiction—it seems that the plans of conquest had already been abandoned when negotiations with the barbarians were discontinued at the outbreak of war. The reasons for this conclusion are that soon after its recruitment legio II Italica was given a stone-built fortress in Noricum (Ločice),[7] and that both new legions had already been assigned to their final base-fortresses, Castra Regina (Regensburg) and Lauriacum, in the last years of Marcus, as if they were no longer intended for newly conquered territories. Presumably, therefore, Marcus had already given up his plans of conquest *c.* 166 or 167, and the conclusion may be drawn that it was the outbreak of the war that prompted him to do so. The plans for *Marcomannia et Sarmatia*, then, somehow arose in connection with the negotiations between the governors and the Danubian barbarians. But how? It is obvious that the nature of the barbarians' demands must be deduced from their subsequent behaviour and intentions, about which Cassius Dio gives some information. It is frequently mentioned during the course of the war that groups of barbarians 'surrendered to the emperor', or offered armed help in exchange for land in a province, or that they specifically tried to settle in the empire—and this Marcus actually allowed some groups to do. The barbarians had probably already come forward with the same demands before the outbreak of war, as well as in the negotiations, and this was apparently an entirely novel factor in Roman–barbarian relations which might well result in a novel reaction on the Roman side. At first sight it does not seem very likely that the barbarians, who during the preceding centuries had taken up arms against Rome several times, and against whom a powerful army had to be maintained on the bank of the Danube, should suddenly wish to exchange their freedom for Roman bondage. We are accustomed to think of the role of an army as offensive, particularly in the case of an empire whose size is somehow logically connected with its continual growth; an empire somehow has to be imperialistically orientated. How, then, could the enemies of the empire demand their own suppression? But in the second century the Roman army was anything but offensive. Contemporary notices from Florus [8] to Tertullian prove the contrary: the profitability of Rome's conquests was questioned, the *limes* was regarded as the wall of the empire, whose purpose was seen as not only to protect the empire from enemy attacks but also to exclude the barbarians from it: *quot barbari exclusi!* Tertullian exclaims.[9] Since the time of Augustus the elaborate system of treaties concluded with barbarian nations on the right bank of the Rhine and the left bank of the Danube formed an organic part of this system of security and defence. The 'exclusion of the barbarians' from the empire apparently became increasingly a problem of foreign policy, and this can easily be understood if we consider the difficult situation of the barbarian frontier people. According to the terms of the treaties made with barbarian princes, the contracting parties had to assist each other with arms: but this, of course, never implied any inclination by Rome to make armed intervention in the hostilities of rival barbarian tribes or princes. The barbarians, on the other hand, had to maintain peace and quiet along the border, and in return those barbarian princes and their followers who were faithful to Rome were allowed to participate in Roman wealth and the splendours of the empire. *Raro armis saepius pecunia* was the slogan of this policy based on treaties;[10] it led the allied peoples sooner or later to the conclusion that it would be better to be the subject of Rome than her ally.[11]

It is clear why this problem had become topical in the sixties of the second century in particular. Among modern historians it has become a commonplace to ascribe the Marcomannic wars in the last analysis to the Gothic migration: the great migrations and displacements of peoples in north and east Europe, begun in the first half of the second century, had an impact on the people on the border of the Roman empire, and caused a north-to-south or possibly east-to-west pressure on the *limes*.[12] This may be roughly correct. But the negotiations which had preceded the war suggest that those Danubian barbarians who were allied with Rome had initially tried to attach themselves to Rome or to win Rome over for a fight against the barbarians on the move.[13] The forces of Rome were, however, tied in the Orient, and for this reason the governors of the Danubian provinces had good cause to delay an armed conflict. Their success proves that the barbarian frontier peoples had not been resolved on war with Rome from the first. And during the negotiations, the possibility of making provinces of barbarian territory beyond Pannonia was presumably discussed too; in the existing situation it could provide a solution welcome to the threatened Danubian barbarians.

This reconstruction is confirmed by the fact that the first enemies to cross the *limes* and attack the empire, as far as we can tell, were not the Marcomanni, Quadi and Sarmatians, but the Lombards, Obii and perhaps also the Vandals, all of whom lived north of them. A hostile movement of the frontier people themselves is first recorded only from 167 onwards; but it must be added at once that the chronology of the Marcomannic wars has not so far been satisfactorily established.

The first attack was made by the Lombards and Obii, 6,000 of whom attacked the section of the *limes* around Brigetio. However, two auxiliary units of the Upper Pannonian army, one of which was stationed in Arrabona, succeeded in repelling them at once.[14] Dio associates this attack with a large embassy which petitioned Iallius Bassus, governor of Pannonia Superior, for peace.[15] The leader of this mission was Ballomarius, king of the Marcomanni, who led as many as ten ambassadors, one from each people. If this mission was connected with the attack by the Lombards and Obii, then one cannot help seeing in it a last attempt by the client-princes to procure Roman armed aid. If other barbarians were responsible for the invasion, why should it have been the Marcomanni who asked for peace? Aggression by their northerly neighbours, however, could well be used as an argument by them. It is very striking that eleven peoples sent emissaries to Carnuntum at the same time and for the same purpose, though they had not previously undertaken joint action; nor is anything recorded of a frontal attack by eleven peoples during the governorship of Iallius Bassus.

It is, therefore, very likely that the first phase of the war passed as a kind of 'drôle de guerre' with Rome's allied neighbours just beyond the frontiers of Pannonia still trying either to gain Rome's armed help or to achieve their admission into the empire. Finally they were left no alternative but to threaten the prospect of a war if they were not admitted into the empire, *nisi reciperentur*.[16] This expression can be interpreted in two ways: admission into the empire by assigning them land within a province, or admission by accepting their territory as a province. It is impossible to guess why Marcus could not bring himself, under such circumstances, to remove the impending danger either by means of armed help or by the *receptio*. Perhaps his experiences over the years with far-

186

from-stable barbarian dynasties decided him against a military alliance; or perhaps the repulse of the Lombard attack gave rise to the illusion that the danger was less imminent than it was represented to be by the allied barbarians.

The government's hesitation resulted in Rome suddenly seeing herself confronted with a great barbarian coalition,[17] capable from about 167 of bringing about a series of catastrophic defeats. What was presumably the first shock was also the most severe. The Marcomanni and Quadi overran the Pannonian *limes*, penetrated northern Italy via the Alpine passes, and, after besieging Aquileia, plundered the small town of Opitergium.[18] The repercussions of this invasion were great, since for more than 200 years Italy had been immune from attack by foreign foes.

The date of this invasion is controversial, and the chronology of the incidents in the first years of the war is also uncertain. The year 168 is one of the few fixed points, when Marcus and Verus established their headquarters around Aquileia, with what was presumably a considerable military force.[19] This was the year, too, that the plague reached the western provinces and Italy, and decimated their population as well as that of the barbarians. In 168 the emperors took the most necessary measures for the securing of Italy;[20] and in this year too they probably established the special command in the approaches of the Julian Alps under the title *praetentura Italiae et Alpium*[21] and started to clear the Danubian provinces of the barbarians who had invaded them.[22] However, the plague forced the emperors to return to Rome at the end of the year; on the way, early in 169, Verus suddenly died. The logic of the—admittedly little known— events supports the dating of the siege of Aquileia and the sacking of Opitergium before the emperors' stay at Aquileia rather than after it. The year 167 was a critical year in Dacia too—as is proved above all by the wax tablets from the gold mines.[23] The clearing of the barbarians from Pannonia and Dacia was directed by Marcus and Verus in person; there are some indications that Verus was active in Dacia and Moesia.[24] It was presumably also in 168 that legio V Macedonica was moved to the fortress of Potaissa in Dacia.

It is unlikely that Marcus stayed long in Rome. In 169 battles were fought with the Germans and Sarmatians on all fronts; details however are unknown, except that M. Claudius Fronto, an experienced general of the Parthian war, who was entrusted with the joint governorship of Dacia and Upper Moesia, fought successfully against the Germans and Sarmatians (Iazyges) perhaps even before the death of Verus, but died in battle in 170 at the latest.[25] The combination of Upper Moesia and Dacia in one command allows us to conclude not only that the Romans had to deal with a frontal attack by the Iazyges from the opposite lowlands but that the two reserve legions, IIII Flavia and VII Claudia, had to be employed as well. The fact that the legions of Upper Moesia could be used elsewhere should indicate that this province was not in immediate danger. Moreover, at the latest after the death of Claudius Fronto the army of the province was reduced by one legion: this is shown by the fact that for some time the governors of this province were of praetorian instead of consular rank.[26] Pannonia Inferior, on the other hand, with one legion—we do not know which —added to her army, was at that time possibly controlled by consular govern- ors.[27] Of course one cannot say that Upper Moesia was spared altogether: at Singidunum a woman *interfecta ab hoste* is mentioned[28] on a tombstone and similar cases are mentioned on several inscriptions in Pannonia and Dacia; while the hinterland of Singidunum in particular was heavily occupied by auxiliary troops under Marcus for the purpose of protecting the newly opened mines of Mount Kosmaj.[29]

The focal point of the war was in Pannonia, and there it remained; Marcus himself was present in this province. In 171 the praetorian prefect Macrinius Vindex was killed.[30] This loss is sufficiently indicative of the critical situation. Nevertheless the defeat seems to have been the turning-point of the war. From 172 onwards an offensive was started under the leadership of Claudius Pompei- anus, Quintilius Maximus, Quintilius Condianus, and last but not least, Per- tinax; Marcus was continually present,[31] and the expeditionary force was formed of vexillations of many legions. Not much is known about these years either. Marcus stayed in Carnuntum for two years and afterwards went for a time to Sirmium.[32] He wrote part of his *Meditations* in Pannonia: twice he mentions at the end of a book the place where he wrote it: one of these was Carnuntum, the other the river Granua (Garam, Hron) in the country of the Quadi.

The sequence of the expeditions which started in 172 leads us to infer the existence of a well-considered plan, which was to concentrate the army on one region and several neighbouring peoples; the first offensive was directed against the Quadi and their neighbours (e.g. the Cotini), who formed the centre of the coalition. The series of reliefs on the column of Marcus in Rome appears to start with this offensive. It can be deduced from the Miracle of the Rain, which is often mentioned in early Christian literature, that it again proved difficult for the Romans to gain a victory. The Miracle of the Rain was not, in fact, experi- enced by the army of Marcus but by that of Pertinax, presumably in the country of the Cotini.[33] The Quadi and some of their neighbours were defeated in the end, and Marcus made conditions devised with regard to the fact that the war was not yet finished: the Quadi had to restore their captives (supposedly thirty thousand), the cattle they had driven away and also, of course, the deserters.

188

They had to recognize a vassal king named Furtius installed by Marcus, and they were denied access to the market-places along the *limes*; this last condition was designed to enable Rome to hold off the still undefeated Marcomanni and Iazyges. But the most important and significant fact is the reception into the empire by Marcus of some groups of Quadi and their settlement in several Danubian provinces and in Italy.[34] The fact that hostile tribes were settled in the theatre of war in Pannonia, Moesia and Dacia, where they could easily make common cause with the enemy, throws light on the character of the whole war: the settlement of barbarians in a threatened province was permitted only when their pacification could be ensured by the very fact of their admission. Similar tendencies are perceptible in the case of the Vandals who were admitted into Dacia. They caused the governor considerable problems, presumably because they had only recently become neighbours of the empire.[35] But of the Quadi who became members of the empire nothing further is heard.

Investiture of a vassal king was not the solution to the problem of the Quadi. When Marcus undertook his expedition against the Marcomanni in 173, the Quadi rebelled and dethroned Furtius, their king by the grace of Rome. Ariogaesus, proposed by the Quadi as their king, wanted to negotiate with Marcus, but, since the success of the Marcomannic expedition was not at issue, the emperor did not enter into negotiations.[36] The conditions of peace which the Marcomanni had to fulfil were not different from those put to the Quadi the previous year, except that, as an additional measure, they had to vacate a broad strip of land along the Danube.[37] Operations against the neighbours of the Marcomanni were also crowned with success. At the present time we only know that the Naristae, a small tribe between the Marcomanni and the Quadi, were defeated and their king Valao killed by the hand of an auxiliary prefect.[38] It can be assumed that similar operations took place against other neighbours of the Marcomanni, as well as those against the neighbours of the Quadi in 172.[39]

The measures that Marcus took for the pacification of the Danubian Germans were, to judge from what has already been said, of varied character. He used the old device of investiture of subject rulers, but combined it with the transfer of groups of barbarians into the empire, a method last used during the Julio-Claudian period. The evacuation of a zone along the border, and thus the establishment of a no-man's-land on the approaches to the *limes*, was presumably a temporary measure, dictated by the critical situations, as was the prohibition of access to market-places along the frontier. It is not quite clear whether the introduction of a stricter control over individual peoples, implemented by army officers, should be counted among the emergency measures. It is in connection with the insurrection of the Quadi during the Marcomannic expedition that we first hear of the Cotini ill-treating a Roman officer; he was the celebrated Tarruntenus Paternus, who was installed as their commander, apparently with some troops at his disposal, after the conclusion of peace in 172.[40] In the second phase of the war every tribe probably received a Roman unit with an officer charged with all aspects of administration. After the Cotini had expelled Tarruntenus Paternus, they were defeated and settled in Pannonia either at once or a little later.[41] At about the same time the Naristae also came to Pannonia.[42]

The end of the year 173 saw only the Iazyges left undefeated. The fact that Marcus intended to leave Pannonia at the end of the year shows that they did not appear too dangerous nor their problem too pressing. But after the Iazyges had invaded Pannonia across the frozen Danube during the winter of 173–4,[43] Marcus transferred his headquarters to Sirmium and began his third expedition from Pannonia. At the same time the emergency measures against the Marcomanni were abandoned or mitigated: they were allowed to visit Roman markets in certain places and on certain days, and the evacuated region along the border was narrowed by half.[44] The Iazyges, on the other hand, had become all the more difficult to handle since their northerly neighbours, the Quadi, had given them help.[45] The king of the Quadi, Ariogaesus, who had also presumably negotiated with Marcus in the meantime, apparently intended to emphasize his demands by supporting the Iazyges. Marcus succeeded—we do not know how or when—in quietening the Quadi, chiefly, perhaps, by managing to take Ariogaesus prisoner.[46] The date of the victory over the Iazyges was probably 175. Their king, Banadaspus, had wished to negotiate with the Romans after their first victory at the beginning of 174,[47] but Marcus was not yet satisfied with Roman success and refused the offer of submission, whereupon Banadaspus was taken prisoner by his own subjects. Marcus entered negotiations with the other king, Zanticus, only later, after a victory of greater importance had been won. The conditions were severe on this occasion too: the Iazyges had to evacuate a zone along the border twice as wide as that evacuated by the Marcomanni, and they had to provide a contingent of 8,000 men. Allegedly they gave back 100,000 Roman captives.[48] According to Cassius Dio, Marcus intended further measures against them but was hindered by the rebellion of Avidius Cassius.[49] Thus the end of the great period of offensives falls in the year 175.

The second phase of the great war started in 177. Marcus himself cannot have considered the severe conditions of peace imposed in 173/5 to be a lasting solution, but it was the barbarians who began hostilities. How Marcus intended to solve the problem of the Danubian barbarians, we do not know, particularly as the new situation after 177 must have rendered new plans necessary.

The chronology of the years 177–80 is likewise not quite clear, but it has been more satisfactorily established than that of the first phase of the war. In 177 the barbarians invaded Pannonia; it remains uncertain whether the Germans acted alone or whether the Iazyges joined them. Although Marcus soon received an acclamation as *imperator*, the praetorian prefect Bassaeus Rufus was killed this year[50] and the other generals were unable to obtain further successes. In the summer of 178 the emperors—Marcus and Commodus—had to appear in person at the theatre of war.[51] In the middle of 179 the Iazyges and the Buri, a tribe in the north-eastern Carpathians, were defeated and peace was concluded with the Iazyges. The conditions were that they had to stay away from the islands in the Danube and that they were not allowed to keep boats on the river. On the other hand they were given certain concessions, above all permission to communicate through Dacia under the supervision of the governor with their kindred on the lower Danube, the Roxolani, and to visit Roman market-places on the bank of the Danube on appointed days.[52]

In the second half of 179 Marcus formally occupied the country of the Marcomanni and Quadi and did not evacuate his troops in the winter of 179/80.[53] According to Cassius Dio, the strength of the Roman forces of occupation was 40,000 men; if this number is reliable, the army equalled in size the provincial army of the two Pannonias. It seems that the whole expeditionary force wintered in barbarian territory. Whether in the last resort this occupation was meant to lead to the formation of provinces out of the countries of the Marcomanni and the Quadi it is not possible to deduce from the events of the spring of 180, when Marcus prevented the Quadi, who were dissatisfied with the Roman occupation—or rather, with the behaviour of the soldiers—from emigrating to the north.[54] On the one hand it should be possible to conclude from this that the Quadi, after bitter experiences such as can be found daily during any occupation, were no longer interested in becoming provincials, whereas Marcus wanted to pacify them as future subjects; on the other hand, Cassius Dio quotes this episode in particular to prove that Marcus merely intended to punish the Quadi, not to conquer their land. He may be right. In the same year the governor of Dacia, Vettius Sabinianus, caused the emigration of the free Dacians to be prevented and assigned land to them in the province of Dacia.[55]

After the death of Marcus on 17 March 180 at Vindobona Commodus continued the war. In August he gained a victory,[56] which, as conclusion of the *expeditio Germanica secunda*,[57] gave sufficient reason for the making of peace. Commodus left Pannonia for Rome in the autumn of 180. The conditions of peace[58] contained only one novelty, permanent control of the internal political life of the Marcomanni and Quadi. This control was exercised by Roman centurions, who probably officiated as *praepositi gentis*.[59] How long this control lasted after 180 is unknown, but it was probably abolished quite soon, and certainly before the death of Commodus. It was based wholly on the intimidation of the defeated people, and could not be sustained without troops of occupation stationed permanently in barbarian territory.

The mutilation of our historical sources hardly permits a satisfactory reconstruction of the new situation. Cassius Dio's fragments contain only episode-like summaries of the negotiations, missions and peace-treaties; yet differences in the emperor's treatment of Germans and Sarmatians can be observed, and we may dare to draw several general conclusions from them. The people who received the strictest treatment during the first phase of the war were the Iazyges, against whom Marcus allegedly intended further punitive measures; they nevertheless were the easiest to quieten during the second phase of the war, and Marcus even allowed some of their old demands to be granted. The Marcomanni and Quadi, on the other hand, were not as easily defeated although Marcus had soon—perhaps as early as 174 or 175—alleviated the severe conditions of the peace of 172-3. Unlike the conditions imposed on the Iazyges, the measures taken against the Germans in 179/80 were severe in the extreme. Considering that during the decade following the Marcomannic war the principal enemies in the Carpathian basin were neither the Suebi nor the Sarmatians but the Vandals who had made their appearance there only recently, it is very likely that the special measures taken against the Suebi originated in the recognition of the delicately balanced situation among the free Germans—which was apparently the result of the newly arrived tribes exerting pressure upon the Germanic neighbours of the Danubian provinces. It is, presumably, not the question here of some irreconcilable German–Roman antagonism, because in that case Marcus would not have admitted some groups of Germans and their small neighbouring tribes (Cotini, Naristae) into the empire, and settled them in Pannonia, Moesia and Dacia. The Marcomanni, Quadi and several other peoples were so hard-pressed by the appearance of the Vandals, Lombards and other Germanic tribes that their only choice lay between submission to Rome and submission to other Germans. The bitter experiences arising from the 'mutual armed aid' stipulated by treaty were presumably the reason for their not readily resuming the old client-relationship with the Romans. The danger therefore existed that they would ally themselves with their new Germanic neighbours. Romans and Germanic tribes thus found themselves confronted with the same alternative: alliance or war, with equally doubtful result. The transfer of several groups from the left bank of the Danube into the empire was merely a palliative, not a solution of the problem. Marcus was presumably well aware of that, but he also presumably appreciated that the incorporation of

Marcomannia in the empire would create new internal problems and new neighbours on the border.

The fact that the Iazyges allowed themselves to be satisfied with the appointment of market days and permission to associate with the Roxolani via Dacia indicates that their problems were of a quite different nature. Nothing is reported of a resettlement of groups of Iazyges within the empire; their commitment in 175 to raise a considerable mounted contingent presumably merely served the purpose of weakening the military power of this nation of horsemen. The Iazyges presumably merely wanted to remove the disadvantages caused by the establishment of Dacia; in other words, to establish a free communication with the Sarmatians on the lower Danube. The control of trade was possibly itself a gain for the Iazyges. For conceivably it was only from this time onwards that they obtained leave to visit Roman markets on the Danube. It is difficult otherwise to explain why all the samian ware found in the Hungarian plain dates from the time of Marcus and the Severi, whereas in the northern half of the Carpathian basin, which was inhabited by Germans and Daco-Celts, a considerable quantity of earlier samian has been found.[60]

I consider it probable that the solution which Commodus adopted originated with his father. Fulfilment of the plan to establish two new provinces, Marcomannia and Sarmatia, cannot have been considered seriously after all that had happened between 167 and 180. It is significant that Cassius Dio, the source nearest in time to the events, who as the young son of a senator must have known personally a number of members of Marcus' supreme command, depends on surmises.[61] The other sources merely repeat the commonplace that Marcus would have annexed the provinces to the empire, had his death not prevented him from doing so.[62] After the interpretation of events given above we may be permitted to doubt whether the establishment of Marcomannia and Sarmatia could still have been seriously considered in 180. Dio corroborates this in an indirect manner not only by what he says about the prevention of the emigration of the Quadi but also in his description of the conditions of peace put to the Iazyges, which constituted a final solution, but should on no account be regarded as preparation for provincialization. On the whole, foreign policy, after Marcus, continued to tread the old paths of client-treaties, and it seems that it proved possible to embrace the new peoples of the Carpathian basin in this policy based on client-relationships. In the peace-treaty it was also stipulated that the Germans were not to wage wars with their neighbours, among whom the Vandals are mentioned.[63] From this we may conclude that Marcus and Commodus succeeded in creating a security system that included the new neighbours as well.

If the Marcomannic wars changed the foreign political situation on the Danube only slightly, their effect on domestic policy was all the greater. First, the Danubian provinces and in particular Pannonia, which had suffered the heaviest losses, had to be rebuilt and economically restored; and, second, it has become manifest that the army of the Danube, representing about one-third of the empire's military forces, and at the same time the nearest to Italy of all the imperial armies, constituted a hitherto unesteemed but potentially important factor in the power politics of the empire. Reconstruction and the awakening of self-confidence in the Danube lands thus went hand in hand. In this, by degrees, the leading role was necessarily taken by Pannonia. The Pannonian provinces contained four legions in the neighbourhood of Italy; and this involved a concentration of economic support. With the economic and social consequences of this favourable situation we shall deal in more detail in the next chapter.

The losses suffered by Pannonia were heavy and various. We learn of wholesale removals of the provincial population into German or Sarmatian captivity. (Allegedly the Germans gave back thirty or fifty thousand, the Iazyges a hundred thousand prisoners.)[64] Cattle are reported to have been driven away and booty of all kinds carried off. The numerous finds of coin-hoards from the years of war, and amongst them such as were hidden during the second phase of the war, are significant as well. Excavations in civil settlements and in forts have revealed time and again the existence of burnt layers of second-century date; it has even become a risky commonplace in archaeology to relate all layers of destruction and traces of burning to the Marcomannic wars. A thorough critical sifting of these traces would presumably show that more damage was done in the civilian settlements than at military sites, where total destruction was an exception. The greatest devastations were presumably effected during the years 167–71 and 177–8, whereas during the years of the offensive, when Marcus himself stayed in the theatre of war, the situation was comparatively normal.[65] No new period of construction, distinctly different from that previous to the war, can be traced in the *limes*-forts. A slightly protruding corner-tower with semicircular plan (Fig. 18, p. 108) and several newly added interval-towers cannot so far be dated accurately; the excavators attribute them to the reigns of Commodus or Caracalla.[66] Of much greater importance were certain other changes in the defence of the frontier which had already been made in the years around 180.

In the first place the transfer of mounted archers from Syria and Africa to the Danube must be emphasized. Already in the first phase of the war a strong unit from the African legio III Augusta was employed against the barbarians of the

Danube, the reason presumably being that the soldiers of this very distant legion were used to the tactics of light cavalry. This African vexillation remained in Pannonia until 180 and part of it was not sent back to Africa but was incorporated in legio II Adiutrix,[67] presumably for the purpose of training the soldiers of the Pannonian legion in the tactics of light cavalry. Under Marcus or Commodus several cohortes Maurorum were brought to the Danube frontier of Pannonia.[68] The cohors I milliaria Hemesenorum equitata civium Romanorum was transferred at the same time to the front facing the Iazyges.[69] It was stationed in the fort of Intercisa, where a small but important Syrian settlement developed. During the Severan period the cohort was further reinforced from Syria.

For similar changes on the Danube frontier of Upper Moesia we have no evidence, but from the interior of this province, where Marcus had stationed many of the cohortes Aureliae raised by him, we have further information. The mines at Mount Kosmaj, south of Singidunum and therefore quite near the Danubian frontier, were presumably opened under Marcus and organized as a *territorium metalli*. For the protection of these apparently very important mines, several auxiliary regiments were transferred from the Danubian *limes* in Upper Pannonia;[70] these were soon replaced by the cohortes Aureliae novae[71] whose fort in Stojnik has already been explored to some extent.[72] The fort's hospital was built in 179,[73] but the stationing of auxiliary troops at Mount Kosmaj is attested almost ten years earlier. Since the troops first transferred to this place were taken from the army of Upper Pannonia, where after the outbreak of the great war all provincial troops would be needed, we must date the organization of the mines and the transfer of troops from Upper Pannonia to the first years of Marcus' reign; and it follows that the military occupation of the mines should be attributed to the disturbance of the internal peace of the province by *latrones*; there was a danger of the *metallarii* making common cause with the *latrones*.

Even during the Marcomannic wars the *latrones* gave trouble to Marcus. Special commandos had to be sent against them. These may also have had to solve the problem of incorporating the *latrones* in the expeditionary force for the Marcomannic war. For the *latrones Dardaniae* this is attested in the *Historia Augusta*,[74] but it would be erroneous to assume that the two cohortes Aureliae Dardanorum recruited by Marcus consisted of former *latrones*. Cohors II Aurelia Dardanorum was assigned to the fort of cohors I Thracum Syriaca in Timacum minus, which lay right in a centre of the *latrones*.[75] The fort of cohors I Aurelia Dardanorum is unknown; it must be looked for somewhere in the valley of the Morava north of Naissus or on the northern slope of the *Scardus mons*, where attacks by *latrones* are also known to have occurred.[76] Both cohortes Dardanorum were recruited from among the Romanized population of southern Moesia (Dardania). Marcus' military measures in Moesia Superior thus, for the most part, concerned the internal situation in the centre of the Balkan peninsula; the decisive factor was probably the realization that the diagonal road leading from Singidunum to Byzantium could become one of the most important military roads of the empire. The towns situated on this road—Naissus, Horreum Margi, Margum—were probably, without exception, promoted municipia under Marcus. The events of the third century were fully to justify Marcus' foresight.

Another change in the defence of the frontier is traceable through inscriptions from Pannonia Inferior. A number of similar building-inscriptions of the year 185 attest new buildings on the bank of the Danube below Aquincum.[77] The inscriptions, which are uncommonly long and complicated in their formulation for this period, report that Commodus had fortified the whole bank of the Danube with *burgi* and *praesidia*; the *burgi* were built from scratch, and the *praesidia* erected to watch places most suitable for the secret crossings of the *latrunculi* (Fig. 35). The word *latrunculi* has caused controversy in that it has been taken to signify smugglers, rebellious subjects of the empire, or simply barbarians, i.e. Iazyges. The very general derogatory diminutive (*latrunculi*) was not intended to provide an accurate identification of those against whom these

Figure 35 Inscription of Commodus mentioning *latrunculi* from Intercisa

praesidia were built; it included all who intended to cross the Danube illegally. Commodus' measures, attested by these inscriptions, were thus concerned not with the better protection of the frontier against barbarian attacks, but with the supervision and regulation of traffic across the border. After what has been said the necessity of these measures can easily be understood, coming as they did a few years after the conclusion of peace with the Iazyges. Marcus had permitted them to visit Roman markets on stated days and had opened the frontier of Dacia to them under the control of its governor; thus under pretence of peaceful and legal traffic across the frontier small groups of barbarians were able to raid the province, and they presumably also tried to trade illegally outside the appointed places and days. It was difficulties of just this kind which presumably necessitated a punitive campaign against the Sarmatians. The date of this expedition is controversial; the *Historia Augusta*,[78] however, connects it with an event which is very characteristic of the altered situation in Pannonia, and which has left traces in the *burgus*-inscriptions of Commodus.

It seems that only some of these *burgus*-inscriptions were actually attached to the newly built *burgi*; others were left in a stonemason's workshop at Intercisa, and were later used as slabs for other purposes. The reason is possibly that the governor of Lower Pannonia, Cornelius Felix Plotianus, had fallen into disgrace in the meantime: his name was erased on those *burgus*-inscriptions that had already been finished. This coincides in time with the overthrow of the praetorian prefect Tigidius Perennis, whose sons were at that time attached as young officers to the army of Pannonia.[79] According to the *Historia Augusta*, Perennis intended to credit his sons with the success against the Sarmatians; and their presence in Pannonia was connected with his further plans. He had evidently realized that in the event of a *coup d'état* the four legions of the Pannonian army were closest at hand for intervention in Italy. A slightly obscure phrase in the *Historia Augusta* (*Pannoniae quoque compositae*)[80] allows us to suppose that the fall of Perennis involved reprisals in Pannonia on a considerable scale, and we may deduce discontent among the Pannonian army against the rule of Commodus.

One German campaign of Commodus, his 'third German Expedition' cannot be dated.[81] It was presumably punitive, like that against the Sarmatians. The client-relationships established by Marcus not only outlasted Commodus but presumably survived in the main until the middle of the third century. The departure of the Pannonian legions for Italy under Septimius Severus in the year 193 did not tempt the barbarians to invade Pannonia; according to Herodian, Severus negotiated with the tribes across the frontier before his departure, and induced them to keep quiet, presumably by the same means as were used under Tampius Flavianus in A.D. 69.[82] Thereafter the main centre of unrest lay on the

Dacian *limes*, where the Vandals and several Dacian groups, above all the Carpi, were the enemies. This unrest endangered the Danubian border of Pannonia only seldom and merely indirectly. Our sources for Severan times are, however, so mutilated and so unreliable that research has had to rely to a large extent on coin-hoards, votive inscriptions and other indirect evidence. The reconstruction of events thus obtained, however, has not only yielded too large a number of wars on the Danubian border but has been thought to reveal barbarian invasions deep into the province as well, and this is difficult to reconcile with the mainly peaceful conditions of the Severan period. Furthermore, we must remember that the Pannonian legions, or at least vexillations from them, took part in almost all domestic and foreign wars under Septimius Severus and Caracalla; they formed, so to speak, the élite troops of the emperor. The reserve legion, IIII Flavia, had to take the place of legio II Adiutrix at least once under Septimius Severus in Aquincum.[83] The Pannonian legions were employed in Dacia as well against the barbarians, and this could not have happened if Pannonia or Moesia Superior had been subject to pressure by barbarians across the frontier at the same time.

The only barbarian invasion into Pannonia under the Severi which is to a certain extent reliably documented took place under Caracalla in 212 or early 213, and was connected with Carpian–Vandal unrest on the northern fringe of Dacia. The invasion affected the north-eastern corner of Pannonia,[84] and possibly was the direct cause of Caracalla's alteration of the boundary between Upper and Lower Pannonia the next year. This line had run from the bend of the Danube south-westwards, and the result was that a barbarian attack, once it had crossed the Danube frontier of Upper Pannonia from the north, could reach Lower Pannonian territory immediately: there troops of the governor of Lower Pannonia had to be employed against it. In 214 the border was therefore shifted further westwards, approximately to the mouth of the river Arabo. As a result Brigetio with legio I Adiutrix joined Pannonia Inferior, and so both Pannonias became provinces with two legions, and the governors of Lower Pannonia were henceforth consulars too.[85] Reasons of state were also presumably behind this decision, since three-legion provinces were abolished by the Severi as having armies of excessive size and therefore governors of excessive power.

The events of *c.* 213 throw a clear light on the continued existence and undisturbed functioning of the system of client-kingdoms on the Danube frontier. According to Cassius Dio, Caracalla boasted of having created hostility between the Marcomanni and the Vandals.[86] According to Herodian's highly imaginative account, Caracalla took pains during his visit to the Danubian provinces to pose as a friend of the Germans.[87] Presumably he tried hard to isolate the Danubian Suebi (Marcomanni and Quadi) from the increasingly hostile Vandals; with the Marcomanni he succeeded. The Quadi, on the other hand, who invaded Pannonia in 212 or 213, he could punish on the basis of the client-treaty without any particular consequences: the Quadi had to surrender their king Gaiobomarus and several nobles of the tribe to him, and these Caracalla simply ordered to be executed.[88]

The Severi were thus successful in maintaining the system of client-kingdoms although presumably against increasing difficulties. The existence of permanent and well-paid interpreters, whose position on the governor's staff was presumably first established under the Severi, shows how much attention was paid to diplomatic relations at that time.[89] In provinces where the native population spoke a dialect not unlike that of the barbarians such interpreters were presumably not necessary, particularly since a large proportion of the soldiers, part of whose duty it was to serve in the *officium consularis*, were capable of understanding the language of barbarian emissaries without difficulties. The fact that in Pannonia interpreters, whose official name was *interpres officii cos. salariarius legionis*, are attested for Lower Pannonia only, presumably has its explanation in the fact that the governor of Upper Pannonia had German soldiers at his disposal. In the *officium* of the governor of Lower Pannonia interpreters are attested for three languages: Sarmatian, German and Dacian. The need to employ Dacian interpreters as well arose from the Dacianization of the Celtic and Germanic peoples on the northern fringe of the Lowlands which had begun long before. The Cotini settled under Marcus in Pannonia around Mursa and Cibalae bore Daco–Thracian names for the most part.[90] Interpreters would not have been necessary if knowledge of the Latin language could have been presupposed among the barbarian princes, and if it had not been necessary to maintain continual contact with the barbarians, whether of a political or economic kind.

The basically peaceful situation in Pannonia was advantageous to the Severi, the main support for whose power was the Pannonian army. Of course, we do not know how much money the treasury had to pay for this peace on the Pannonian frontier. The emperors must probably have attached special importance to the Pannonian legions being always in a state of preparedness, and for that reason grudged no pains to create a *modus vivendi* with the Danubian frontier peoples. Caracalla, the 'friend of the Germans' in Herodian, played his part in this aspect of the foreign policy of the Severi. We have already drawn attention, in connection with the overthrow of Perennis, to the ability of Pannonia in the event of a *coup d'état* to play a major part because of its four legions, and even more because of its nearness to Italy. This fact had already become obvious during the Year of the Four Emperors, but the internal situation of the empire did not give opportunity for another century for the Pannonian army to exploit the *arcanum imperii*. The Marcomannic wars were the first to demonstrate how quickly Italy could be reached from Pannonia. This was presumably realized by the Pannonian soldiers, and, in his numerous *adlocutiones* during the Marcomannic war, the Emperor Marcus obviously did not conceal the fact. The key position occupied by Illyricum during the third century has already become a commonplace. [91]

The attempt of Tigidius Perennis to stir up the Pannonian army against Commodus was not successful; but seven years later, when the governor of Pannonia Superior, L. Septimius Severus, was proclaimed emperor on 9 April [92] in Carnuntum, Rome and the empire found themselves confronted by an army marching with astonishing spirit and discipline behind its candidate. Only in the case of legio X Gemina is there a suspicion that it did not take part in the proclamation of the new emperor; [93] but it is extremely unlikely that this legion offered resistance, stationed as it was in the immediate neighbourhood of Carnuntum. At any rate, X Gemina did take part in the struggle against Pescennius Niger. [94] After the troops of the Danubian provinces had marched into Rome with Septimius Severus the new praetorian guards were chosen from amongst their men, [95] mainly from Pannonians, but to a lesser but still considerable degree from Moesians, Thracians and Dacians. The future recruitment of the praetorian cohorts, too, was secured from these Danubian provinces, service in the guards serving at the same time as the first step to a higher military career. Legio II Parthica, stationed not far from Rome, also consisted of Danubian soldiers. Besides the Pannonian legions, the legions of the other Danubian provinces too were from the beginning supporters of the regime of the Severi. For this reason the collective name Illyricum began to be used immediately after Septimius Severus' seizure of the throne. Gradually this name gained special political meaning. As early as Herodian we find the word Illyricum in use as an inclusive term for the Danubian provinces; [96] its origin is not clear. In the second century this term was used for the Danubian provinces only in the customs administration (*publicum portorium Illyrici*), and in Appian's *Illyrike* the statement is found that the word Illyricum has no linguistic-ethnic content. [97] It is therefore quite possible that the name of the customs district was transferred to the Danubian provinces in general. Modern historiography, much too sensitive to national and racial differences, is inclined to attribute the part played by the Danubian military to a national self-assertion of the Illyrian race; but we shall see that this Illyrian military consisted of Celts, Thracians, Illyrians, even of Romanized city-dwellers of mixed origin, and of other constituents of the empire who migrated to Illyricum. From Septimius Severus and his dynasty onwards till the middle of the third century, the representatives of Illyricum were not even people from the Danube but mostly senators from other parts of the empire; people who were Illyriciani by birth came into power only in the last third of the third century.

The many thousand Danubian soldiers serving in and around Rome were able to exert a not inconsiderable pressure on politics in Rome. They behaved, indeed, as representatives of the interests of their homeland, expressing these interests from time to time vigorously. Best known is the case of the historian Cassius Dio, who in 229 was prevented by the threats of the praetorians from attending in person the inauguration of his second consulship; their reason was that as governor of Upper Pannonia Dio had previously implemented unpopular policies.[98] Similar occurrences soon produced an attitude of reserve towards the Illyriciani among senatorial circles and the representatives of various other provinces. This attitude had a welcome pretext in the lack of *humanitas*, in other words the rural roughness and low level of education, of these soldiers. In the previous century Fronto had contrasted the clumsiness of the Pannonians with the sophistication of the Syrians.[99] Herodian attributed the seizure of power by Septimius Severus to the stupidity of the Pannonians;[100] and the urban Roman population, plebs as well as senators, watched with horror the presence of these barbarians who did not even understand Latin well.[101] During the first years of the dynasty of the Severi, the fear of Illyricum was presumably employed as an element in power-politics; it was thus used, for instance, by Septimius Severus in 196 when, on setting out against Clodius Albinus, he committed his son Caracalla to the care of the Danube army after having him proclaimed Caesar in Viminacium in the summer of that same year.[102] The young Caesar received the news of the victory over Albinus in Pannonia.[103] In 202 the emperors, coming from the Orient, travelled through the Danubian provinces, and by journeying along the *limes* took special care to visit the troops of Moesia and Pannonia in their forts.[104] It is, moreover, not unlikely that the journey from Carnuntum to Rome was arranged in such a manner as to enable Severus to be in Carnuntum on the *dies imperii*, 9 April, and to enter Rome on the day of his state entry nine years before. By 202 the great days of the Illyriciani were already over; Pescennius Niger and Clodius Albinus were defeated, the war against the Parthians finished. The legions were in their fortresses again: II Adiutrix had perhaps returned with the emperors from the East—and Septimius Severus perhaps wanted to demonstrate before the troops as well as before the empire at large that Illyricum was to remain the most favoured part of the empire.

Sovereignty, however, cannot be based on so limited a social foundation. The interests of the Illyriciani came more and more to be considered as sectional and there are a few indications that as early as Alexander the Pannonian troops were not quite satisfied with the government.[105] Between foreign and internal politics there existed a connection in that the Danubian troops did not have to advocate concentration of the empire's resources in Illyricum as long as the Danubian border was quiet—and under Alexander, especially, the peace seems to have been undisturbed. But a serious problem arises from the fact that during the following decades the Illyriciani wished to make use of the resources of the empire to protect their homeland. Such a demand may perhaps have been made for the first time during Alexander's Persian war, when the Rhenish and Danubian troops desired to return home on the news of a war with the Germans.[106]

The outbreak of this war against the Germans on the Danube is not well authenticated, and it is by no means impossible that the German danger in Illyricum was a false rumour. Coin-hoards with latest coins of the years 228 and 231 respectively suggest on the contrary a Sarmatian attack, crossing the Danube border of Lower Pannonia.[107] It is not mentioned in written sources. Perhaps this Sarmatian attack was the first beginning of the Sarmatian war, fought by Maximinus Thrax from 236 to 238. During these years the imperial residence was at Sirmium,[108] so used for the first time since Marcus had stayed there in 174–5, but destined to be by no means the last. We must dwell on this point a little, since Sirmium and its surroundings played an increasingly important part in power-politics and strategy from the middle of the third century onwards. We have already seen that much earlier the Scordisci owed their temporary hegemony to the circumstance that they were in possession of the area around Sirmium, a country lying at the focus of routes connecting Italy and the west with the Balkans and the Orient (p. 10). After the Goths made their appearance on the lower Danube, and at a time when the Rhine and Danube legions had to be employed repeatedly against the Parthians and the Persians, deployment of troops through Pannonia and the Balkan peninsula became necessary. As we have seen, this had already been recognized by Marcus, and it was presumably not by chance that he chose Sirmium as headquarters against the Sarmatians. In the third century Sirmium gradually became one of the great headquarters of the empire.[109] The results were twofold: on the one hand economic prosperity prevailed in the area, but on the other a clique of Sirmians arose who often succeeded in obtaining empire-wide power. All Pannonian emperors originated from the Sirmium region, and some emperors although demonstrably from other Illyrian provinces have been taken for Sirmians by some of our sources.[110] The first emperor from the Sirmian region was Decius; the last was Valentinian, who came from Cibalae.

It is not unequivocally established that the enemies in Maximinus Thrax's Sarmatian war were indeed the Sarmatians of the Lowlands: Dacia, not Pannonia seems to have been the province to be exposed to the heaviest pressure by barbarian attacks until the middle of the third century, and the attacks on Pannonia thus often came from barbarians living on the northern fringe of the Lowlands and in the neighbourhood of Dacia, as for instance the Vandals and the Carpi. Emperors assumed the title Sarmaticus whenever the *ripa Sarmatica* was attacked, irrespective of the defeated peoples being really Sarmatians or not. More recent research has endeavoured to fill the gaps left in the direct sources by certain hypotheses, which might enable us to reconstruct the struggles that took place on the Danube *limes* in the middle of the third century. It has been assumed that imperial surnames granted to the troops (e.g. legio II Adiutrix Philippiana) were distinctions for martial successes, and accordingly that barbarians had to be fought on the Pannonian Danube *limes*, especially in the northeast part of the province, in the reigns of Gordianus, Philippus, Trebonianus Gallus, etc. If this hypothesis based on the imperial surnames is correct, the question still remains to be answered whether the troops concerned were given their titles for wars in Pannonia, or because of their part in an expedition outside Pannonia. The Pannonian and Moesian troops presumably quite often took part in the wars in Dacia and Moesia Inferior, although this can rarely be proved because of the lack of suitable information in our sources.

Coin-hoards from the time of Trebonianus Gallus certainly allow the conclusion that after 252 the Quadi became restless and attacked the *limes* around Brigetio.[111] This conclusion is not unreasonable, for the reason that, to all appearance, both Marcomanni and Quadi were among those responsible for the great catastrophe of the years 258–60. But the first certainly attested struggles in Pannonia are those which occurred under Gallienus; these had vast consequences. Although the situation in the provinces of Illyricum became increasingly dangerous after the forties of the third century, at first a situation of chronic war developed only in Dacia and on the lower Danube, where the Goths and the peoples under pressure from them were an increasing danger to the *limes*. The representatives of Illyricum in the imperial court presumably demanded energetic measures, and when Philippus in 247 personally took part in the difficult war against the Carpi, he had the opportunity to convince himself that extraordinary measures must be taken to avert the impending danger. The obvious step was to unite the military forces of Illyricum under a supreme command of their own—thus in effect following the example set in the years 117–19 or 168–70, when Marcius Turbo and Claudius Fronto had been entrusted respectively with the simultaneous governorship of several adjacent Danubian provinces. Possibly as early as his return march, but certainly at the beginning of 248, Philippus appointed Ti. Claudius Marinus Pacatianus as commander of all Danubian troops or at least of the troops of Pannonia Inferior and Moesia Superior.[112] The constitutional standing of this special commission, which henceforth was given frequently, is not clear.[113] Probably the supreme commander was not simultaneously governor of all the provinces concerned, but merely a military commander without administrative authority in the civilian sphere. His seat seems to have been Sirmium; this, too, could be indicative of the intention not to impair the governor's position, for Sirmium was neither the residence of a governor nor a legionary fortress.

It seems that the Illyrican soldiery had only waited for a supreme commander of their own to proclaim him emperor as their representative, and it must not be forgotten that they could count on the aid of the praetorians and the other Illyriciani in Rome, and thus had good reason to believe that they would be able to repeat the march of Septimius Severus on Rome. A devastating attack on Lower Moesia by the Goths and many other peoples prevented Pacatianus from turning against Philippus, and his troops soon felt induced to dismiss him, presumably because of his failure or his inability. A decisive factor in their decision, however, was the dispatch of Decius to Illyricum: when the news of the seizure of power by Pacatianus and other similar news reached Rome, and when Philippus had let his intention to abdicate be known, the senator C. Messius Traianus Decius, a native of Sirmium, was sent to Illyricum against Pacatianus.[114] Decisive for Philippus' choice of Decius was presumably the assumption that as a Sirmian he would be welcome to the Illyriciani. Decius, who apparently knew his countrymen better and was presumably also better informed of the intentions of the leading Illyriciani, had warned the emperor not to entrust him with the supreme command in Illyricum; but he had in the end to set out against Pacatianus. Pacatianus was killed, probably as a direct result of the news that the Sirmian Decius had been sent out as new supreme commander, and Decius managed to fight successfully against the Goths. But his successes gave rise amongst the soldiery of the Illyriciani to the idea that Decius would be the right man for the realization of their plans. Against his will Decius was proclaimed emperor in the summer of 249 and was forced to march against Philippus.

The circumstances of the proclamation of the first Pannonian to become emperor show that the Illyriciani were keen on finding a suitable representative capable of mobilizing against the increasing danger from the Goths not only the forces of Illyricum but those of the whole empire as well. The chief problem to be solved by each supreme commander was victory over the Goths, the imperial throne being merely a means thereto. After the death in battle of Decius the procedure remained the same. In the place of Decius, Trebonianus Gallus, governor of the most seriously threatened province, Moesia Inferior, was proclaimed; but as soon as a general capable of dealing successfully with the Goths emerged in the person of Aemilius Aemilianus, supreme commander in Illyricum, the purple was given to him. Aemilianus was forced, as Decius had been before him, to set out against Italy, but he did not succeed in maintaining himself there, and after having defeated the Emperor Trebonianus Gallus he succumbed to Valerian.

After 253 there was a sudden calm as far as the Illyrican soldiery was concerned. Since the government of Valerian and later of Gallienus was certainly somewhat anti-Illyrican we must not ignore the possibility that they had removed the representatives of the Illyrican soldiery from the most important offices. It is at any rate striking that after 253 no supreme commander is attested in Illyricum. We may deduce either that none was appointed or that he successfully resisted the intentions of the Illyriciani. Nothing is said about larger attacks by barbarians at this time; there was thus no immediate inducement to proclaim an emperor in order to fight them. This is confirmed by events from *c.* 256 onwards. About this year or at the latest in 257 the younger Valerianus, son of Gallienus, was appointed supreme commander of Illyricum; this shows clearly that the situation on the borders had become aggravated to such an extent that the appointment of a supreme commander—while perhaps at the same time placating the demand of the Illyriciani—had become imperative. However, the commander sent was a youth, who although in the confidence of the emperor cannot be considered a general.

The chronology of events between 257 and 260 is far from clear or attestable in sources. We do not intend to add to the vast number of chronological hypotheses, nor do we wish to follow any of them. Everything is as yet unfixed and will remain so, until new finds give new information. Presumably it is better to enumerate the individual facts without attempting to relate them chronologically.

In 257 Gallienus took the titles Dacicus Maximus;[115] it follows that the province of Dacia, continually overrun since Philippus, bore the brunt of this new attack as well. Pannonia was probably invaded somewhat later, but then all the more heavily; the number of coin-hoards—over a dozen—datable to the years 258–60 illustrates the seriousness of the attack. The coin-hoards are evenly distributed over the whole area of the province,[116] and attacks by Suebi and Sarmatians are alike attested.[117] The attacks also affected part of the Moesian border along the Danube, where coin-hoards are also known.[118] Up to this time Upper Moesia had been spared heavy barbarian attacks; the province owed its immunity solely to its favourable geographical situation. There are, in fact, a few coin-hoards from Upper Moesia—some of them huge—which were concealed during the reign of the first emperors of the 'Military Anarchy'.[119] But we believe that these should rather be attributed to internal disturbances, in particular to concentrations of Illyrian troops under the leadership of usurpers: the generals who were proclaimed emperors by the troops of Illyricum mostly fought on the lower Danube and in Moesia Inferior, and from there they had to make their way to Italy via Upper Moesia, and in particular along the Naissus–Singidunum–Sirmium road.

In Pannonia's critical situation pretenders proclaimed themselves once more. Probably they were supreme commanders of the Illyrian army as Pacatianus, Decius and Aemilianus had been. The first one was Ingenuus,[120] who was perhaps appointed to replace the younger Valerianus now deceased or killed. Characteristically enough for his own position and the part played by Sirmium, he was proclaimed emperor in this town. Ingenuus was supported by Pannonian and Moesian troops, but he was soon defeated by Aureolus near Mursa.[121]

The second usurper was the senator P. C(. . .) Regalianus, who was also proclaimed in Sirmium, and by Pannonian and Moesian troops.[122] We know more about this man than we do about his predecessor; perhaps we can deduce from this fact a comparatively longer reign. It is reported that he had fought successfully against the Sarmatians. Towards the end of his reign he permitted coins to be struck, which circulated in the north-western part of Pannonia (Fig. 36) and above all at Carnuntum.[123] These coins were struck in great haste and—for want of raw material—from older coins overstamped. The *denarii* of a senatorial lady, Sulpicia Dryantilla, overstruck in the same way, belong also to this group of numismatic rarities. This Sulpicia Dryantilla accordingly was either the mother or the wife of Regalianus. But it is curious that more examples of the coins of Dryantilla are known than of those of Regalianus. From the distribution it could be deduced that Regalianus had to withdraw to the north-western corner of Pannonia towards the end of his reign, where he then had the irregular coins struck—possibly for mere propaganda purposes.

During these periods of unrest Gallienus had to go to Pannonia at least once. One episode, otherwise not datable, can be attributed to these years by the fact that it presupposes Gallienus' presence in person: according to this story the emperor became infatuated with the daughter Pipa of the Marcomannic king, Attalus, and therefore transferred a part of Pannonia to him.[124] This surrender or transfer of a piece of land of the empire must, of course, be understood in the

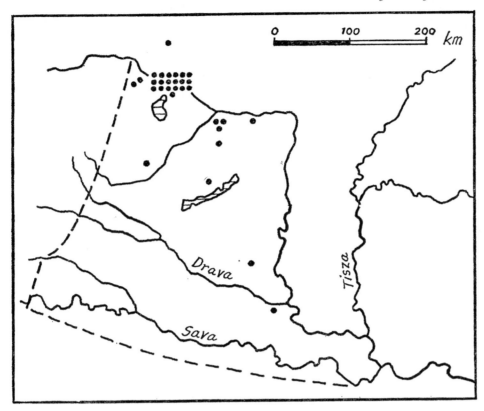

Figure 36 Distribution of coins of Regalianus and Dryantilla

sense of a group of Marcomanni being settled by Gallienus in Pannonia. And since the Marcomanni had taken part in the war of 258–60 we may place the settlement of some groups of them in Pannonia within these years.

For the further events, possibly the most significant fact was that Gallienus—presumably through his general Aureolus—brought a contingent of probably considerable strength, consisting of vexillations of the British and German provincial armies, to Pannonia and stationed them at Sirmium.[125] They remained in Sirmium till after the years of crisis 258–60: this is attested by their taking part in the battle against Macrianus in the Balkans near Serdica (*bellum Serdicense*) whence they returned to Sirmium.[126] Furthermore, they could not be sent home because of the Gallic empire. By their presence in Sirmium they held the Illyriciani, whose leading clique was domiciled there, in check, and at the same time they occupied the most important place in the Danube region.

Regalianus was killed by his own soldiers, if this information in the *Historia Augusta* is to be trusted. Such a murder was in itself by no means unusual, since the emperors who were proclaimed by the Illyriciani were treated as the tools of a junta whose political intentions were not decided by personal points of view.[127] It is striking that the end of the usurpations in Illyricum coincides with the important military and political reforms of Gallienus, which were directed towards the organization of a mobile central army. Possibly the British and German vexillations were one of the reasons which induced Gallienus to establish this central army. Gallienus was intelligent and wise enough to have the Illyriciani represented in the newly organized army; at least the so-called legionary coins of this emperor, probably referring to the new central army, allow us to conclude that the *legiones Illyricianae* too had provided detachments.[128] Under such circumstances the junta of the Illyriciani had no choice but to make sure themselves that their interests were understood in this new concentration of power, and therefore they gave up the tactics of local usurpations. The fall of Regalianus possibly belongs to the period of these events.

During the remaining years of the reign of Gallienus the interior of the Danubian provinces remained calm, nor were there any conflicts of importance on the *limes*. Pannonia had, however, suffered so much during the crisis years 258–60 that this temporary quiet under a government faced with the most severe difficulties both in foreign and in home policy did not make any reconstruction or consolidation possible. Moreover, a further circumstance, not to be underrated in importance, has to be considered here: for obvious reasons the government of Gallienus did not consider it a major duty to make use of the resources of the empire for the reconstruction of Pannonia: first, because these resources were not sufficient anyway; second, because several other provinces, among them Moesia Inferior and Dacia in particular, needed reconstruction just as much; and, third, because the experiences of the last two decades had provided ample reasons for not considering the demands of the Illyriciani. One important measure, however, has to be mentioned: the establishment of the mint in Siscia, which was to become one of the principal mints of the empire and the most important one in the Danubian provinces. Thereby, the supply of our provinces with inflationary coinage was secured on the spot. The raw material was provided by the mines of Dalmatia and Moesia; the reason for choosing Siscia for the mint was presumably that this town, situated fairly near to Italy, had good connections with the mines.[129] At the same time the mint of Viminacium, the only place to produce colonial coinage of the area of the middle Danube, was closed. This mint of Viminacium, which had been in production from the time of Gordian to the time of Gallienus, in addition to the bronze coinage of *Col(onia) Vim(inacium) P(rovinciae) M(oesiae S(uperioris)* had also struck the bronze coins of the Provincia Dacia.[130]

208

There is no need, I trust, to prove in detail that the appearance of the Goths on the lower Danube and of the Gepidae in the Carpathian basin created an entirely new foreign situation which was soon to destroy the old client-relationships (which had functioned well during the time of the Severi) on the lower Danube and in the eastern fringe of Dacia as well as in the Carpathian basin. The frontal attack on Pannonia in the years 258–60 has quite a few features in common with the first phase of the Marcomannic wars (167–71), when all neighbouring peoples possessing a client-relationship with Rome entered upon a period of unrest. And the reasons for this attack were presumably the same, too. At any rate we must interpret the settlement of Marcomanni in Pannonia in the same way as the settlements by Marcus of Suebi and Cotini in Pannonia after the first war: the old neighbouring people wished to obtain the *receptio* by force, and this *receptio* was granted to a few groups. This time the peoples under pressure were no longer only the Suebi and Sarmatians, but the Vandals and Carpi as well who were exposed to the pressure of the Goths and the Gepidae. After 260 heavy fighting between Gepidae, Vandals and Sarmatians took place in the Carpathian basin, outside the provinces Pannonia and Dacia. The Gepidae who had penetrated into the Carpathian basin soon after the middle of the third century[131] wished to drive away the Vandals, who since the time of Marcus had settled on the northern fringe of Dacia; whilst from the South a new Sarmatian tribe, the Roxolani, invaded the Lowlands. The Roxolani, who were forced to evade the pressure of the Goths, migrated through Dacia to their kindred the Iazyges, presumably by the route that Marcus had permitted them to use. There are some indications that they had already appeared on the border of Pannonia *c.* 258–60.[132] The most seriously endangered area was therefore Dacia, which had the Carpi and Goths as neighbours on the east, the Gepidae and Vandals on the north and Vandals and Roxolani on the west. Under such circumstances the evacuation of part of Dacia under Gallienus cannot be passed over lightly. Under Gallienus the two legions of Dacia, V Macedonica and XIII Gemina, appear with their officers and their administrative ranks in Poetovio, where they richly furnished a Mithraeum with sculptures and votive inscriptions.[133] This shows that after 260 the army of Pannonia had been increased by two legions, apparently for the protection of Italy but at the expense of Dacia, which after the appearance of the Goths and Gepidae had become a cause for war rather than a strong-point for the protection of the Danube frontier. It is possible that Dacia had already been the subject of negotiations between Goths and Romans, and as a result of these had been partially evacuated already at the beginning of the fifties.[134] This episode is, however, little understood as yet, and does not come within the scope of this work.

The relative calm on the Danube after 260 presumably has its cause partly in the mutual struggles of the Gepidae, Vandals and Sarmatians and partly in the preparations of the Goths for the invasion of the Mediterranean by sea. These Gothic attacks on the coastal regions of Asia Minor, Thrace and Macedonia sooner or later made it clear that the preservation of Dacia not only had no strategic advantages for the empire—in particular since the Goths could advance on the heart of the empire—but that it had become a burden on the imperial army. The final proof was given by the great attack of the year 268, when the Balkan provinces were laid waste from the south by the Goths, who did not even spare Athens. They invaded Moesia from the south and advanced along the Thessalonica–Scupi–Ulpianum road as far as Naissus; it was only there that Gallienus, who in the meantime had arrived with the mounted central army, could defeat them. Soon afterwards, Gallienus had to turn against Aureolus, and he left the war against the Goths in the Balkans in the hands of his general Marcianus.[135]

Now came the great moment of the Illyriciani, who, as *protectores* in the entourage of the emperor and in the supreme command of the mounted central army, had gradually got the upper hand during recent years. The exclusion of the senatorial order from the highest posts of the state and army by Gallienus was a favourable circumstance for the Illyriciani. They soon succeeded in getting rid of Gallienus as well as of Aureolus and in proclaiming their candidate, the Dardanian M. Aurelius Claudius, emperor. The leading men of the Illyrican soldiery at that time were the praetorian prefect Heraclianus, the later emperor L. Domitius Aurelianus, and the Emperor Claudius, probably all of them born in Moesia Superior.[136] Whether an expression of the local interests of the province Moesia Superior, which had suffered much, especially from the Goths, is to be seen here may be left undecided. At any rate Claudius soon went to the scene of the struggle in the Balkans, where he remained until his death at Sirmium in the spring of 270.

Claudius died from the plague, which was then breaking out as it had a hundred years before. The highest representatives of the Illyriciani, assembled in Sirmium, did not follow the choice of the Senate in Rome; they refused to accept Claudius' brother, Quintillus, as emperor, but proclaimed Aurelianus the second highest man in their clique. This man, regarded by later tradition as an 'honorary Pannonian', a Sirmian,[137] was to be the man who restored the unity of the empire. To him is due much of the reputation of the Illyriciani as the 'saviours of the empire', as *satis optimi rei publicae*.[138] Among the regional juntas the Illyriciani, indeed, appear to have been the only one to have a feeling for the unity of the empire. Reared in a strategically important part of it, which was strongly occupied by the military, but one whose economy was little developed and which was incapable of maintaining its huge army on its own resources, they were interested in employing the wealth of the economically better developed provinces for the maintenance of the army of Illyricum.

The clearance of the Goths from Illyricum was probably not completed before 270. Aurelian, who in the meantime had gone to Italy, had to return soon to repel a dangerous attack on Pannonia by the Vandals.[139] Possibly the Vandals desired to settle in the empire after having been expelled by the Gepidae. This invasion, costing the emperor a heavy battle and expenditure on food supplies for the retreating Vandals, seriously damaged the province of Illyricum, which during the previous decade had been able to enjoy peace and which towards the end of the reign of Gallienus had showed signs of a consolidation.[140] The next year (271) when Aurelian marched through Pannonia and the Balkans towards the east against Zenobia he had to fight with the Goths who had invaded Illyricum and Thrace. He gained a great victory,[141] but was presumably able also to convince himself finally that disturbances beyond the frontier caused by the Goths and Gepidae could be brought to an end only by radical measures. He now decided to undertake the difficult operation of abandoning to the Goths and Gepidae the bulwark of Dacia, which had long ceased to play its part in the defence of the empire. This was not a sacrifice made by the Illyriciani on the altar of empire. The abandonment of this province, as we have seen, had been seriously contemplated for a certain time already, and when Aurelian ordered the evacuation presumably few troops and a much diminished civilian population remained to be transferred to the right bank of the Danube. Thus Illyricum was able to obtain temporary relief from the great pressure of the barbarians; the empire as a whole gained only an indirect advantage. It was indeed time to solve the problem of the Danube frontier. The provinces of Illyricum—Pannonia, (particularly in 258–60 and in 270), Moesia Superior (in 268–70), Moesia Inferior and Thrace (strictly speaking since the forties of the third century until 270)—had had to suffer so much that there was a danger of losing Illyricum with its irreplaceable man-power once and for all.

The abandonment of Dacia diverted the pressure which had hitherto weighed on the Danubian and Balkan provinces. The Goths and Gepidae began to quarrel about the evacuated province; the Goths and some of their allies emerged victorious.[142] The Gepidae had to give way to the pressure of the Goths by moving westwards and south-westwards, which created a tension in Sarmatian territory. The consequences of this new situation, however, showed up only rather later, when the wars against the Sarmatians began along the *limes* of Pannonia.

The other result of the evacuation of Dacia was the establishment of a new province on the right bank of the Danube.[143] This province was formed by cuts from the territory of Moesia: the north-eastern part of Upper Moesia and the western part of Lower Moesia were detached to create it. Legiones V Macedonica and XIII Gemina which had previously been moved out of Dacia were given fortresses there, and the Romanized population of Dacia was evacuated to settle in it. The fact that the lower Danube *limes* was given two new legions shows that danger from the Goths was reckoned with for the future. The new legionary fortresses (Ratiaria and Oescus) were chosen so as to bar the two most important approaches to the Balkan peninsula: the road leading from Ratiaria to Naissus and on into part of the Balkan peninsula, and the way leading from Oescus through the Isker valley to Thrace and the Aegean.

Chapter 7
The second age of prosperity: rise and collapse

The towns of Upper Moesia

A fresh period of active provincial policy had already begun under Marcus. As was shown in the previous chapter, the rich mines on Mount Kosmaj south of Singidunum were opened under him, probably even before the outbreak of the Marcomannic wars, and were garrisoned by troops from Upper Pannonia. Municipal status was given by Marcus to all the places of importance in the valley of the Morava in Upper Moesia along the road from Byzantium to Singidunum (Fig. 37, p. 220). This road was later to become one of the most important trunk roads of the empire. The exact date of these town foundations is unknown, and so we cannot decide whether these measures of Marcus were induced by experience resulting from the war, or whether he had certain plans, independent of the crisis, which he then was able in part to realize, despite the difficulties created by the war. In Pannonia the measures taken seem without exception to have their origin in the experiences of the war, but it is uncertain whether they were initiated by Marcus or only later by his son.[1] The opening of the mines at Mount Kosmaj fits well with the systematic development of natural resources in Upper Moesia, which had been gradually carried forward since the time of Trajan.

Marcus' programme of municipalization has a number of details in common with the Flavian foundation of municipia in the valley of the Save (Neviodunum, Andautonia, Siscia, Sirmium). The tendency to create bases and route-centres for traffic, on a road the importance of which was recognized, is unmistakable.

The towns of Naissus and Margum were certainly creations of Marcus; the foundation of the third town in the valley of the Morava, Horreum Margi, cannot at present be dated accurately, but its foundation by Marcus would be the most likely possibility.

Naissus was situated not far from the confluence of the Morava and the Nišava,[2] in the area where all important roads through the Balkans of necessity meet: first, the Singidunum–Byzantium road, the principal route in the Balkan peninsula, second, the road from the Adriatic coast at Lissus via Ulpianum to the valley of the Morava, and, third, the road connecting Naissus via the Timok valley with the Danube; of this the terminal points were Ratiaria on one bank, and Dacia on the other. The Lissus–Naissus road was joined by the one from Thessalonica in the Vardar valley via Scupi to the Danube. The successor of Naissus, modern Niš, is also one of the most important route-centres for Balkan traffic. Under Septimius Severus, Naissus is named as the birthplace (*origo*) of a praetorian in Rome, and it must therefore already have been a town with municipal status.[3] From Naissus very few *decuriones* are known, but from the fact that they were Aurelii[4] it can be concluded that the council was composed of native-born new citizens. Nor is there a large number of recorded foreign settlers in and around the town. It lay on the northern bank of the Nišava in the region of the Turkish fortress (Kale) of Niš; one quarter, however, lay on the southern bank and was connected with the centre of the town by a bridge across the Nišava. The bridge was situated in the same place as the modern bridge, near the southern gate of the Kale. Otherwise we know very little of the town. Under the high mound of the Turkish fortress the Roman layers are buried several metres deep, and the excavations so far have done no more than establish the fact of Roman settlement. The *territorium* of Naissus was remarkably large, probably the largest of the *territoria* in Upper Moesia. It included not only the valleys of the Nišava and of the two arms of the Upper Timok (Svrljiški and Trgoviški Timok) but also the middle valley of the Morava between the Klisura and Horreum Margi. To the west the valleys of the Jablanica and the Toplica belonged to Naissus, too.[5] No urban centres sprang up in all these regions, with the exception of Remesiana (Bela Palanka) in the Nišava valley, which was of special importance as the seat of the *Concilium provinciae*, the place where the *Ara Augusti Provinciae* was situated[6]—and later as a bishop's see. Hitherto, however, we have no evidence for the municipal status of Remesiana. Discussion of the problems connected with the birthplace of the Emperor Justinian is beyond the scope of this work. This place in the valley of the Jablanica near Leskovac was chosen for the seat of the vicar of Illyricum and was developed into a town under the name Justiniana Prima. Recent Jugoslavian work has solved the controversy about Justinian's birthplace by excavations on the Caričin Grad near Leskovac.[7] For problems of earlier times, however, information from the time of Justinian is of first importance, since this provides the only means of determining the territorial division of Upper Moesia in the third and fourth century. The size of the territory of Naissus, also, can only be established by means of early Byzantine sources, and below we shall have to refer to these sources frequently. At the latest under Septimius Severus, a road-station (*praetorium*) was established at Naissus for imperial officials travelling on official business; this station was administered by the *stratores* of the governor of Upper Moesia. These *stratores* have left an imposing number of altars.[8]

A *strator* is also attested in Horreum Margi.[9] This town is situated at the point where the road down the valley of the Morava crosses the river. From Naissus down-stream the road lay on the right bank of the Morava, but near Horreum Margi (Ćuprija) it changed over to the left bank, and then, gradually leaving the valley, it crossed the hilly region of north-west Serbia to Singidunum. The crossing of the Morava was by means of a bridge, the piers of which are still partly preserved in the bed of the river. Horreum Margi, too, was a road junction: the great diagonal road across the Balkans had to make a detour in order to reach the seat of the provincial administration of Upper Moesia, the legionary fortress of Viminacium. The Horreum Margi–Viminacium road also left the swampy valley of the lower Morava soon after Horreum Margi, crossing over to the valley of the Mlava.[10] As a nodal point, Horreum Margi had obviously played a certain role from very early on: at least its name shows clearly enough that the place had been a supply-base for the Danubian army at an early date. Of the town itself hardly anything is known. A rectangle measuring *c*. 300 by 350 m, whose walls are, however, late Roman or early Byzantine, could possibly be the successor of earlier walls.[11] The small area (11 ha) enclosed by these walls would correspond well to the size of a small town. The persons known from the few inscriptions from Ćuprija were retired soldiers or civilians, all of them originating from the native population. For the existence of a municipium at Horreum Margi we have two pieces of evidence, one from the time of Caracalla and the other dating to the year 224. The foundation thus falls before Caracalla, either in the time of Marcus or in that of Septimius Severus.[12]

The valley of the lower Morava is very boggy even today. Because of its swampiness the roads which followed the valley left it near Horreum Margi and ran, as we have seen, through the hilly region to the Danube. Nevertheless, a route-centre arose at the mouth of the Morava on the actual bank of the Danube, which had a certain importance as early as the second century. Margum (Orašje near Dubravica) was the seat of a *conventus civium Romanorum* in the second century, for its *curator* is mentioned on an inscription.[13] This shows that Roman citizens early settled at the mouth of the Morava, presumably in response to the growing use of the river for the transport of goods, after the mines of Upper Moesia had been put into operation. The raw material of the mines could be transported to the Danube only via the Morava. Recent excavations have disclosed a small part of the town on the bank of the Morava,[14] but we have no knowledge of its size and appearance. Whether it had an area exceeding 60 ha, as older records would have it, can be neither proved nor disproved. The date of its foundation can be deduced with certainty from its name, municipium Aurelium Augustum.[15] The very small body of inscriptions from the site are of foreigners, mostly freedmen or slaves, and especially of *Augustales*, and this agrees well with the fact that the settlement, as seat of *cives Romani*, had already been of some importance earlier as a centre of merchants and business people.[16]

Mines in Upper Moesia

The opening of the mines on Mount Kosmaj,[17] too, was the work of Marcus, and presumably dates from before the outbreak of the Marcomannic wars. We have already seen that, for the protection of this mining territory, troops were transferred from Upper Pannonia. The inscription of a tribune of cohors I Ulpia Pannoniorum[18] dates from the year 168 at the latest; somewhat later, these troops were replaced by the cohortes Aureliae, newly recruited by Marcus. The mines on Mount Kosmaj show an exceptional feature among the mines of Upper Moesia, which suggests that a new policy was employed towards the mines of the province which may have been the work of Marcus. This feature is the fact that the inhabitants of this mining territory were to all appearance wealthier than those of others, as is strikingly demonstrated by the monumental remains of the Kosmaj area. From the other mines of Upper Moesia either no inscriptions at all (Aelianum, Aurelianum) or very few only (Ulpianum, Dardanicum) are known, whereas the Kosmaj area has yielded a great number of inscriptions which differ in some respects from the body of inscriptions of the province. These differences, also, allow some deductions.[19]

First, the whole body of inscriptions from the Kosmaj area belongs to the period from Marcus roughly to Severus Alexander, though the majority probably belong to the early years of the Severi. In this period very good stonemasons worked there, who had come either from Viminacium or from some town of the Save valley and who produced richly ornamented, probably very expensive grave-reliefs of the south Pannonian–Norican type. This at once

216

allows us to infer a suddenly developed prosperity which enticed good stone-masons to come there. Second, practically without exception, foreigners only are named on the inscriptions, e.g. Dalmatians,[20] Thracians,[21] south Moesians and some from the western provinces,[22] too, a colourful medley which otherwise can be proved to have existed on the Danube in the Flavian–Trajanic period only. There is a remarkable number of peregrini among them, considering that in Upper Moesia in particular peregrini very rarely set up inscriptions. Third, the inscriptions are amazingly full of mistakes. There exists hardly any text that is not incorrect in one way or another, and several are entirely incomprehensible. All this can only mean that, under Marcus or a little later, excellent opportunities for profit were available at the mines of Mount Kosmaj which attracted a colourful mixture of foreigners. The prosperous conditions attracted hither good stonemasons, too; but the patrons—part of them peregrini—were either uneducated people or people who did not understand Latin. The erection of inscriptions had become a fashion in which some even indulged who were incapable of reading and understanding their own texts.

The reasons for this economic blossoming are not known. One might suggest that the output of the newly opened mines was abundant, or that Marcus granted the miners particularly favourable conditions. Neither reason seems unlikely, since a new mine with rich resources of lead and silver situated not far from the Danube and from the road-centre at Sirmium was apparently considered by the government to be an economic asset of first importance; and in that case it will not be surprising to find these particular mines protected by auxiliary troops.

The economic prosperity of the Kosmaj area we can easily trace back to local factors. A similar prosperity, observable in the Danubian provinces in general, has causes that are far more complex and to be interpreted only when studied on an imperial scale. These causes have already been discussed in the previous chapter. Reconstruction and the awakening self-confidence of the Danubian military, who played the leading part in the struggle for power under Septimius Severus, led to an economic upsurge in the Danubian provinces, and especially in their border area; this depended not merely on the prosperity of the local economy but also on resources drawn from the wealth of the empire as a whole. We are as yet far from being able to define at all accurately the mechanism of this process, but must now attempt to characterize its main features.

Urban development in the third century

The first certain indications of the intention of the Severi to follow an exceptional policy towards Pannonia can be seen soon after the accession of the first

Severan emperor. Carnuntum and Aquincum, the two seats of provincial administration in Pannonia, were granted the rank of *coloniae* in the year 194,[23] a distinction previously bestowed only very rarely; Hadrian, for example, found it difficult to understand why towns should apply for this rank. Whether the title of *colonia* in the case of Carnuntum and Aquincum involved also a grant of the *ius Italicum* or something similar, we do not know. One very important change did, however, take place with the granting of colonial rank, and this change should be interpreted as a favour to the settlements outside the legionary fortresses. As far as can be seen, it did not affect Carnuntum and Aquincum alone but all settlements outside the fortresses of the Danube in so far as municipal self-government existed among them already.

As we have seen in Chapter 5, at those legionary fortresses which were also the seats of governors there had existed since at least the time of Hadrian two autonomous communities side by side: the settlement around the fortress itself (*canabae*) and distinct from it, and situated a little distance off, the municipium Aelium (Fig. 22). Both had local autonomy with the necessary corporation and magistrates. Under Septimius Severus the *canabae* ceased to exist as separate autonomous units and were either given a municipal type of constitution of their own, as at Dacian Apulum, or they were incorporated into the municipium. Henceforth only one municipal community existed at these legionary fortresses and this incorporated the settlement around the fortress. Only at Aquincum and Carnuntum, the seats of the two Pannonian governors, is it attested that the community thus enlarged was at the same time given the rank of *colonia*; the other municipia still retained their old rank for some time, like Viminacium, which became a *colonia* only under Gordian III.[24] It is remarkable that at Viminacium, according to building-inscriptions of Septimius Severus and Caracalla, the *canabae* continued to exist; these emperors ordered the *canabae* of Viminacium to be rebuilt.[25] The stone slab on which this is recorded is, however, remarkably small, and from this we may infer that the newly built *canabae* were not comparable with the very extensive second-century settlement *ad legionem*. This conclusion is supported by the fact that we have no epigraphic proofs at all of the existence of *decuriones* or magistrates of the *canabae* for the time of the Severi. The *canabae*, therefore, ceased to exist as a type of community, and only a very small settlement, possibly consisting of only a few houses, continued under the control of the legion.[26]

It is not possible to say why this change was necessary. It may hold true in general that Septimius Severus diminished, or even practically abolished, those lands in the frontier provinces which were under military control, the reason being that these *territoria* were no longer able to secure the provision of the army. This emperor, being an innovator in the field of army provision, introduced the *annona militaris* as a special tax mainly on natural produce, which was very oppressive on the civilian; this must imply radical reform of the old system of army supply.[27] But these reforms did not necessarily involve the municipalization of the *canabae*, since even under Hadrian they had only a partial and indirect place in the provision of the army. Presumably, other reasons were decisive; we may suggest the permission to marry, which was the legal recognition of a situation which had long existed *de facto*. As long as the *canabae* remained as a special type of community, the members of soldiers' families were in theory citizens of another community, and possibly one where not even their grandparents had been living. The municipalization of the *canabae* abolished this complex legal status of the soldiers' families which did not correspond to the actual situation. Thereafter the legal status corresponded to the true one: the soldiers' families were citizens of the community in which they had settled in former days. At the same time the municipalization meant an improvement of position, since the *canabae*, despite their pseudo-self-governing type of constitution, were not properly autonomous, in that they could not be the *origo* of Roman citizens.[28] It is possible, too, that between the two communities which existed after Hadrian in the vicinity of the legionary fortresses tensions arose from various reasons. As communities under military control, the *canabae* were deprived of the glamour of independence, but in some respects they were privileged, and among their inhabitants there were a few rich people who should have offered their riches in the neighbouring municipium for purposes of that community.

Before entering further into the foundations of towns by the Severi, mention must be made of some measures taken by Septimius Severus in Pannonian towns, which suggest favour towards Pannonia. The colonia Siscia was henceforth called colonia Septimia Siscia;[29] this suggests some significant measure there. No details are known, but we might suppose some reform of the finances of the town, the grant of certain privileges or some similar act. Something analogous was undertaken in Savaria, where a *Kalendarium Septimianum* is attested, administered by a *vilicus*, who was a municipal slave.[30] The word *Kalendarium* perhaps denotes a particular kind of book-keeping, presumably of special funds donated to the town of Savaria by Septimius Severus.

Further foundations of towns by Septimius Severus are not unequivocally attested; but this by no means indicates that he did not create any new municipal communities in his favourite country. After the second half of the second century, however, mention of the *tribus* of a citizen on inscriptions becomes exceedingly rare, and the imperial by-name of the town, too, is mostly omitted from

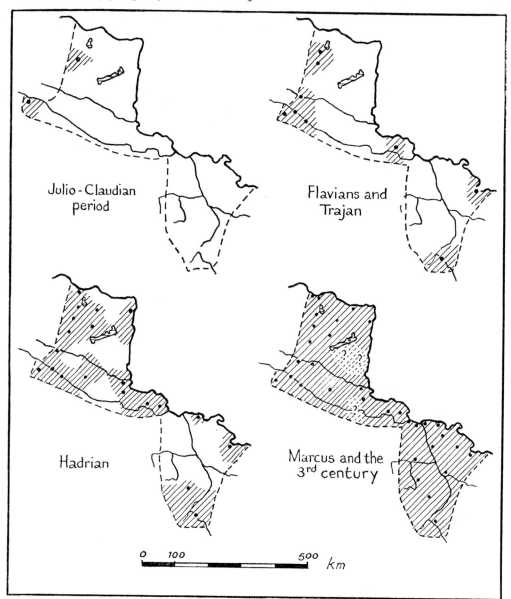

Figure 37 The progress of urbanization

inscriptions. The foundation date of a town can be learnt from its being named on a dated inscription. Since these dates, too, merely provide a *terminus ante quem*, it is best to enumerate here all foundations of the third century, because everything suggests that the provinces of Pannonia and Moesia Superior were entirely municipalized by the end of the third century—that is to say, their entire area consisted of municipal *territoria* (Fig. 37). Since the majority of the new towns are named as municipalities for the first time under Caracalla or a little later, it is very probable that they were promoted to the rank of municipium under Caracalla, presumably as a result of the *Constitutio Antoniniana*. As a result of this enactment the *civitates peregrinae* as such vanished but continued to exist as *de facto* municipal communities; the moment at which they became a municipium in the formal sense presumably depended only on the ability to assemble a town council (*ordo*).[31] At any rate we do not know of any non-municipal communities after Caracalla,[32] and everything suggests that sooner or later—but most of them during the time of the Severi—all civitates were converted into real municipia with *decuriones* and magistrates. The incorporation of the *canabae* in the adjacent municipium or colonia, whichever it might be, already illustrates a tendency to abolish differences between types of communities, in other words to create a common denominator for the various territorial authorities. This trend led to the very simplified system of the late Roman period, when all autonomous communities were called civitas, and words like municipium and colonia became so devoid of meaning that they were hardly used any more, for instance in the *Codex Theodosianus*.[33]

In Pannonia Superior, Vindobona and Brigetio, the two other places with legionary fortresses, probably did not share in the improvement of status under Septimius Severus. Brigetio is named *Municipium Brigetio Antoninianum* on an inscription,[34] and a foundation by Caracalla can be inferred. Certainly by 217 Brigetio was a municipium.[35] At that time, or possibly later still, Vindobona also became a municipium; of its *decuriones* and magistrates, only one is as yet known.[36] In Vindobona as well as in Brigetio the duality of settlement, described in Chapter 5, was probably completely developed before the end of the second century, but the settlement lying further away from the fortress was given municipal status only in the third century; the grant, however, will have also included the *canabae* in this improvement of rank. The pseudo-autonomous organizations of the *canabae* at any rate ceased to exist in Brigetio, too, and some of the *decuriones* of the municipium Brigetio had themselves buried in the former *canabae*, and furthermore they erected buildings there; from this it must follow that they were citizens and inhabitants of the settlement around the legionary fortress.[37]

Further municipal communities of Upper Pannonia which obtained this status probably only during the third century were Jovia on the Drave east of Poetovio and the municipium Faustinianum somewhere east of Siscia. If these towns are entered on a map together with the rest of the earlier foundations (Fig. 37) it can be seen that only one area of this province does not contain any municipium known to us: this is the area south of Lake Pelso where we located the Cotini in Chapter 3. Here only Tricciana (Ságvár) is a possible candidate, but up to the present no proofs of its municipal status are available. Furthermore it is not certain that Tricciana lay in Upper Pannonia.

Jovia, probably known originally as Botivo, is mentioned in the *Itinerarium Burdigalense (Hierosolymitanum)* as a *civitas*,[38] showing that it must have become a town at the latest under the Tetrarchy. No more is known about it, except that it was a bishop's see in the fourth century.[39]

The municipium Faustinianum is mentioned as the home of a praetorian and an *eques singularis* in Rome;[40] in addition, its *scriba* is named on an inscription in Siscia.[41] The find-spot of this inscription itself is sufficient to suggest that this municipium must be sought somewhere in the vicinity of Siscia. And since the valley of the Save above Siscia was already completely municipalized, only the valley below Siscia comes into question as its *territorium*. Siscia incidentally furnishes a further illustration that in the small Pannonian towns the *scriba* was of high social position and as the *de facto* director of municipal affairs was presumably fairly rich, too. The *scriba* of the municipium Faustinianum was at the same time Augustalis in Siscia.

In Pannonia Inferior, the only new town to be mentioned is Mursella on the Drave west of Mursa. This town is referred to on a single inscription only,[42] which mentions a *decurio* M. Aurelius. However, its foundation is not easily understandable: it was only ten or twelve miles from the Hadrianic colonia Mursa, and—possibly for this very reason—it is not recognized as a town in the *Itinerarium Burdigalense*, which specifies the ranks of places (*civitas, mansio, mutatio*). Moreover, there is a possibility that the *decurio* on the inscription was in fact a town-councillor of the town of the same name in Upper Pannonia. If so, Mursella (Mursa minor) in Lower Pannonia should not be counted among the municipia.

The map of Lower Pannonia shows more gaps than that of the Upper province. In the area of the civitas Breucorum no urban centre is known as yet, nor is a municipium known in the area which was formerly the civitas of the Hercuniates. The municipium Volg (. . .) could be claimed for the Hercuniates, but it is entirely unidentifiable, since it is only named on an inscription from Intercisa.[43] As a town with municipal status, Sopianae (Pécs) on the southern slope of the Mecsek mountains comes into consideration. As early as the time when the Italians started to extend their trade-connections in Pannonia, it had been of importance as a route-centre,[44] and under Diocletian it became the seat of the civil administration of the province of Valeria.

In Moesia Superior the civil settlement of the legionary fortress of Singidunum was also made a municipium, presumably under Caracalla.[45] The topography of the Roman settlements under modern Beograd is not yet altogether clear.[46] It is known that the legionary fortress lay on the site of the later fortress at the confluence of Danube and Save (Kalemegdan, Tvrdjava). Traces of a civil settlement on the southern and eastern outskirts of the fortress have already been established, while uncertain traces further east still may perhaps indicate a second civil settlement. Here, too, the municipium presumably comprised both settlements.

Of the remaining municipia of Upper Moesia, two can be dated to the period of the Severi and presumably to the reign of Caracalla, the only basis of the dating, however, being the fact that their *decuriones* carry the name M. Aurelius. But as only one *decurio* of either of these towns is so far known, even greater caution is needed in determining the foundation date of others. The two towns where *decuriones* are known are the municipium Dardanorum or Dardanicum in the valley of the Ibar[47] and the municipium Celegerorum in the valley of the western Morava. The identification of municipium Dardanorum, whose name was long known in the abbreviation *mun. DD*,[48] with Sočanica in the Ibar valley became possible only recently. Here had lain the seat of the administration of the *metalla Dardanica*. Of the municipium Celegerorum hardly anything is known. The tombstone of its only attested *decurio* was discovered long ago, but its findspot (Ivanjica) poses a number of problems, since the valley in which it lies opens only towards Dalmatia, and it would have been difficult in this secluded area for a town to come into existence. The *decurio* may have had his estates in the valley of the Moravica, but municipium Celegerorum must be looked for somewhere in the valley of the western Morava (perhaps at Kraljevo).

The nomenclature of these two Upper Moesian towns recalls the Pannonian municipia Latobicorum and Iasorum. We might therefore deduce that both had been formed from *civitates peregrinae* (of the Dardani and the Celegeri). But this is true only of municipium Celegerorum, for municipium Dardanorum goes back to the *metalla Dardanica* and this is the reason why we cannot exclude the possibility that its real name was municipium Dardanicum.

The remaining three Upper Moesian towns were also based on former mining districts. South of Viminacium in the valley of the Mlava the *Itineraries* indicate a station named Municipium; in this area had lain the *territorium* of the *metalla*

223

Aeliana Pincensia.[49] In the region of the *metalla Aureliana*, however, we have
knowledge of a place called Aureliana,[50] which was presumably the centre of an
otherwise unattested municipium Aurelianum. The assumption that a municipium
existed in the area of the *metalla Aureliana* is justified by the fact that in early
Byzantine times a separate χώρα (chora) with its centre Aquae existed here,[51]
and that the area of this χώρα consists of parts of the province of Upper Moesia,
which cannot be attributed to any of the known urban *territoria*.[52] Since early
Byzantine territories (of the time of Justinian) are directly derived from the
urban territories of imperial times, a precursor of the *chora* of Aquae has to be
presupposed. By analogy with municipium Ulpianum (=*metalla Ulpiana*) and
municipium Dardanorum (=*metalla Dardanica*) we can assume that the *metalla
Aeliana* became a municipium Aelianum and the *metalla Aureliana* a municipium
Aurelianum; supporting evidence is given by the facts that a municipium is
attested in the area of the *metalla Aeliana* and that Ulpiana, the late antique version
of the name Ulpianum, corresponds in form exactly to the place-name Aureliana.
From the order in which the place-names of the *chora* of Aquae are given it can
be inferred that Aureliana was situated somewhere near the confluence of the
White and the Black Timok; this was the natural centre of the *chora* and is where
today the most important place of the Serbian Krajina, Zaječar, lies. Not far from
Zaječar is a well-known Roman site, Kostol, where a large, palace-like house has
already been excavated (Fig. 49) and where older notes report a fairly large
Roman settlement (Fig. 31) (of *c.* 16 ha).[53] The municipium Aelianum was
situated in the valley of the Mlava, near the modern Kalište, Boževac and Batuša.[54]

The third town of late foundation, also based on the municipalization of a
mining district, was Aureus Mons (Seona) on the bank of the Danube east of
Singidunum. It is listed as a *civitas* in the *Itinerarium Burdigalense*.[55] Quite near to
this otherwise unimportant town lay the mines of Mount Kosmaj, which most
probably were incorporated in the territory of Aureus Mons. We do not know
when this occurred, but it was certainly only after the time of the Severi, since no
trace of a municipal administration is extant on the numerous inscriptions from
the period when this district of mines enjoyed its greatest prosperity.

As we have seen, the municipalization of Upper Moesia was based to a considerable extent on the mines. At first glance it does not seem probable that the
mining districts (*territoria metalli*) became municipalized, as this involved removal from direct imperial administration; but a law of the late Roman period
refers to the Upper Moesian mines in particular as being administered 'according
to ancient custom'[56] by the *curiales* of the towns. On an inscription the *metalla
Dardanica* are called *Metalla Municipii*, proving that the municipalization of these
mines had already occurred in the third century.[57] Whether this reform of ad-

ministration has to be attributed to a decline in the production of the mines cannot be decided, but it is noteworthy that the undeniable prosperity on Mount Kosmaj lasted only to the municipalization.[58]

It is not improbable that the settlements which had sprung up near the mines were the only urban settlements in the areas in question, or more accurately, the only centres on which a municipal autonomy could be built. This is perhaps the easiest way to explain the municipalization of the mines. Moreover, the town foundations of the third century were not able to produce towns in the true sense of the word whether in Pannonia or in Upper Moesia. The difficulty encountered in trying to determine the number or even the sites of these towns is caused by the absence of any epigraphical or archaeological traces worth mentioning. We have only to recall our ignorance of a single *decurio* at a considerable number of these towns (Aelianum, Aurelianum, Aureus Mons, Horreum Margi, Jovia), and the fact that the remaining towns have mostly produced only one *decurio* on inscriptions (mun. Dardanorum, mun. Celegerorum, mun. Volg . . ., Mursella, municipium Faustinianum); whereas from towns which truly deserve the title of 'town' a considerable number of *decuriones* can be listed.

In this connection it is a little obscure why the centres which already existed in some districts had not obtained municipal status. To all appearance, Aureus Mons was much less important than the mining settlement on Mount Kosmaj, and in the area of Aurelianum Timacum minus, a civil settlement of importance at the junction of the Naissus–Ratiaria and Naissus–Aurelianum roads had developed near the fort of cohors II Aurelia Dardanorum.[59] In the same way, not one of the more important *limes*-forts on the Danube was chosen as the centre of the municipalized civitas Hercuniatium; instead municipium Volg(. . .) was possibly selected, whose position though unknown was certainly not on the Danube, since it is not listed among the places on the *limes*-road either in the *Antonine Itinerary* or in the Peutinger Table. We might possibly conclude that the government tended to grant municipal rights only to civil settlements, but not to settlements round forts. Aureus Mons was in fact probably situated near an auxiliary fort, but even more troops were stationed at Mount Kosmaj.

As well as new foundations, promotions of towns to higher rank must be mentioned. These were apparently based on entirely political considerations. As already mentioned (p. 218) Septimius Severus granted the provincial capitals of the two Pannonias the rank of *colonia*; during the time of the Severi, Cibalae and Bassiana also became colonies,[60] then Viminacium[61] under Gordianus in 239, and Singidunum and Brigetio[62] at an unknown date towards the middle of the third century. Whether further towns were granted the rank of *colonia* is an open question; at any rate the promotions thus far clearly illustrate motives based on power-politics. First, the provincial capitals, whose troops had helped Septimius Severus to the throne, were distinguished; not much later the towns Cibalae and Bassiana in the Sirmian region—a proof that the significance of this area was clearly recognized by the Severi; then followed the capital of Upper Moesia, Viminacium, and finally the towns with legionary fortresses, Brigetio and Singidunum. And this means that only the towns with legionary fortresses and towns in the surroundings of Sirmium were distinguished with the rank of *colonia*; all of them were towns of particular relevance to the basis of imperial power in the third century.

Town foundations, of course, only created the administrative and juridical frame of a communal life, but the latter could never be created solely by the formal procedure of foundation. Although in the third century some communities of the empire—above all the towns proper who could demand full autonomy—had a preference for emphasizing that they formed a *res publica*— a community of highest jurisdiction—and although some Africans, Spaniards or Orientals may have cherished the utopian view that the empire was a conglomeration of towns equivalent, autonomous and possessed of equal rights,[63] nowhere had these ideas less foundation than in some parts of the Danube provinces, where urban life with all its requisites could never arise. Even in those towns which as military *coloniae* (e.g. Ratiaria) or as *canabae* (e.g. Aquincum) produced an active communal life and became important centres of trade and industry, it can be noted that the municipal upper class and in particular the *decuriones* often owed their wealth and social position not to participation in local economic life, but to the circumstance that they represented some branch of the administration of the empire. In Ratiaria the customs-farming families of Sabinii, Julii and Antonii played a leading part in the *ordo* of the *colonia*,[64] and the customs-farmer Titius Antonius Peculiaris was one of the richest decurions of Aquincum and had connections with the municipium of Singidunum as well.[65] It is obviously one of the most characteristic traits of urban life on the Danube that wealth and prosperity were not based on the sudden growth of the local economy but rather on participation in the central power or simply in the administration. Wealth was derived from empire-wide sources and not from those of local production. We shall refer to this feature of the Danube provinces more than once in this chapter; here a short summary was necessary, because one important characteristic of town life is particularly connected with this feature. As we have seen, the majority of the newly founded towns were incapable of producing a municipal life. The decurions of these towns had neither the material nor the cultural background to support either the wish or the ability to immor-

226

talize themselves on inscriptions; only the *canabae* of the legionary fortresses figure as an exception to this rule. There was more than one reason why municipal status was first granted to them only under the Severi, but even before this they had been able to produce an active communal life.

Syrians in Pannonia

During the time of the Severi a very extensive oriental immigration becomes perceptible in the towns in the true sense of the word which has not yet found a final explanation. The possibility has frequently been considered that this immigration was part of the programme of reconstruction after the Marcomannic wars; organized resettlement has also been suggested as a possibility on the supposition that the Orientals had not voluntarily come to the Danube. In general historians are in agreement that the Orientals contributed significantly to the economic prosperity of the Danube provinces in the time of the Severi. Intercisa, the fort of cohors I Hemesenorum, is generally referred to, where, as a result of reinforcement of the regiment from Syria, a Syrian enclave came into existence; and this enclave, according to the abundant body of finds and monuments,[66] created a prosperity unparalleled at other sites with auxiliary forts on the Pannonian *limes*. Thus arose the attractive picture of the reconstruction of a Pannonia laid waste in the Marcomannic war: with Marcus already transferring oriental troops to Pannonia, and the civilian Orientals then following, induced to migrate to Pannonia by the government in order to employ their long-renowned business acumen and their economic ability for the benefit of a province that had suffered severely. And it is true that Orientals are much more often attested in Pannonia than in Upper Moesia, which had been much less exposed to barbarian attacks.

To this picture some corrections must be made. First, oriental immigration is already detectable in the second century. The *decurio* of Carnuntum, Domitius Zmaragdus, who had the amphitheatre built at his own expense around the middle of the second century, came from Antioch;[67] the freed woman of a Syrian named Bargates can be traced at the Danube bend as early as the turn of the first century;[68] Syrian soldiers served on the Danube even before the time of Marcus, and so on. It is well known that the large-scale colonization by Trajan of newly conquered Dacia brought (or rather lured) a great number of Orientals to that province, and this oriental immigration left traces in the two Moesias, too, where Orientals are as well attested before Marcus as afterwards.[69] The Orientals who reached Pannonia in the time of the Severi were following a route trodden by Orientals for at least a century, and we must not forget the part played by

Aquileia: this port, like her successor Venice, was the greatest gateway to the east in Europe.

In the second place, it is presumably no coincidence that the Orientals did not settle in the towns which needed 'economic aid' most urgently: they are absent in the small towns that had been founded a short time previously or were then just about to be founded, but they gathered all the more in those towns that were flourishing centres of the province anyway.

It cannot be determined when the actual immigration of the Orientals into Pannonia began at this period. The earliest immigrants are attested under Septimius Severus, excluding, of course, the soldiers and their families belonging to the cohors Hemesenorum, who made their appearance at Intercisa under Marcus or Commodus. There is also the possibility that immigration on a larger scale started only when the legions of Pannonia were employed in the Orient against Pescennius Niger and then against the Parthians. It is known that the legions of Pannonia, and those of Upper Moesia too, brought oriental soldiers with them on their return from the East.[70] These men presumably brought their families with them, and thus a longer period of immigration could have begun.

There was however an undoubted immigration of completely civilian elements, above all Jews, inhabitants of Asia Minor and especially Syrians. Sporadic epigraphic evidence proves the existence of larger or smaller oriental settlements in Sirmium, Mursa and Savaria. This evidence is for the most part confined to tombstones,[71] but in Mursa a Jewish house of prayer is attested, which was reconstructed in the reign of Septimius Severus,[72] and a synagogue must be assumed to have existed in Intercisa, where the head of the *synagoga Iudaeorum* was a customs official named Cosmius.[73] Remarkable is the sporadic occurrence of Jews in north-eastern Pannonia in the area round Aquincum,[74] though inside the town they are less well attested. A small Syrian enclave was formed near the auxiliary fort Ulcisia Castra, where in the period of the Severi a Syrian cohort was stationed.[75]

As one can deduce from the epigraphic evidence, these enclaves lived lives which were rather isolated from the local population. The Syrians in Intercisa married only among themselves, their sole outside links being with Syrian families in Aquincum; the reason for this is mainly that some soldiers' sons from the circle of the cohors Hemesenorum were enlisted in legio II Adiutrix. The Syrians in Ulcisia Castra, too, had connections only with the Syrians of Aquincum. This isolation is all the more striking since the Syrians in Aquincum and especially in Brigetio had worked their way up to membership of the *ordo decurionum*,[76] and at the latter town became, to all appearance, the richest and most important decurions of the town. Aquincum and Brigetio were the two largest centres of oriental immigration, where newly arrived Orientals can also be proved in the later years of the Severi.[77] Nor did immigration cease after the Severan age; even in the fourth century fresh groups can be proved to have arrived.[78]

The markedly rural origin of the Syrian immigrants is noteworthy. On their inscriptions they often state their origin with full details, not only of the city *territorium* but also of the native village within it.[79] The possibility might therefore be entertained that these Syrian inhabitants of villages had been soldiers who after their discharge had made their way through a municipal career, and as such no longer recorded their military service. This possibility, however, can be confidently dismissed, since the Syrian decurions of Brigetio did, in fact, sometimes state that they were veterans,[80] and the circumstantial statement of the native place of origin would not be imaginable without an equally circumstantial statement of the military career. Thus it seems more probable that some Syrians came to Pannonia as civilians, but it remains to be explained why they came in such great numbers, and why during the time of the Severi in particular.

As we have stated above, Orientals, and among them Syrians, were already represented in the municipal upper class of Pannonia and other Danubian provinces during the second century. Relations with the Danube area were thus already developed, and Syrian merchants were also to be found more and more frequently in the provinces of western Europe. An emigration from Syria was therefore going on, but it certainly cannot be attributed to some decision of imperial policy. Furthermore, we must note that Orientals appeared in great numbers under the Severi in Pannonia in particular, but not in other Danube provinces. In these a sudden increase of immigration during the time of the Severi can be proved only inasmuch as this period has yielded most of the inscriptions, and therefore epigraphic evidence for Orientals is virtually confined to it. Only in Pannonia is a preponderance of Orientals in the *ordo* of certain towns or more generally in the population of some of the settlements established for Severan times. Thus we may conclude that the Syrians, who knew very well how to win the markets of the empire and to exploit economic possibilities offered by some of the provinces, found in Pannonia a particularly favourable field for activity and were free to develop it. Since we are here concerned with Pannonia in the period of the Severi we must associate oriental immigration with the economic upsurge which occurred there at this time, and which lured businessmen, and above all of course the Syrians who were the most active of them all, to Pannonia.

Thus we must interpret the oriental (Syrian) immigration into Pannonia in the context of a sudden general development, but we cannot recognize in them the cause or motive power of this boom. Other immigrants are, of course, attested too, especially Thracians, Africans and, sporadically, elements from other parts of the empire. These were, however, less significant both numerically and socially, and indeed their arrival has a close connection with certain troop-movements. Africans were brought to Pannonia as early as under Marcus, and at the same time some Italians entered the legions.[81] The Thracians were for the most part incorporated in Pannonian units when the latter marched through Thrace on their way to the east.[82] The Thracian element in the military was very strong in the time of the Severi.[83]

Prosperity in the early third century

Immigration in the Severan period is only one of many symptoms of the economic advance which created the second and greatest period of prosperity in the Danube provinces, and in the province of Pannonia in particular. The causes of this prosperity, in so far as they lay in central power-politics, have already been discussed in the last chapter. It now remains to set out the further evidence.

The most conspicuous phenomenon is the exceedingly large number of inscriptions from the time of the Severi. During the Severan period the fashion of erecting inscriptions was widespread throughout the entire empire. We must not therefore attach too much significance to the mere preponderance of inscriptions at this time. The fact that the large number of inscriptions was, nevertheless, a symptom of economic prosperity emerges only from their varying frequency from territory to territory. Since inscriptions have been collected or at least copied for centuries, whereas the systematic archaeological investigation of the ground in some parts of the Danubian area is a comparatively recent development, the epigraphic sources have a higher value as evidence for the identification of regional differences. And the composition of the epigraphic material is also significant: in contrast with pre-Severan times, votive, building- and honorific inscriptions are very frequent, permitting deductions of several important changes in public life, in religion and in the social structure.

On the whole it is true to say that the majority of Roman inscriptions in the Danube provinces belong to the Severan period. But this verdict is by no means valid for towns like Poetovio, Savaria, Siscia, Sirmium, Ulpianum and Scupi whose inscriptions for the most part date from earlier times.[84] In Brigetio, Aquincum, Singidunum, Viminacium, Timacum minus, Ratiaria, Intercisa and in some parts of the border zone of the Danubian *limes*, however, inscriptions

230

were first erected in large quantities in Severan times; while in several small towns in the interior of the province of Moesia Superior the setting up of inscriptions was begun for the first time only under the Severi, for instance at Horreum Margi, Naissus, Municipium Celegerorum or in the Kosmaj area. Inscriptions were set up regularly only after a community life in the Roman manner had sprung up, and present and future appreciation of them could be relied upon. The formation of autonomous Roman municipalities in the interior of Upper Moesia thus provides sufficient explanation for the appearance of the fashion for inscriptions. But in settlements where community life had already been consolidated during the first or second centuries the sudden increase of inscriptions will have to be explained in different terms.

To start with it is no mere chance that the accumulation of inscriptions particularly attests prosperity in those settlements which, because of the presence of a fort, had military inhabitants. In the towns in the interior of the provinces, containing an entirely civil population, this economic boom cannot be traced, at least not with the aid of inscriptions; and this is another indication of the character of the boom. Here we come to the second question, which faces us when we consider the composition of the body of inscriptions. As already mentioned, the epigraphic material of the Severan period is remarkably varied: it consists not only of inscriptions on tombs—though these were in the majority in the period before the Severi—but of a considerable number of altars, building-inscriptions and various other kinds of inscriptions as well, all of which were rare in the pre-Severan period. In earlier times altars and other votive inscriptions were, in fact, erected only exceptionally: those setting them up were mostly town magistrates, governors and the personnel of the *portorium*. This fact allows the conclusion that these *vota* were official monuments. The deities honoured on these inscriptions were gods and goddesses of the official Roman pantheon; local traits did not enter the dedications or were official too, such as Danuvius, Diana and other gods honoured by imperial officials. In some towns this situation remained unaltered in the Severan period, for instance at Scupi.[85] Elsewhere, however, particularly in the towns along the Danube *limes*, altars for gods not previously attested in the province become frequent; the pantheon that can be traced in the inscriptions becomes very colourful and displays local traits too, while among those setting up altars practically all social layers are represented: there are huge altars and imposing bases of statues and very small primitive altars, presumably produced by the person setting the altar up himself, with scratched writing of cursive character. The building inscriptions point in the same direction. Most of these are dedications from temples and shrines[86] built by a very varied selection of the population. Their concentration in the time of the Severi must be partly attributed to a contemporary fashion for setting up inscriptions. A respectable number indicate not new buildings but reconstructions of buildings, most of which are presumably pre-Severan in origin. These buildings had originally lacked an inscription; building-inscriptions of pre-Severan date are very rare. Nevertheless, to some extent it is possible to deduce a lively building activity from the large number of Severan inscriptions, and this is corroborated by direct archaeological evidence.

The inscriptions of Severan times thus attest an undeniable prosperity among the provincial population of the frontier regions. A large part of the population, including simple soldiers without rank or office, were able in this period to afford the erection of grave-inscriptions (Fig. 38), and they were capable, too, of displaying building activity and a certain munificence; but the fact that this prosperity did not extend over the whole territory of the province follows from there being hardly any traces of it at all among inscriptions from the inner parts of Pannonia and Moesia. Moreover, the suddenly increased quantity of inscriptions shows traits that allow deductions about important changes in society.

To start with we have to draw attention once more to the votive and building-inscriptions. As already stated, dedications and buildings of private character became frequent. A consistent distinction, however, between private and official dedications is neither possible nor desirable, for there exists an imposing number of votive inscriptions, set up for the welfare of the emperor or some regiment or corporation, the erectors of which were private persons or ordinary soldiers. All votive dedications have something public and impersonal about them, and so, if votive and building-inscriptions become frequent in general, an intensive participation of wide classes of society in public life can be deduced. From this point of view, too, it will not be possible to consider that their concentration in the frontier area is entirely accidental: the politically favoured soldiery of the Severi were stationed in this zone which, as we have shown, was also possessed of material privileges. The fact that this rapid development can be demonstrated much better in Pannonia than in Upper Moesia proves the economic character of its causes. Among the Danubian armies the Pannonian military forces were the most privileged since from the view-point of power-politics they possessed the greatest importance. In the undeniably large body of inscriptions from Singidunum or Viminacium, ordinary soldiers or simple private persons are much less represented than in Pannonia.[87]

We must now consider aspects of religious history which are revealed by the dedications of the Severan period. To the economical and political prosperity of the frontier area is due the appearance of local traits in religion. The earliest indications of these date back to the first year of Septimius Severus' reign, when

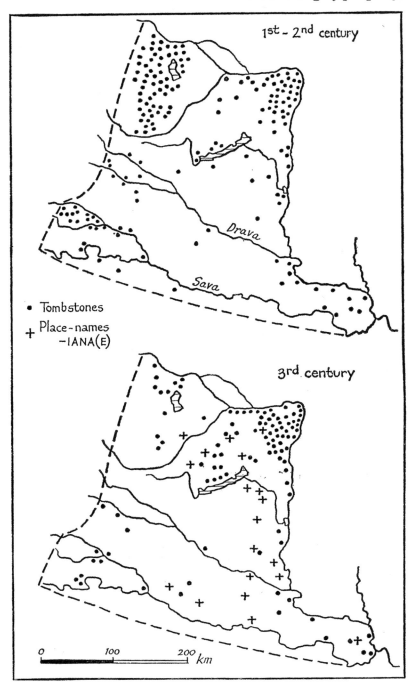

1st - 2nd century

Drava

Sava

• Tombstones
+ Place-names
  −IANA(E)

3rd century

0   100   200 km

Figure 38 Distribution of tombstones in Pannonia

the Matres Pannoniorum were honoured at Lugdunum, and the Pannonian augurs announced the victory of the new emperor.[88] The increasing political activity of the Pannonian military thus also finds its expression in the numerous votive inscriptions. The strictly Roman form is always used, and this is connected with further characteristics of the inscriptions of the Severan period.

As we have already seen in Chapter 5, grave-inscriptions are known in great numbers from some parts of Pannonia and from the Metohija. They were erected by native inhabitants who mostly carried local, non-Latin (Celtic, Illyrian or Thracian) names. In the area of distribution of these early-imperial funerary inscriptions grave-inscriptions are equally frequent in the period of the Severi; with negligible exceptions, however, the personal names are colourless Latin ones. A few Celtic names,[89] which exceptionally turn up among the families which erected these tombstones, show that these were descended from the native families which had earlier been named on tombstones at the very same place. But now the local traits of native culture, which had been deliberately cultivated and preserved in the second century, were abandoned; a Latinization of names was taking place which went hand in hand with the abandonment of further local customs, such as of wagon-graves, *tumuli* and such peculiarities in the sculpture of tombstones as astral symbols, in the same area. Since nomenclature was only rarely and sporadically Latinized in the second century, we must recognize in its sudden Latinization an important process, which, in the last resort, was based on the desire of the indigenous population consciously to adapt itself to Roman forms.

A comparison of the zones where inscriptions of the first and second centuries have been found, with those in which those of later date are distributed, immediately shows (Fig. 38) that they are roughly identical, the difference being that in the south and west, that is in the interior of Pannonia, the find-spots become more scattered, whereas in the frontier areas they are either equally or possibly even more frequent than they were before. But it is curious that the stations established for *beneficiarii*,[90] whose network was developed precisely in the early Severan period, can be demonstrated only in those areas of the country where no or very few inscriptions were erected (Fig. 39). The security service was presumably most necessary in those places where it was to be feared that inner tensions would lead to movements of unrest and robbery—both called *latrocinium*.

The prior condition for a larger number of tomb-inscriptions is a social class that is strong in numbers and in wealth. Where grave-inscriptions are rare or altogether absent, the most prosperous upper social layer was small in number; in other words, society was in its economic aspect strongly polarized. It cannot well

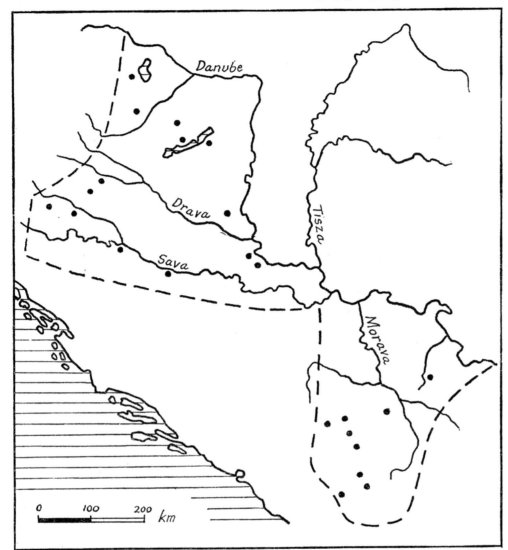

Figure 39 Stations of *beneficiarii*

be accidental that those parts of Pannonia, where in the first and second centuries numbers of grave-inscriptions were set up and none in the third century, were, all of them, areas where either an early colonization or the development of great estates in the second century can be demonstrated. Among these must be included the surroundings of the town of Savaria, the district to the north of Poetovio, the territory of Scarbantia, and to a certain degree also the neighbourhood of Neviodunum. North of Lake Pelso grave-inscriptions do, in fact, occur in the third century too, but in contrast to the previous period, when it was alien settlers for the most part who set up inscriptions, these are erected by soldiers buried in their own native villages.[91] These tombstones recall those from the villages of the frontier area, which are mainly family gravestones of soldiers of an adjacent garrison.

Accordingly we must attribute the cessation of tombstones in inner Pannonia to the establishment or further development of great estates; and the large number of funerary inscriptions in the frontier area and north of Lake Pelso must be attributed to the comparatively favourable material position of the soldiers. One might even hazard the conclusion that a certain prosperity, or at least a state of affairs that expresses itself in the number of inscriptions, can be proved only in places where recruitments were made from the local population. Admittedly, the towns of inner Pannonia continued to provide recruits for the legions and auxiliaries in the third century but these soldiers are either not traceable on inscriptions, or they left behind inscriptions in the towns only and not in the country. It is therefore perhaps not unlikely that the soldiers who originated from inner Pannonia were mostly inhabitants of towns. A final answer is not yet possible, but the privileged position of the frontier areas which even in the second century had provided the greatest number of recruits seems to be certain.

In Upper Moesia the situation shaped a little differently. Until the Severan period the government refrained for security reasons from making use of the population of the whole of the province. In the second century recruitments were made only in the two military *coloniae* and only for the legions. Marcus was the first to form some auxiliary troops from Dardanians; approximately at the end of the second century, recruitment, for the legions as well as for auxiliaries, started in other parts of the province though possibly still not all. By means of soldiers' records of their native place, and by means of the family gravestones of soldiers returned home or buried in their native villages, recruitment, which probably took place regularly, is attested in the valley of the Morava (Naissus, Horreum Margi) and in the Dardanian area (Ulpianum). It is however very characteristic that at Horreum Margi[92] and in the territory of Naissus[93] erection of grave-inscriptions began only with the recruitments, and the inscriptions are, with few exceptions, all gravestones of soldiers' families.[94]

To sum up: the epigraphic material, the majority of which dates to the time of the Severi, is mostly concentrated in the frontier area, proving an economic prosperity which is closely connected with the military. The composition of the body of inscriptions allows us to deduce increasing participation by the population in public life; this activity caused an adaptation to Roman forms and formalities, and the abandonment of local, non-Roman customs in nomenclature, and in burials and the like.

Economic prosperity left manifold traces in the archaeological finds of the frontier areas. Although excavations have, so far, been made in few of the towns on the bank of the Danube and despite the fact that the stratigraphy of the sites was not always satisfactorily dealt with, one can say in general that stone buildings and certain elementary requisites of cultivated Roman life obtained general currency in towns no earlier than Severan times. The most striking ground plans in the parts of Aquincum laid open so far and presumably of Carnuntum, too, are Severan in date, and this applies in particular to the private buildings and small shrines (Figs 28, 30). In the *canabae* of Viminacium the best-built houses seem to originate from the same period.[95] The decurions of the Danubian towns were remarkably generous in the first third of the third century; a large number of inscriptions attest various buildings, donations and endowments of public character, bestowed by members of the municipal upper class.[96]

This same class presumably introduced the luxuries of the town houses too. The mosaic floors in the settlements on the Danube originate, practically without exception, from the time of the Severi; the exceptions, like the mosaics of the governor's palace in Aquincum (probably of mid-second-century date) belong to a socially quite different category.[97] It is not impossible that demand in some large centres of the frontier area caused the appearance of workshops for the production of mosaics. In Aquincum a workshop produced several mosaic floors with a composition of large figures, rather crudely executed (Pl. 28b); the compositions go back to known iconographic originals in Hellenistic art.[98] But also at Aquincum a mosaic floor came to light recently,[99] of which the central picture, showing the scene with Hercules, Deianeira and the centaur Nessus (Pl. 28a), was certainly imported and inserted into the floor made by the local masters. Wall-paintings, occasionally with large figural compositions too, were also in demand, and walls and ceilings were often adorned with stucco.[100] To all appearance there were no differences between the two parts of the Danubian towns: the luxury houses are found equally in what had been the municipium and in the former *canabae*.

The luxury of the upper class showed in burials as well: in the whole of the Danubian area burial in sarcophagi became fashionable towards the end of the second century (Pl. 29); in some towns such as Viminacium it was possible to establish that from the end of the second century onwards rich citizens no longer used tombstones but greater grave-monuments or subterranean burial-chambers, in which the often richly decorated stone sarcophagi were placed.[101] This fashion appeared earliest perhaps at Ratiaria where several sarcophagi, possibly going back to the middle of the second century, are derived from garland sarcophagi of Asia Minor.[102] The spread of this custom along the Danube can easily be followed.[103] The change to burial in sarcophagi meant, of course, at the same time a change to inhumation and abandonment of the rite of cremation. This change in rite took place first among the rich inhabitants of the towns; whether it originated from the Orientals, as has been frequently maintained, cannot be decided. It was at any rate almost invariable that the richest people in the towns on the Danube in the third century did not set up tombstones but sarcophagi or great funerary monuments. The fact that many Orientals were among them will not be surprising in view of the part played by this ethnic component, which has already been described.

In the period of the Severi the decurions of the Danubian towns to all appearance lived inside the towns where they displayed their building activity and where they were buried. In the rural surroundings of these towns no great villas that could be interpreted as the centres of larger estates and as seats of estate-owners are so far attested. The only exception seems to be the area south of Carnuntum, where the estates of the Boian aristocracy lay. A few smaller villas of this region which have been investigated were abandoned, according to the stratigraphical evidence, in the second half of the second century,[104] presumably as the result of the estates being incorporated in larger estates. However, the largest of the villas, the palace of Parndorf (Fig. 32), seems to have been lived in and used without interruption. It was the centre of a presumably very large estate or even a *latifundium*. In the remaining parts of the frontier zone no large estates can be demonstrated, and this is the reason for our venturing to draw the conclusion, already hinted at above, that the wealth of the upper class in this area was not based on the concentration of agrarian estates.

The existence of a well-to-do class in the country cannot, nevertheless, be denied. Apart from monuments, such as the huge burial-vault from Brestovik (Pl. 30), east of Singidunum, which was built on the model of grave-chambers in Thrace and Asia Minor and adorned with wall-paintings, and which was at the time apparently considered exceptional,[105] all settlements and cemeteries uncovered in the border area so far suggest economic prosperity and the spread of the Roman way of life. In the civil settlements of the auxiliary forts of Pannonia it can be observed that with few exceptions stone houses do not occur before the time of the Severi. This is the case not only at Intercisa, where the Syrian enclave created a situation which must not be generalized, but also at those of Ulcisia Castra, Albertfalva and Matrica (Fig. 54), of which the civil settlements have been investigated to a certain extent.[106] The same is true of the non-military settlements of the frontier area and of the hinterland of the *limes*. As has been stated in Chapter 5, no villas of early imperial date—nor any stone houses—are so far known in north or eastern Pannonia.[107] In the surroundings of Aquincum, however, where excavations have been carried out in many places, smaller villas and stone houses could be examined; their construction has been dated, or can be dated, to the Severan period. These houses are heated by hypocausts, and have painted walls and the like too, but their size is mostly smaller than that of the villas of early imperial date in the western and interior parts of Pannonia. These villas are particularly frequent in the immediate surroundings of Aquincum.[108] A group of fairly humble houses not far from Aquincum had a cemetery of its own in which the grave-inscriptions of the inhabitants came to light.[109] From the name of a woman, Bithynia Severa, it can be inferred that the owners were Orientals. As we have seen, there are some indications that the wealth of the upper class in the frontier area did not derive from agricultural estates. The origin of its prosperity must be looked for elsewhere. Our starting-point must be the circumstance which is perhaps most characteristic of society in the frontier area, that no strict distinction can be made between the municipal (civil) upper class and the military. One of the reasons was that decurions and garrisons were closely interwoven: there are cases where the son of a decurion entered the legion,[110] and in some towns the veterans formed a continual replenishment for the *ordo* of the town.[111] Many private soldiers seem to have been no less wealthy than the prosperous civilians. We have for example a fair number of sarcophagi set up by active or discharged soldiers of the legions.[112] The great number of family gravestones belonging to legionaries not only attests regular recruitment from the immediate neighbourhood of the garrison but, even more, close interconnection between the local population and the military. The village inhabitants probably provided recruits readily, and military service yielded a regular and not negligible income for small family units. Apparently this was the reason why the formation of large estates did not take place in the border area. The small estates of the village population in the frontier zone were able to compete and could hold their own because they could count on a regular income from outside which came to them through the military. The 'normal' development in the direction of town-based estates and finally of large estates— these latter already outside the limits of municipal economy—was thus impeded in the border area, and remained so for as long as the military had at its disposal an income from outside the economy of its own province.

This hypothesis finds its confirmation in a comparison between the circumstances of the border area and those of the interior of the province. As we have seen, a concentration of estates in the interior of Pannonia was already going on during the second century. A comparison of the distribution of tomb-inscriptions in the ordinary countryside with those areas where villas and—by means of villas and other indications—large estates can be assumed to have existed shows immediately that these two distributions are more or less mutually exclusive. Where villas or large estates can be demonstrated, tomb-inscriptions are lacking, and vice versa. Since grave-inscriptions in the open country mostly represent families which possessed at the most only moderate prosperity and were very often closely connected with the legions in the vicinity, we can distinguish two types of social and economic development. These are the aforementioned 'normal' development towards the concentration of estates on the one hand and on the other, development associated with the prosperity of small estates in the border zone which were closely connected with the military. The place-names of inner Pannonia furnish further proofs: there is a small number of place-names formed from a native name or from a *cognomen*, such as Mariniana, Maestriana, Bassiana. These names can be considered as adjectives to the word *villa*, e.g. villa Mariniana = 'Villa of Marinus', and thus may go back to villa establishments of early or middle imperial date. Their distribution is confined to the areas of 'normal' development (Fig. 38, p. 233).

Social development in Upper Moesia in the third century

The interior of Pannonia belongs as a whole to the zone of normal devloepment. In Upper Moesia, the situation was probably altogether different. We must first notice that the mines formed a factor not present in Pannonia, and secondly, that recruitment was late in starting throughout the whole of this province, while the areas which provided most of the recruits (Scupi, Ratiaria) did not lie in the border section—which was anyhow short. Moreover, the army of Upper Moesia drew to a much larger extent on the population of the un-armed adjacent provinces. For this reason and also because tension between original inhabitants and aliens was always perceptible, the military could not act as a connecting link between the provincial upper class and the native village population. Apart from Scupi and Ulpianum, the towns of the interior of Moesia were all creations of Marcus and the Severi. They were never really able to de-velop a municipal life of their own, despite the obvious tendencies of the govern-ment to include the local upper class in the municipal administration. In the earlier towns, management had been in the hands of foreign settlers and business-men who were rather isolated from the native population. Ulpianum, whose aris-tocracy—unique in the province—consisted, in the second century, of Dardani-ans, had the normal development, as is shown by the absence of tombstones of military families; in the third century it did, in fact, provide recruits.[113] It is precisely in the *territorium* of Ulpianum that one of the few *latifundia* of the Danubian area can be demonstrated: this was the great estate of the senatorial families of the Furii and Pontii on the northern slope of the Scardus Mons, the slave personnel of which left behind quite a number of inscriptions in the Severan period.[114] A freedman of these senators, Furius Alcimus, is known, too, as the recipient of a *responsum* from the jurist Ulpian.[115]

Except at the two legionary fortresses, tombstones of soldiers' families are attested in Upper Moesia only at Timacum minus, where lay the fort of cohors II Aurelia Dardanorum and in the valley of the Morava (Horreum Margi and Naissus).[116] The families of soldiers at Singidunum, Viminacium and Timacum minus were of mixed foreign provenance, continually settling near the garrison and forming a hereditary military profession. The sons of soldiers in Timacum minus frequently entered the legion even as early as the second century. These families had had practically no connection at all with the local population of the district. The fact that from the end of the second century onwards recruiting for the legions of Upper Moesia was carried on outside the military *coloniae* of Scupi and Ratiaria is proved, apart from some statements of origin by soldiers from the city of Rome, only by family tombstones from Horreum Margi and from the territory of Naissus. It might be assumed that the formation of a social class consisting of small peasants and soldiers—after the pattern of the Pannonian frontier districts—took place in this area, but suspicion is roused by the circumstance that the prosperity attested by these tombstones had come into being along the very road that played such an important role in the third century. The causes of prosperity were thus not necessarily of a politico-military kind. The seeds of a developing prosperity cannot, however, be denied in some parts of Upper Moesia. The difference between Pannonia and Upper Moesia is probably based on the fact that, owing to the late beginning of local recruiting, a population of military complexion was unable to come into being until very late, if at all. The military of this province was composed, much more than that of Pannonia, of a foreign class of hereditary soldiers, who were isolated from the population of the province. The causes had deep local roots which were discussed in Chapter 5: for reasons of security local recruiting could not be spread over the whole of the province, only few Roman self-governing cities were founded before Severan times, and the original population was thus excluded from the leadership of the local life of the community.

The Illyriciani

What has been said gives us some clues to the social characteristics of the
Danube army at the culmination of its political activity, as well as to its historical
role. A first, very typical component was formed by the resident families in the
villages of the frontier area, who had an economic connection with the provincial army, in that the income of their members, serving in neighbouring units,
rendered the family estates competitive. This social class was neither a peasant
soldiery nor a hereditary military cast. It had the opportunity of rising into the
higher military ranks as well as into the civilian urban upper class, which was
not strictly isolated from it. It was, however, continually reinforced from the
circles of the local native population. A retrospective characterization of the
Illyriciani in Aurelius Victor apparently had this peasant class in mind: *his sane
omnibus Illyricum patria fuit, qui quamquam humanitatis parum, ruris tamen ac militiae
miseriis imbuti satis optimi rei publicae fuere* ('Almost all these were natives of
Illyricum; though lacking culture they were reared on the hardships of rural
and military life and were therefore good servants of the state.')[117]

A second component was the military families, long resident in the vicinity
of the garrisons; with them military service had practically become hereditary.
These families presumably provided most of the centurions and higher posts
in the army and provincial administrations, for the officers of the Danube army
in the third century still seem to have been mostly people with Italian *nomina* or
with older imperial family names who cannot have come from the rural population of the frontier region.[118] Reinforcement for this hereditary profession of
soldiers was ensured by recruits from the unarmed Danubian and Balkan provinces (Dalmatia, Thrace and Macedonia), and from the eastern provinces.
Recruits from those interior regions of military provinces which were not
occupied by the army must be reckoned as belonging here too, since it may be
considered unlikely that the inhabitants of areas of 'normal' development sought
to join the army unless they were poor or had lost their estates. For the inhabitants of the frontier area, however, service as soldiers was a means of maintaining the prosperity of their estates and other undertakings. The non-agrarian
enterprises in the border area in particular were directly dependent on the
army, as was the economic and community life of the border area as a whole.

For the same reason the civil population of the towns in the frontier area was
closely dependent on the army, too. As we have shown, large estates were unable to develop just because the rural small estates could draw from the sources
of the military economy. The same was presumably true of commercial and
industrial undertakings, since it is a significant characteristic of economic life

in the frontier provinces that hardly any industrial centres capable of achieving something for the economy of the empire arose in them. We shall return to this problem later in this chapter. Trade was thus chiefly an import- or transit-trade across the frontier, and economic life was based much more on consumption than on production. Under such circumstances a social contrast between rural soldiery and urban citizens, such as was imagined for instance by Rostovtzeff, could not arise, at least not in those areas where the army was stationed—or rather, not only stationed but so intertwined with local society as to give this society its most significant characteristic.

A certain democratic tendency in the aspirations of the Illyriciani is, nevertheless, undeniable. The soldiers of the Danube army were in origin peasants— indeed, peasants belonging to an agricultural structure that knew neither slave labour nor coloni. The most influential social element was presumably formed by them, not by the soldiers from the hereditary military families in the garrison-towns, whose sole basis of existence was their affiliation with the army. As professional soldiers their sole interest was presumably that their existence should be secured by the imperial government. Thus they were excellent tools in the hands of any military movement that wished to count on their help. It is therefore justifiable to speak of an Illyrican soldiery.

Unfortunately the percentage figures of the two component elements of the Danube army are unknown. The share of the professional soldiers presumably varied a little from legion to legion; it was, for instance, presumably much larger in legio VII Claudia than in the legions of Pannonia Inferior (I and II Adiutrix), where rural soldiers' families are attested in the border area by many tombstones. The participation of the professional soldiers must also have been large in the Upper Pannonian legions X and XIV Gemina, since large estates, developed from the Boian estates, lay in the very hinterland of Carnuntum; and elsewhere in this province, too, there are a number of proofs of 'normal' development.

Economic developments of the third century

The Danube region's economic and political centre of gravity lay in the frontier zone. A certain wealth and prosperity in the interior of the provinces cannot be denied, as we have already seen, but stronger social polarization there set a barrier to this prosperity. This polarization is attested, above all, by the large estates, the full development of which falls into the Severan period. Those villa establishments in the interior of Pannonia which had originated in an earlier period continued in use, and some of them were very luxuriously appointed. Most famous are the large mosaic floors from Baláca, north of Lake Pelso, which have recently been dated to late or post-Severan times.[119] It is very probable that, as a result of the concentration of estates, some smaller villas were abandoned; this could be demonstrated archaeologically at several sites in the territory of the Boii near Lake Neusiedl (Lake Fertö in Hungarian). The wider surroundings of Sirmium, that is the country between Danube and Save, must have been a region of large estates too. The Senator C. Messius Decius, proclaimed by the Illyriciani as the first Pannonian emperor, was born in Budalia, a village west of Sirmium, on the Sirmium–Cibalae road,[120] presumably in a family villa. The Emperor Probus, too, had estates around Sirmium:[121] it was on their account that he was killed by his soldiers. In this fertile country around Sirmium objects of luxury and sculptures of good quality have come to light as unassociated finds in large numbers; they allow us to infer richly appointed rural seats.[122]

Possibly during the third century some of the large estates passed into imperial possession. Some imperial estates occur even earlier, for instance one somewhere near Ratiaria,[123] and another not far from Sirmium[124] where Herodes Atticus stayed in 174. The place-name Caesariana on the Savaria–Aquincum road, in the region of large estates north-east of Lake Pelso, also points to an imperial estate. It can be assumed at once that during the course of the third century a number of large estates passed into imperial possession as members of the aristocracy were killed or exiled in the course of political upheavals. The family estate of the Emperor Decius apparently became such a domain, but a change in private ownership in the third century has to be assumed for the palace-like villa of Parndorf in the country of the Boii. From the beginning of the fourth century onwards we have an imposing number of demonstrably imperial domains, in Pannonia as well as in Upper Moesia.

The owners of large estates in the interior of Pannonia were descendants of the municipal aristocracy of early imperial times. A certain regression of municipal life, shown in the small number of building- and votive inscriptions of decurions, can presumably be attributed to increasing withdrawal by prosperous members of the *ordo* from municipal affairs. Of course, new buildings are not wanting in the towns even after Marcus; for instance in Savaria a great shrine of Isis was built, possibly on the site of an earlier temple (Fig. 41, Pl. 27b).[125] In Savaria, too, mosaics of the Severan period are attested. The prosperous middle classes of the population of both town and country, however, have disappeared. The tombstones which are characteristic of this social class become rarer even in those places where they had been exceedingly frequent in the first and second centuries; as at Savaria, Poetovio and in the region of Lake Pelso. Imperial endowments and other measures, like the *Kalendarium Septimianum* in Savaria (p. 219) or the *colonia* (*nova*) in Siscia, under such circumstances presumably served the purpose of preventing regression of municipal life. These attempts, however, have left no trace of any notable success. One of the reasons for this lack of success may have been the ability of wealthy members of the municipal aristocracy to use the prosperity of the times to further their own interests; they did not represent a social stratum which for political reasons could be, or had to be, treated to its disadvantage. As early as the second century this class had given capable officers to the empire, like the famous M. Valerius Maximianus from Poetovio, and a number of equestrians, some of whom are attested in various positions of the *tres militiae*.[126] Though the notorious self-made men of the Severan period mainly originated from military circles in the frontier area, like Aelius Triccianus, who started his career in the officium of the governor,[127] later representatives of the Illyriciani were Sirmians and so presumably men of wealth, or else the majority of them possessed family names which allow us to infer a citizenship of older origin; they cannot therefore be taken to represent the original inhabitants of the frontier area.[128] Decius had attempted to avoid supporting the cause of the Illyriciani, but the confidence which the latter had in him proves that a conflict of interests between the military and the prosperous upper classes of the areas of 'normal development' did not exist.

No considerable changes in the trade-relations of Pannonia and in the focal points of its economic life took place at the end of the second century. Imports, demonstrated by finds, mostly come from the western provinces, with the difference that the areas on the upper Danube, being situated nearer to Pannonia, come more and more into the foreground as suppliers. After Marcus, finds of samian ware are almost exclusively Raetian (Westerndorf ware).[129] So-called Raetian pottery is also well represented in Pannonia, but the majority of the products was presumably manufactured on the spot.[130] The same should be true of the glassware, which, though in shape based on products of the Rhenish factories, was for the most part produced in Pannonia.[131] Imports from the west included enamelled bronze brooches and other types of enamelled bronze ornaments; these were also imported by Pannonia for the transit-trade: enamelled bronzes were very popular among the Sarmatians of the Hungarian plain.[132] Rhenish motto-beakers, bronze mounts, carvings in jet, etc., are found on all the larger sites.

Importation in equally large quantities from the eastern provinces cannot be proved. A few striking objects do however come from workshops in Syria, Egypt or Asia Minor. Without exception they are luxury goods or small works of art such as bronze statuettes,[133] sculptures in ivory[134] and richly ornamented silver or bronze vessels[135] produced in the workshops of Alexandria. A considerable number of pieces of gold jewelry presumably also comes from the East, as well as the glass-paste imitations of cut stones, often showing Greek mottoes.[136] Fine textiles were apparently brought from the Orient; they can only rarely be traced among our finds.[137]

At this period oriental imports presumably no longer came to Pannonia by way of Aquileia, but already on the Byzantium–Naissus–Singidunum road through the Balkans or to an even greater extent by water, on the Danube. And because of this trade-route on the river we may assume that oriental wares played an even greater part in Upper Moesia's import trade. Because work on small finds has not progressed very far proofs are not yet available, but oriental influences show fairly directly in stone sculpture and they may even perhaps be represented by imported pieces such as the garland sarcophagi at Ratiaria and possibly at Viminacium too.[138]

In general, finds of Severan date are much more uniform, much less various, than those of earlier times, the chief reason being that the local production of everyday objects managed to displace imports almost everywhere, even where the quality of the local goods was inferior to that of the imports. Imported goods were soon imitated on the spot. It has for instance only recently been discovered that so-called Raetian pottery was also made in Pannonia.[139] In the same way, it becomes increasingly probable that most of the glassware was produced in Pannonian workshops. From not uncommon discoveries of moulds for casting knee-brooches, buckles, mounts and similar objects[140] it is believed that the simple bronze objects of use in all centres of the province were locally produced.

Nevertheless, industry in Pannonia and Upper Moesia did not produce any speciality of significance in the trade of the empire. So far, research has not succeeded in demonstrating that objects of Pannonian or Upper Moesian origin exist in large numbers in other parts of the empire. Identifiably Pannonian products were of hardly any significance economically, and it cannot even be shown that they were distributed by the normal processes of trade; they could just as well have owed their dispersal to troop-movements. These Pannonian products were small votive tablets of the so-called Danubian Rider-god or of Mithras, made from the soft yellowish crystalline limestone of the Alma Mons (Fruška Gora).[141] Writers of the early imperial period inform us, furthermore, on a number of other Pannonian exports, such as medicinal herbs, wild animals for the amphitheatre, hunting dogs and possibly also horses.[142] In their mountain farms the Dardanians of South Moesia produced a cheese speciality.[143] It cannot be said of course how these exports affected the trade-balance of our provinces. Of greatest importance, presumably, was the export of silver, lead and copper from Upper Moesia; but as has already been seen, it contributed to the prosperity of only a limited number of towns and settlements.

In the time of the Severi export trade across the frontier was to all appearance very significant. This export is confirmed by indisputably Roman goods among the finds from sites outside the empire as well as by the influence exerted by the provincial industry on local barbarian pottery production. Barbarian–Roman trade-relations are demonstrated, too, by several typical kinds of objects found in settlements outside forts along the Danube, for instance special shapes of vessels of Sarmatian type.[144] For the centres on the bank of the Danube this trade presumably formed a component of prosperity, but from an imperial view-point it may have brought but little gain, especially since the treaties with barbarian border peoples arranged for their regular subsidy. Roman exports thus tended to be bought by barbarians with Roman money. No evidence at all exists for any important import-trade from the barbarians into Upper Moesia or Pannonia.[145] The presumably busy trade across the border along the Danube, though a local source of gain for the border areas, was quite expensive for the empire. Therefore the encouragement provided by the border trade for the economy of the frontier region must be counted among the phenomena of the period of prosperity.

Standardization of local culture in the third century

Uniformity of industrial production went hand in hand with a standardization of local culture. In all aspects of culture there becomes evident a fading of the local characteristics which in the second century in particular had shown up very significantly. This standardization was so intensive as to cause the almost complete disappearance of local, non-Roman nomenclature in Pannonia; specifically local rites of burial (*tumuli*, wagon-graves), local burial symbolism (astral symbols) and much else besides disappeared as well. This comparatively swift process which took place at the end of the second century has been held to mean that the native population had been so reduced by the Marcomannic wars as to render necessary the oriental and other immigrations for the purpose of populating the country again. But the war under Marcus, although undeniably catastrophic, did not cause such an ethnic vacuum. Celtic names sporadically continue to occur after Marcus, particularly in those regions where epigraphic evidence for local nomenclature in the first and second centuries is densest. These sporadic names, now appearing as *cognomina* in the three-names system, are carried by persons whose families were already changing to the Roman system of nomenclature.[146] This Roman nomenclature is very colourless: the only feature which is to some extent characteristic is the preference for certain *cognomina* such as Decoratus, Candidus, etc. A few further names are successful Latinizations of Celtic or Illyrian ones, such as the *cognomen* of Aelius Triccianus, formed from the Celtic Triccus.

In Severan Pannonia the widest retention of non-Latin nomenclature presumably occurred among the barbarian groups settled under Marcus in the province. But these groups have left behind hardly any epigraphic monuments. A group of Cotini were settled under Marcus in the territories of Mursa and Cibalae; already under the Severi they were entering the legions, and some even rose to positions in the praetorian guard. In Rome, these are listed in the *laterculi Praetorianorum*, with *cognomina* that suggest an intense Dacianization of the Cotini.[147]

Energetic Latinization of nomenclature in Pannonia can, of course, be proved only with reference to the elements named on inscriptions. However, the fact that this 'epigraphically traceable' social class all of a sudden gave up their non-Latin nomenclature, to which as far as we can tell they had clung tenaciously during the second century, is a symptomatic and important phenomenon. As we have seen in Chapter 5, these epigraphically traceable Celts and Illyrians were the chief components of the local aristocracy which enjoyed the favour of the government, and they retained certain traditions of their culture despite or because of their Romanization. If this attachment to tradition on the part of the local upper class was a conscious attitude in the second century, then the abandonment of this tradition must be considered as equally conscious. Their increasing share of local authority, even more their growing political importance in the empire, apparently rendered an adaptation to Roman forms necessary.

The relinquishment of *tumuli* and wagon-burials probably also belongs in this context. Hitherto, no *tumuli* indisputably datable to the third century have been found. A few cart-graves may possibly date to the early Severan period, but in the third century this expensive rite fell completely into disuse, and certainly long before the economic crisis reached the Danube; the disappearance of the fashion cannot therefore be attributed to economic reasons. Among typical local representations of tombstone symbolism, astral symbols and representations of chariots, too, disappeared soon after the end of the second century; the only iconographic composition to continue in use was the one which was least un-Roman: this was the so-called scene of sacrifice with the tripod and sacrificing figures standing on either side (*camilli*). In the period of the Severi this scene is found on the tombstones of non-local families too.[148]

Among the monuments of Upper Moesia, such a recession of native tradi-

248

tions cannot be traced, mainly for the reason that the native population as a result of insufficient adaptation to Roman influence had never hitherto produced anything specific. It is only from the beginning of the Severan period that we can speak of adjustment which affected the whole province. It was then that most of the self-governing towns were founded. The consequences of this late consolidation became apparent only towards the middle of the third century. With the exception of the inhabitants of the *territorium* of Ulpianum, the original population of Upper Moesia only very rarely set up inscriptions: it was done by a few soldiers' families in the valley of the Morava and some very few decurions of the towns of late foundation in the interior of the province. This epigraphically traceable stratum of the native population, however, already carried Latin names. Particularly instructive in this respect is the extensive list of 169 recruited and 195 discharged soldiers of legio VII Claudia;[149] it enumerates one age-group without gaps and thus renders possible a comparison with the composition of the epigraphically traceable legionaries. On this list, a fairly large number of Upper Moesian soldiers with Thracian *cognomina* are named (from time to time Illyrian *cognomina* occur, too), whereas on the other inscriptions of the province legionaries with Thracian (or Illyrian) *cognomina* are excessively rare. From this fact it is possible to conclude that a large mass of already fairly Romanized natives are not traceable on inscriptions such as grave-inscriptions and altars. Only after the middle of the third century do they become traceable here and there, and on inscriptions they often appear with Thracian names—unlike contemporary Pannonians. A certain Thracian self-assurance, too, is seen when a town-born Roman soldier, a citizen of Scupi, gives his origin as Bessus.[150] This indication of origin has caused considerable difficulty, since the clan of the Bessi were not settled in the *territorium* of Scupi, nor even in Upper Moesia at all. In late usage, however, Bessus simply signifies someone speaking Thracian; presumably by the name Bessus the soldier of Scupi merely expressed his membership of the Thracian-speaking inhabitants of the empire.

It cannot be denied that this developing self-assurance is somehow connected with the self-assurance of the Illyriciani. Particularistic self-assurance and emphasis on genuine Romanity were equally characteristic of these soldiers, and their Danubian origin often gave an opportunity to show genuine old-Roman traits. Thus his origin in Dardania became a proof that the Dardanian in question (the Emperor Claudius II) was in the last resort a Trojan—in other words an original Roman of the best kind.

Religious cults in the third century

When the Pannonians suddenly stepped on the stage of history in 193, and put their official cult to the service of the Severan party, they did not intend to enlarge their local pantheon and its local names, but chose Roman deities, or at least laid emphasis on deities and cults that could join the Roman cult without difficulty. Pannonian augurs are reported, who possibly go back to Celtic usages, but who were as augurs nevertheless acceptable to any Roman;[151] the Matres Pannoniorum, too, could easily be understood by all, although they may possibly descend from the Nutrices of Poetovio.[152] Later on in Pannonia itself cults made their appearance which, although they may be taken to represent real Pannonian cults, do not show truly local, non-Roman traits. Possibly the most essential trait of these cults was that everything about them—the name of the deity, appearance, iconography—was Roman, or at least, they did not show any local traits not elsewhere attested in the empire. This particularity of the cults and of the religion of Pannonia can be understood only when seen in connection with the process we have just described: the sudden fading of local, native peculiarities of the culture in favour of a standardized, colourless Roman attitude, which, in the last resort, was based on Danubian–Roman self-assurance.

In the cult of Silvanus[153] in particular quite a number of difficulties are removed if it is interpreted in this connection. These difficulties had arisen because those investigating the cult wished simply to see in Silvanus an Illyrian deity, if not the chief god of the Illyrians. This conclusion reached long ago was based more or less only on the circumstance that an exceedingly large number of monuments in Pannonia was dedicated to the god Silvanus; in some settlements the number of altars to Silvanus exceeds even those to Juppiter Optimus Maximus; and in general Silvanus is the god with the greatest number of altars after Juppiter. The floruit of the Silvanus-cult falls into the Severan period. (The earlier monuments of the cult are found only in the towns of the Amber Route and have connections with the cult of Silvanus Augustus in Aquileia,[154] who cannot be considered as a native god.) A further point in favour of an interpretation of Silvanus as a renamed Illyrian or, more generally, native Pannonian deity, is the connection of some aspects of his cult with pre-Roman fertility-cults, especially with ithyphallic representations of Iron Age date.[155] No doubt Silvanus is from time to time represented in an ithyphallic style on Pannonian reliefs too,[156] and the fertility concept was not alien to the Silvanus-cult as a whole. These traits are, however, too generally Roman to be taken as a local specialized trait. As was pointed out several decades ago[157] the Pannonian

Silvanus shows no traits of character that are not attested in the Italian cult of Silvanus also. It is therefore quite probable that for their favourite native god the Pannonians had selected an Italian deity; the fact that this god was not simply based on a pre-Roman deity of like nature follows from the introduction of his worship only from the third century onwards, and then only in an unspecific form.

It is not to be denied that in Pannonia Silvanus was the most popular, most frequently worshipped deity in the period of the Severi (Pl. 31a). He is represented only very rarely on altars with other gods; he is mostly represented alone or in company with other deities of nature, related to him in character, like the very colourless Silvanae (occasionally named Quadriviae and, when in the singular, Diana). With the gods of the classical pantheon Silvanus is not brought into connection, and even less with oriental mystery religions, though these during the time of the Severi were widespread too. On some altars Silvanus is actually named in company with the Capitoline Triad or with Juppiter Optimus Maximus,[158] which bestows a certain official character on him. It is not impossible that as early as the beginning of the second century Silvanus was officially depicted as the symbol of Pannonia.[159] But if we take into consideration the fact that a god named Silvanus Augustus was worshipped in Pannonia at an early date, then the conclusion suggests itself that Pannonia's figure of Silvanus was in fact derived from outside. This is not at all surprising since elsewhere too the *interpretatio Romana* arose from the Romans putting some native cults, which they worshipped as *genii loci*, on a par with their own gods. Silvanus, however, is not an example of an interpreted local god, or at least not of a single god, since he has no known corresponding deity with a native name. He has a few non-Latin titles;[160] but each of these is only attested once, and already their numbers allow the deduction that one single Illyrian chief god, equated with Silvanus, never existed. At one of the places dedicated to Silvanus—it seems that this god rarely had shrines or temples—altars for an otherwise unknown god Vidasus and for a goddess Thana came to light.[161] This single case cannot be taken, for the very reason of its being unique, as an argument in favour of an interpreted Illyrian god.

The reasons why it was Silvanus who became the most popular god of Pannonia, and why this did not happen before Severan times, can be most clearly deduced from the characteristics of the god himself. As we have already made clear, Silvanus did not become the god of a political idea. At the most he embodied the province Pannonia in the symbolism influenced by literature, but the Pannonians themselves were not interested in establishing Silvanus as the symbol of their country. The most frequent *cognomen* of the god was the exact contrary of everything official: *Silvanus domesticus*, and most of the representations are of this aspect. They show a simply dressed bearded man with the knife of a wine-grower or gardener, with fruit, with a tree and a watch- or sheep-dog.[162] In this context the *cognomen herbarius*[163] also belongs, as do the association of his worship with that of the Lares,[164] and the connections with fertility. All these traits are taken from the Italian Silvanus, as is also the iconographic type of Silvanus' representations. *Silvanus domesticus* is the protector of the garden, of the estate, of the fruit and the harvest (*Silvanus Messor*), and of fertility in general, as far as agriculture, horticulture and cattle are concerned. As a god of decidedly private character he is mostly worshipped on tiny altars in the house and on small reliefs, and often, too, on altars that do not state the name of the donor, since they were meant for domestic usage.[165] As a basis and prerequisite for the distribution of this conception of the deity the family estates of small peasants and the small properties of house, garden and field apparently served.[166] As we have already explained in this chapter, the floruit of these family-estates in the frontier area exactly coincided with the Severan period, and the monuments of the cult of Silvanus, too, mostly originate from this area. The forms of the cult's rites are purely Roman: altars for daily domestic worship and domestic sacrifices; small reliefs, also presumably for domestic shrines or for the Lararia and so on. Although the cult itself is closely connected with agriculture, it was hardly practised in the country; most of the monuments come from the military settlements on the Danube.

The remaining aspects of Silvanus are Roman as well; best known was the god Silvestris, a rural parallel for Domesticus. Silvestris often appears together with the nymph-like Silvanae, with Diana and other classical goddesses of Nature; and once he is explicitly considered the god of hunting.[167] This aspect of Silvanus was mainly worshipped by the higher levels of local society and has a certain literary tinge—that is to say it is provided with classical traits and associated deities taken from the classical pantheon.

That Silvanus is not an Illyrian god is conclusively shown by the fact of his being worshipped much more in the northern frontier zone than in the Illyrian–Pannonian south of the province.[168] It is at any rate of the greatest significance that the popular god of the Pannonians, and particularly of that part of the population most influenced by the army, was a god who in all his characteristics and in all the forms of his cult is purely Roman if not purely Italian. At the outset this cult was presumably spread by the Italians, who were glad to discover their own Silvanus—the *tutor finium*—in this land of small peasants, and in the end they created a cult which was taken up by peasant soldier circles.

The remaining gods, who are often traced back to native deities in the

252

specialist studies, were also the products of introduction from abroad. It has often been emphasized that the goddess Diana (Pl. 31b) was an Illyrian or at least native deity (goddess of fertility and of the forest) in Roman attire; but it should be remembered that in the literature of the early empire Pannonia already figures as a densely wooded country rich in game,[169] and foreigners anxious to placate the divine powers of a strange land would therefore naturally worship Diana on the spot as a goddess particularly present in Pannonia. It would be idle to enumerate all deities in whom local (Celtic and Illyrian) traits have been discovered, for the most important circumstance is that natives, and especially those carrying local names, only exceptionally set up altars or cult-reliefs.[170]

Gods described by their native, non-Latin names are to be considered as merely the by-products of a cult which in form and contents was entirely Roman. Genius Ciniaemus[171] and Minitra[172] are probably the only gods attested so far who are known nowhere but in Pannonia, and it is significant that they are mentioned only once each. Sedatus, Juppiter Optimus Maximus Teutanus, Epona, Mars Latobius, etc., who also enjoyed worship in Pannonia,[173] were introduced by foreigners, and they exhibit hardly anything suggestive of local roots. These divine figures, originating and given their character elsewhere, belong in Pannonia to the Roman pantheon, as do the Mystery religions, practised by town-dwellers and the army. Our conception of cult and religion in Pannonia is controlled by what is preserved by means of classical Roman forms —altars, reliefs and sanctuaries. The most characteristic trait of the religion of Pannonia was that it did not allow a synthesis of Roman and local conceptions to arise; everything was expressed in Roman terms and the expression itself was not original. The fact that even this local religion does not become really traceable before the Severan period may be attributed in the last resort to the fact that active participation in Roman civilization by the Pannonian army and its closely linked civilian population only began to develop after the reign of Marcus, as the result of the political activity of the Illyriciani.

This interpretation makes sense of the even more colourless religion of the province of Upper Moesia. The gods themselves and the forms of cult do exhibit a greater variety, but only because Upper Moesia lay on the border between two areas of culture: the province lay between the Latin and the Greek worlds, or more accurately between the Latin-speaking world of Italian and western ideas and that of Thracian and oriental ideas expressed in Greek. The 'native' gods of Upper Moesia are in part purely Thracian deities whose names, forms of cult and iconography were developed in Thrace. They spread in the districts adjacent to that province. The other influence came from the north-west, bringing south Pannonian, Italian and general Roman cults.[174] A Thracian god of heaven, weather, and of everything supernatural, seems to have been the Juppiter who was worshipped under the titles Culminalis, Epilophios, Paternus, Capitolinus and so on.[175] Epilophios being a Greek name points to a Greek formulation in Thrace. None of these cults, however, won a distribution as wide as that of Silvanus in Pannonia, although they and presumably several others (Nemesis at Ulpianum,[176] Mars in the Timok valley,[177] etc.) are characteristic of some smaller areas, for instance the title Paternus or Patronus in the north-western parts,[178] and the Greek formulations in the south-eastern. The only really local, specific-ally Upper Moesian (Dardanian) god seems to have been Andinus, named once only, on an altar dedicated by a *beneficiarius*,[179] i.e. by a foreigner. The names Andinus, Andia, Andio, derived from that of the god, turn up quite often, especially among Dardanians.[180]

There is one particular form of cult with pronounced characteristics, which developed in Pannonia or in Upper Moesia: this is the so-called Danubian Rider-god (Pl. 34a) or, more accurately, pair of Riders, after the pattern of the Dios-curi.[181] It is very likely that its floruit fell within the later years of the third century. His monuments are small reliefs in lead or marble, with a complicated crowd of different gods and symbols, composed in an order which is formalized by tradition. The reliefs, of which several hundred are known, are distributed in the middle Balkan area, Dacia and Pannonia, but the centre of the distribution lies in southern Pannonia and northern Moesia, spreading outwards from these areas. Nothing original was produced by this cult: all gods and symbols, and even the iconographic settings are well-known types of classical, oriental and other religions. An interpretation as a superstitious pantheistic synthesis of the gods most worshipped towards the middle of the third century is the best sug-gestion; in this Sol, the gods of the stars, the Dioscuri, the symbols of Mithras, Epona, and much else were mingled; the only original feature is its strictly organized eclecticism. The representations on small reliefs may well be assigned to the lowest levels of religious expression; a few of them reached distant border areas in the empire, apparently among the belongings of soldiers.[182] I consider it quite possible that this Rider-god, who must not be confused with the Thracian Rider-hero, arose from pantheistic speculations in the circle of the Danubian army; we must not forget that his centre was the same as that of the Illyriciani—the surroundings of Sirmium.

In this superstitious and primitive (because too undifferentiated and mechani-cal) synthesis of theological ideas, elements of the Mystery religions were also included. They constituted the most important religious movement in the Danubian area in Severan times, though their diffusion had occurred earlier. Leaving Isis aside, since her worship in western Pannonia goes back to the north

254

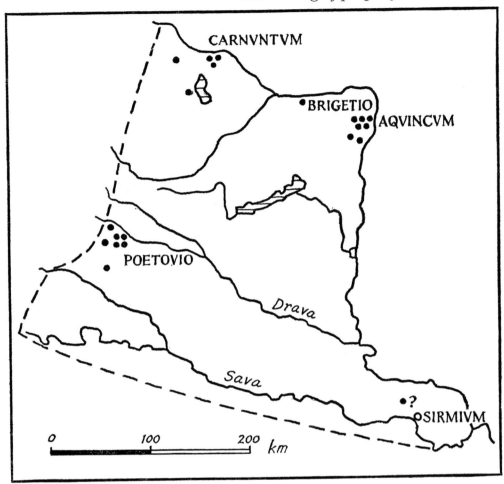

Figure 40 Mithraea in Pannonia

Italians of the early period, we find that Mithras and Dolichenus made their appearance in Pannonia and Upper Moesia no later than the first half of the second century. Their first followers were soldiers from the east and the personnel of the *publicum portorium Illyrici*. The earliest temples of Mithras in both Pannonia and Upper Moesia were built by the staff of the customs, by the slaves of some *stationes* of the *portorium Illyrici*.[183] The distribution of the Mithraea so far uncovered or attested by inscriptions suggests that the cult of Mithras was spread by regular missionary activity:[184] these temples are attested only in certain towns and in their immediate surroundings (Fig. 40). The fact that this missionary activity proved successful only after the Marcomannic war is presumably explicable, not only in the presence among the newly arrived Orientals of no

255

doubt numerous followers of oriental religions,[185] but also in the change which occurred towards the end of the second century in the whole religious and spiritual atmosphere of the Danube provinces. The abandonment of local traditions and the adaptation to general Roman forms of civilization presumably proved propitious circumstances for the spread of oriental religions, too.

Juppiter Dolichenus was worshipped here as elsewhere mainly in military areas, although noteworthy monuments of his cult have come from entirely civilian settlements, for instance Savaria.[186] The cult communities seem to have been well organized: the priests of Dolichenus throughout the whole of Lower Pannonia assembled for an unknown reason and immortalized this event on a votive altar.[187] Possibly this occasion was the visit by Septimius Severus in 202, for this emperor seems to have been a promoter of the cult. Whether the spread of the cult among the Pannonian army was the result of Severan policies or whether on the contrary Severus favoured the cult because of its popularity there, is not known; but it is worth noting that one of the earliest, if not the earliest, monument to Dolichenus in the western provinces was found at Carnuntum where the first of the Severi was proclaimed emperor. Two sanctuaries of Dolichenus are known in Pannonia, at Carnuntum and at Brigetio (Fig. 41). They were small rectangular rooms, strikingly narrow and modest, and they stood in the vicinity of other cult-rooms in an unpretentious temple-quarter.[188] Recently a Dolichenum with the whole of its cult-equipment came to light at Egeta, a fort-site on the bank of the Danube in Upper Moesia.[189] This sanctuary was an oval building, small and lacking architectural features. It is strange that it also contained a dedication to Mithras. The Dolichenum of Brigetio was adjacent to a Mithraeum: there are no indications that the eastern Mystery religions were mutually competitive.

It is unnecessary to describe the numerous Mithraea of Pannonia in detail as they do not differ from the usual type. The cult-relief of the chapel is a special feature, however, for it belongs to a distinct Danubian iconographic type. The most important centres of Mithraic religion, from which missionary activity started (Fig. 40), were at Poetovio (possibly the earliest centre in Pannonia), Carnuntum and Aquincum. In these towns several Mithraea have been excavated, and they also occur in the wider surroundings of these towns, whereas in some of the other towns such chapels can be demonstrated neither archaeologically nor epigraphically and are absent in the neighbourhood too. The Mithraic communities mostly consisted of a group of people who were associates in public life: employees of the customs' administration, family and domestic staff of a wealthy man, soldiers in a regiment. Moreover, the topographical distribution of Mithraea at Aquincum and Poetovio suggests the mutual separation of their

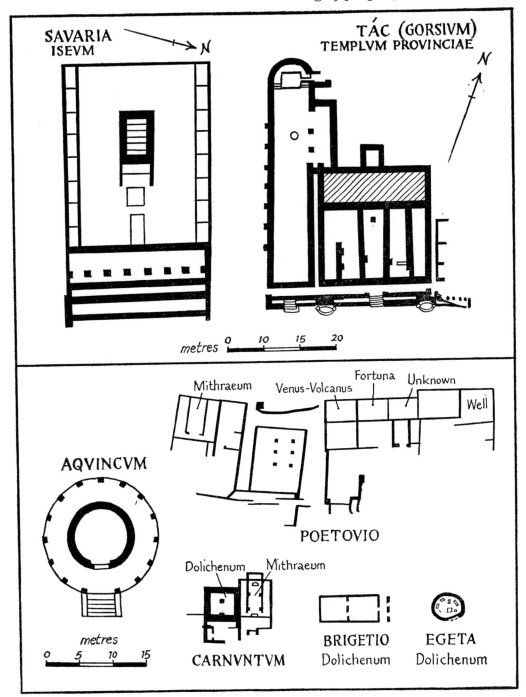

Figure 41 Plans of temples and shrines

spheres of influence rather in the manner that Christian communities were separated. A few more centres of the cult must be supposed in the less-investigated valley of the Save. The existence of a Mithraeum in the surroundings of Sirmium has been proved only recently.[190] The cult-centres in Upper Moesia cannot at present be accurately determined. The rather larger number of monuments in the Danube Valley round Singidunum and Viminacium point to Mithraic communities in these towns.

The remaining oriental Mystery and religious movements found much less response on the Danube. The cult of Sarapis has a number of monuments in southern Moesia and Pannonia;[191] its connection with the religious policies of Caracalla is sometimes very probable.[192] The Egyptian cults to all appearance survived only till the early years of the Severan period, but these marked only the start of Mithraic prosperity. There are, of course, also, no small quantity of monuments relating to cults from Syria and Asia Minor, but these often lack clear definition in the Danube area: we cannot even be sure that they came directly from the Orient. Dea Syria for instance possesses a number of monuments in Brigetio, Aquincum and elsewhere, but her cult is missing in the very settlement where the inhabitants had most preserved their Syrian traditions, at Intercisa. The distribution of all oriental religions in Pannonia and Moesia Superior is a phenomenon that we must interpret in relation to the religious history of the Roman empire as a whole rather than to the special circumstances of these provinces themselves (for instance as the result of local oriental immigrations). Orientals by no means absented themselves from these cults, but the part they played in the mission was no greater than anywhere else in the empire.[193]

Only the cult of Dolichenus seems to be somehow closely connected with the Pannonian army, and can possibly therefore be related to particular traits of Pannonian development. As already noted, Septimius Severus was a promoter of this cult. It is therefore presumably not coincidental that, of all the provinces of the empire, Pannonia has most monuments of Dolichenus, and that the spread of the cult in the western provinces is often attributed to the missionary activity of Pannonian soldiers in particular.[194] The puzzling disappearance of the cult in the second half of the third century is suspicious enough to suggest political reasons as its cause. To the Christians at the beginning of the fourth century Dolichenus was no longer an enemy worth fighting. Thus the polemic writings of the Christians, which enumerated all sorts of pagan beliefs and superstitions, do not mention Dolichenus at all. It is possibly rash to draw conclusions from the circumstances in which the Dolichenus sanctuaries were found, but it is a remarkable fact that all their furnishings have been preserved, whereas in the shrines of Mithras, for instance, small finds belonging to the equipment of the cult are very rare. It seems therefore that the Dolichenus sanctuaries were destroyed by a single violent act, and that at a time when they were still in full use. The members of the cult-community were unable to save the sacred furnishings in the way in which other pagan cult-furnishings were saved from the Christians in the fourth century—by hiding them in the ground. The seemingly sudden fall of the Dolichenus cult is thus possibly to be connected with political conflicts during the period of military anarchy. The so-called Mithraeum III at Poetovio,[195] luxuriously furnished by the Dacian legions under Gallienus, or the Sol-cult busily propagated by Aurelian, to name only two examples, are significant for the entanglement of politics with religion.

Christianity probably did not make its appearance in the middle Danube area before the time of Gallienus. The first fairly secure indications of the existence of communities of Christians originate from this time, but this of course does not exclude the possible presence of individuals earlier. During the persecution under Diocletian, a number of martyrs during their trial or when speaking to believers referred to their having been brought up as Christians; but the possibility cannot be excluded that they were born as Christians not in one of the Danube provinces but somewhere in the east, particularly since the Christians earliest attested on the Danube mainly came from the Greek linguistic area. Before the Tetrarchy, Christianity seems to have been strongest in and around Sirmium, and this was presumably connected with the increasing number of foreigners present there from the middle of the third century. The leading Danubian soldiers and politicians were all pagans with an often marked hostility towards Christians, which was evidently rooted in their old Roman attitude. In the third century Christianity was not yet a political factor in the Danube area.

Latin and Greek elements in provincial culture

In the civilization of the Severan period, as already mentioned, the local traits produced by the Romanization of the second century yielded to a colourless, undifferentiated, empire-wide culture. That this culture was Latin-Roman in almost all its aspects and that the oriental elements were represented in it only as far as these had penetrated Roman civilization in the western provinces in general, is due to the fact that Upper Moesia as well as Pannonia belonged to the so-called Latin provinces of the empire. Latin and Greek provinces to all appearance were strictly separated, and this separation probably did not correspond to the linguistic situation in real life—first, because the local languages (Celtic, Illyrian and Thracian), as we shall see below, were not given up by the original population, and, second, because Greek was probably much more widely distributed as a colloquial language than appears on the inscriptions, which are with few exceptions in Latin.[196] So we may be justified in supposing that the separation came from above, and was official. The demarcation line between Latin and Greek (often called the Jireček line after the Balkanologist K. Jireček who first mapped it) ran along the southern and eastern boundaries of Upper Moesia. Macedonia and Thrace were Greek, Upper Moesia—and even more of course Pannonia—Latin provinces. The fact that this line coincided with the frontier of the provinces clearly proves that the official language of each province had been decided some time during the early imperial period. In this respect it is noteworthy, too, that the border station where the most important route in the Balkans, the great diagonal Byzantium–Singidunum road, crossed the Jireček line was called Latina.[197] Those coming from 'Greek' Thrace entered 'Latin' Moesia at this point.

Upper Moesia and Pannonia were Latin provinces in the sense that the language of public life and of administration—both at provincial and at municipal level—was Latin. Hence the language used for the inscriptions, whether official dedications by imperial officials and municipal dignitaries, or appearing on private votive and funerary monuments. This of course by no means implies that only Latin was spoken in towns and fort-settlements. In the larger settlements a presumably considerable part of the population spoke Greek for certain; this is clear above all from Greek graffiti on tiles and on the walls. The non-Latin elements, however, erected Latin inscriptions in those places, too, where the public apparently understood another language better. The Syrian families at Intercisa and Brigetio set up neither Syrian nor Greek inscriptions, and at Scupi, where the population of the town was probably better versed in Greek than Latin, the language of the grave-inscriptions was Latin too. In Scupi the Greek language turns up only in unintended slips, for instance on some inscriptions where the Latin text was rendered in Greek letters,[198] evidently because it was dictated to somebody capable of writing only in Greek. On a number of inscriptions where the essential part—the name and age of the deceased—was in Latin, the subjective additions (verses, expressions of grief, etc.) were added in Greek for the reason that they could not be produced in Latin by the members of the family who had hardly any command of that language.[199]

It is very likely that Greek was particularly to be found in the largest towns, the very places in which Latin was most used as the colloquial language. A broad class of the population speaking only Latin or mainly Latin can be supposed in towns only where the population consisted of a homogeneous mass of

260

the descendants of Italian immigrants, or else of a heterogeneous mixture of immigrants who could make themselves mutually understood only by means of a neutral language. And for this reason Latin never really made progress where the great majority were natives; the reason was that the language understood by all was already provided. Thus it will not be possible to suppose too wide a distribution of Latin in the country and in the less important towns, especially those of late foundation; still less a section of the population with Latin as its native language. The fact that we nevertheless do not possess any inscriptions in local languages simply means that writing, like many other attributes of higher civilization, was an inseparable part of the externals depending on and produced by Roman rule.

The fact that the representatives of the Illyrian soldiery who had their origin in the native Pannonian population understood only one of the two cultural languages of the empire, namely Latin, is unequivocally attested by a source.[200] But the majority of Pannonian soldiers did not even speak Latin. Grave-inscriptions set up by Pannonian soldiers of the Rome garrison are often so full of mistakes that they form eloquent testimonies to the defectiveness of their command of the Latin language.[201] All the more attention deserves to be paid to a special group of Pannonian monuments in stone, attesting a knowledge of themes from classical mythology. Mythological subjects were popular on gravestones in the second century in western Pannonia[202] and at Viminacium[203] too (Pls 6c, 11b); but in west and south Pannonia and in Upper Moesia they were not so widely distributed as in east and north Pannonia, where from the middle of the second century onwards they formed a very specific and numerous group (Pls 32, 33).[204] The reliefs were mostly parts of larger, sometimes huge grave monuments. The mode of representation and composition goes back to well-known and widespread themes of Hellenistic–Roman iconography, and the variety of themes is remarkably large: it extends from the old-Roman legends of Aeneas and of the foundation of Rome to little-known stories from Greek mythology. In north-eastern Pannonia it had already been a fashion before the Marcomannic wars to furnish grave monuments with stone slabs ornamented with reliefs; these earlier reliefs, however, took their themes from the everyday life of the deceased, and a few very lively and attractive popular representations resulted from it;[205] the most delightful is a representation of a woman in native dress, holding a pig's head on a wooden plate (Pl. 13a).[206] It seems very characteristic of the culture of the Severan period that this popular art was given up in favour of a classical one. In part the classical themes are closely connected with the cult of the dead and with conceptions of the other world, in part they can be attributed to the self-assurance, mentioned before, of the inhabitants of the

Danube countries, which expressed itself in old-Roman cults and practices. The apparently very popular scene of Aeneas fleeing from Troy (Pl. 32b), the scene of Rhea Sylvia with Mars or of the she-wolf with the twins at any rate fit well into the framework of this theory. Perhaps we may mention here further Trojan scenes such as Priam before Achilles (Pl. 33a), the dragging of Hector (Pl. 33b), or the story of Iphigenia; but the majority of these scenes can hardly be explained in terms of old-Roman self-assurance. Of course, the question has also to be left open whether all these scenes, which are sometimes even represented in an abbreviated and simplified manner, were understood and recognized by the customers and public at all. It is quite possible that the scenes were simply taken from the pattern-books of the stonemason's workshops without any scholarly understanding. Perhaps we should only stress the supersession of scenes from everyday life by those from classical art; this new choice of themes presumably indicates conscious adaptation to classical culture, an adaptation presumably based on political and sentimental motives but by no means implying the spread of deeper knowledge. In this respect, too, society in the frontier region wished to inflate its image; it is certainly not coincidental that the classical themes were most widespread precisely in the border area, where the population drew most of its gain from prosperity based upon military policy.

The existence of schools capable of providing more than reading, writing and arithmetic can be supposed only in the capital towns; some evidence suggests that the sons of the small-town aristocracy had to go to school in Carnuntum or Aquincum (Pl. 29b).[207] Instruction in law can also perhaps be assumed, to provide the elementary knowledge necessary for administrative staff. Since, however, the staff of the provincial administration were soldiers, it is not absolutely necessary to suppose elementary legal training among civilians. There is much to suggest that knowledge of the law and a certain interest in the standards and terms of Roman law were derived from the military.[208]

A few quotations from Vergil among scribbles on tiles[209] point to texts of classical authors possibly being read in the higher forms in schools, but wider deductions can be drawn neither from these Vergilian quotations nor from the verses on graves which contain manifold Vergilian reminiscences.[210] These poems were mostly only adapted on the spot and the Vergilian flourishes curiously derive almost entirely from books I and VI of the *Aeneid*; the authors of these verses, not necessarily natives of the Danubian area, thus had a schooling not much superior to that of their clients in the Danube provinces.[211]

Illiteracy can be demonstrated by means of incorrect, often even unintelligible, grave-inscriptions. Since those who set up the grave monuments undoubtedly belonged to the upper class of provincial society, education and social position must be treated as not inseparable. It is known that it was not necessary for soldiers to know how to read and write. The knowledge of writing was indispensable only for those who were employed in the administration and in business life, and that is why the relatively best-educated people must be sought among the municipal *scribae*, the non-free employees of the financial administration and the administrative grades among the troops. A freedman of the Furii at Ulpianum, Furius Alcimus, put a legal question to the most famous jurist of his time,[212] though the decurions of Ulpianum themselves were not necessarily literate people. This point must be stressed in order to prevent the large number of Severan inscriptions being considered as proof of the spread of Latin civilization—writing and language.[213] An essential, if not the most significant characteristic of the Romanization of the Danube areas was that it took on forms which did not correspond to its real content, and that in all its manifestations it exaggerated its intensity. This, however, was only possible in the special circumstances of a politically based prosperity. Certainly the Romanization of the second century, which had created in many areas of our provinces a special culture of local colouring, was much more honest and started on a development which later on might have contributed something of value to the civilization of the empire. But when the soldiery suddenly rose to political importance they wished to stress their Roman identity at the expense of these special traits, and therefore created the colourless Danubian Romanization which, although drawing much from the culture of the empire, lacked a deeper understanding of it. This very superficial and in many respects only make-believe Romanization could not hold its own against the storms which were soon to break out, and collapsed much more quickly than the Gallo-Roman, African and other cultures in the empire which more successfully preserved their local traits.

Third-century collapse

To speak of a collapse in the true sense of the word is justified, not only because the barbarian invasions and wars in the second half of the third century were truly catastrophic for the Danube provinces, but because there must have been further, deeper causes responsible for Pannonia recovering so slowly after the crises. To point to the general crisis of the empire is too easy a way out. But, in the last resort this general crisis does seem to have been the cause of many of the barbarian invasions. As we have previously mentioned, peace on the Danube cost the empire a presumably considerable sum of money. This sum was regularly paid by the Severi, because the availability of the Danubian army for wars in other parts of the empire was a consideration of the first importance with them, and the loyalty of the troops themselves also depended to a considerable extent on government willingness to raise the sums necessary for securing peace along the Danube frontier. Although it goes beyond the authority of our sources, it can be considered probable that the elevation of pretenders of their own by the Danube provinces coincided with inability to pay the annual sums due to the barbarians, and this could occasionally lead to barbarian invasions. The war under Regalianus against the Sarmatians could be explained in this way; and, vice versa, if the central government with the financial resources of the empire at its disposal for some reason or other did not make the annual money available, the Danubian troops will have tried to put their own candidate in control of the empire. The appearance of the Goths on the lower Danube must not be forgotten here, for this was an entirely new factor in Roman foreign policy which further aggravated the potentially dangerous situation.

The war under Gallienus (259 or 260) in Pannonia and the war against the Goths in 268 in Upper Moesia probably resulted in devastations at least as great as those which occurred during the Marcomannic war under Marcus: at any rate the very large number of coin-hoards from all parts of Pannonia suggests a general catastrophe. But though the situation in Pannonia returned to normal again fairly soon after Marcus, nothing can be said of an early reconstruction or consolidation in the last third of the third century in Pannonia and in Upper Moesia. The sudden decrease in the number of inscriptions well illustrates this, and it is also striking that in the whole of the archaeological material we have no types or groups of finds which can be dated indisputably to this period. This presumably indicates not merely an absence of much building activity and reasonable luxury during this time, but decline of industrial production to a minimum. Nor can it be said that an import-trade worth mentioning existed.

We have inscriptions from the time of Gallienus to that of the Tetrarchy, but these are mostly either votive or building-inscriptions set up by persons of high rank (such as legionary prefects and governors) or else milestones. The building-inscriptions attest a certain amount of reconstruction at military installations. But inscriptions set up for private purposes and grave-inscriptions are practically missing altogether. The few exceptions are marked out by strikingly primitive or simple execution, which when compared with the good standards of stonemasons in the first half of the third century attest a sudden decline of the workshops and of the artistic ability of the stonemasons themselves. Demand must have lessened considerably. Sarcophagi and fairly large grave monuments were not being commissioned; the only attested type of grave monument is the gravestone (stele) which was in most cases crudely executed.[214]

The general decline had far-reaching consequences for the further development of the provinces. The damage suffered by both people and land was not, it seems, put right during the period from Gallienus to the Tetrarchy.[215] The reason appears to be twofold. First, individual emperors were not able to concentrate the forces of the empire in the Danube provinces—and if they could, the problem of Dacia had first to be solved. Second, it was easier to disregard the particularistic view-point of the Illyriciani—and in this the 'Illyrian' emperors themselves set an example, although they did attempt to achieve something in Pannonia. It is presumably significant that Probus, who had been born in Sirmium, compelled his own soldiers to take part in the economic reconstruction of Pannonia and Upper Moesia.[216] This was a new policy, entirely different from the economic policy pursued by the Severi. It caused the death of the Emperor Probus, but the idea of reviving the economy of the Danubian provinces under their own efforts was a step in the direction of the methods which were to be used by Galerius. Decline and the end of the economic boom were closely connected, in that the central government in which the Illyriciani themselves had a part abandoned the parochial outlook of the Danube army. In Pannonia and to a certain extent in Upper Moesia too, this new policy spelt bitter decades of misery and insecurity. It is, of course, a different question, and one which the historian cannot answer, whether it was because of the general economic crisis or by a gradually achieved political insight that the Illyriciani came to, or were compelled to, lay aside their long-established egoism.

Chapter 8
The Danube frontier in the late Roman period

Aurelian fell victim to a court conspiracy whose effect was to bring the aged Tacitus to the throne. To him the Illyriciani stayed faithful while he lived,[1] but on his death soon afterwards the troops once again proclaimed an emperor from the Danube provinces. Probus had been born at Sirmium and belonged to the immediate circle of Aurelian, and he continued the political programme of his predecessor, a programme which now had nothing in common with the parochial views of the older generation of Illyriciani. The political outlook of Probus showed the same tendency towards universalism as that of Aurelian had done, and in his policies he could presumably reckon on the co-operation and understanding of the Illyrican junta; the political thinking of this group, as had become increasingly obvious from the last years of Gallienus, had grown to include a concept embracing the whole of the Roman empire. The hostility towards the Christians shown by Aurelian fits into the framework of this concept, as does his intention to end the economic passivity of the Danube provinces, so that imperial policies should not have to depend on the parochial outlook which controlled the demands of the Danube army. The mutilation of our sources forbids description of this economic policy which we can trace only in a few events, as for instance in the accounts of drainage activity and the planting of vines around Sirmium and on the Upper Moesian Danube under Probus.[2] These were presumably part of a larger economic plan, since they were demonstrably continued by Galerius on a large scale. A curious votive stone[3] of about the same period from Pannonia records extensive private planting of vines; the fact that they appear on a commemorative stone shows that activities of this kind were considered at the time as politically important.

266

The men of the Danube army probably could not rise to this new policy. It was not only a personal tragedy for Probus that he was murdered by his soldiers on account of it: the soldiers on the Danube gradually ceased to share in the control of affairs; after Probus' murder the initiative for choosing a successor did not come from the Sirmian troops but from the army of Raetia, and an emperor was elected whom even the *Historia Augusta* could hardly call an Illyricianus. When Numerian and Carinus, the sons of this short-lived new emperor, had been killed by members of the Illyrican junta, the new regime of the Jovii and Herculii no longer relied on the Danube army. The decisive victory over Carinus took place near Margum in Upper Moesia[4] in 285, and won for the Dalmatian, Diocletian, undisputed rule over the empire. Although the co-rulers, whom he soon called in, were all men from the Danube provinces —Maximianus from Sirmium, Galerius from Romulianum in the new Dacia, Constantius from Naissus—as if all Danubian provinces were to be represented in the Tetrarchy—the policies which they were able to bring to fruition corresponded with the imperial outlook of the new generation of Illyriciani, and not with the old ideas of the Danubian army, which could be deprived of its political influence by means of the new economic policies. However, Sirmium managed to keep its leading role and experienced its greatest period of prosperity as one of the principal imperial capitals. It is therefore not surprising that natives of Sirmium and its surroundings continued to rise to the highest administrative posts of the empire and that the dynasty of Constantine was finally replaced by them, when Jovianus from Singidunum and Valentinian from Cibalae were successively proclaimed emperor.

The evacuation of Dacia did relieve the situation beyond the frontier to a certain extent, but the results were tangible only on the lower part of the Danube *limes*. The Goths who had taken possession of Dacia after its evacuation no longer exerted pressure on the *limes* of Moesia and the new Dacia. But the diversion of their forces in the direction of the old Dacia created an increasingly serious situation on the *limes* of Pannonia. We are very badly informed on the wars that occurred under Probus, Carus, Carinus and Numerian; the accounts in our sources can best be interpreted in the context of a process leading up to the numerous Sarmatian wars of Diocletian.

The Sarmatians came under pressure, as already mentioned (pp. 209 ff.), not only from the north, after the Vandals' migration, but also from the east and north-east, from the Gepidae and Goths. This may have been the main reason for their renewed attacks on the *limes* of Pannonia. The first attack, of which we are informed only indirectly,[5] happened under Probus *c.* 278; it is possible that the Vandals also took part in it. A few years later Carus gained a victory over the Sarmatians,[6] and about the same time a victory over the Quadi was celebrated.[7] Since this victory can be dated to the year 284, and as Diocletian took the title Germanicus on one more occasion than his colleague Maximian, it is likely that this war against the Quadi continued perhaps into 285 and was brought to an end by Diocletian acting alone.[8] The Danubian Suebi afterwards caused few problems,[9] the Sarmatians many more: the latter were to give the Tetrarchy as well as the Constantinian dynasty plenty to do.

Unfortunately we have only a very rough idea of what took place in the Sarmatian lowlands after the evacuation of Dacia. At first the Sarmatians were able to offer resistance to pressure by the Germanic tribes, but their attacks on Pannonia prove their renewed attempts to obtain armed assistance from Rome or the *receptio*. Leaving aside doubtful archaeological finds, good evidence for the *receptio* of a few smaller groups of Sarmatians [10] is afforded chiefly by place-names, such as *Mutatio Sarmatarum* in the Morava valley,[11] and by the mention of a few Sarmatian units in the *Notitia*;[12] but of these, some, at least, were probably admitted into the empire later.[13] Because of the Sarmatian wars Diocletian had to spend most of the period 289–94 at Sirmium,[14] even after he appointed Galerius ruler of Illyricum (on 1 April 293).[15] Only the outline of the chronology of the Sarmatian wars is at all certain;[16] according to it, victories over the Sarmatians occurred in the winter of 289–90, in 292 and in 294, when an important victory may have been won; then in 299 when the Marcomanni had to be fought too; and three further Sarmatian wars of Galerius are attested in the years 299–305. Unfortunately we do not know whether the new assumptions of the title Sarmaticus were the results of actual military successes or whether smaller skirmishes on the bank of the Danube and mere diplomatic steps were sufficient to justify them. At any rate the victory of 294 received special celebration,[17] and a few other events in the same year suggest important measures and successes on the part of the Romans. In the autumn of 293 Diocletian inspected the *limes* of Pannonia Inferior; his presence at Lugio, one of the most important stations on the Sarmatian front, is attested by two edicts issued on 5 November.[18] Since Galerius had then been Caesar and the real ruler of Illyricum for six months, the presence of the Augustus—his senior in rank—in Galerius' dominions must be connected with particularly urgent and important measures; Diocletian visited Pannonia on only one further occasion, for the conference in 308, when he was constitutionally no longer emperor.[19] The Sarmatian war of 294 was probably conducted by Diocletian in person: he remained at Sirmium from September 293 till August 294 almost without interruption. The preparations made in the autumn of 293 were evidently intended for the war planned for the following year, but the measures being taken were quite novel

268

and so demanded the presence of the senior Augustus. These new measures are recorded for us in an entry for the year 294 in the Consularia Constantino-politana, which reads 'This year forts were built in Sarmatia opposite Aquincum and Bononia' (*his consulibus castra facta in Sarmatia contra Acinco et Bononia*).[20] A series of coins with the new reverse type showing a fort-gate, which was to be repeated often in the future, may also be related to this achievement.[21] The two forts are also mentioned in the *Notitia Dignitatum* as garrisons *in barbarico* [22] which is all the more curious since counter-fortifications on the left bank of the Danube and on the right bank of the Rhine are never otherwise called forts on barbarian ground in this source. In the case of these two forts, therefore, *in barbarico* must mean that they were built not on the *limes* but deep in the country of the Sarmatians.

The fort-sites have not so far been found; but a few indications, among them a large votive inscription of Diocletian and his colleague from the Lowlands reading *ob devictos virtute sua Sarmatas* [23] (according to R. Eggers's restoration of the text), strongly suggest the existence of Diocletianic fortifications. Furthermore, though research on the Roman frontier has made considerable advances in Hungary it has not so far succeeded in proving a Diocletianic period of construction on the Danube *limes*, although large numbers of late Roman frontier buildings, as we shall see, have been discovered and investigated. Finally we find on the Sarmatian bank of the Danube, the *ripa Sarmatica*, a type of fortification which itself suggests either that Roman troops were permanently stationed in Sarmatian territory or that constant transport of supplies thither was necessary.

This type of fortification[24] has so far been attested at about seven crossing-places, on the right as well as on the left bank. It is a rectangle surrounded on three sides by walls and towers and on the fourth open towards the Danube (Fig. 42); the walls run down into the water. This type of plan can be interpreted as a fortified landing-place, and since a special tile-stamp, which is attested on the Sarmatian *limes* only in these fortifications, suggests that they were built at one time, we must either connect them with a planned offensive or preferably with a phase of foreign policy which necessitated regular intercourse between Pannonia and the Sarmatians. The obvious conclusion is to connect these landing-places with the construction of fortifications that occurred in 294. If this rather bold reconstruction of events is correct, then the absence of a Diocletianic building period on the Pannonian *limes* can be explained. Diocletian and Galerius presumably recognized that the Sarmatians under pressure from the north and east had no choice but to force a *receptio* or armed aid by continual attacks. Under such circumstances, reconstruction and further

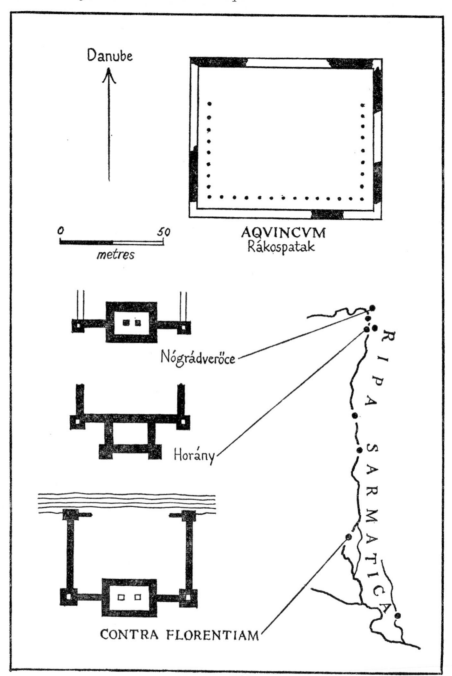

Figure 42 Fortified landing-places

fortification of the Danube *limes* could only have been a palliative, and for that reason construction on the *limes* was abandoned and an entirely new measure in the field of foreign and military policy was introduced: the partial occupation of Sarmatian country by Roman troops. The aim was not conquest or formation of a province but simply to ensure security and peace on the Danube. It could also be interpreted as armed aid for the barbarians. Events on the *ripa Sarmatica* in the first half of the fourth century thus lead us to the conclusion that the Sarmatians were bound to Rome by special treaties and that the disturbances which broke out from time to time were caused not by fundamental hostility on the part of the Sarmatians towards Rome but by pressure on them from their neighbours.

It would be tempting to date the so-called Devil's Dyke (Csörszárok) in the Sarmatian lowland to this period (Fig. 43). This earthwork, which has only

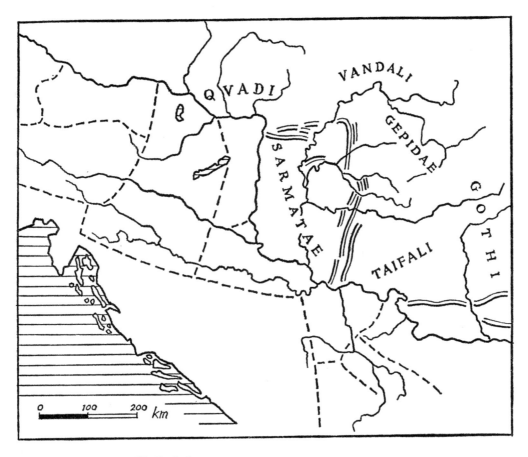

Figure 43 The 'Devil's Dyke'

recently been mapped and investigated,[25] enclosed the very area inhabited by the Sarmatians in the third and early fourth century. We already possess certain clues for dating it, notably a *terminus post quem* in the third century: the earthwork in some places cuts through traces of settlements of the second and third centuries and therefore cannot be dated before the end of the third century at the earliest. Other finds prove that it must have been built before the ninth century.[26] As a fortification it had hardly any value at all, but as a visible boundary line against the peoples on the fringe of the Lowlands, the Vandals, Gepidae and Goths, it could play a certain part, provided it could be somehow guaranteed by a third power. As long as its exact date remains unknown it is wiser to leave the question undecided. A slightly later date was recently proposed by the archaeologist investigating the earthwork.[27] The dyke must have been in use for a long time, since its course was altered a number of times and some sections showed evidence of reconstruction.[28]

That the new government of Galerius at Sirmium considered the war of 294 and the measures taken after it as a solution of the Sarmatian problem is clear from the fact that he soon afterwards set out against the Carpi and other enemies on the lower Danube *limes* and settled part of the defeated people in Pannonia. Moreover, these settlements were connected with large-scale drainage, and soon after they occurred the Diocletianic administrative reforms came into effect. All this suggests that after 294 Galerius considered the situation in Pannonia sufficiently consolidated for far-reaching reforms. Only 299 saw further war, and this probably took place on the northern border of the Sarmatians and in the territory of the Marcomanni.[29] Otherwise Pannonia, Upper Moesia and the new Dacia were able once more to enjoy an assured peace.

In the years after 294 Galerius maintained the economic policies on the middle Danube which had been started by Probus. Drainage was carried out in the interior of Pannonia by canalizing the natural outlet from Lake Pelso to the Danube (i.e. by building what is today the Sió canal) and this, like the drainage around Sirmium undertaken by Probus, was designed to increase the area of cultivatable land.[30] At the same time the Carpi were settled in Pannonia after their defeat in 295 on the lower Danube. Possibly these barbarian groups were given land in the very areas that had been drained, as had happened with the Cotini, who had been settled by Marcus around Mursa and Cibalae, in the marshy region of the *Hiulca palus*. Possibly the improvements themselves may have been carried out by the newly settled barbarians. A gift of land to the Carpi in eastern Pannonia north of the Drave can be established from a passage in Ammianus Marcellinus.[31] The Sió canal flows through this area, and it was this part of Pannonia which was created a new province under Galerius.

The late Roman division of provinces and separation of civilian and military administration was applied only gradually in our area. The first step had been taken by Aurelian when he established the new Dacia. Of the towns of Upper Moesia, probably only Ratiaria and Aurelianum at first belonged to this province. After 294 under Diocletian, or more accurately under Galerius, the great reform was set in motion. As we have stated, the north-eastern part of Pannonia, that is to say the northern half of Pannonia Inferior, was constituted a new province. The date at which this occurred can be established from the fact that the new province was called after Valeria, Diocletian's daughter and wife of Galerius, and from the association of naming and drainage.[32] Thus the province of Valeria came into existence after the victory over the Carpi, i.e. in 295 at the earliest.[33]

A few uncertain indications[34] suggest that the remaining late Roman provinces of Pannonia and Upper Moesia were organized not then but later, and it is not impossible that the administrative reforms of the period from Diocletian to Constantine the Great were modified to some degree. In the absence of direct information, and also of thorough research into these questions, it will be better to describe here the final situation as it had come to exist roughly under Constantine.

In Pannonia, four provinces came into being as a result of the reforms (Fig. 44): Valeria was formed from the northern half of Pannonia Inferior, and Pannonia Secunda from the southern half of the same former province, while Savia was constituted from the southern, and Pannonia Prima from the northern part of Pannonia Superior. The border between Upper and Lower Pannonia from now onwards is that between Prima and Valeria, and Savia and Secunda respectively; the frontier between Savia and Prima followed the Drave, but that between Valeria and Secunda ran north of the Drave, since a piece of land north of Mursa—presumably the *territorium* of this town—belonged to Secunda.[35] The seats of the civilian administration were the towns Siscia in Savia, Savaria in Pannonia Prima, Sopianae in Valeria and Sirmium in Pannonia Secunda. Sirmium also became the centre of a unit of administration, being variously the capital of a diocese and of a prefecture. The governors of the Pannonian provinces were not equal in rank, but their status was altered during the fourth century, and all that is certain is that, when Roman rule came to an end, Prima was under a praeses, Savia under a corrector and Secunda under a consularis.[36] In the course of the administrative reforms a town of Pannonia was ceded to Noricum: under Constantine, Poetovio no longer belonged to Savia.[37]

In the area of Upper Moesia the following provinces were established (Fig. 45): Moesia Prima and Dardania lay wholly within the former Upper province;

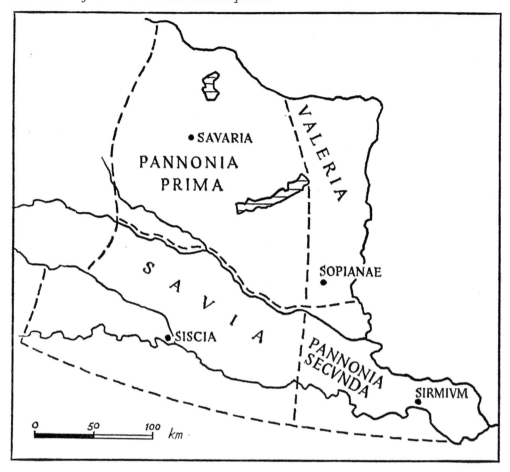

Figure 44   The late Roman provinces: Pannonia

Dacia Ripensis and Dacia Mediterranea, however, extended into parts of Moesia Inferior (the remaining area of which was named Moesia Secunda) and into Thrace. It can, however, be shown with a large degree of probability that Aurelian's new province of Dacia and Dacia Ripensis had the same extent. At what date a second province named Dacia was established is not certain, but possibly not before Constantine, when Dacia Mediterranea was created out of parts of Dardania and Thrace. For Naissus, which certainly belonged to Dacia Mediterranea from the time of Constantine onwards, is still mentioned as a town of Dardania only a few years earlier.[38] This, incidentally, was the reason why the family of Constantine could be traced back to the Trojan Dardanos. Dardania and Moesia Prima were constituted by dividing the remaining area of Moesia Superior, probably under Diocletian. Later, the territory of Naissus was attached

Figure 45 The late Roman provinces: Upper Moesia

to the new-founded province of Dacia Mediterranea, and this arrangement was maintained until the time of Justinian. Dardania included the territories of Scupi, Ulpianum and municipium Dardanorum; the parts of the two Dacias that formerly belonged to Upper Moesia were the territories of Ratiaria and Aurelianum and of Naissus respectively; the rest was given to Moesia Prima. The seats of the provincial administration were at Viminacium, Ratiaria, Serdica and Scupi. In the *Notitia Dignitatum* the governors of the provinces, with the exception of the consularis of Dacia Mediterranea, are all praesides.[39]

The ten years until the abdication of Diocletian in 305 probably passed without any notable hostilities on the Danube frontier. The written sources on the reigns of the last pagan emperors are very incomplete and distorted—presumably this is no coincidence—but Christian tradition would not have suppressed wars of any importance conducted by Galerius. The economic reconstruction under Galerius, which is briefly alluded to by later historians, also suggests that the situation at that time was probably peaceful and stable. The victorious title Sarmaticus, which was frequently adopted by Galerius and his successors, must be understood in the same way as the titles *semper victor*, *triumphator*, etc. The adoption of the title Sarmaticus could be the result of diplomatic moves or simply of the emperor's success in maintaining peace on the *ripa Sarmataca*; in this sense he would have been the conqueror and master of the Sarmatians. Moreover, Galerius had to fight several battles in the east and, after 296, was not personally present most of the time on the Danubian front.[40] A war against the Sarmatians is attested during the second half of 305, in which the young Constantine, accompanying Galerius as Caesar, distinguished himself.[41] The scale of this war, whose only recorded episode took place in a marshy area and therefore presumably not far from the Danube, cannot be established.

The long-drawn-out crisis of the Tetrarchy from the abdication of Diocletian to the assumption of sole rule by Constantine affected our province only towards its close. Galerius, the only consistent representative of the Illyrican junta among the Augusti and Caesars, was undisputed ruler on the middle Danube, and when the emperors assembled at Carnuntum in November 308[42] Illyricum once again came under the rule of an Illyricianus. However, the conference of the emperors at Carnuntum was indicative of the changed balance of power in the empire inasmuch as the choice as venue for the negotiations fell not on Sirmium, the stronghold of the Illyriciani, but on what was by now a rather unimportant town in Pannonia which was not even the seat of a praeses. When choosing the place for the meeting of the *religiosissimi Augusti et Caesares*—as they called themselves on the altar to Mithras (Pl. 34b) from Carnuntum[43]— Sirmium was evidently deliberately avoided.

In November 308 the Pannonian and Upper Moesian provinces were assigned to the rule of Licinius, a fellow-countryman of Galerius.[44] At first Licinius probably resided for most of the time at Sirmium, but his court had to leave there at the end of 314. Constantine reached Pannonia with his army in September of that year, defeated the more powerful army of Licinius on 8 October near Cibalae and pursued Licinius down the Balkan peninsula as far as Thrace.[45] As a result Constantine became the virtual master of our provinces, though the leading circles of Illyricum were presumably opposed to him. At Cibalae, Danubian soldiers had been defeated by west European troops, and a curious group of hoards from Pannonia and Moesia suggests that the Licinian party continued strong and that Constantine found himself confronted with considerable internal political difficulties in the region of the middle Danube. Silver dishes with the punched inscription *Licini Auguste semper vincas* (Pl. 35b), which were produced on the occasion of the tenth anniversary of Licinius' accession in 318, apparently in the court workshop, and then were sent to persons of high rank filled with gifts, have come to light in several places in Hungary, Jugoslavia and Bulgaria.[46] They prove two things: first, a busy propagandist activity on the part of Licinius in the dominions of Constantine, and, second, a persecution, sooner or later, of the followers of Licinius; for the dishes have been found in fairly large numbers only because they had to be concealed in the ground.

Between 315 and 324, that is until the elimination of Licinius, Constantine stayed practically continually at Sirmium; the interruptions were short visits to north Italy (315–16, 318) and in particular to the middle Balkan area, where he is attested mostly in Serdica and in his birthplace, Naissus (316–17, 319, 320, 321, 322, etc.).[47] Serdica gradually became a rival to Sirmium. In Sirmium, Constantine presumably met much opposition from the Illyriciani, among whom there were many supporters of Licinius. His long stay at Sirmium was presumably connected with this, since there were no reasons of foreign policy for his staying: the Sarmatians became hostile only from 322 onwards. The alleged intention of Constantine to choose Sirmium as the capital of the empire was possibly only a rumour spread for tactical reasons, with an eye to the Illyriciani. After the removal of Licinius and especially after the foundation of Constantinople, Constantine's visits to Sirmium were limited to the occasions of passing through on his way from the Balkans to the west.

In May or June 322 Sarmatian troops under the leadership of a certain Rausimodus invaded Valeria.[48] The chance of tradition allows us to identify the site of the battle: Rausimodus besieged the Danubian fort of Campona not far from Aquincum and set the wooden superstructure of the stone fort on fire. Constantine soon appeared from Sirmium, attacked Rausimodus near Campona and pursued him beyond the Danube, where he defeated the Sarmatians in their home territory and Rausimodus met his death. This was an opportunity for Constantine to undertake an expedition through the whole of Sarmatia. He entered imperial territory again near Margum in Moesia and thus presumably traversed the whole of their land from north-west to south-east. The reasons for this seemingly considered undertaking are not quite clear. Constantine spent hardly more than a month beyond the frontier and soon went to Savaria, where he is attested by the end of July.[49] His whereabouts in summer and autumn are not recorded, but at the end of the year he was in Serdica. It is not impossible that the journeys of the emperor in the summer were connected with the preceding operation in Sarmatia. According to Optatianus Porfyrius, Constantine brought back much booty, especially captives who were distributed at Bononia.[50] It is quite possible that in 322 Sarmatian groups were settled in the Danube provinces and that a new treaty was made with their compatriots beyond the frontier.

This suggestion finds its confirmation in the struggle which broke out ten years later, for this allows an inference about events in the Danube–Tisza plain between 294 and 332. Since parts of Sarmatian territory, as we have seen, were occupied by Roman troops, and since these garrisons, according to the *Notitia* were permanent, we cannot regard the action of Rausimodus as a simple plundering raid, but rather as an indication that peace on the *ripa Sarmatica* could no longer be secured in the manner successfully used in 294. The attack by Rausimodus must have been caused by increasing pressure from the Goths, for the Roman victory over the Goths in 332 within Sarmatian territory[51] can only mean that before this date the Goths had already invaded an area which was officially recognized as Sarmatia. The testimony of a further source allows us to infer that in 332 the Sarmatians were allies of Rome,[52] evidently as a result of the treaty renewed in 322. Since we are further informed that after 332 a huge mass of Sarmatians was settled in the Balkans and in Italy,[53] we must interpret these events in the light of the old alternative: armed aid or *receptio*, the root cause being increasing pressure exerted on the Sarmatians by the Goths and other Germanic tribes. When in 322 Constantine marched across the country of the Sarmatians he might well have convinced himself that they were caught between two fires and that their problem could be solved only by effective armed aid or by receiving some of their groups into the empire—i.e. by *receptio*. To evacuate Sarmatia altogether and thus to sacrifice it to the Goths, Gepidae or Vandals was apparently a solution that could not be seriously considered. Constantine therefore had no choice but to transfer Sarmatian groups into the empire and at the same time to provide effective armed assistance for those that remained outside. The Devil's Dyke which cuts off Sarmatian country towards the north and the east will have been recognized by the Romans as a frontier, and very possibly it was under the formal supervision of Roman garrisons in Sarmatia; this produced a commitment which Roman policy could no longer avoid in the old manner.

This is the reason why in 332, when the Goths penetrated into Sarmatia, the Romans made a decisive and energetic intervention on the Sarmatian side and 'defeated the Goths in the country of the Sarmatians'.[54] This success, aided by large-scale fortifications on the lower Danube, could have forced the Goths to keep quiet for some time, if the Sarmatians had remained content with Roman help. Instead they took a desperate measure that brought its own immediate punishment, by arming their subjects, the Limigantes. After the victory over the Goths an uprising of the armed Limigantes broke out, and part of the Argaragantes, the Sarmatian ruling class, had to flee into the empire.[55] Another part of the Argaragantes were given refuge by their northern neighbours the Victohali.[56] Then followed two years of war with the insurgent Sarmatians, and Constantine's own presence there in 334.[57] A huge coin-hoard at Campona suggests heavy losses among troops on the frontier.[58] In a description of much later events in Ammianus we learn[59] that the Limigantes were organized under Roman supervision in the border section of Pannonia Secunda and Moesia Prima. They were divided into two groups, named after two Roman forts: the Amicenses after Acumincum (Slankamen near the mouth of the Tisza) and the Picenses after Pincum (Veliko Gradište); it seems that control over them was in the hands of the commanders of these forts.

The plain, which had become somewhat depopulated as a result of these wars, and particularly owing to the emigration of the Argaragantes, became a bone of contention among Germanic tribes; the wars between individual tribes are reported—with many gaps and in a mutilated form—only by Jordanes. Despite the Roman victory over the Goths, the latter, and perhaps the Gepidae gradually too, probably became masters of large stretches of the plain. The Vandal king Visumar, who likewise laid claim to the depopulated country of the Sarmatians, was defeated by the Goths and fled to the Roman empire.[60] In the eastern half of the plain, particularly in the area north of the river Maros and east of the Tisza, Germanic influence becomes perceptible in the finds of the fourth century,[61] and this suggests that the Devil's Dyke gradually lost its role, at least in that part facing the original Dacia now occupied by the Goths and Taifali.

It is very likely that the transfer of Sarmatian groups into the empire went far wider than reception of the expelled Argaragantes, and was deliberately and systematically undertaken by Constantine. For we have records of a mass of people in unprecedented numbers, who were settled in Thrace, Scythia Minor, Macedonia and Italy, whilst more than twenty years later we hear of Argaragantes who had remained beyond the frontier.[62] The war of 332–4 thus displays certain similarities with the Marcomannic wars: the enemy was defeated but *receptio* was also employed. Constantine had presumably recognized that force and conclusion of treaties were not by themselves sufficient when the peoples in the border lands outside the frontier came under pressure from beyond.

The combined method of *receptio* and military intervention now resulted in a longer period of peace on the middle Danube. From *c.* 335 to 356 no disturbances worth mentioning took place on the *ripa Sarmatica*. Constantine presumably did not neglect to negotiate with the Germanic tribes who had invaded the Lowland; the treaty with the Goths on the lower Danube no doubt had its parallel on the *ripa Sarmatica*.

All our evidence so far supports the view that late Roman fortifications on the *ripa Sarmatica*, new buildings and reconstructions alike, were first begun under Constantine. We have already pointed out the absence of a Diocletianic period of construction in Pannonia, and we have connected this with a new frontier policy inaugurated by Diocletian, with which the Devil's Dyke on the plain may be associated. If this hypothesis—over-bold in the present state of research on the earthwork—is correct, it follows that the very marked Constantinian period of construction on the Pannonian *limes* means a return to traditional frontier policies of the pre-Diocletianic period. The events of the years 332–4— if their interpretation is correct—may throw light on this return from a different point of view.

The periods in the history of the Upper Moesian *limes* are less clear. The very intensive new work on the *limes* in the Djerdap has produced a number of new fort-plans, but a definitive identification of their periods must be reserved for future work (Fig. 46). The situation is rendered more difficult by the further fact that fortification of the Iron Gates was also undertaken under Justinian. Furthermore, this particular section of the *limes* was least accessible from the left bank, and so the absence of certain types of tower by no means proves their absence on other parts of the *limes* in Moesia Prima and Dacia Ripensis. Moreover, a kind of Devil's Dyke is also known in the country opposite Dacia Ripensis: this is the Brazda lui Novac, which runs eastwards from the Iron Gates and is designed against an enemy from the north.[63] It could equally well be attributed to a time when places on the left bank of the Danube were not yet given up after the evacuation of the old Dacia and therefore had to be protected, as to the activities of Constantine who had bridges constructed and roads built

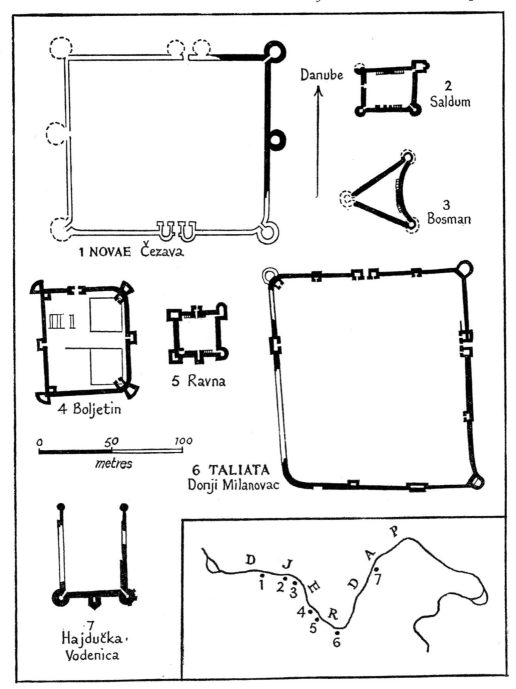

Figure 46   Late Roman forts in the Djerdap

on the other side of the Danube in that particular section. It would be premature to suppose that the Devil's Dyke in Sarmatia and the Brazda lui Novac must be of the same date. But the evidence hitherto is in favour of a Constantinian date for the building programme on the Upper Moesian section of the *limes*.

The fact that construction work on the Danubian *limes* was not restricted to existing forts can be seen in the fort-lists of the *Notitia Dignitatum*, which contain many new place-names. Nor is there any lack of late Roman fortifications revealed by excavation on the bank of the Danube, but their exact date of construction has only occasionally been established. In default of sure dating evidence a number of fort-plans must be omitted from any reconstruction of the history of the frontier.

Among Pannonian fortifications on the Danube uncovered so far (Fig. 47), fresh buildings of the late Roman period occur in considerable numbers.[64] A common characteristic must be pointed out: without exception they were built on elevations and, if the ground allowed it, on plateaux with steep sides. Further common features were the irregular outline normally adapted to that of the plateau, the small number of gates (usually one only) and the great thickness of the walls compared with those of earlier forts. These common traits, however, do not furnish any clues for dating, nor are the small finds, coins and pottery sufficient for closely fixing the date of construction. But there is one type of tower which is certainly of Constantinian date; it was preferred both in new buildings and also in reconstruction of earlier forts. This is the 'fan-shaped tower'. It has several variants which were probably not the result of typological development but local variations of one and the same type. In new fortifications these towers were built in such a way that their walls were radial extensions of the line of the fort-walls at the corner, or they were left open towards the interior of the fort.[65] In fortifications which already existed the towers were added to the rounded corner, and the curving part of the fort's corner was presumably pulled down in order to open the tower towards the interior.[66] In practically all forts on the *limes* a regularly recurring sequence can be observed at the angles (Fig. 18, p. 108); an internal trapezoidal corner-tower dating from the first stone period in or after the middle of the second century; next, a small rounded external tower dating from the end of the second or the beginning of the third century, and finally a long projecting fan-shaped tower of the time of Constantine. The dating of the fan-shaped tower is based on the following facts: at the newly built, roughly triangular hill-top fort of Visegrád (Pl. 14b) monetary circulation starts only under Constantine; second, the renaming of Ulcisia Castra as Castra Constantia[67] is presumably connected with the rebuilding associated with this type of corner-tower; and third, when the fort of Campona

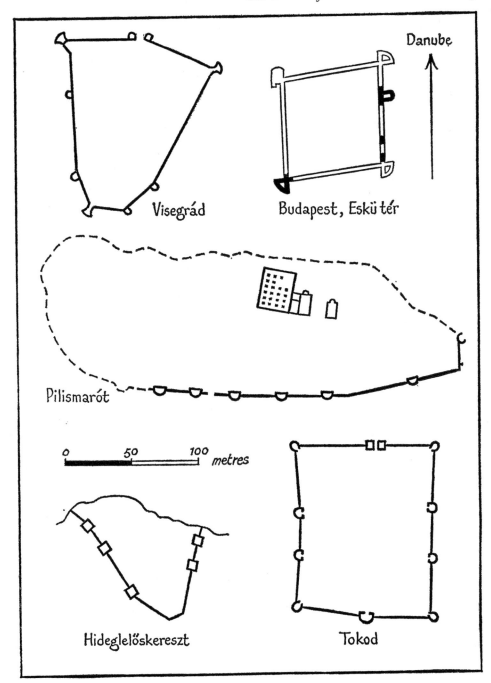

Figure 47 Plans of late Roman forts in Pannonia

was destroyed in 322 and had to be reconstructed it was given fan-shaped towers.

A further change must be connected chronologically with this type of tower. So far, in at least three *limes*-forts, it has been established that the *portae principales* or the *porta decumana* were cut off from the outside by a huge curving wall (Fig. 18, p. 108); it had the effect of converting the former gate into a single complex of three towers.[68] From now onwards the *limes*-road had to make a detour, as can be seen particularly well at Campona in the behaviour of the modern road. This alteration has a decidedly defensive character and has its parallel in the newly built fortifications with only one gate or two at the most. The ditch was also retained in the fourth century, and sometimes it was even doubled. At some forts the ditch ran remarkably far from the walls, which had the effect of providing a very wide berm between ditch and wall.[69] The projecting towers therefore had the evident purpose of allowing an enfilade of the besiegers in front of the walls.

The earthen ramparts which had supported the fort-wall from the inside were sometimes removed and buildings added to the inner face of the wall. This can be demonstrated at the legionary fortress of Carnuntum[70] or at Campona, and in all new fortifications the bank was omitted;[71] presumably it was for that reason that they had thicker walls than the earlier forts of the *limes*.

Further types of towers, as yet not accurately datable, are semicircular projecting interval-towers which are characteristic of new late Roman building;[72] horseshoe-shaped interval- and corner-towers[73] which, although sometimes possibly of Constantinian date, are probably in general characteristic of more than one late Roman period; and lastly circular corner- and interval-towers which probably come late in the history of late Roman construction: they occur as subsequent additions on late fortifications.[74]

In general we may conclude that the system of strong-points and forts on the *ripa Sarmatica* became substantially denser during the fourth century. It is less probable that a comparable amount of construction was going on on the front of Pannonia Prima which faced the Germans. At present frontier research—especially in the fairly recent excavations of S. Soproni—has shown that the Danube bend in north-east Pannonia was provided with exceptionally strong defences (Fig. 48). Whether the compact row of fortifications is characteristic of this section of the frontier only, or whether the results obtained there may be generalized to apply to the whole of the *ripa Sarmatica*, cannot yet be decided. It is equally uncertain whether certain types of tower can be taken as indicative of simultaneous building activity along the whole of the bank of the Danube or only within shorter sections of the frontier. The distribution of very typical late tile-stamps bearing the names of officers, units or places to some extent supports the idea that the construction of fortifications was differently organized in each province: some types of stamp are limited to the *limes* sector of a single province; those of the *limes* in Moesia Prima and Dacia Ripensis illustrate this particularly well.[75] If the expansion of the letters DRP on a group of tiles whose main distribution lies in the area of Dacia Ripensis, as D(acia) R(i)P(ensis) is correct,[76] then we have in these tiles a *terminus post quem* for extensive building activity on this stretch of the bank, for Dacia Ripensis as the name of a province is attested only from Constantine onwards.

As can be seen from walled-up gates of forts and newly-erected hill-top strong-holds, Constantinian building activity on the Danubian *limes* was based on a defensive strategy. The representation of fort-gates on coin-reverses, frequently repeated since the time of Diocletian, is proof that this defensive policy was deliberately published as political propaganda.[77] Peace on the Danube, of course, had been secured not by these fortifications but by the martial and diplomatic successes of the emperor, and not least by the mutual animosities of barbarian neighbours. Nor is it likely that the title Sarmaticus assumed in 339 by Constans referred to clashes of importance on the Danube. Contemporary literature says nothing of a war which could justify this title, and we might rather conclude that totally unimportant frontier skirmishes gave a welcome opportunity to the young emperor to award himself the title of Sarmaticus after the example of his father.[78] Constans is, indeed, attested in Pannonia in 339: in the spring he was at Savaria, soon afterwards at Mursella north-east of Savaria, and only in September can he be shown to have been at Naissus.[79]

After the death of Constantine the Great, his sons assembled at Viminacium[80] to divide up the empire. The area of the middle Danube fell to Constans, who had been placed in the guardianship of his eldest brother, a guardianship which he soon energetically rejected. Constans stayed in our area mainly at Naissus until the spring of 340, that is until the defeat of the brother who was his rival. Afterwards he twice again returned to the Danube from the west: on both occasions he is attested for a short time at Sirmium (in 342 and 346).[81]

It was only in 350 that our provinces became the scene of important events once again. Magnentius had murdered Constans in 350 and had assumed the purple in the West; he was on the point of seizing the Danube region without difficulty. It was the idea of Constantia, a daughter of Constantine the Great, to set up a counter-emperor in the Danubian area, in order to save the dynasty. Vetranio, the old man chosen for this purpose, was in no way a representative of the Danubian area. Although only a tool, he was remarkably quickly pro-claimed in Mursa and soon afterwards in Sirmium, too;[82] this gives a hint of the survival of certain old Illyrican tendencies. Vetranio was anything but capable—presumably it was for that very reason that he had been selected—and after deplorable hesitations he finally moved out at the end of 350 to meet Constantius II who was making a leisurely approach from the east; he greeted him at Serdica and on 25 December abdicated at Naissus at a ceremony arranged in as festive a way as can be imagined.[83]

Constantius II spent the winter presumably at Naissus, or more accurately not far from Naissus at the large villa owned by the dynasty at Mediana,[84] and in the spring of 351 advanced to Sirmium where Gallus was created Caesar.[85] The following months brought a heavy trial for Constantius II and once again made southern Pannonia the scene of war. Magnentius appeared in Pannonia from north Italy with a very large army, consisting in part of Rhenish barbarians.[86] The whole of the summer was taken up with a gradual retreat by Constantius along the major road leading from Emona via Poetovio and Mursa to Sirmium. A first fight near Atrans left Magnentius the victor. The latter did not wish to pursue his enemy who in the meantime proposed negotiations, but advanced up the valley of the Save in the direction of Sirmium, took Siscia and attempted the siege of Sirmium. The army of Constantius in the meantime remained at Cibalae. Apparently the adversaries hesitated long before they at last joined battle south of Mursa on 28 September.[87] Magnentius had to withdraw to Italy and Constantius entered Sirmium where he stayed till May 352.

At the battle of Mursa the empire suffered an enormous loss of men fit to bear arms. Nevertheless the situation was not exploited by the barbarians near the frontier of Pannonia; along the border quiet reigned, a further proof that the wars with the barbarians cannot be explained by the simple formula of Roman–barbarian confrontation. Since the development of the client-system in the first century, great attacks by the barbarians were caused by the situation beyond the frontier rather than that within the empire, and the wars that Constantius II had to wage on the Danube a few years after his victory over Magnentius can also be interpreted only in the light of this consideration.

In 356 whilst the emperor was in Rome, the Quadi, who had not been heard of for many decades, overran the province of Valeria, while the Sarmatians attacked Pannonia Secunda and Moesia Prima, that is the *limes* on the southern border of the plain.[88] How far these invasions penetrated into the provinces we do not know;[89] it by no means follows from the later negotiations for the return of prisoners of war[90] that the invasions were comparable to the great disasters under Marcus or Gallienus. The brevity with which Ammianus Marcellinus dealt with this attack suggests that it was itself of no great importance, though its consequences for the future on the *ripa Sarmatica* were far-reaching.

Constantius II arrived in Pannonia in the summer of 357 and entered negotiations with the Quadi and Sarmatians.[91] The following years were taken up with various operations and negotiations on the *ripa Sarmatica*, the emperor meanwhile staying continuously in Sirmium. He left Sirmium only in the middle of 359, to go to Constantinople on the news of the fall of Amida.[92] Our only source, which is however very detailed and occasionally well-informed, is Ammianus Marcellinus, though his record contains a few inconsistencies.[93] For instance, the information that the Quadi and Sarmatians had attacked the empire during the emperor's stay in Rome can presumably be taken to be a duplication, since the same is reported for the winter 357–8 in similar words, and according to Ammianus the emperor travelled from Rome to Illyricum, whereas no traces of such a journey are to be found in the well-attested itinerary of Constantius. It was in the summer of 357 that Constantius first came to Pannonia.

The fact that the emperor, though present in Sirmium in 357 and receiving emissaries from the Quadi and Sarmatians, did not start his great expedition before 358, suggests that the situation in 356–8 resembled a *bellum suspensum*; the barbarians made demands which were then the subject of negotiations. Only the attack of the Quadi and Sarmatians in the winter of 357–8 induced the Romans to interrupt these negotiations which were being carried on within the setting of the presumably very complicated client-system, and to proceed to war. The report of Ammianus on the expedition of Constantius II is written very much in the vein of the late Roman ideal of the triumphant emperor. Therefore we learn details only from the homages paid by defeated peoples and little is said about which of the many small kingdoms were in fact defeated, which of them were merely intimidated, and where the fighting actually took place. Ammianus gives a fairly long list of Sarmatian and Quadan kings, each of whom apparently possessed an alliance with Rome of his own and who were independent of each other. They are listed, however, only after the emperor's victorious advance, whereas the description of the expedition, which is in any case short, merely talks of Sarmatians in general.

The expedition started in April of 358 when Constantius II led his army from Pannonia Secunda over a pontoon bridge against the Sarmatians. Evidently great store was set by the Roman attack achieving surprise. After their country had been plundered far and wide, the Sarmatians retreated into valleys which can only have lain among the Dacian mountains on the eastern edge of the Lowlands. From this it can be deduced that the expedition advanced towards the north-east. Another army advanced from Valeria and looted and burnt in a similar fashion in the northern sector of the Lowlands. The emperor's army then effected a junction with these other troops and the united army invaded the territory of the Quadi.

This is the extent of our knowledge about the expedition. Ammianus continues his report with the negotiations for peace and with the homages paid. First Zizais, a youth of royal birth, approached the emperor, and then, encouraged by the mild treatment received by Zizais, came Runo, Zinafer and Fragiledus. Among other things they all proposed to put themselves and their people under Roman rule: in other words they offered themselves for *receptio*. After a client-treaty had been made with these chieftains, Araharius and Usafer, kings of north-westerly Sarmatian and Quadan peoples, made their appearance.

A particular problem was posed by the Argaragantes who appeared only after the others. As we have seen, these had once fled to the Victohali. They believed that now their time had come. They came forward with the demand for a guaranteed freedom. They were given Zizais as their king.

All these negotiations were carried out by the emperor in person. In Ammianus' manner of description, they have the appearance of a continuous ceremony in conquered country rather than peace-negotiations with peoples all equally defeated. The possibility cannot be excluded that the imperial expedition in the spring was nothing but a demonstration for the purpose of intimidation, in order once again to convince neighbours of the frontier under pressure from beyond that Rome intended neither to conquer them nor to accept them into the empire.

The negotiations with the Sarmatians took place somewhere in the country beyond Valeria. From here, court and army proceeded to Brigetio, the base of one of the expeditions into the country of the Quadi. The Quadan kings—Vitrodurus, Agilimundus and others—were ready at once to make a new client-treaty with the Romans.

As we have seen, the Romans were only interested in maintaining the system of client-kingdoms along the border. The first king of the Sarmatians with whom negotiations were entered was Zizais, who was instituted as king of the Argaragantes. From this we may suggest that in the last resort the unrest had started with the Argaragantes, the problem of whom had not been solved in a sufficiently radical manner by Constantine. In the war of 332–4 Constantine had been satisfied with receiving into the empire some of the Argaragantes who had been expelled by the Limigantes, and letting the rest of them disappear from Roman sight, and he laid greater store on obtaining control of the Limigantes, the new masters of the lands beyond the borders of southern Pannonia and Moesia. Although the exiles—*exsules populi* in Ammianus—found hospitality among the Victohali about 334, they probably bore the main responsibility for the ultimate assumption of arms by the Quadi.

That the initiative for the Sarmatian–Quadan attack in 356 and 357 came from the Argaragantes also follows from a phrase used by Ammianus.[94] The expedition of Constantius II was thus aimed at peoples made restless by the Argaragantes. The fact also enables us to understand the demand of the Argaragantes that the Romans should grant them their freedom, for they had had bitter experiences from the armed assistance given by Rome in 332: Constantine though supporting the Sarmatians against the Goths had not reconstructed the old-accustomed social stratification.

The attitude of the Limigantes during the expedition against the Argaragantes and the Quadi is very obscurely described by Ammianus. It is stated that they had felt induced by the activity of the Argaragantes to attack the Roman frontier, compelling Constantius II to punish them as well in the end. Since the Danube border of Moesia lay opposite the Limigantes and since according to Ammianus this section of the frontier had also been attacked in the winter of 357–8, it is probable that the Limigantes did indeed undertake an action of their own. But we cannot exclude the possibility that the new treaties with the other Sarmatians had arranged for common Roman–Sarmatian action against the Limigantes: for it was Zizais in particular who took part as a Roman ally in the campaign against the Limigantes, and according to Ammianus the action was aimed from the beginning at the resettlement of the Limigantes. In his summary of the outcome of the war, Ammianus stresses two results: the restoration of their land to the people who had been expelled, and the establishment of a new king, that is one recognized by treaty as a Roman client. Thus one can perhaps count the solution of the problem of the Limigantes among the main aims of the expedition. Ammianus devotes more attention to this part of the expedition than to the first part. In the campaign against the Limigantes, the Romans, the rest of the Sarmatians under Zizais, and the Taifali proceeded in alliance. The revolt of 332 was revenged with unprecedented cruelty, but it was not possible to reconstitute the old social order. The Limigantes had to evacuate the country opposite Pannonia Secunda and Moesia Prima, which they had occupied, and in it the Argaragantes were permitted to settle again. The Limigantes received land elsewhere and also a client-treaty.

After these achievements Constantius II victoriously entered Sirmium in June 358. There he spent the winter, during which news arrived that the Limigantes had left the land assigned to them and that there was danger of disorder once more in Sarmatian territory owing to their raids. In April 359 the emperor therefore again prepared an expedition. This time he moved with his army into the province of Valeria and from there sent two tribunes to the Limigantes to call them to account. The Limigantes answered that they desired to be accepted into the empire and to be given land in some part of the *orbis Romanus*. Constantius considered this a conclusive solution to the problem of the Limigantes and prepared for a ceremony of which the scene chosen was the bank of the Danube not far from Aquincum. The only precaution to be taken was the placing in readiness of a unit of the fleet on the Danube. The Limigantes, however, on being admitted to the emperor's presence attempted a surprise attack on him; in the ensuing tumult Constantius barely escaped with his life, and the Sarmatians were only put down after a hard struggle. But the massacre secured quiet on the *ripa Sarmatica* and peace lasted till 365.

Constantius II soon afterwards left Pannonia for his last visit to Constantinople and the East.[95] The change of government in 361 affected our provinces only inasmuch as Julian, descending the Danube in secret, had to make a short stop at Sirmium and afterwards stayed in Naissus till the end of November, where the news of the death of Constantius reached him.[96] In the autumn of this year massing of troops did take place in southern Pannonia and on the diagonal road across the Balkans;[97] but on the whole the change of government was peaceful and there are a few indications[98] that paganism in Illyricum was still strong enough to give occasional expression to loyalty towards Julian. On the occasion of the proclamation of the Moesian Jovian and of the Pannonian Valentinian, an Illyrican junta of the old sort hardly played any decisive part. The Pannonian origins of the Valentinianic dynasty gave rise to repeated controversies between Illyrian and non-Illyrian cliques—some of this can be seen in the attitude of the *Historia Augusta* towards Sirmium and the Pannonians, and Ammianus also was distinctly anti-Pannonian—but Valentinian himself was anything but representative of an Illyrican trend. He did not reside in Sirmium, and on his visit to Pannonia in 375 he showed little feeling for the complaints of his fellow-countrymen.[99] It cannot, of course, be denied that the choice fell on the Danubian officers Jovian and Valentinian because the officers making it had some awareness of the prestige of the old Illyriciani. Both emperors, who successively entered into the heritage of the Constantinian dynasty, came from the surroundings of Sirmium: Jovian from Singidunum[100] and Valentinian from Cibalae.[101] They presumably owed their military careers simply to the circumstance that their family homes lay in the vicinity of an imperial residence. A number of other important personalities of this period were Pannonians too, and they obtained swift promotion for the reason presumably that they were fellow-countrymen of the emperors. But in the second half of the fourth century Sirmium had already lost its central role. In the division of the empire between Valentinian and Valens, Sirmium happened to fall in the periphery of the two areas; the centres of gravity of east and west then became finally fixed, and the emperors stayed in the new centres.

Valens and Valentinian had shared court, power and empire at Mediana, the large imperial villa outside Naissus.[102] In June 364 they were still attested at Naissus, and in the summer they were both in Sirmium.[103] Next year Quadan and Sarmatian attacks are reported,[104] which were probably dealt with by the *receptio* of some Sarmatian group; for not long afterwards the Gallic poet Ausonius writes of the 'recently settled Sarmatians' in the valley of the Mosel.[105] The barbarian attacks—which are attested on practically all frontiers of the empire[106]—were presumably connected with the change of government. Possibly a parallel can be seen in the 'Sarmatian war' of Constans, which broke out soon after the death of Constantine. It may have been more or less regular for barbarians beyond the frontier to consider their treaty as void on the death of the emperor, or to fear that the new emperor would make a less favourable treaty. However, the death of an emperor could also give rise to the hope of a more favourable treaty.

The attack of the Quadi and Sarmatians in 365 could be dealt with easily and briefly. The centre of unrest shifted into the area of the lower Danube, the *ripa Gotica*, where fortification building was started energetically that year.[107] But building activity on the *limes*, which receives good notices in the contemporary literature, was not caused by the immediate situation, for it was soon taken up on the *ripa Sarmatica* and on the *limes* of Pannonia Prima as well. It is one of the most serious problems of Pannonian frontier research that this activity, which was probably extensive, cannot as yet be sufficiently clearly distinguished from the other periods of construction, which are themselves only roughly determinable. For a long time an apparently reliable clue was provided by tiles bearing the stamps of the *duces* Frigeridus and Terentianus—both in Valeria—datable to the year 375. And since these tiles are mostly found associated with other late Roman tiles, many other tile-stamps were considered to be also certainly of Valentinianic date. But it has recently been shown that some stamps— especially those of the so-called OFAR group, which are attested in the fortified landing-places of the *ripa Sarmatica*—cannot be of Valentinianic date, and that the *dux* Frigeridus cannot have functioned in Valeria in 375.[108] The Valentinianic period of construction on the *limes*, which is certainly important, therefore cannot be clearly defined. The most important question is still undecided: did Valentinian mainly repair existing forts and, apart from that only order a great number of *burgi* to be built, or are the fairly numerous new late Roman fortifications attributable to him? If so, this would mean that a colossal amount of

291

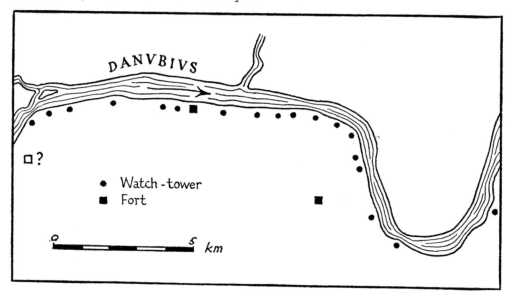

Figure 48   Late Roman forts on the Danube bend

work on the *limes* was carried out by Valentinian. A fortunate discovery made recently shows that the very numerous tiles with the stamp *Frigeridus vir per-fectissimus dux*, known from many places in Valeria, must be datable to 372 at the latest.[109] The tiles in question, however, were roof-tiles, which attest either the completion of the buildings from which they came, or else mere reconstructions and repairs: they do not prove new buildings. The only certain new buildings of Valentinianic date are a few *burgi* and one fort, which are named on building-inscriptions.[110] But these very inscriptions give rise to the suspicion that Valentinian ordered new construction at only a small number of places: it is certainly curious that small watch-towers, never exceeding 8 by 8 m in size, were given a long, pompous inscription. The problem becomes even more complicated when we take the distribution of stamps of the Frigeridus group and the find-spots of the building-inscriptions into consideration as well: practically all come from the Danube bend, where recent excavations have indeed discovered an astonishing number of *burgi*, all of the same size (Fig. 48).[111] Does this indicate that Valentinianic building activity was restricted to the section of the *limes* above Aquincum and below Brigetio?

The director of construction was Aequitius, *magister militum*, named on all Pannonian building-inscriptions. Work probably started soon after the accession of Valentinian, as the construction of a 'fort with towers' was completed in 367 if not earlier. Two *burgus*-inscriptions are dated to 371 and 372 respectively; the latter was recently found in a small *burgus* at Visegrád (Pl. 36b). The other *burgus* was built for the supervision of trade across the frontier. Unfortunately the find-spot of the building-inscription relating to the fort is not known accurately; it can be related to one of several forts in the vicinity of Solva (Esztergom).

Buildings on other sections of the Danube frontier are even less easily identifiable. A building-inscription—unfortunately fragmentary—came to light in Carnuntum;[112] officer-stamps of the Frigeridus group are met with in Valeria in other places too, and small *burgi*, corresponding in every detail to that of Visegrád, are proved in some places, though never so numerous as on the bend of the Danube, where the hilly bank and the meandering of the river (Pl. 14b) necessitated a compact row of towers.[113]

By 373 building activity was extended to the land beyond the frontier. Ammianus reports that Aequitius had a fort erected on Quadan territory,[114] and a lucky find has recently revealed the presence of a *burgus* of Valentinianic type in the country of the Sarmatians.[115] This *burgus* stood on the northern border of the Lowland, about 60 km away from the Danube, and so deep inside Sarmatian territory that the security of the small garrison could only have been guaranteed by treaty and/or by the presence of Roman troops in Sarmatia. There is further evidence for the stationing of Roman soldiers there in the *Notitia Dignitatum*, which lists the two forts built in Sarmatia in 294 as lying *in barbarico*.[116] They were thus still occupied in the second half of the fourth century. There is no mention of Roman posts in the country of the Quadi before Valentinian, and from Ammianus' description we may be sure that his construction of forts there was indeed a new experience for the Quadi, since the events of the years 374–5 were caused by this very work. Before Valentinian the Quadi had not been dangerous, as—amongst other evidence—the results of fighting on the approaches to Pannonia testify. Throughout the fourth century they always appear merely as easily pacifiable allies of the Sarmatians and were never of more than secondary importance on the Pannonian *limes*, as is also emphasized by Ammianus.[117] Valentinian's desire to occupy Quadan like Sarmatian territory with watch-towers and garrisons is either based on a new strategic conception or is due to increasing tension in the Carpathian basin. Whether one should attach importance to Ammianus' expression: 'as if the country of the Quadi were already under Roman rule',[118] I would prefer to leave undecided. It is, however, strange that Ammianus speaks of only *one* fort in the country of the Quadi, and that Valentinian's punitive expedition of 375 did not advance into the centre of the country but attacked their eastern border, the hilly region opposite Aquincum. It follows from the negotiations of Constantius II in 358 that the centre of Quadan territory was near Brigetio. It is therefore quite possible that the fortification was in fact a link in the chain of occupation of Sarmatian country, and was to be constructed somewhere in the Sarmatian–Quadan border area. Since Roman posts could not be erected in a barbarian country without a previous war or treaty we must assume that the Quadi lodged a protest with Aequitius, on the ground that the fortification had been started without a previous Roman–Quadan agreement; and that the reason why it had been undertaken without this previous agreement was because the place in question was in a border area, presumably the northern border of Sarmatia.

This interpretation of events is confirmed by the fact that Aequitius immediately ordered construction of the fortification to be stopped, when the Quadan emissaries had protested against it. Because of the delay in its building, Aequitius suffered an accusation, and the young Marcellianus, who was the son of Maximinus from Sopianae in Valeria, was made *dux Valeriae*. The young man ordered the construction of the fortification to be resumed, and committed the catastrophic blunder of inviting Gabinius, the Quadan king, under the pretext of negotiations, and then killing him.

The Quadi decided to avenge this outrage. This time they were the ones to call in the armed assistance of the Sarmatians, and both peoples invaded Pannonia in July 374. The invasion found the provincial population entirely unprepared; the country people engaged in harvest were slaughtered or dragged off, and the daughter of Constantius II, on the way to join her bridegroom Gratian, was almost taken prisoner. From the large number of coin-hoards ending with coins of Valentinian[119] from all parts of the province, various invasions, by several different groups, can be inferred, though new study shows that some finds which were traditionally associated with this war must be excluded.[120] The imperial city of Sirmium was in greater danger than it had been for more than a century. Probus the praetorian prefect had the neglected fortifications of the town repaired: money from the city treasury, which had been collected for building a theatre, was made use of for this purpose. Two legions, which together could have got the upper hand against the Sarmatians, were not coordinated because of staff rivalries and were defeated. Only the *dux* of Moesia Prima, the later emperor Theodosius, succeeded in defeating the plundering barbarians. They retreated and declared themselves ready to comply with the demands put to them: they were all the more ready to do so, since news of the arrival of troops from Gaul diminished their chances.

So far the account of Ammianus: he also gives a hint that it was the Sarmatians who did the larger part of the damage in Pannonia even though the

Quadi had begun the war. When Valentinian left Trier for Pannonia in the spring of the next year, Sarmatian emissaries did in fact come to meet him, with a variety of promises.

Valentinian arrived at Carnuntum about the end of April or the beginning of May.[121] He was accompanied by the imperial family with the court and army and established his headquarters near Carnuntum. From information given by Ammianus it can be deduced that the court lay at the Villa Murocincta, 100 Roman miles from Brigetio and in the vicinity of 'deserted and ruined' Carnuntum, i.e. probably in the huge villa of Parndorf.[122] The fact that Valentinian had to remain there for three months—the whole of the summer—is evidence that the purpose of his visit was not merely the preparation and carrying through of an expedition, but that a number of affairs required putting in order in Pannonia also. Even the preparations for war, of course, met with difficulties: since the war of the previous year had happened during harvest, provincial stocks were presumably not sufficient for a large expeditionary force. The tax policy of Probus, the praetorian prefect of Illyricum, moreover, had given rise to general discontent, giving the emperor much to do.

At the end of the summer the Roman army at last departed and attacked the Quadi from Aquincum. A pontoon bridge was built quickly and unnoticed north of Aquincum, and the imperial troops surprised the Quadi who did not dare to fight in the open. The cavalry under Merobaudes had already started an advance from a different point. Valentinian plundered and burnt the country of the Quadi, and at the beginning of November returned to Aquincum. Quadan emissaries were received at Brigetio, but during this reception the last Pannonian to occupy the imperial throne was overtaken by death on 17 November.

The next days were occupied in Brigetio with negotiations inside the supreme command, since it was to be feared that the army would demand an emperor. In order that an imperial election might be conducted in the camp in the least damaging manner, the boy Valentinian II was fetched from Murocincta and presented to the soldiers. The proclamation fell on 22 November, whereupon mother and son went to Sirmium, which was chosen for them as their residence.[123] Valentinian II stayed in Sirmium until 378; he was the last emperor to reside in this famous Pannonian town.

How the problem of the Quadi and Sarmatians was solved we do not know. Presumably the client-treaties had to be renewed, and possibly the annual payments increased. At any rate we hear nothing during the following years of any difficulties from the Quadi or Sarmatians on the Pannonian *limes*. But on the lower Danube the situation became all the more aggravated; here, pressure by the Huns had caused movements of the various Gothic groups. In 376 the *receptio* was, in fact, given to some of the Gothic princes, and this caused a chain-reaction of catastrophic events. These incidents in the Balkan peninsula, in which the area of Upper Moesia was also affected, opened a new period characterized by the settlement of *foederati* in the Danube area.

The middle Balkan area, that is the provinces of Moesia Prima, Dacia Ripensis, Dacia Mediterranea and Dardania, had enjoyed a peace which had lasted from the Tetrarchy until the death of Valentinian; its peoples owed it above all to their geographical situation. The frontier district of Moesia Prima formed part of the *ripa Sarmatica*, the centre of which was in Pannonia. The frontier of Dacia Ripensis was the western part of the *ripa Gotica*—where barbarian pressure on the Lower Moesian *limes* was concentrated. The middle Balkan area was of importance only as the deployment zone between the Sarmatian and Gothic fronts, and became the scene of fighting only when barbarian troops stood on Roman soil. When once the Goths had crossed the frontier of Lower Moesia, however, the centre of war shifted to the Balkan peninsula, and the strategic importance of Upper Moesia showed up clearly. As deployment zone it was used much less frequently for the settlement of *foederati* than were the remaining Danubian provinces, but from the end of the fourth century onwards it had as much to suffer as its neighbouring territories.

Pressure by the Huns was thus at first advantageous to Pannonia, where after the Gothic invasion of the Balkans the crisis was relaxed. The Sarmatians and Quadi were presumably easy to appease after 375, because Gothic pressure on the Lowland had decreased as the result of the Balkan invasion.

Chapter 9
The final period of prosperity

The late antique period in the Danubian provinces is often characterized by historians as one of gradual decline and disintegration. Responsibility for this rather one-sided generalization must be laid partly on the sources themselves and partly on the limitations of archaeological research. As far as the sources are concerned, it is often pointed out that late antique and early Christian inscriptions are rarities in the provinces;[1] gravestones were rarely erected any more, and the absence of building- and votive inscriptions is due simply to the poverty of the formerly rich and votive-loving municipal aristocracy. Stone sculptures became so rare that one might doubt the continued existence of stonemasons' workshops. As far as archaeological research is concerned, it long restricted itself, particularly in our provinces, to the *limes* and to the towns on the bank of the Danube, where it is true that appreciable prosperity cannot be proved. It is only recently that doubts have been expressed whether it is legitimate to extend the symptoms of decay, which undoubtedly existed along the Danube, to the interior of the country and to all social classes as well. The decline of inscriptions and the poverty in the frontier area are in contrast to a number of written sources, which report large and flourishing towns in Pannonia and Moesia, and count our provinces among those which were wealthy or at least self-supporting.[2] The excavations of the last decades in certain of the towns furnish unequivocal proofs that no decline can be seen at least in the towns in question. A careful consideration of the decline of Roman civilization in the Danube provinces is therefore opportune.

As we have mentioned (pp. 266, 272), several imperial measures towards the end of the third century indicate an intention to make agriculture in Pannonia and Upper Moesia profitable. The Emperor Probus ordered vines to be planted at the *Alma Mons* (Fruška Gora) north of Sirmium and at the *Aureus Mons* east of Singidunum, and these plantations he 'put at the disposal of the population'; furthermore, it is reported that he ordered the marshy area east of Sirmium to be drained 'in order to render native farmland fertile and to increase its extent'. A road-station east of Sirmium carried the name Fossis,[3] and the modern Serbian place-name is Jarak, which means 'ditch'. Improvements were also carried out by Galerius in the province of Valeria by canalizing the natural outlet of Lake Pelso in order to gain arable land. These works, ordered on the highest level, must have been considered of great importance at the time, and therefore they have been stressed in the concise imperial histories of the late fourth century.[4]

At the same time private persons were also presumably called upon to produce money and man-power for this 'economic plan': a curious altar from the north-west corner of Pannonia Secunda is intelligible only against the background of this economic policy enforced from above. This altar was dedicated to Liber Pater by a certain Aurelius Constantius and his son Venantius on the occasion of vines being planted on a very sizeable piece of land 400 arpennes in area (= 200 iugera = *c.* 100 acres). The kinds of wine are enumerated on the altar: Cupenis, Terminis, Vallesibus, Caballiori. Unfortunately these names create great difficulties. But it is interesting that the size of the piece of land is expressed in Gallic units of area. Whether ground in Pannonia was generally measured in arpennes we do not know, and we are therefore not able to decide whether Aurelius Constantius may have come from Gaul. The recording of a vine-planting on an altar we must suppose was due to the importance attached to this activity at the time.[5] The dating of the altar does not offer much difficulty, for the names Constantius and Venantius are typically late antique though the pagan vow suggests not too late a date. Attention should be drawn to the circumstance that the name and rank of Aurelius Constantius are erased; evidently the estate-owner fell into disgrace and it will be permissible to draw the conclusion that his estate passed into imperial possession. And this conclusion brings us to the period of internal unrest under the later soldier-emperors or to the later years of the second Tetrarchy.

Not far from the find-spot of this altar a mountain is attested in the *Itineraries* called *Aureus Mons*—which is not to be confused with the place of the same name in Upper Moesia [6]—where the vineyards of Constantius can be looked for. This mountain, like the Alma Mons and Aureus Mons in Upper Moesia which had been planted with vines by Probus, lay on the bank of the Danube; the river provided the transport of the wine. Viticulture in Pannonia is attested by finds such as vinedressers' knives,[7] which are also represented in the hand of the deity Silvanus (Pl. 31a), and by a few wine-presses;[8] Dio talks of Pannonian wine, however, in a very disparaging manner,[9] though at the time the surroundings of Aquileia are praised by Herodian as the wine-garden of Illyricum.[10] Pannonia was thus dependent on imports of wine even in the Severan period, and we have to take the vine-plantings by Probus as a measure aimed at securing economic independence for the province. Nevertheless, as late as 383 it is reported that Pannonia supplied corn to north Italy and was given wine in exchange.[11] It would be very interesting to learn what effect the vine-plantations on the Danube had on the trade-balance of Pannonia and Moesia, and how north Italy reacted to them. It is, of course, quite possible that the vineyards in Pannonia Secunda and Moesia Prima mainly served to supply the population of Sirmium.

Villas and their estates in the fourth century

The plantations and improvements towards the end of the third century presumably went hand in hand with the growth of large estates. In some parts of Pannonia and Moesia large estates made their first appearance at that time. This process led to Pannonia soon being counted among the provinces exporting corn and cattle. Late antique sources describe Pannonia as a rich and cheerful land.[12] The *Expositio totius mundi* (*c.* 350), which attests the agricultural exports of Pannonia, also mentions that Dardania supplied Macedonia with cheese and lard, and gives the information that Moesia was only capable of supporting itself. So Dardanian hill-farming had become capable of exporting, and the agriculture of Moesia too enjoyed a comparable growth.

The predominant type of estate in the late Roman period seems to have been of medium and of large size. The estates can be assessed with some probability from the size of the villas. The large villas of former periods continued in use in the fourth century; they were even enlarged and furnished with mosaics, as for instance the villa of Deutschkreuz near Leitha[13] and in the same area the huge villa of Parndorf, to be discussed below. It is, however, much more important that we already have a very large number of villas and of partly explored dwelling-houses which were first built in the fourth century.[14] Some of them were decidedly luxurious, as shown by stucco decoration, architectural stone fragments and wall-paintings. Often the bath was in a building of its own, and this also suggests that the villas were large.[15] In areas where even in earlier times large estates had been the predominant form of land-ownership, such as north of Lake Pelso,[16] new villas were erected, but large and small villas can also be demonstrated in areas where similar buildings are not previously attested. In this context the province of Valeria requires special mention: an imposing number of late Roman villas have been discovered within its borders.

As far as size and ground plan of late villas are concerned, we find a variety based on the introduction of ground plans which had not hitherto been employed in the province (Fig. 49). Apart from villas with a central peristyle or a central room,[17] there are large, sometimes even huge, palaces with corner-towers or wings,[18] other buildings erected around a large rectangular inner courtyard,[19] and also a number of smaller dwelling-houses which may or may not belong to an agricultural estate.[20] These smaller dwelling-houses are distributed particularly thickly in the near vicinity of Aquincum,[21] but this does not imply that elsewhere only bigger villas were built.

The exact dating of the late villas and rural dwelling-houses is as uncertain as that of the late Roman *limes*. For a few a date in the Constantinian period has been proposed, but it is not impossible that others were erected even later. The process of concentration of estates therefore probably continued during the fourth century; in any case we have evidence that Pannonian estate-owners were living on their estates till the end of that century.[22]

Within what was formerly Upper Moesia we have so far only a few ground plans of villas, for instance that of Remesiana, which belongs to the type of huge estate-centre with a large central court;[23] a palace from Kostol near Zaječar[24] lying in the area of the supposed municipium Aurelianum; and among others a dwelling-house with bath-suite[25] in the mining district on Mount Kosmaj. The date of construction of these villas and palaces is not known, but it is striking that from this area we have a few hoards and isolated finds of the fourth century which unequivocally point to extremely rich, high-ranking owners.[26] The state cameo from Kusadak,[27] south-east of Singidunum, was certainly in the possession of a fourth-century man of high rank, and the silver hoards including silver dishes (Pl. 35b) commemorating the *decennalia* of Licinius[28] (p. 277) attest rich people, too. We know much more of the imperial estates and palaces of Upper Moesia, about which the evidence is available in our sources and from the results of excavation.

Among the villas were some that belonged to imperial estates, and were furnished as palaces. It is *a priori* probable that Pannonian estate-owners, who were involved in the political struggles of the military anarchy and Tetrarchy, did not always emerge as members of the successful party, but often lost influence, wealth and even life, and suffered the confiscation of their estates. Moreover, emperors who originated in the Danube area presumably did not omit to erect a palace or summer residence at their birthplace; these palaces were, of course, at the same time centres of large imperial estates round about.

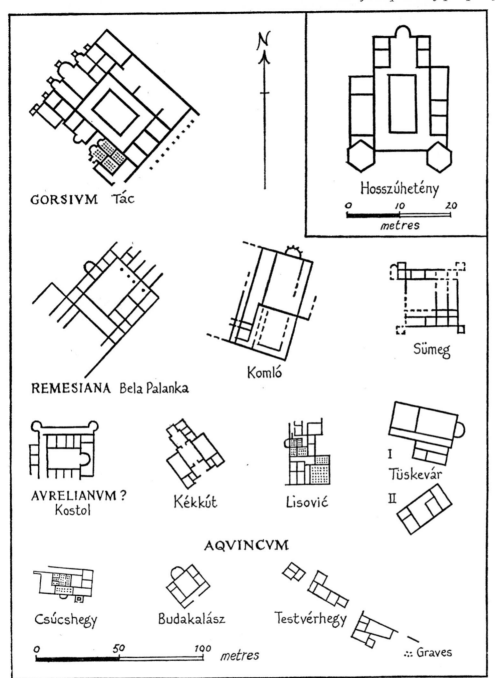

Figure 49 Plans of late Roman villas and houses

Maximianus Herculius ordered the erection of a palace at his birthplace not far from Sirmium,[29] Galerius had a palace in his birthplace in Dacia Ripensis, named Romuliana after his mother Romula,[30] and the Constantinian dynasty owned an immense seat at Mediana,[31] a few miles east of Naissus, where it is possible that Constantine the Great was born; it is presumably justifiable to maintain that influential representatives of the Illyriciani at the imperial court had luxurious country-houses constructed for themselves in their homeland.

Of the imperial palaces mentioned, that of Mediana has already been partly investigated.[32] The portion uncovered consists of state rooms provided with mosaic floors (Pl. 43b), walls encased in marble, and rich plastic decoration. Not far away household rooms with huge storage-containers and wine- or oil-presses came to light. The area enclosed by the wall of the Pannonian villa of Parndorf[33] was almost as large as a legionary fortress, and behind the palace long barrack-like buildings were found (Fig. 32, p. 170). At Parndorf it is not only the size of the site which suggests an imperial owner, but even more so the great *aula* or audience hall added about the end of the third century, and the luxurious furnishing of the palace: almost all the larger rooms were laid with mosaics about the beginning of the fourth century. It is not therefore rash to follow Saria in the belief that this villa was the scene of the conference of emperors at Carnuntum in 308. It is even more likely that the family and court of Valentinian stayed in this villa in 375: the summer-residence of Valentinian, the villa Murocincta, lay after all in the neighbourhood of Carnuntum, 100 miles away from Brigetio. All this fits well with Parndorf.

There were of course imperial palaces in the towns too, but they often lay in the suburbs, not in the centre of the town. A palace furnished with mosaics north of Poetovio was excavated in the nineteenth century; it too is mentioned by Ammianus.[34] He also writes of the *regia* at Savaria,[35] but we may expect the largest residence at Sirmium. The palaces of Poetovio, Parndorf and a few others to which we shall return later were equipped with mosaics which in ornament, style and execution are strikingly uniform and which can be attributed to a central workshop, possibly even to the same school that produced the early Christian mosaics in Aquileia.[36]

Within the rectangular walled enclosure of the villa of Parndorf a huge *horreum* (store-building) was uncovered (Fig. 32), proving that Murocincta was the centre of a large estate. This brings us to a question which so far has found no answer, but whose solution could probably be attempted in this context. In the interior of Pannonia a few late Roman fortified settlements are known, which are strikingly uniform in size, method of construction and ground plan.[37] The best investigated and best known is the fortification of Fenékpuszta[38] on the south-western corner of Lake Pelso (Fig. 50), not far from the place where the Sopianae–Savaria road, which played an important part in late Roman times as part of the road from Sirmium to Trier, crossed the river Sala. The walls of the fortification enclose a rectangle measuring 392 by 348 m, which is not compactly built up and shows no indication that the settlement had a military character. Inside the fortification a large *horreum*, a Christian basilica with many periods of construction (Pl. 38a), and a number of larger and smaller buildings were exposed, among them a house which in size and ground plan can be taken for a palatial seat. The defensive walls were very thick (2·60 m, with an even broader foundation); the corner-, gate- and side-towers, altogether about forty-four in number, were of large completely external circular plan, and the two gates, on the north and south sides respectively, were also fortified with an inner court-yard. To the fortification belonged extensive and comparatively rich cemeteries, in which a Christian tomb-chapel was also excavated.[39]

The lay-out of Fenékpuszta, whose ancient name was probably Valcum, has a good parallel in the fortification of Gamzigrad (Fig. 50) in Dacia Ripensis, the best-preserved and most imposing Roman monumental building in our provinces (Pl. 37).[40] Gamzigrad, too, is a rectangle (*c.* 300 by 230 m) enclosed by mighty defensive walls, which was only loosely built up. Up to the present, only one building has been uncovered inside Gamzigrad, a large palace appointed with all imaginable luxury: mosaics from the same school as the late Roman mosaics of Pannonia, rich architectural plastic art, sculptures of marble and porphyry, a large number of expensive objects of use and so on. Specially noteworthy is the theme of the mosaics in the largest room in the palace: this mosaic floor is divided into hunting-scenes, with the hunters wearing a shield-shaped badge on the upper arm like the hunters on the hunt-mosaic at the imperial villa of Piazza Armerina. The immensely strong fortification of this seat also suggests that its owner was the emperor. The defences were similar to those of Fenékpuszta, very thick walls (3–4 m wide), huge round towers (twenty in all), two gates and so on. Established as it is not far from a famous thermal bathing establishment, this fortification can perhaps be interpreted as an imperial hunting-castle, built in a fairly remote but pleasant region for the leisure of the emperors. Though it falls outside the scope of this work, it must be mentioned that Gamzigrad was rebuilt under Justinian and chosen for the bishop's see and administrative centre of the χώρα (chora) of Aquae. The name Aquae is evidently connected with the thermal springs of Gamzigrad.

Gamzigrad was also presumably the centre of an imperial estate, probably consisting mainly of forests and hunting-grounds. Its relationship with Fenék-puszta is striking and must surely be considered in any interpretation of this

303

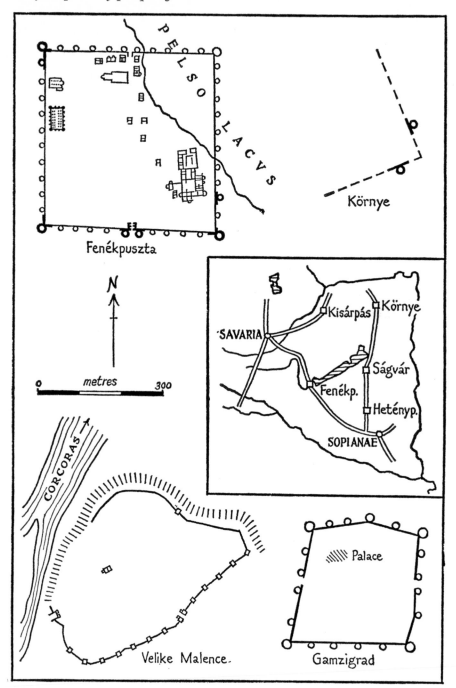

Figure 50   Late Roman fortified settlements

Pannonian fortification. The great palace of Fenékpuszta has unfortunately not yet been explored in an expert manner; the finds so far made in the whole of the settlement are however poorer than those of Gamzigrad, and do not suggest imperial luxury. What do qualify the fortification of Fenékpuszta for the category of imperial possession are, first, the size, which is excessive for anything of a private or military kind, and, second, the buildings inside the rectangle enclosed by the walls, which suit neither a large civilian settlement nor a military camp. Fenékpuszta was not a town, nor was it a fort for some branch of the late Roman army. The large *horreum*, which in itself might be imagined as belonging to a large late Roman city or to a large garrison, must accordingly be interpreted as indicating that Fenékpuszta was the centre of a *latifundium*. The fact that this *latifundium* cannot have been a private one is shown by its not being unique but representing a distinct type of late Roman fortification in Pannonia.

Apart from Fenékpuszta, large late rectangular fortifications with thick walls ranging from 2·30 to 2·60 m in breadth and with many round towers are known from four further places, all in the interior of the province and lying on the diagonal roads that connected south-east and north-west Pannonia. On the Sopianae–Brigetio road, about half-way between Sopianae and Lake Pelso, lies the fortification of Heténypuszta near Gölle (ancient name, Jovía);[41] and on the same road south of Lake Pelso is found the fortification whose ancient name is very likely to have been Tricciana.[42] This fortification (modern Ságvár) measures 292 by 268 m; so far only its cemetery has been investigated, and this is one of the richest and largest late Roman inhumation cemeteries of Pannonia. Here also Christian grave-chambers have been discovered.[43] Approximately on the line of the same road lies the fortification of Környe (ancient name unknown); although this also has thick walls (2·30 m) and large round towers, it was probably of smaller size than the others, perhaps *c.* 150 by 200 m.[44] Lastly the fortification of Mursella must be mentioned, which lies on the Savaria–Arrabona road; here too, only the cemetery has so far been investigated.[45] A noteworthy characteristic of all these fortifications is that they seem to have been carefully designed. Not only are the straight walls and round towers (which are always tangential to the wall) identical (Fig. 50), but so are the corners which form exact right-angles. This latter characteristic requires especial emphasis, for town- and fort-walls elsewhere rarely and indeed exceptionally show walls at exact right-angles to each other.

We may justifiably doubt whether such uniform planning in the lay-out of fortifications could have been carried through on private estates. Walls of such extraordinary width, exceptional even in late Roman forts and fortifications, are not likely to have been used in private buildings, for the reason that fortification

of private villas towards the end of the fourth century was still only exceptionally permitted. And, as we have pointed out, they cannot be considered as
military constructions or as newly built urban settlements either. It is all the more
noteworthy that we have dated imperial edicts from Mursella and Tricciana;[46]
and this allows the inference that when the occasion arose emperors did reside
within these fortifications. And since the edict of Constans from Mursella is
dated to 339, we may date their construction either to the Tetrarchy or better
still to the reign of Constantine the Great. The coins found seem to start under
Constantine.

At Gamzigrad, the fort-like construction of an imperial hunting-castle can be
fairly accounted for because late Roman imperial seats were often regarded and
furnished as *castra*. In this respect Diocletian set an example when he had his
palace at Salona (Split) built as a luxurious fort. The defensive character of sites
such as Fenékpuszta, which do not seem to have been built specifically as
imperial residences, is less understandable. To point out the increasing danger
from barbarians in the border provinces of Pannonia is not convincing, particularly if these fortifications had really been built in the first half of the fourth
century. It is, of course, by no means impossible that large centres of imperial
latifundia already existed under Diocletian or Constantine and were given defensive walls later on when faced with the increasingly dangerous foreign situation
on the Danube. The possibility also exists that the estates of which these fortifications formed the centres were intended for the settlement of barbarians from
the left bank of the Danube, who were transferred under Diocletian or Constantine to Pannonia, and that the centres of the estates were heavily fortified in
order to intimidate the barbarians who had been accepted into the empire. It
might be relevant that Fenékpuszta and Ságvár were situated at Lake Pelso,
which was regulated by Galerius just after 295 when the Carpi had been give
land in Pannonia. A settlement of Carpi is moreover attested around Sopianae,
in the general area of which Heténypuszta lies. It is hoped that the excavations
started at Fenékpuszta, Heténypuszta and Ságvár will be able to provide an
answer to these questions.

It is quite possible that further agricultural centres existed, now unknown, of
the type of Fenékpuszta: apart from the latter, which has been well known for a
century, fortifications of this type have come to our knowledge only during the
last few decades. So we cannot exclude the possibility that the imperial estate
with a fortified centre constitutes an estate-type in Pannonia. Even the sparse
epigraphic material from late Roman times does yield one piece of evidence for
imperial estates: this is the *silvae dominicae* in the surroundings of Savaria[47]—but
an administrator of the *res privata* is attested only in the province of Savia.[48] A

306

number of place-names of varying age can be attributed to imperial estates too, for instance Caesariana, which is to be sought in the surroundings of the large villa of Baláca, and a number of places with the names Jovia or Herculia, of which however only some can be localized (e.g. Heténypuszta).

Fortifications of Fenékpuszta type have a further important relevance for Pannonian history. Behind their strong walls in the early Middle Ages there survived remains of the Roman population, and other ethnic elements which are not easy to define. These were sometimes responsible for an archaeological 'culture' of special type, such as the so-called Keszthely culture of the inhabitants of Fenékpuszta.[49] To all appearance the fortifications were places of refuge much in demand during the Dark Age, and were therefore often captured and chosen as residences by successive masters of the country. We shall later return to this aspect of them.

As has been seen, the development of large estates partly in private but mainly in imperial possession was a significant trend in the Danube provinces. Up to the middle of the third century large estates played a certain role only in the interior of Pannonia and Upper Moesia, in the wider hinterland of the *limes*. But even in some territories of the hinterland, as for instance at Scupi, municipal medium-sized estates were preserved down to the time of the Severi,[50] and the special circumstances of the mines of Upper Moesia also resulted in tendencies that are different from 'normal' development. For these reasons the establishment of large estates over most of Upper Moesia can be regarded as a process caused by late Roman development.

If our hypothesis about the conditions necessary for the preservation of the estates of small peasants in the border area is correct, that is if local military recruitment in the Severan period did indeed maintain the prosperity of the small estates, then it must be concluded that the soldiers' living conditions underwent far-reaching changes towards the end of the third century or a little later, and that these changes contributed their share to the formation of large estates, or were even responsible for it. It is beyond the scope of this work to outline the military reforms from Gallienus to Constantine; their essential character is presumably well known even if their details and in particular the stages of their development are still obscure. These reforms were doubtless as effective in Pannonia as elsewhere; as a result the frontier troops lost their old political role and also suffered material handicaps. It was the newly created mobile troops in the larger centres in the interior of the empire who received political and material privileges: these places were also centres of administration and so no distinction can be made between military and civilian prosperity.

In the fourth century the centres of prosperity and wealth were no longer the military areas along the Danube, but the agricultural estates in the countryside and the centres of administration, which also lay in the interior of the province.

The towns in the fourth century

The development of large estates was presumably the most important cause, or at least one of the most important causes, of the decline of municipal life. Till the middle of the third century the curial upper class of the towns had remained a social group prosperous, active, prepared to make sacrifices, and—in the frontier-zone—also closely interlinked with the military: this class is hardly heard of in the fourth century. Under the Tetrarchy the decurions or higher magistrates of the towns had indeed erected a few dedications for the welfare of the emperors[51] in conformity with duty, but after this they disappear from the inscriptions; so do the soldiers, who had been the other social class of the province to be active epigraphically. The loss of their old role by the curial class and finally by munici-pal administrations is attested by an inscription from Savaria, which records that the *horrea* of the town were replenished under Constans and the supply secured.[52] On this inscription, dealing with a purely municipal affair, neither the municipal magistrates nor the colonia Claudia Savaria are named. On the late Roman inscriptions of the province only officers of the imperial administration and dignitaries of the new aristocracy are named: soldiers occur but rarely, and municipal magistrates not at all.[53] Possibly this is to be accounted for by the custom of Christians to omit mention of the worldly position of themselves and their relatives from grave-inscriptions (Figs 55, 56), but it is more likely that holders of the traditional offices no longer erected inscriptions: they were, after all, the social class most tenacious of traditions—and of the old religion amongst others. Those who set up inscriptions in the late Roman period were, as has been said, members of the new aristocracy, and the distribution of inscriptions suggests a considerable social displacement, for late inscriptions mainly come from the interior of the province (Fig. 51), not from the border area. The inscriptions themselves, of course, do not allow any further conclusions, especially in view of their general rarity. The important points are where and by whom inscriptions were still erected; the answers to these questions point to a social displacement.[54]

The curial upper class did not of course cease to exist. At the town of Emona, which at that time had long belonged to Italy, the decurions went out to greet Theodosius in the traditional toga,[55] and even towards the end of the fourth century the administration of the mines of Upper Moesia was entrusted to the charge of the *curiales*, or, more correctly, to procurators chosen from among the

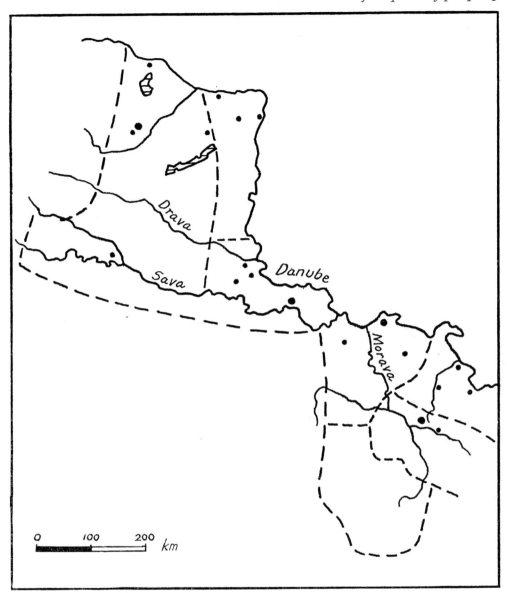

Figure 51 Distribution of Christian inscriptions

number of the *curiales*.[56] This office presumably formed one of the numerous burdens that the *curiales* had to bear in the late Roman period. It is even possible that some self-governing towns were not founded before the beginning of the late Roman period, for instance Jovia, formerly named Botivo, east of Poetovio on the Drave, and possibly some of the late Pannonian and Upper Moesian towns such as Remesiana, the see of a bishop in the fourth century; Sopianae, seat of the praeses of Valeria; Aurelianum or Aureliana, etc. The prerequisite for founding a town was still the existence of a prosperous social class capable of becoming the class of *curiales*.

It was presumably to these curial estate-owners that the smaller but often quite luxuriously appointed villas in the surroundings of Aquincum belonged, which were of fourth-century date.[57] The neighbourhood of Aquincum is quite well investigated, which enables us to perceive a number of other symptoms as well. The emigration by the rich inhabitants of the town into the surrounding territory could be nicely proved. The late antique cemeteries of Aquincum are very poor, but there are rich individual burials, including mummified interments and sarcophagi, in the wider circumference of the town.[58] It could also be observed that building activities at Aquincum continued only till about the middle of the fourth century; after that, not only were no new buildings constructed but the older buildings were either not kept in repair or they continued in use in a very poor way. The hypocausts were no longer employed, the rooms were heated by primitive hearths and ovens, the drains that had collapsed were no longer repaired and so on.[59] The houses and the graves in the surroundings, on the other hand, attest a certain wealth, though this was no longer concentrated in the town but in the countryside instead.

The results of research in Aquincum can probably be generalized for the other towns of the border area. Carnuntum is described by Ammianus as desolate and ruined, and excavations there have not so far disproved him.[60] However, his description of circumstances in Pannonia at the time of Valentinian's visit in 375 has given rise to controversy, since to begin with it is not certain whether the symptoms of decay resulted from a longish process of decline or from the taxation policy of Probus the praetorian prefect. Some details of Ammianus' description[61] suggest that in Savaria and even more in Sirmium everything was still in good order, even though the town-wall was in a state of neglect as a result of the long period of security. When the Sarmatians broke in during 374, the walls and ditch of Sirmium had to be repaired in great haste—the ditch was filled with stones that had fallen from the walls—and for this work money was used from the municipal treasury which had been collected for building a theatre. As far as Savaria is concerned we learn that the gates were in a tumbledown state at the time of Valentinian's visit. In my opinion, we cannot consider this the result of long peace, because normal urban life cannot be imagined without busy traffic through the gates. There is much to support the view that at Savaria at least the symptoms of decay are traceable to the taxation policy of Valentinian, or rather to its cruel implementation by Probus. From Ammianus himself it is clear that at that time the interior towns of Pannonia were affected by a temporary crisis whereas that of the Danube towns was chronic.

As already shown, the rich inhabitants of Aquincum—presumably the *curiales*—moved out of the town and had country-houses built for themselves in the surrounding territory, where they also had themselves buried. The towns on the Danube, formerly centres of prosperity and focal points of politics, became impoverished and presumably depopulated too, first because the richest and most active leaders withdrew from urban life, and second because the soldiers in the border zone no longer consisted of troops of the first quality. The process of concentration of agricultural estates accordingly extended over the whole of the area of the province. Nevertheless there were towns able not only to preserve their role but to experience a new period of prosperity. This is all the more curious since the towns in the interior of Pannonia had already lost much of their original importance, mainly as the result of the concentration of estates. Now these very towns became flourishing centres once again. Not all of them, though; if we take into account late Roman buildings known so far, the wealth of burials and the occurrence of inscriptions, we can suggest that prosperity is only characteristic of towns that for some reason were privileged by the government (among them towns in the vicinity of Sirmium, and a few special cases such as Constantine's birthplace, Naissus), or else that were centres of the provincial administration, such as Sopianae, Savaria, Siscia or Viminacium. It will be better to speak of a further displacement of the centres of influence than of a gradual decline of the towns; this displacement caused the decay of some regions, but not of the whole area.

This shifting of the focal points probably did not happen immediately under the Tetrarchy. We have a few indications that life started to move again in the old ways in the time when consolidation was beginning. A few grave-inscriptions, which are more crudely executed and poorer than tombstones of before the middle of the third century, must presumably be dated to the time of the Tetrarchy.[62] And we have a few altars and building-inscriptions dating at the latest from the beginning of the fourth century that still retain the traditional form and formulation, and the majority of these come from centres which had played a leading role in Severan times.[63] In the area of the new Dacia, to which the legions of old Dacia had been transferred, and where the Romanized civilian population evacuated from Dacia had been settled, there exist sculptures and inscriptions later than the time of Aurelian; some of the ornamental details and over-all design of the stone sculpture from Dacia Ripensis have suggested that evacuated Dacian stonemasons continued to work in the new Dacia.[64]

Under the Tetrarchy some towns will have been provided with town-walls. So

311

far this is certain only for Scarbantia, where the wall—which remained in use virtually without alteration during the Middle Ages and is preserved to the present day (Pl. 20b)—could be dated in fairly recent excavations.[65] The town-wall of Scarbantia enclosed a roughly oval polygon. An area of similar shape was surrounded by the town-walls of Bassiana too; it is only known from air-photography.[66] Certainly late Roman are the town-walls of Ulpianum, Horreum Margi and Remesiana [67] and some others hardly yet investigated [68] though known to exist. However, when we remember that from Diocletian to Valentinian our provinces were fairly peaceful and that under Valentinian the walls of Sirmium were in a neglected state and those of Savaria tumbling down, it may not be rash to conclude that the walls were built under the Tetrarchy. But the proof is available only at Scarbantia.

It is only under Constantine that a radical change took place. After Constantine all archaeological, epigraphic and other evidence combines to show urban prosperity in the centres of administration. Among these the bishops' sees presumably have to be included; but most of these lay in the same centres as those occupied by the bureaucracy. At any rate, wealth is often expressed in an abundance of Christian monuments and in particular by richly furnished Christian cemeteries of which the date must be Constantinian or later.

The imperial city of Sirmium was given most of its monumental buildings under the Tetrarchy (Fig. 29, p. 163). This conclusion is based not only on the disturbed conditions preceding the accession of Diocletian which would not allow of large-scale building, but above all on the choice of Sirmium as a permanent imperial residence by Galerius. Licinius ordered baths to be built there, for which the column-capitals were fetched from a quarry on the Dalmatian island of Brač.[69] Whether these *thermae Licinianae* can be identified with the truly huge remains of baths (Pl. 39a) uncovered by the new excavations[70] is as yet undecided. These baths at Sirmium were constructed in building-tiles, as were the other large buildings recently discovered at Sremska Mitrovica;[71] among them were a large *horreum*, part of a palace-like building (Pl. 39b), a hippodrome probably connected with the Imperial Palace and, in the south-west corner of the town, a row of shops which possibly belongs to a market-hall. The remaining excavated buildings are dwelling-houses, among them one of a distinctly upper-class character. Buildings constructed of tiles are very rare in Pannonia, and this is all the more striking since in many areas no building-stone is available. There are large districts where stones for building had to be fetched from quarries 50 km away from the building-site. Presumably the reason why, exceptionally, tiles were used in Sirmium was that the very pretentious technical execution of the public buildings was only possible in tile. Of great significance are the Christian cemeteries of Sirmium (Fig. 52). These were excavated in the nineteenth century, unfortunately with little care (but perhaps with too much imagination). They have yielded a large number of grave-inscriptions and an extraordinary variety of subterranean buildings.[72] Below we shall return to this cemetery: here it need only be mentioned that similar cemeteries are equally characteristic of the other administrative centres.

Naissus experienced its first real prosperity only in the fourth century, not because it was an administrative centre but because it formed an important nodal point in west–east communications and was also the birthplace of Constantine the Great. Stephanos Byzantios was probably correctly informed when he considered this town to be the 'foundation and the home of Constantine'. Although Naissus had become a town as early as the reign of Marcus its real founder was evidently its greatest son, to whom a bronze statue was set up not far from the bridge over the Nišava.[73] Of the cemeteries which extend over a very considerable area east of the town some parts have been laid open in several stages.[74] The large cemeteries of the late Roman town are very uniform. Here, too, subterranean tombs whose walls are built of tile or stone, grave-chapels above ground, even funerary basilicas and grave-chambers with frescoes are frequent and prove not only the prosperity of the town but its size also.

There exists a very close relationship between two painted grave-chambers from Naissus[75] and Sopianae,[76] which was also the seat of a provincial administration. From Sopianae come the most important early Christian monuments so far discovered in Pannonia (Fig. 57, Pl. 40). The grave-field (Fig. 53) is as rich and shows as much variation as the cemeteries of Sirmium and Naissus.[77]

Finally we must mention the cemetery of Savaria,[78] the capital of Pannonia Prima: it has not so far been investigated at all but is unequivocally attested, and to judge by its inscriptions was probably as rich as the others. In Savaria, too, the remains of a large building were uncovered thirty years ago, containing a large hall with an apse; its floor had been adorned with first-class mosaics. The walls were covered with marble slabs, and the whole hall was provided with a hypocausted heating system. The mosaics can be attributed to the workshop which at the beginning of the fourth century produced a number of mosaic floors for the palaces and Christian cult-buildings of Illyricum and Aquileia. The excavation of this building in Savaria was from the start carried out on the assumption that it was the Christian basilica, built under Constantine for the martyr-bishop Quirinus. The doubts voiced during the last decade about this identification have only recently been systematically argued and at the same time the hypothesis has been suggested that the hall with its mosaics was the *aula* of the imperial palace in Savaria. This palace lay in the immediate vicinity of

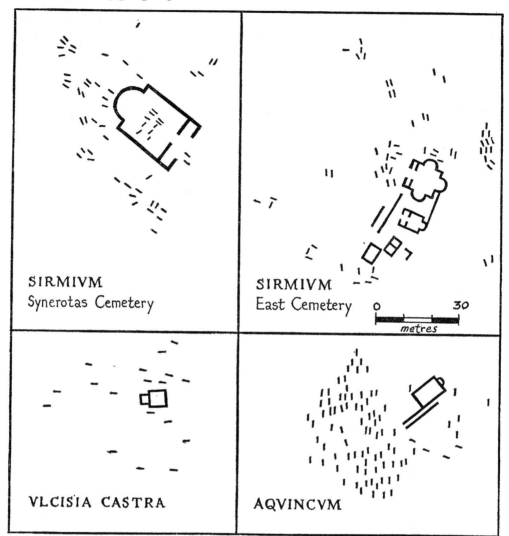

Figure 52 Plans of Christian cemeteries

the forum of the colonia, as is shown by fragments of the statues of the Capitoline Triad which were found in the southern neighbourhood of the *aula*.[79]

Apart from Sirmium, Sopianae and Savaria which were seats of the provincial administrations, and apart from Naissus, the birthplace of Constantine, other towns in the interior of Pannonia and Moesia display the uniform characteristics of late Roman municipal prosperity. The recently explored Christian cemeteries of Ulpianum (which was already called Ulpiana)—the quarters of a *legio pseudo-comitatensis* according to the *Notitia*[80]—sporadic burials around Scupi which often employ older tombstones as packing,[81] the grave-chambers in the surroundings of Remesiana[82] and the graves of Viminacium[83] all have exact parallels in Pannonia. There are no differences in the rite or the combination of grave-goods, but only in the wealth of the burials, and in this respect, too, the difference lies not between Pannonia and Upper Moesia but between the interior of Pannonia and of Upper Moesia on the one hand and the frontier zone of both provinces on the other. In the towns of the frontier zone the burials are poorer; the graves are mostly simple earth-graves or graves with tiles, burial-vaults are rarer, and structures ornamented with paintings or mosaics even more so. This is equally characteristic of the towns and of the settlements at military sites on the frontier; whilst in the hinterland, as well as rich cemeteries at the towns, a number of wealthy single graves are also known from non-urban settlements.[84]

Through the archaeological indications of prosperity in the towns in the interior of the province a further important difference can be traced. We have already pointed out that the majority of towns of late foundation were not towns in the proper sense of that word, for the social, economic—and often the geographical conditions as well—for implanting regular urban life were not available. And these 'towns' were the ones to display no traces of prosperity in the late period, even though they may have been situated in the interior of the province. The reason was not simply that for the seats of provincial administration, and presumably for camps of the field army, traffic-centres and thus the sites of genuine towns were chosen. In this respect it is characteristic that all three pseudocomitatensian legions in Upper Moesia were placed in important places of the country: Timacenses, Ulpianenses and Scupenses.[85] And the development of large estates must also be considered an important factor. It is possibly not without importance that one of the great fortified *latifundia* centres, discussed above, was built in a municipium of Hadrian. The distribution of late Roman–early Christian inscriptions (Fig. 51) is limited either to towns (Savaria, Siscia, Mursa, Cibalae, Sirmium, Viminacium, Ratiaria, Naissus) or to areas where there had been no regular towns even in earlier times: at the latter these inscriptions come to light in unimportant small places, evidently burial-places of owners of large estates, or in the estate-centres.

As we have already stated, late Roman burial-rites are strikingly uniform. The same tendency towards uniformity can perhaps also be discerned in other aspects of daily life. One characteristic is particularly striking and can be further established by excavation: as examples we may take the settlement at the Pannonian *limes*-fort of Matrica (Százhalombatta) and the large settlement of Gorsium (Tác) in the hinterland of Matrica.[86] In both settlements a succession of layers extending from the second to the fourth century came to light. In the

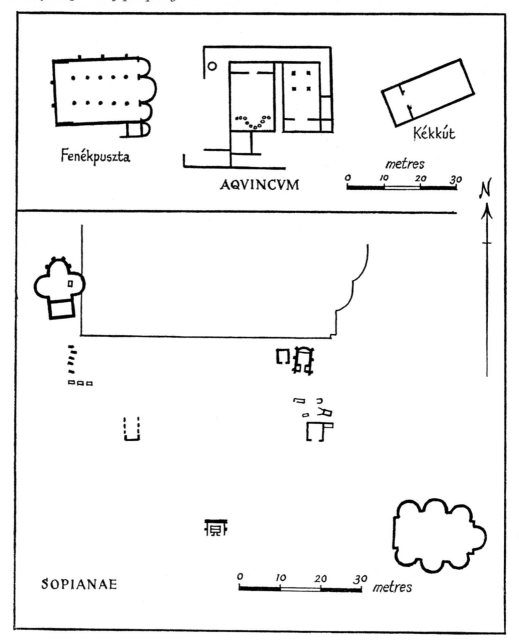

Fenékpuszta

AQVINCVM

Kékkút

metres

0   10   20   30

N

SOPIANAE

0   10   20   30   metres

Figure 53 Christian *basilicae*: funerary monuments in Sopianae

second century both settlements consisted mainly of dwelling-houses and buildings not constructed from stone or tile. Stone buildings appear first in the third century but become general only in the fourth, so much so that one could reckon that all dwelling-houses were stone-built at that period (though the worsening quality of the stonework cannot be denied). At Matrica (Fig. 54) remains of several small houses were discovered whose walls had no foundations, nor were the stones set in mortar. Dry-stone walls without foundation disappear only too easily, and so they cannot always be traced in excavations. This is presumably the reason why late Roman village settlements or, in general, simple dwelling-houses are only seldom traceable archaeologically. The dwelling-houses of the earlier periods were often dug partly into the ground and settlements of early imperial date are marked by the great number of pits. Settlement layers of the fourth century have left no such obvious traces behind. The number of stone-built houses of fourth-century date established by minor excavations, however, is not negligible,[87] and we can therefore assume a general change in building-method in the fourth century.

The barracks in late forts also probably had dry-stone walls without foundations. Various reconstructions and additions in the left half of the *praetentura* in the legionary fortress at Carnuntum[88] and the interior buildings of the recently investigated late Roman fort of Tokod (Pl. 36a),[89] consisted of dry-stone walls. In urban living-quarters, for instance at Aquincum, dry-stone walls also become frequent. Bonding courses of tiles can also be considered a technical innovation in construction; these are characteristic of late Roman military buildings and presumably civilian architecture derived them from that source. The courses, two to four tiles thick, were built into the walls at intervals of about 1 m, in order to strengthen the rubble masonry.

This change in the way and standard of life presumably follows certain changes in the social structure. But investigations so far undertaken are insufficient for full comprehension of these, nor are our sources suited for the answering of such questions. As the epigraphic sources, which yielded rich information up to the middle of the third century, dry up, greater reliance has to be placed on purely archaeological results, and this we have attempted to do. We have been able to show the probability that the municipal upper class which had such close connections with the army was replaced by an aristocracy of owners of large estates; these men may have been partly derived from the former municipal upper class, and in the fourth century they did include *curiales* in their number; but in general they were much more closely linked with members of the new bureaucracy than with the military on the *limes*. And among owners of large estates we may assume that the emperor himself had by far the largest posses-

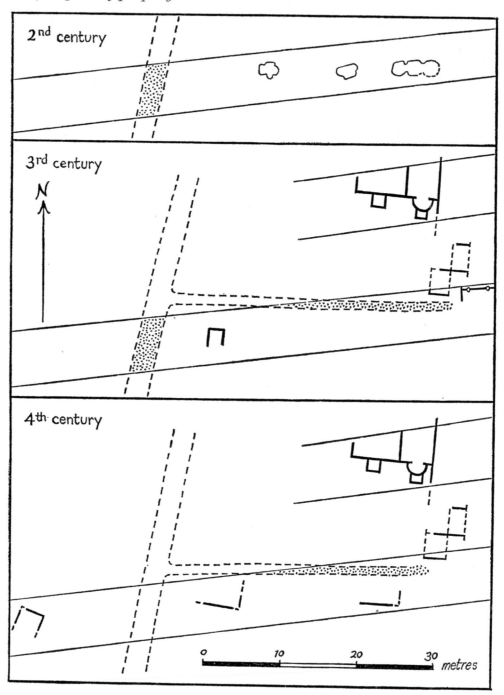

2nd century

3rd century

N

4th century

0   10   20   30
metres

Figure 54 Plan of the *vicus* of the auxiliary fort at Matrica

sions in land. But a social structure of that kind would necessarily result in a social polarization such as had hitherto been unknown in the provinces; yet to judge from the standardization of the way of life which we have described, this polarization does not seem to have occurred.

Economic and social developments in the late third and fourth centuries

Nevertheless it may be suggested that there did exist a sharper polarization in our provinces than in former periods. In the first place it is significant that hoards composed of objects of precious metal and of works of art are mainly fourth-century or were buried at that time.[90] The rich graves of the fourth century, that is those furnished with precious grave-goods, are wealthier than graves considered rich in the earlier periods. The luxury of the wealthiest dwelling-houses increases, yet at the same time these luxury houses are met with more rarely. Furthermore a gradual process of polarization can be traced by analysis of late Roman cemeteries; fourth-century graves exist in a quantity sufficient for statistical investigation. It has been discovered only recently that a gradual pauperization can be traced in late Roman cemeteries, not in the sense that all graves became increasingly poor, but because poorly furnished graves and those without grave-goods became more and more frequent, whilst the wealth of the few rich graves remains the same. There are, however, marked differences in wealth between individual cemeteries; in general, cemeteries in the interior of Pannonia are richer than those of the settlements on the *limes*, while outstanding wealth is seen in the cemeteries of administrative centres and those belonging to the centres of fortified imperial *latifundia* of the type of Fenékpuszta.[91]

The standardization of the way of life is thus not based on a levelling of classes. It shows itself in characteristics indicating increased industrial production rather than actual wealth. In Chapter 7 it was pointed out that local characteristics in the pottery and bronze industries, in burial-rites and in costume and so on, suddenly faded out in the Severan period. Provincial culture became colourless and undifferentially 'Roman', and this is clearly seen in different fields and manifestations of culture, from pottery to the art of the stonemason. This process continued in the fourth century. It has also been shown in studies of late Roman cemeteries that with very few exceptions the grave-goods were produced on the spot; this could be proved only by mapping the minor peculiarities. This may be illustrated by a few examples. Bone bracelets are among the most frequent grave-goods in virtually all cemeteries. At first glance they are all very similar. But the simple decoration—incised circles and dots—shows local characteristics. Jugs and beakers of pottery are also frequent grave-goods. The jugs mostly have a handle and a bulbous body, and the beakers are of types that are closely interrelated. There are however hardly any types that have a distribution over the whole of the province. A fine typology demonstrates that each type is characteristic of a limited area only. Moreover, in cemeteries round which no suitable clay was available, pottery is rare, but in others the majority of grave-goods are ceramic. When classified closely, differentiation of grave-goods increases during the fourth century, though if an insensitive typology is employed the material looks very uniform. The production of local workshops was everywhere able to satisfy demand for cheap objects of use; the wares produced were very simple and mass-produced, attesting uniform poverty among the population.[92]

The wares not produced on the spot are very few in number. Chief among them must be considered the cross-bow brooches (Zwiebelknopffibeln) which are also very uniform and which were probably produced in central workshops. There are some indications that this type of brooch was reserved as a badge for certain ranks in the administration. From central workshops, too, came the bronze mounts decorated with figures, used on small caskets (Pls 44b, c); some of these representations refer directly to the emperor, and possibly had a certain official character. Of the remaining representations some are decidedly Christian in character, the remainder taken from pagan classical symbolism.[93] The remaining imports are mostly luxury goods, such as glass vessels with cut ornament probably from Rhenish factories,[94] objects of jet also from the west,[95] and finger rings with intaglios, buckles decorated in niello, and so on. The contribution of the east in these luxury goods was presumably larger than can be proved by available parallels; inscriptions where they occur on such objects are strikingly often in Greek.

Some of the objects which reached Pannonia from distant sources were officially provided for the use of certain ranks and cannot be included among evidences of genuine trade. Import-trade was restricted to luxury items the categories of which were probably much more limited than those handled by the busy commerce of Flavian to Severan times, when import in bulk is proved above all by samian ware (*terra sigillata*).

What has been said of course applies to Pannonia only, since no studies exist of excavated late Roman cemeteries in Upper Moesia, which are in any case few in number; nor do we possess massive publications of finds from excavated settlements. It will, however, be permissible (with certain reservations) to extend the results obtained in Pannonia to Upper Moesia too; this is because the cemeteries that so far have come to our knowledge do not differ, in rite and in the grave-goods, from those in Pannonia. A uniformity covering both provinces can therefore be considered likely. This was, however, a uniformity of broad sections of the population possessing in the main a rather poor standard of living. The small upper class was richer than the larger municipal and military upper class of former times, but, perhaps precisely because of its small numbers, it was the bearer and representative of the uniform upper-class culture of the empire which now rarely yielded any local characteristics.

Because the finds do not admit an import in bulk, we are faced with the problem how to explain the most striking peculiarity of late Roman monetary circulation in Pannonia and Upper Moesia. It has long been known that coins from Balkan and eastern mints form a very considerable part of all the hoards and isolated finds, whereas Italian and western mints are hardly represented at all; not even Aquileia has a percentage worth mentioning.[96] Perhaps it was the high eastern proportion in the coinage of Pannonia that induced historians to suggest that late Roman Pannonia enjoyed an increasing import-trade from the east. But the present state of knowledge offers no cause for that. The east had always played its part in imports to Pannonia, and in the fourth century, as already mentioned (p. 320), there are still a number of articles which must be derived from the eastern provinces. For example, the few finds of textiles from fourth-century graves, with silk and linen represented amongst them, presumably came from the east.[97] But there is no archaeological evidence that the east achieved predominance in the trade of Pannonia, and we must therefore assume either that oriental imports were composed of organic material and are therefore untraceable archaeologically, or that Pannonia was paid in coin for exports to the east; for this second supposition the coin-finds can be cited in evidence.

Here we would have let the question drop were it not that a remarkable source offers grounds for further hypotheses. This source is the *Expositio totius mundi*, a mainly commercially orientated survey of the then known world, written *c.* 350 by a Syrian. The author knows hardly anything of the trade of the western provinces; even provinces such as Gaul are dealt with in insipid general remarks and political details. It is all the more striking that he considers Pannonia to be 'a country in all respects rich', having a busy trade in corn, cattle and 'in part' in slaves too. We must attach particular importance to this information because the countries on the lower Danube (Moesia and Dacia Ripensis) though famed in the empire for their agriculture, according to the Syrian author were only self-supplying; that is, they did not export. Whether the renewed attacks by the Goths from the middle of the third century onwards and the subsequent Gothic settlements were responsible for the decrease of agricultural exports, I am not able to decide; it is however not unlikely that the reason for Pannonia being able to enter the trade of natural products was that the provinces which had a better geographical position, commercially speaking, were no longer able to export. The prerequisite of production for export was presumably the large estates.

The Syrian author of the *Expositio* names Noricum as supplying garments (*Norica vestis*) in the same chapter with Pannonia; this might perhaps suggest that the Norican export went via Pannonia, and this makes sense only if it travelled down the Danube towards the east. In the same way, Dardania, exporting Dardanian cheese and lard, is mentioned only in passing in the chapter on Macedonia; the reason being that it had been Macedonia, in fact, that had traded these Dardanian goods.

Apart from corn and cattle, slaves are named as exports from Pannonia. Where these slaves came from we do not know; primarily we might imagine that internal unrest in the Lowlands beyond the Danube, the frequent struggles between Sarmatians, Gepidae, Vandals and Goths,[98] might generate a slave-trade on the frontier through the sale of prisoners by the Sarmatians into the empire. Regular trade on the bank of the Danube is attested not only by late Roman small coinage, which is very frequent in the Hungarian Lowlands, but also by the curious inscription from the time of Valentinian, reporting that a *burgus* was erected for the purpose of border trade.[99] In the fortified landing-place on the left bank of the Danube near Nógrádveröce pieces of raw amber were found[100] attesting the continuance of the amber-trade through the barbarian approaches of Pannonia in the fourth century.

Accordingly it is possible that Pannonia in the fourth century exported agrarian products and slaves, and that the eastern provinces were interested in this export. The only relevant traffic-route was the Danube, on which the goods were transported either to Raetia and Noricum or to the Black Sea and ultimately to Asia Minor. Even southern Pannonia had access to the trade of the empire only via the Save and the Drave which ran eastwards, and so we need not confine our search for the large estates responsible for the export to the vicinity of the Danube in Valeria and Pannonia Secunda.

The fact that Dardania exported the products of its mountain farming—cheese and lard—to Macedonia will not be surprising seeing that the Axius was a convenient and at the same time the only water-route connecting Dardania with the Mediterranean.

Religion in the late third and fourth centuries

The trade-relations with the east were presumably partly responsible for the fact that oriental immigration did not cease after the crises of the third century. On Christian grave-inscriptions Christians newly arrived from the eastern provinces are repeatedly named (Fig. 55); oriental Greek participation in the earliest Christian communities seems to have been very great. The large majority of Christians known by name had Greek names;[101] so did the bishops, the majority of whom had Latin names only from the middle of the fourth century onwards.[102] But even among Christians with Latin names we may assume there were Greeks; one of the earliest Danubian bishops, Victorianus, was a Greek.[103] The earliest Christian communities on the middle Danube were small groups of people from the eastern provinces. They are found mainly in and around Sirmium[104] where Orientals attracted to the headquarters of the Illyriciani and later to the imperial court had gladly settled. The legendary list of bishops of Sirmium goes back to the disciples of the apostles. The real circumstances and prerequisites of the Christian mission in our provinces can presumably be recognized in the attraction exerted by Sirmium, despite the fact that until the end of the Tetrarchy the policy of the Illyriciani was decidedly hostile to Christianity.

Paganism, indeed, maintained itself tenaciously in our provinces, and when Constantine stayed in Sirmium in the 310s and 320s he probably had to meet religious opposition from the Illyriciani of the Licinian party. After Constantine's victory over Licinius the pagans in our provinces were silenced, as is best illustrated by the sudden ceasing of pagan dedications on inscriptions. Since conservative pagan elements were especially distinctive of the circle of the *curiales*, one of the reasons for the disappearance of epigraphic evidence for the *curiales* may have been the disinclination of pagan *curiales* to express their beliefs either on votive inscriptions or on tombstones. At any rate the drying-up of inscriptions cannot be simply accounted for by a change in taste or by economic decline. As we have already stated, pagan dedications, and grave-inscriptions too, are still found even though not in large numbers under the Tetrarchy after the consolidation; in their sudden disappearance under Constantine, we must recognize political and religious reasons. For at this same time the Christians start to set up gravestones in certain towns;[105] it is inconceivable that none of the *curiales* should have had the material means of setting up a tombstone for himself. The distinctly luxurious burials however are indisputably Christian, and grave monuments and grave-inscriptions are erected only by Christians; and this proves not only the gradual Christianization of the wealthy upper class but also the assured and even challenging character of the new religion. Since the attitude of the Illyriciani up to the time of their defeat was hostile to Christianity, we must assume that the ruling class in our provinces was pagan up to the time of Constantine, and attribute their conversion to his reign; no doubt political considerations played the main part in the change, as also in the choice of the new aristocracy. Before we enter on the ecclesiastical history of our provinces it will therefore be worth while to set out the evidence concerning the end of paganism.

Under the Tetrarchy paganism was still flourishing in Pannonia and Upper Moesia. Official votive inscriptions set up by municipal magistrates are, in fact, rather colourless—dedications to Juppiter Optimus Maximus are the most frequent—but there is good evidence that sanctuaries of Mithras were still used and maintained; indeed, the cult of Mithras may have enjoyed the greatest popularity of all. This at any rate is suggested by the action of the imperial participants in the conference of Carnuntum in 308, who dedicated to this particular god a large stone (though in fact re-worked from an old altar (Pl. 34b)), and who restored a Mithraeum at Carnuntum.[106] To all appearance the Mithraic sanctuaries fell victim only to the destructive rage of the Christians; this can be deduced from the fact that often all the trappings of the cult had been collected and hidden in the ground,[107] showing that they were still in use at that time. It is well known that some pagan rites and beliefs managed to survive still later, and it is significant that a small relief of the Danubian pair of Rider-gods was found in a layer of Constantinian date, and a similar small relief even came to light in the imperial villa at Mediana.[108]

Finds of course exist that prove the use of pagan cult-places till late in the fourth century. Least significant are the coins thrown into springs and wells,[109] which often go right down to the end of the Roman period. Magical texts and similar superstitious finds do after all turn up even in Christian contexts,[110] and this sort of superstition is behind the offering of coins to water. But as late as the reign of Gratian a Mithraeum seems to have been the scene of religious rites not far from Scarbantia.[111] As far as can be established, pagan sanctuaries all suffered violent destruction, from the *capitolium* of Scarbantia to individual altars standing in the open, which were often buried in the ground[112] or walled-up.[113] The smashed pieces of the colossal statues of the Capitoline Triad of Scarbantia (Pl. 27a) were also walled-up, and the statues of the Triad at Savaria also show traces of destruction by force. It is, however, not always possible to decide whether the smashed remains of a sanctuary were buried by Christians in order to get rid for good of the idol in this way, or whether they were concealed by pagans for reasons of reverence. But there are collections of cult-objects that have come down to us unharmed, as a treasure hidden away; in such cases the pagans obviously managed to hide away their cult-images and the sacrificial implements in time. These destructions and concealments cannot be dated; on the whole the process will have lasted roughly from the time of Constantine to that of Julian.[114]

Under the Tetrarchy the government probably took official steps for the promotion of the old religion and cults. It has been assumed that it was for an anti-Christian occasion that the priests of Dolichenus of Lower Pannonia assembled.[115] But this assumption is untenable for the reason that Dolichenus, as we have seen, disappeared from the late pagan pantheon too early. A sensational new find proves, on the contrary, a deliberate and fully detailed pagan religious policy under the Tetrarchy.

At Brigetio a small group of graves was found, dating from the beginning or first half of the fourth century; most of them were in sarcophagi. In the centrally situated and richest grave lay a man who held a silver augur's staff (Pl. 44a) in his right hand. The very pretentious goldsmith's work of the staff points to the end of the third century as the earliest possible date: this date of manufacture for the staff is likely to be significant. That it accompanied the augur to the grave—he was evidently a municipal dignitary of Brigetio—is a striking illustration of the religious circumstances of the time. The *lituus* was intended to serve several generations of augurs; the man who took it to the grave with him was clearly a pagan, and his friends were evidently convinced that he would be the last augur of Brigetio.[116]

Christianity and the persecutions

At the period when the augur of Brigetio could still practise his cult without disturbance, there already existed in southern Pannonia small groups of people, mostly Greek-speaking, who confessed themselves Christians. Their local tradition presumably did not reach back much further than the middle of the third century. The first Pannonian martyr is indirectly established as the victim of Valerian's persecution, and under Gallienus some Christians, either during their trial or during conversations with fellow-believers, referred to their having been brought up in Sirmium as Christians.[117] Some of the Pannonian martyrs, therefore, were Christians of the second generation in Pannonia. As can be deduced from their names, however, they were mostly immigrants—Orientals or people from the Greek-speaking provinces of the empire.

The first certain indications of Christianity on the Danube can be traced with the aid of the Acts of the Martyrs. The earliest martyr in our provinces seems to have been Eusebius, bishop of Cibalae, whose Acts indeed have not come down to us, but whose memory has been preserved in local tradition. The Christian lector Pollio of Cibalae, who was executed in 303, referred to his example among other reasons because his own martyrdom fell on the same day as that of Eusebius (28 April). According to Pollio, Eusebius met his martyr's death during an

325

earlier persecution (*persecutio superior*), and this can most probably be identified with the persecution under Valerian.[118] It is not impossible that at that date organized Christian communities existed not only at Cibalae but in a number of other towns in south-east Pannonia—above all in Sirmium—and that these also had to suffer in Valerian's persecution.

However, the life of the Christian communities was probably undisturbed and peaceful during the last decades of the third century, as can be deduced from the writings of Victorinus, bishop of Poetovio.[119] At this time there already existed a few heresies within the communities, and it is not impossible that the Catholics in Sirmium had a separate community from that of the heretics.[120] The appearance of Christian communities with their own bishops probably occurred in the period from Gallienus to Galerius; by 303, bishops and priests are already attested in several towns in southern Pannonia. The execution of the four or five stonemasons, who later enjoyed a special veneration in Rome under the name of *Sancti Quattuor Coronati*, was probably an isolated instance, not part of a systematic persecution. Since the Acts of these martyrs prove the personal intervention of Diocletian, the martyrdom, which took place on 8 November, should probably be dated to the year 293 when Diocletian was present in Pannonia and—specifically in November—made a journey along the *limes*. He ordered the stonemasons, who worked in a porphyry quarry, to be executed because they refused to produce a statue of Aesculapius.[121]

The great persecution of the year 303 throws a sudden clear light on the Christian communities of our provinces.[122] And one thing must not be forgotten here: to all appearance the persecution was directed towards the politically most important area of Sirmium, where the procedure was truly extreme. In other parts of Pannonia and Upper Moesia either no martyrs are known or we have information only about the bishops. This means that in general only the most prominent leaders of the communities were punished, and in some remote areas not even these were consistently persecuted: we know almost two dozen names of persons later revered as martyrs from the surroundings of Sirmium, whereas no martyrs are known from the provinces of Pannonia Prima and Valeria, and the statements on martyrs in Dardania, Dacia Ripensis and Dacia Mediterranea contain hardly more than a few doubtful names and commemorative days: Florus and Laurus in Dardania, Hermes, Aggeus and Gaius on 30–31 December in Bononia or Ratiaria; at Naissus there are not even martyrs' names. The only martyr in the province of Savia was the famous Quirinus, bishop of the provincial capital at Siscia. Quirinus was executed in Savaria, presumably in order to intimidate the Christians of that town.

The possibility, however, cannot be excluded that the transmission of the names of martyrs and the survival of their cults depend on the later history of the Christian communities in question. Sirmium preserved its role for a very long time after the collapse of the western empire; in the gradual emigration from Sirmium, the relics of martyrs and their cults were taken along and a new cult established in the new home. One of the most important saints of the Greek-Orthodox and Slav churches was the Sirmian Demetrius, whose relics were transferred to Thessalonica. The emigration to Italy in the beginning of the fifth century played a decisive part in the survival of a number of other martyr-cults, for instance that of Quirinus and the four Coronati. But the martyrs and cults of some towns in which their Christian communities died out early in the fifth century could easily have been forgotten. Nevertheless, the north Italian churches have preserved the memory of a long list of Danubian martyrs and commemorative days, and since these were for the most part Sirmians, it is quite probable that most of the martyrs came from Christian communities in and around Sirmium. But there remains the possibility that the martyrs of less important centres in our provinces were much more easily forgotten.

The long list of Christians executed in Sirmium contains some names that received especial glory in the cult of the medieval church, above all Demetrius, deacon of Sirmium (the Saint Demetrios (Dmitrij) of the Eastern Church) and Irenaeus, bishop of Sirmium; the gardener Synerotas and Fortunatus and Donatus had a special cult in Sirmium; their places of rest became the object of pious veneration and focus of a large Christian cemetery there.[123] Synerotas was presumably not executed in the same year as the others; his commemorative day is 23 February whilst most of the others died their martyr's death in March or April. The commemorative days, which are concentrated in the spring, allow a tentative reconstruction of the progress of the persecution. In 303 all was quiet at Sirmium at first. At the end of March and the beginning of April the heads of the communities at Sirmium and Singidunum were arrested and executed: these were Bishop Irenaeus and the deacons or priests Fortunatus, Donatus, Demetrius, Romulus, Montanus (together with his wife Maximilla) and presumably many more, for instance the Seven Virgins whose individual names are not known. At the end of April the persecutions extended to the neighbouring towns: at that time Pollio, lector of the Christians at Cibalae, was executed, and probably not much later the bishop of Siscia, Quirinus, was arrested. Tradition speaks of Montanus fleeing from Singidunum to Sirmium and being therefore convicted there. But it is probable that Christians from neighbouring Singidunum were brought to Sirmium for trial and judgment.

Possibly for different reasons, Bishop Quirinus was brought to Savaria. After having been arrested on his flight he was led 'through the towns', and the trial was held at Savaria, and there in the theatre, that is, with full publicity. From the Acta we learn that at that time Savaria too already had a Christian community; but of Savarian martyrs tradition knows nothing. The ostentatious management and the delay in the criminal proceedings can therefore be explained in the following manner: the Christians of the provinces of Savia and Pannonia Prima were not so strictly and consistently pursued as those of Pannonia Secunda, and the example of Quirinus was intended to suggest the possibility of future persecution. On 8 June Quirinus was flung into the small river flowing through the town of Savaria. In the place where his body came to land an oratory (*locus orationis*) was later built, and under Constantine his relics were deposited in the basilica erected for him.

In Sirmium the persecution continued longer, presumably until 304. The commemorative days of the remaining martyrs, among them the gardener Synerotas, are spread over the months July to February; nothing is said, however, about their being dignitaries of the church hierarchy, and many women were also included in their number. This was the stage in the persecution when laymen too were falling victims. It is very probable that smaller, more local persecutions continued to occur under Licinius too; for instance, it was probably under him that Hermyle and Stratonice of Singidunum were thrown into the Danube there. The martyrdom of Synerotas, too, can be dated to the later years, *c.* 305–7.

As we have seen, the persecutions under the Tetrarchy suggest that in the larger centres of our provinces organized Christian communities existed under bishops. The same assumption applies also to towns which had neither martyrs (or more correctly martyrs that were later venerated) nor Christians mentioned for any other reason in our sources. The next occasion when we learn something about the organization of the church on the Danube is the council of Nicaea, at which the Christian communities of Pannonia and Upper Moesia were represented. The Danubian bishops named among the fathers of the council mostly signed as representatives of a province—*Domnus Pannoniae, Dacus Dardaniae*—and, considering the distance from the Danube to Asia Minor, this possibly means that they attended as the emissaries of a whole province, since not all bishops of the province could take part in the council.

We learn some details of the dioceses in our provinces only from 343, when some otherwise unknown bishops took part in the council of Serdica. At this council, too, not all Pannonian and Moesian bishops were present; the majority were bishops of towns situated nearest to Serdica, such as Horreum Margi, which is attested as a see only this once. There are, in fact, hardly any towns in Pannonia or Moesia whose fourth-century lists of bishops are handed down without gaps. But it is probably legitimate to assume that by the first half of the fourth century all towns (*civitates*) in the Danubian provinces had bishops, even where these are not named in the sources. The following towns of our provinces are attested as bishops' sees:

PANNONIA PRIMA: none.
VALERIA: none.
SAVIA: Siscia, Jovia.
PANNONIA SECUNDA: Sirmium, Mursa, Cibalae.
MOESIA PRIMA: Viminacium, Singidunum, Horreum Margi, Margum.
DACIA RIPENSIS: (western part) Ratiaria, Aquae.
DACIA MEDITERRANEA: (western part) Naissus, Remesiana.
DARDANIA: Scupi, Ulpianum.

A complete bishops' list we have only for the fourth century and only for Sirmium. Of the remaining sees only some bishops are known by name; Aquae and Horreum Margi are mentioned merely in the list of the council of Serdica, and the bishopric of Margum is attested only by the fact that a bishop, not mentioned by name, had an embarrassing experience with the Huns.[124] The bishop of Jovia appears only once, on an inscription,[125] and lastly, certain Moesian sees are attested only for the fifth or the sixth century. By and large our knowledge depends on a variety of chances: the geographical situation of the town in question, the part individual bishops played from time to time in the general history of the church (for instance Valens of Mursa and Ursacius of Singidunum, etc.). Taking all these chances into consideration, it would not be right to maintain that there were no bishops in Pannonia Prima and in Valeria; we must remember that in other provinces, for instance Moesia Prima, almost every town is attested as an episcopal see. The evidence of early Christian monuments so far discovered suggests a bishopric at any rate in the provincial capitals of Savaria and Sopianae; and with the organization of the early church in mind we can say that it is quite probable that there were bishops in the remaining late Roman civitates as well.

We are, of course, best informed on the church of Sirmium, although its bishops did not play a part of first importance in the dogmatic dissensions in the middle of the fourth century. The leading roles were played by the bishops of two towns near Sirmium: Valens, bishop of Mursa, and Ursacius, bishop of Singidunum, were the leaders of the Arian party in the western part of the empire, and exerted a decisive influence on the government of Constantius II which was well-disposed towards the Arians. With their appearance the ecclesiastical history proper of our provinces begins; until the disintegration of the

Roman administration and provincial organization this was determined by the controversy between the Catholics and the Arians. The bishops of Sirmium were as dependent on the powerful influence of Valens and Ursacius as most of the bishops of our provinces were; the only exception was Photeinos, bishop of Sirmium, who was repeatedly in exile or being reinstated: he represented a heretical trend of his own.

The activity of Valens and Ursacius began in the 330s when they made their appearance as eager followers of Arius (who had been exiled to Illyricum) and succeeded in obtaining the bishoprics of Mursa and Singidunum respectively. At the same time they succeeded in sending the Catholic bishop of Sirmium, Domnus, into exile, and in establishing in this important bishopric the Arian Eutherius. Valens' attempt, in 341, to become bishop of Aquileia failed, and this incident caused a temporary decline in Pannonian Arianism. The bishops of south-western Pannonia seem to have been under the influence of the north Italian churches, always on the side of Catholicism, and in the council of Serdica the Moesian bishops also took sides with the orthodox Osius. Even Eutherius came over to the Catholic side, and this left Valens and Ursacius no alternative but to profess Catholicism themselves.

In the same year, 343, however, Eutherius died, and the new bishop of Sirmium, Photeinos, only weakened the unity which had been so difficult to establish; as early as 346 Valens and Ursacius, who most of the time travelled in affairs of church politics, appeared against Photeinos in Milan. A synod in Sirmium not much later condemned Photeinos, and in 351, in the presence of the Emperor Constantius II another synod was convoked at Sirmium which sent Photeinos into exile. The new bishop, Germinius, was an Arian, for Valens had in the meantime succeeded in establishing his influence over the emperor by prophesying to Constantius at his prayers his coming victory over Magnentius near Mursa. As a church-politician Bishop Germinius was hardly more than the mouthpiece of Valens.

The last years of Constantius II were under the influence of Arianism, but already by 358 there were indications that western Catholicism was on the point of regaining Sirmium. Two synods at Sirmium (357, 358) showed Arianism already in retreat, and at Rimini Valens and Ursacius were defeated. The decisive impulse came from north Italy, particularly when Photeinos was recalled under Julian and the Sirmian church thus fell victim to a different type of heresy. Although in 365 the Arian Germinius was allowed to recover his see, from then onwards Valens and his irresolute follower Germinius gradually lost ground; and when Valens and Ursacius, by now probably old men, disappeared from the scene, Ambrosian Catholicism succeeded at last in establishing a bishop of

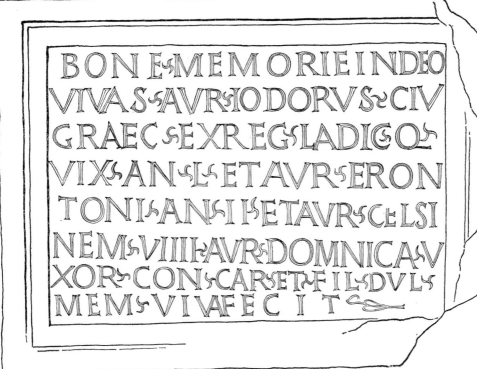

BON E·MEM O RIE I NDEO
VIVAS·AVR·IODORVS·CIV
GRAEC·S·EXREG·LADIG·O·
VIX·AN·L·ETAVR·ERON
TONI·AN·I·I·ETAVR·CELSI
NEM·VIII·AVR·DOMNICA·V
XOR·CON·CAR·ET·FIL·DVL·
MEM·VIVAFE C I T·

Figure 55   Christian tombstone of Aurelius Iodorus, a Greek from Laodicea, found at Savaria (Hungarian National Museum)

orthodox views in Sirmium. Under Valentinian the church of Sirmium with its bishop Anemius was already under the influence of north Italian Catholicism. An event which would have been impossible only a quarter of a century earlier happened in 378: western Roman bishops assembled in Sirmium and unanimously condemned the east Roman Arian heresy.

It is to be hoped that this survey, despite its brevity, has shown that any part played in general church history by the Christian communities of Pannonia and Upper Moesia was due to their proximity to the imperial court at Sirmium and the opportunity this afforded for involvement in high politics. The bishops of towns that lay further from Sirmium either played no part at all or entered church politics only as secondary followers of this party or the other; and this may have been the reason for their being named only in exceptional cases and by chance in the sources.

After all that has been said, it would seem idle to ask whether the Christians of our provinces professed themselves Arians or orthodox. Not even the leading personalities of the communities created or represented a dogmatic trend of their own; on the contrary, a sometimes unattractive manoeuvring was characteristic of the most important bishops. The Arianism of a Valens or a Ursacius seems to have been mainly dependent on the general situation of ecclesiastical politics and not on the opinion of their own community or its priests. Although they were the most eager exponents of eastern Arianism in the west, their influence and their successes rested all too much on the favour of the Court. The broad mass of believers were presumably little affected by politically conditioned dogmatic subtleties, and were all the more ready to leave controversies to their priests and bishops, since they were mostly still involved in the simple dilemma of Christianity versus paganism. That Arianism was able under such circumstances to gain the upper hand was probably mainly due to the fact that the earliest, and therefore dogmatically best orientated, Christians were, as we have seen, Greeks and Orientals, whilst it was only in the first half of the fourth century that the native population began to be converted. The oriental component in the Christian communities can be demonstrated everywhere where early Christian inscriptions have come to light (Figs 56, 57), and even the language of many early Christian grave-inscriptions is Greek,[126] whereas the Greek element in the epigraphic remains of earlier periods was very small. Whether, in the Arianism of the Pannonian bishops, we are observing tactics of an Illyrican kind for securing Danubian influence in the policies of the western church, cannot be proved though it is not improbable.

Early Christian monuments, of course, can only prove the existence of Christians: they cannot indicate the proportion of Christians to pagans. But there are some monuments whose interpretation is questionable. A number of ground plans, the results of old excavations, have been interpreted as Christian cult-rooms, and in some cases modern re-excavation has proved that these interpretations are incorrect. Apsidal rooms are often considered to be Christian basilicas, although the ground plan itself admits other interpretations. For instance, it has recently been shown that the so-called basilica of Quirinus at Savaria was in fact the Aula of the imperial palace and not the resting-place built under Constantine for this bishop of Siscia.[127] Despite this, the Christian monuments of Pannonia are very numerous and varied, particularly in the better-explored northern part of the province, and a similar wealth of monuments can well be expected in Upper Moesia too, where excavations at Ulpianum (Pl. 38b) and Naissus suggest large Christian communities with extensive cemeteries.

Indeed, the most notable remains of the Christian communities are their cemeteries. It is extremely likely that in the fourth century Christians instituted cemeteries of their own, separated from those of the pagans, at least in those

Figure 56 Christian tombstone of Artemidora buried near St Synerotas at Sirmium

settlements where they were numerically powerful. On an inscription from the Christian cemetery of Savaria a *custor cymiteri* is named,[128] and in all larger cemeteries so far investigated it could be established that the centre was the cult-chapel of a deceased Christian who was particularly revered in the community. A few written sources mention martyrs' resting-places, and epigraphic evidence[129] makes it certain (Fig. 56) that the focus of the Christian cemetery at Sirmium was the burial-place of the martyrs Synerotas and Fortunatus, who are of course otherwise also well attested.

The centre of the Synerotas cemetery at Sirmium was a *Cella Trichora*; this type of cemetery-chapel has also been found in Sopianae and Aquincum (Fig. 53).[130] The remaining chapels in undoubted Christian cemeteries are distin-

Figure 57 Painted vault of the *cella memoriae* at Pécs (Sopianae)

guished by their variety.[131] A simplified variant of the *Cella Trichora* was the chapel with two apses in which, for the most part, sarcophagi were placed; this type is attested at Ulpianum and in the cemetery of Tricciana (Ságvár).[132] The sarcophagi were either buried in the ground or they stood on the floor of the chapel (Pl. 38b); for the most part they were simply executed in the so-called house type, without decoration. The Belgrade sarcophagus, re-worked from an earlier sarcophagus with the representation in relief of the Good Shepherd and of Jonah, is unique,[133] as is also the sarcophagus from Szekszárd in the province of Valeria,[134] though this carries Christian symbols only on its lid. In the grave-chapels the dead were mostly buried below ground, quite often in a subterranean grave-chamber. These grave-chambers were often painted, in a rather unpretentious, simple, but very engaging and straightforward style. The best-known monuments of this kind are Grave-chapel I at Sopianae (Pl. 40b) and the closely related example at Naissus, both of which show the Chi-Rho monogram between Peter and Paul in the main scene;[135] this iconographic design is also found in the apse-mosaics of basilicas in Rome, datable not much later. At Sopianae, there are also portraits(?) in medallions and scenes from the Old Testament (e.g. Jonah, Noah, etc.) represented on the vault (Fig. 57). A further common trait

334

on murals at Naissus and Sopianae is the fence of a garden (Pl. 40a), intended to symbolize the garden of Paradise—or so it is interpreted, presumably correctly. In a simplified manner the motif of the garden is painted in a very small stone-built tomb, also in Sopianae.[136]

Exceptionally, grave-chapels also had mosaic floors, for instance at Ulpianum, but otherwise they were probably decorated in a rather simple fashion. The larger chapels contained many burials, the dead being walled in next to each other in the chamber. In the largest chamber that has so far come to light, found in Sopianae, there was room for twenty dead.[137] Apparently there existed no connection between the size of the chapels and their artistic or architectural equipment. The painted chamber I at Sopianae had a rather modest super-structure, whilst the unique *Cella Septichora*[138] at Sopianae (Fig. 53) had no subterranean tomb. Modest chapels with one apse or even with only a rect-angular ground plan are attested in many cemeteries, but in the huge late Roman cemetery at Naissus cemetery-basilicas of often considerable size were uncovered. It is quite probable that the crypt of the eleventh-century Cathedral of Pécs (Sopianae) was an early Christian cemetery-basilica too, especially as it has a basilican plan with nave and two aisles and as the cathedral is built in the area of the early Christian cemetery—which itself is presumably not accidental. The *Cella Trichora* of Sopianae was rebuilt in about the ninth century and its walls repainted.

The early Christian grave-inscriptions of Pannonia and Moesia are always small tablets, which were built into the walls of grave-chapels or served to close the *loculi*. These alone are sufficient to prove the rich architectural achievement of the early Christian cemeteries: in some of them such grave-tablets came to light in great numbers, for instance at Savaria and Sirmium. And this is evidence that cemeteries with rich furnishings plentifully supplied with buildings above and below ground are Christian. There are of course a large number of extensive fourth-century cemeteries which although fairly rich in grave-goods are fairly uniform in burial-rite and decidedly poor in architectural elaboration; pagan burials are probably to be sought in these poorer surroundings.

The centres of Christianity, however, can probably be looked for in the larger settlements and in the large estates. This is suggested not only by finds of Christian grave-goods but also by the cult-buildings hitherto certainly identified. As we have seen, the cemeteries that can certainly be established as Christian are municipal cemeteries, or at least those belonging to large settlements and to fortified centres of imperial *latifundia*. It is in such places that Christian cult-buildings are to be expected, and in addition in some larger villas. Christian basilicas (Fig. 53) have been proved at Ulpianum (Pl. 38b), Remesiana—both episcopal sees in the fourth century—Aquincum, and at Fenékpuszta (Pl. 38a), and also in the area of a late Roman villa north of Lake Pelso, where broken tiles with the Chi-Rho monogram suggest a Christian cult.[139] Marble altar-slabs have come to light in the villa of Donnerskirchen, at the villa of Örvényes—which has also yielded other pieces of architectural sculpture from a cult-room—and from Csopak in the vicinity of Örvényes;[140] the last two find-spots are situated on the north shore of Lake Balaton (*Pelso*), where great estates had long been the predominant agricultural type.

Art in the late Roman period

If Christianity appears to some extent as a religion linked to urban centres, the administration and owners of estates, this is even more so with late Roman art in the Danubian area. The first two periods of prosperity under Roman rule had, as everywhere else, produced a provincial Roman culture of a special character, traceable for us in a local style of sculpture in the grave monuments, in the minor arts partly derived from pre-Roman tradition, in a few local cults and in other similar characteristics. Although this provincial Roman civilization was the culture only of the ruling upper class of provincial society, this class had a sufficiently numerous membership and was present practically everywhere, as either a military or a municipal élite, or in earlier times, as a tribal aristocracy. In times of peace this class had richly ornamented grave monuments set up with inscriptions in Latin, donated cult-buildings, cult-statues, reliefs, votive gifts and altars, adapted its way of life more and more to Roman standards, and was keen in all its lasting manifestations to appear as a Roman upper class. Though it declined in numbers and in wealth under the Tetrarchy the class continued to exist and left behind many traces of its earlier outlook. But soon it reached complete stagnation, and this has to be attributed to far-reaching social changes which we have already described as far as possible. The artistic monuments of late Roman times stem now from a new aristocracy, which was in part an official class tied to the administration of the empire and in part a necessarily small group of great landowners. It was, on the one hand, numerically much weaker than the earlier military–municipal aristocracy, and therefore was not able to sustain a local artistic production; but on the other, as part of the imperial aristo-cracy, it was rich enough to patronize imperial art and to cultivate centres of art and artists beyond the province. Early Christian grave-inscriptions which for the most part had been set up by the middle classes were for that reason tablets of no artistic pretension, lacking plastic execution and carved with clumsily chiselled letters: the customers had to apply to local stonemasons who were no longer accustomed to plastic works that a stonemason of Severan times would still have been able to execute without difficulty. The new aristocracy, on the other hand, sent for masters from Aquileia when they wished to adorn their villas with mosaics; they acquired works in silver (Pl. 45a) or in gilded bronze from the important workshops and had sculptors invited to attend them.

But we must not underrate the proportion of late Roman artistic monuments in Pannonia and in Upper Moesia which are attributable not to this aristocracy but to the emperor himself, who was either the owner or the donor of these works of art, and often the object of the representation too, as was the case with the imperial statues of Naissus, for example. The evidently very rich sculptural decoration of the villa at Mediana, or the statues in porphyry found here and there—and significantly in Gamzigrad in particular—were commissioned for a high official if not the highest; finds like the gilt-bronze portrait of a young emperor from Sopianae,[141] or the state cameo from Kusadak[142] in north-western Moesia evidently derive from court workshops—and from the view-point of local art and culture it is irrelevant that some of these court workshops were centred at Sirmium and Naissus. The gilt parade-helmets of high-ranking officers found in the fort opposite Aquincum and at Berkasovo,[143] not far from Sirmium (*Frontispiece*), are the products of factories listed in the *Notitia Dignitatum*. The silver dishes from various late hoards were produced in workshops at Naissus and were probably intended to contain imperial gifts and presentations for dispatch to persons of high rank.[144] The portrait-bust from Sopianae, already mentioned, can also be interpreted as an imperial present, being fixed either to the tip of a consular sceptre or to the middle of a dish. The richly ornamented silver tripod from Polgárdi[145] in Valeria (Pl. 45a) could equally well have been found in one of the famous late silver-hoards such as those of Augst or Traprain.

Even some of the less valuable products of the applied arts should perhaps be attributed to central workshops. Possibly the most characteristic group of late Roman small artistic objects in Pannonia is the bronze mounts of small boxes with ornamentation in chased relief (Pls 44b, c).[146] The themes are very varied, but are related to imperial art and have a certain official character. These small boxes, too, can be interpreted as official presents—of so far unknown destination. Lastly, possibly the most striking example must be mentioned: the technically skilful, complicated cross-bow brooches so uniform in the manner of their production, and in their ornamentation, that they must certainly originate from state-owned factories. Among them are gilt, nielloed, even gold examples, notable products of a very exacting goldsmith's art. From an earlier period, too, there exist brooches that were the products of a highly developed artistic industry. But they are less uniform than the late Roman cross-bow brooches and for typological reasons must be attributed to local masters; for instance the so-called Norico-Pannonian wing-brooches of the native Celtic population are notable products of local metal-working and of local taste. But there are no types of finds and groups of finds of late Roman date which show local character or a locally limited distribution. Costume and way of life had become largely standardized, not only in high social circles, which though weak in numbers were closely interlinked with the aristocracy of the empire, but also among the common people, whose way of life became uniform presumably as the result of poverty. Social change in the late Roman period resulted in the former middle classes, composed of the military and the municipal aristocracy, being either promoted into the imperial aristocracy or—presumably the more common alternative—sinking back into the common people. This polarization finds its expression in the history of the settlements, in the growth of large estates, in the material of the small finds, in burials, in the arts and in the last resort in religion too. The process must be summarized and emphasized once again at the end of this chapter, since it had far-reaching consequences for the fate of our provinces after the end of the fourth century. Roman culture and presumably the Latin language as well were confined to those who by office or by wealth were closely connected with the government and with the administration; they no longer formed a military–municipal social class mediating between superiors and inferiors, but a class that was rather isolated from the people, and which was not strong enough to maintain itself under the changed circumstances after the end of the fourth century, in the way in which the senatorial estate-owners of Gaul, Spain and Africa were able to maintain themselves. The special situation of our provinces as frontier-provinces may have been responsible for this: the incorporation of masses of barbarians from the left bank of the Danube affected the Danubian provinces most of all, and the largest estates here probably belonged to the emperor; particularly so since these estates had come into existence only lately, as the result of politico-military changes and of a systematic economic policy.

Chapter 10
The beginning of the Dark Age

In the autumn of 376 hordes of Goths under the leadership of Fritigern crossed the lower Danube and were given the *receptio* into the empire by a *foedus*, which allowed them to settle in the provinces of Dacia Ripensis, Moesia Inferior (Moesia Secunda) and Thrace. This date is a turning-point in the history of the Danubian and Balkan provinces, since after it barbarian groups never ceased to be found within the frontiers of the empire. Late Roman historians fully appreciated the significance of this turning-point; Jerome, by birth a Dalmatian, starts the final misfortunes of the Romans at Illyricum at this date,[1] contemporary and later writers do not tire of speaking of the fall of Illyricum, and a chronicler of the eastern empire took the year 377 as the beginning of the rule of the Huns in Pannonia,[2] and not wholly without reason, as will be seen.

The Goths who had been settled on imperial territory started the same year to become restless, and plundered far and wide in the provinces that had been assigned to them. The immediate cause of these raids was that the Goths were not, or not regularly, given the provisions which they had stipulated and which they should have obtained as *foederati*. The situation was further aggravated by the appearance of a new group of Goths under the leadership of Farnobius from across the Danube. The first Roman counter-measure, taken in 376, merely succeeded in diverting the plundering Goths from Gratian's part of the empire. This was due to Frigeridus (possibly a former *dux* of Valeria), who, on Gratian's orders, appeared in Illyricum with western, including Pannonian, forces. The next year, Gratian sent more troops under Richomeres to Illyricum, and these made contact with the eastern troops of Valens. In the meantime, however, there had also been an invasion of Taifali from Trajan's Dacia into Dacia Ripensis; they placed themselves at the disposal of Farnobius. Frigeridus succeeded in defeating them south of the Haemus (in the Balkan mountains), but this did not change the fact that the Goths were in rebellion on imperial territory and neither western nor eastern troops could either throw them back over the frontier or force them to be peaceful.

In spring 378, therefore, both emperors set out for the Balkan peninsula with their field-armies. Valens came from Antioch and Gratian from Trier. For Gratian the organization of the campaign was so urgent that he went by boat down the Danube and stopped only at Sirmium for a short halt. From Sirmium he also probably went by boat, and did not set foot on land before the eastern edge of Dacia Ripensis, where he found himself faced by a group of Alans. Near Castra Martis he suffered a minor defeat at their hands, which however contributed considerably to his decision not to advance any further. Valens is attested in Constantinople at the end of May,[3] and it will have been somewhat later that he appeared in Thrace.

The group of Alans that opposed Gratian near Castra Martis belonged to a Gothic–Alan group whose leaders were the Goth Alatheus and the Alan Saphrac. A band of Huns which had deserted the main body of Huns also entered this alliance; a circumstance which was to be of a certain importance in the event. This mixture of people under Alatheus and Saphrac entered the empire probably in 377 and at first remained in Dacia Ripensis. In 378 the two chieftains got in touch with Fritigern and contributed considerably to the catastrophe near Adrianople on 9 August. After the battle, like the other barbarian groups, they went on raids in the Danubian and the Balkan provinces, and finally came to Pannonia.[4]

Gratian remained at Sirmium during the winter of 378–9 where he created Theodosius an Augustus on 19 January.[5] Soon afterwards he left Illyricum, leaving the problem of the Goths to the new emperor. Theodosius went to Thessalonica where he made preparations to fight the barbarians. Only after a number of measures had been taken did he leave Thessalonica for the purpose of clearing the valley of the Axius and Dardania; how far north he managed to advance, we do not know.[6] The chronology of this year is very confused. We are informed of a military enterprise the leader of which was the *dux et praeses* of Valeria, Maiorianus, who was soon afterwards appointed *magister*.[7] Possibly he fought against the groups led by Alatheus and Saphrac which had penetrated into Pannonia. A victory over Goths, Alans and Huns was announced in Constantinople on 17 November 379,[8] while Theodosius had suffered a number of smaller defeats south of Pannonia at the hand of Fritigern.[9] The Alans, Goths and Huns, however, who were defeated in the autumn may be identified with the peoples of Alatheus and Saphrac. The victory over them (? in Pannonia) could have no wider results because of Theodosius' lack of success, and we hear, in fact, that the towns of Pannonia had to live through difficult times in the winter of 379–80, and some settlements were even destroyed. The commander of the Pannonian troops, Vitalianus, was powerless against the barbarians, who were attacking on a wide front.[10]

Gratian had therefore to appear again in Illyricum. He sent a number of troops ahead of him, who undertook clearing-up operations in Dacia Mediterranea and Thrace,[11] and at the end of the summer came to Sirmium, where Theodosius, who had been impeded by illness, also arrived not much later. At the meeting of emperors in Sirmium, which was probably regularly repeated during the following years at either Sirmium or Viminacium, some principles for regulating the situation in Illyricum, which had fallen completely into disorder, could be worked out: in the following years negotiations were set going with the Goths of Illyricum, leading, in 382, to a new treaty with the Goths, according to which they were given extensive areas for settlement in the provinces of Dacia and Thrace. In 380 the peoples of Alatheus and Saphrac were settled in Pannonia, presumably in the areas of which since their plunderings in the winter of 379–80 they were the *de facto* masters.[12]

We have no information on the size and position of the areas designated by the treaty with Alatheus and Saphrac. The only fairly secure clue is given by the grave-inscription of a bishop, who was buried in Aquileia approximately at the beginning of the fifth century. In another source, this Bishop Amantius is attested in 381 as bishop of the small Pannonian town of Jovia in the valley of the Drave, east of Poetovio, and his grave-inscription praises him among other things for having 'emigrated' (presumably from his native town of Aquileia) to Jovia, where he undertook a blessed activity among the 'two peoples' and developed good connections 'with two *duces*'. Whether the two peoples were the Romans of Jovia and the barbarian *foederati* or whether they were the subjects of the two chieftains, cannot be decided; it is certain, however, that the *gemini duces* must be Alatheus and Saphrac.[13] From this we may infer that these barbarian chieftains lived in the diocese of Amantius, or at least in the vicinity of his diocese. Jovia lay in the valley of the Drave, of which the largest Pannonian town, Mursa, was troubled by barbarians in the winter of 379–80. Jovia was, moreover, situated in the province of Savia, on the southern boundary of which lay Stridon, the birthplace of Jerome, which also underwent a barbarian attack in 379–80. A barbarian tribe—called *antiqui barbari*—who were possibly the descendants of these *foederati*, are mentioned in the province of Savia in the sixth century.[14] Thus it is suggested that the settlement area of the *foederati* of Alatheus and Saphrac lay in the valley of the Drave in the province of Savia.

In the meantime the situation beyond the Danube had changed, too. We do not know exactly, of course, what happened during these years in the approaches of Pannonia and Moesia, but Ammianus mentions under the year 378 that all the peoples on the left bank of the Danube, from the Marcomanni and Quadi to those living by the Black Sea, had started to move, and, driven from their homes, were threatening the *limes*.[15] Presumably they wanted to be given the *receptio*—but the reception of Fritigern in 376 did not serve as a good precedent to induce the Romans to accept further groups of barbarians. Hoards, hidden in the ground in the Hungarian Lowlands around 375,[16] suggest unrest in the country of the Sarmatians; the Taifali, under pressure from the retreat of Athanarich into the southern Carpathian mountains, broke into Dacia Ripensis, as we have seen, and in 381 even Athanarich had to ask for the *receptio*, presumably as the result of Fritigern's Goths having had to retire over the Danube into the old Dacia a short time previously.[17] The reason why we cannot establish the effects of unrest beyond the frontier on the *limes* of Pannonia and Moesia is probably that the attention of contemporaries was too much focused on the situation inside the empire. An invasion of Sarmatians occurred at an unknown date before 384,[18] but nothing is otherwise recorded of fighting on the *limes*; this is partly due to the fact that pressure by the Huns at that time only affected the lower Danube and because the situation in the approaches to the *limes* had been relieved by the emigration of large masses of Goths. Gothic pressure on the Sarmatians therefore decreased, and the disturbances of *c.* 375–80 seem soon to have been settled.

There are, indeed, some indications for a certain amount of consolidation on the middle Danube in the 380s. Bishop Amantius at Jovia has already been mentioned. St Ambrose reports that in north Italy Pannonian corn was traded for wine in 383,[19] and Valerius Dalmatius, a governor of the Gallic province of Lugdunensis III, at the earliest towards the close of the 380s, retired to his estates in the south of the province of Valeria.[20] Since, however, the province of Lugdunensis III was created by the usurper Magnus Maximus, and its governor was therefore probably one of his supporters, we must suppose either that the area controlled by Maximus included Valeria for a time, or—less probably— that the former supporter of Maximus was exiled to Valeria. It is not impossible that Magnus Maximus had considerable influence with the *foederati* in Pannonia, since in the same year as Maximus appeared in Pannonia, and in Savia in particular, barbarians plundered the province.[21] And Maximus got in touch with the Goths on the lower Danube, which enabled him to hinder Theodosius' advance against him.[22]

The towns of Savia, among them the important base of Siscia, fell into the hands of Maximus in 387. Theodosius however was able to take the field against him only in the summer of 388. He advanced from Thessalonica through Dardania and Moesia I to Pannonia, and defeated the usurper, first near Siscia, then near Poetovio.[23] A factor of importance for an estimate of future events must be mentioned here: the mint of Siscia, which with its four *officinae* was the largest in Illyricum, ceased production in this time.[24] The great mints of the west (Arelate, Lugdunum, Treviri) ceased production mostly towards the end of Theodosius' reign, and at that same time the other Pannonian mint at Sirmium (which was of little importance) closed down.[25] The closing of the western mints *c.* 395 is evidently the effect of general economic and political conditions—in the last resort the change to an economy based on natural produce; the drying-up of monetary circulation under and after Theodosius cannot, therefore, be simply connected with decay of 'the Roman way of life'. The closing-down of the mint of Siscia broadly, of course, belongs here too, but the date, almost a whole decade earlier, demands special reasons. In my opinion two motives must be considered here: first the *foederati* of Alatheus and Saphrac, as we have seen, were settled precisely in the province, or preponderantly in the province, of which Siscia was the capital. In this way the surroundings of the *Moneta Sisciensis* had become a danger area, and it presumably seemed advisable to the government to shut down this important mint. Secondly, it has recently become increasingly clear that it was not under Theodosius that money circulation in Pannonia— above all north of the Drave—ceased, but earlier, *c.* 375: in other words to all appearance considerably earlier than in the other European frontier provinces. Siscia was able to provide Pannonia with the small change of the years 368–75 in abundance.[26] The sudden decrease[27] is certainly due to serious causes. The reasons for the early cessation of monetary circulation will best be connected with the settlement soon after 375 of a very considerable mass of barbarians in Pannonia, who paralysed economic life and the functioning of the administration over large areas of the country, and probably in those areas in particular which were the most important links between the Danube frontiers and the South. The *foederati* in Pannonia were wedged between northern Pannonia (Pannonia I and Valeria) and the south (Italy and the Balkans), and this situation must be significant.

The political activity of the *foederati* and their employment in various élite units show that their presence in Pannonia and Dacia by no means implied a peaceful coexistence of Roman town-dwellers and of partly barbarian inhabitants of the countryside. Moreover they had chronic difficulties of supply and established themselves as parasites on the provinces, which in any case were economically not up to standard. In 391 Theodosius had to fight against the Goths in the Balkans; he defeated them in 392 near the Rhodope mountains.[28] It was probably the Pannonian *foederati* who attempted an attack on north Italy in the same year,[29] and this may suggest that the situation was even worse in Pannonia, or that Pannonia itself had little left to offer them. Jerome, observing events in his homeland with understandable attention, reports now and then a few details which because of the emotional terseness of his style may appear mutually contradictory. Historians have gone to either extreme: either everything had already perished in Pannonia—or rather in the Pannonian–Dalmatian border region around Stridon—by 380, or everything was still nice and peaceful *c.* 410. Jerome reported that Stridon had been destroyed by the barbarians in the year 380;[30] then under 392 he wrote that 'all perished';[31] a few years later, however, he wished to sell his father's estate in Stridon.[32] An expression of pathos and despair such as *cuncta perierunt*, of course, cannot be contrasted with the sale of 'half dilapidated little farms', as Jerome characterized his father's estate. *Semirutae villulae quae barbarorum effugerant manus* can be presumed to be a factual description of the situation in Pannonia, while general laments on the *damna Illyrica*, the *Illyrica vastitas*, etc., have a long-term justification. Otherwise these expressions would not have been used so quickly and so widely by contemporary historians. Apart from Jerome, they are encountered in the works of St Ambrose, Orosius, Pseudo-Augustine and others,[33] and the fact that this phraseology occurs later, too, in Sidonius Apollinaris or Ennodius,[34] does not affect its truth. The first occasion for these laments was provided by the Gothic movements before Adrianople; their repetition at the end of the fourth century and in the fifth was caused by new attacks and by continual internal unrest in the provinces concerned. The stages of decline were often bewailed in contemporary literature with conventional expressions and metaphors, but the underlying truth must not be doubted despite its guise of literary platitude. The situation in Pannonia is described by Claudian in retrospect in 399 as a continual siege;[35] the inhabitants of the towns do not dare to leave the town-walls, and the peasant cannot cultivate his fields. Certainly, this also is an exaggeration, particularly since Claudian wished thus to emphasize the merits of Stilicho. But times of crisis were repeated only too quickly and consolidation was hindered because the *foederati* could not adapt themselves as producers in the economic life of the country. Their most important sources of income were their position among the élite troops of the central army, and their exploitation of the provincial population, partly by means of the food-rations they drew and partly, no doubt, by straightforward plundering.

Alatheus and Saphrac contributed important contingents to the army of Theodosius in the battle at the Frigidus in 394, in which the Visigoths under Alaric also took part. The troubles that broke out in the following year, which were partly connected with dissensions in the camp at the Frigidus, affected not only the lower Danube but probably Pannonia too.[36] Whether barbarians broke into Pannonia cannot be proved for certain, but Stilicho had to fight with barbarians soon after he had crossed the Julian Alps.[37] Claudian tells us nothing of successes by his patron in this connection, but wide-scale plundering by the Goths and the first appearance of the Huns on the lower Danube probably offered the Pannonian federates an opportunity to fish in troubled waters. Archaeologists in Pannonia have connected many burnt deposits and destruction layers with the plunderings of the year 395; whether they are right to do so cannot be determined, if only because the dating evidence of coins, as we have seen, becomes very rare after 375. However, from Vindobona, where the existence of thick burnt layers can be established, a comparatively large coin-hoard has come to light of which the final coin dates from 395.[38] The *foederati* presumably did not limit their raids to the area assigned to them, but whether they were responsible for the burning of Vindobona cannot be definitely stated. What makes it less likely is that the immediate neighbours of Vindobona, the Marcomanni, are heard of at just this time.

In Paulinus's biography of St Ambrose we read of a Marcomannic queen named Fritigil[39] who, under the influence of the great bishop of Milan, was converted to Christianity, and persuaded her husband to surrender to the Romans. It follows from this that part of the Marcomanni crossed to Pannonia, and was given the *receptio*, and an entry in the *Notitia* confirms this settlement from another angle.[40] We read there of a *tribunus gentis Marcomannorum* in Pannonia Prima, and it is much more likely that this *gens* was the people of the husband of Fritigil than the descendants of the Marcomannic king, Attalus, settled in Pannonia under Gallienus. At the same time this entry seems to be one of the last in the Pannonian chapter of the *Notitia*.

Rome's old neighbours on the Danube frontier, the Marcomanni, Quadi and Sarmatians, and the comparatively new ones such as Vandals and Gepidae are otherwise hardly mentioned at all towards the end of the fourth century. For the most part they only appear in lists of people that had started to move and were threatening the Danube provinces, and we cannot therefore decide whether they were entered in these long lists merely for the sake of completeness or whether they had actually taken part in the fighting,[41] as subordinate partners of the Goths and Huns. The Sarmatians are named fairly frequently by Claudian,[42] but generally only to describe a geographical area or the section of the Danube

known as the *ripa Sarmatica*. These passages in Claudian are at the same time the
only proof that the Court of Ravenna considered the Danube frontier as one still
held by Roman occupying forces. From the *Notitia*, which is an enigmatic source
for late Roman military history, we cannot disentangle which of the troops
enumerated in the respective chapters were the latest ones and when they were
stationed on the Danube. The lists of troops in Valeria and Pannonia Secunda in
particular seem to be rather old-fashioned in comparison with the lists of other
provinces, in that they are in several layers and contain a long succession of fort-
sites enumerated twice and three times over.

The problem of the end of the Roman *limes*, however, is not special to the
Danube provinces. As long as historians assumed a single 'collapse' of the *limes*
as the result of a single barbarian attack, a date was sought before which the
limes existed as a state frontier with regular garrisons, and after which it simply
no longer existed as a frontier and military zone. However, from the end of the
fourth century we must distinguish between formal state frontiers, frontiers of
actual spheres of influence and strategically important lines occupied by the
army. We hear practically nothing of the Danubian frontiers having played any
part in the protection of Roman territory. The territory occupied by *foederati*,
and later by barbarians who were not federates, extended into former areas of the
empire, but this did not prevent the Danube remaining the formal frontier of the
Roman empire. Building periods datable to the fifth century cannot be proved
on the *limes*, and the few, technically inferior and otherwise unimportant repairs
in the forts of the *limes* which must be of post-Valentinianic date can only prove
continued use of the stone buildings by various ethnic or social elements but not
regular use of the *limes*-buildings by regular Roman troops. The settlement of
foederati on the border gradually deprived the garrisons of the *limes* of their
strategic importance; it was less and less necessary and possible to maintain
discipline, to fill up the units by regular recruitment and to secure provisions. A
gradual decline, however, can only rarely be followed in detail and divided into
periods. The process of this decline started under Theodosius and ended in about
the 420s, when the central administration ceased to attempt to keep the lists of
troops in the *Notitia* up to date.

In the period that we have been discussing the Danube *limes* had no significance. It lasted longest on the border of Pannonia Secunda and Moesia Prima,
where no federates were settled. But this section of the *limes* was the least
important on the middle Danube. It gained importance only when the centre of
the Hunnic empire was shifted to the Danube–Tisza plain, when it was this
section of the *limes* which was most exposed to pressure from the Huns. But then
it ceased to be a militarily occupied frontier.

In 399 or 400 Stilicho is praised as *Rheni pacator et Histri* by Claudian [43] and in the same poem a number of Stilicho's successes in Pannonia are listed. [44] The context does not make it certain whether Stilicho visited Pannonia, and the panegyric exaggeration of his successes does not permit recognition of the true state of affairs. According to Claudian the peace of former times returned to Pannonia; but the inhabitants were not able to enjoy the results of this *pacatio* for very long. In 401 the Vandals marched from their home in the north of the Hungarian Lowland through Pannonia to the west, [45] and this migration particularly affected that part of the province which up to that time, though by no means spared altogether, had not been settled by *foederati*. In the same year Alaric also departed and moved with his people to the west. The migration from Dacia Ripensis through Pannonia to Ravenna could be accomplished by only one route, through the valleys of the Timok, Morava and Save, and then over the Julian Alps. Alaric arrived in Italy in 402, and after he had to retreat out of it he was given Pannonia by Stilicho, [46] but probably not as a federate territory, and this presumably aggravated the situation of the province still further. We do not know whether Alaric's brother-in-law and successor, Athaulf, had already settled in Pannonia at that time as well; but a few years later he was certainly in western Pannonia. [47]

This second great settlement of barbarians in Pannonia neither lasted long nor was able to achieve consolidation. In 405 Radagaisus' huge band of barbarians moved through northern Pannonia, probably on the route used by the Vandals when going west. [48] This time it was probably the Visigothic control of southern Pannonia near the approaches to Italy under Alaric and Athaulf which was able to prevent their crossing the Julian Alps. This migration no doubt had a catastrophic effect on the areas of Pannonia which lay in its path. A few years afterwards Jerome counts the *hostes Pannonii* among the peoples that had poured into Gaul, adding a sigh over the *lugenda res publica*. [49] The fact that the *hostes Pannonii* are named at the end of the list, so to speak as a climax, with the added lament, suggests that Jerome meant the non-barbarian inhabitants of Pannonia, though presumably not the provincial upper class. A few statements in our sources attest a flight of the socially higher-ranking inhabitants of Illyricum to Italy and presumably to the Dalmatian coast too. Some of these are to be dated after 403, others before 408, and thus are best related to the march of Radagaisus. These statements are the first to mention Pannonian refugees who from the beginning of the fifth century onwards appeared in various places in the west and in the Mediterranean area.

We can probably recognize Bishop Amantius as one of the earliest fugitives. He is attested in 381 as bishop of Jovia, and after officiating for twenty years in the Christian community of this small town on the Drave, he returned, some time in the beginning of the fifth century, to Aquileia, where he died in 413. The inhabitants of Scarbantia fled from an *incursio barbarorum* to Italy, and took the relics of Quirinus, the martyr bishop of Siscia, from Savaria with them. In 402/3 Quirinus was not yet buried in Rome,[50] and so we may attribute the panic caused by the barbarians in north-western Pannonia to the invasion of Radagaisus. The fact that the bones of the martyrs were taken from their graves and accompanied the flight is significant enough for the situation: the martyrs' graves were the focus of cult-life, they were preserved in a *Cella Memoriae* in the Christian cemetery (as in Sirmium), or in a basilica (as in Savaria). By the translation of the relics, the Christian communities lost their most important treasures.

The flight from Radagaisus seems to have set in motion a fairly large mass of people. In 408 a decree was promulgated, which suggests a large number of refugees from Illyricum.[51] A number of fugitives who did not have the appropriate proofs of identity fell into an awkward situation by being seized as slaves; the decree refers to this situation having already become chronic, and speaks of the whole of Illyricum, both fugitives and those remaining on the spot. It follows that in 408 the situation was still unconsolidated, and this conclusion can also be drawn from other decrees on captives and refugees. Probably Alaric, too, took captive Romans with him to Italy.

About 408 at the latest the Senator Flavius Lupus was active in Pannonia Secunda. His duty was to control the land-tax, whose proceeds were used by imperial decree for construction work on the walls of the towns.[52] The towns of this particular province for the most part presumably already possessed their town-walls, like Sirmium, for instance, whose walls are mentioned by Ammianus under the year 374. But an earlier date for the activity of Lupus in Pannonia is equally feasible.

What happened in Pannonia, after the Gothic kings Alaric and Athaulf had migrated to Italy, is obscure. The only record we have for two decades following 408 is of a military command under a certain Generidus, who, according to Zosimus, was the commander of the troops in Dalmatia, Raetia, Noricum and Upper Pannonia (i.e. Pannonia Prima), about 409, and who proved himself a brave commander.[53] In the *Notitia*, the ducates of Noricum Ripense and of Pannonia Prima are, in fact, united, and the apparently accurate enumeration of Generidus' command in Zosimus may therefore possibly suggest either that, of all the provinces of West Roman Illyricum, it was only in these that troops remained any longer, or else it was only in these that Generidus was in a position to organize a proper occupation. Possibly it was under him that Noricum Ripense and Pannonia Prima, which in the *Notitia* have a common *dux*, were united.

The double mentions of forts, which suggest that the list of troops of the province in question is a compilation, hardly occur in the Norican–Pannonian ducate, and this may mean a relatively late validity for this chapter.[54]

We do not know how long the command of Generidus could be maintained. In the second and third decades of the fifth century the area controlled by the empire of the Huns presumably moved westwards; as a result Pannonia soon found itself neighbour to the Huns and probably in the third decade parts of the province actually came under Hunnic rule. About 408 the centre of the Hunnic empire, of course, still lay on the lower Danube, probably opposite Dacia Ripensis. But the migration through Pannonia towards the west by Vandals, Goths and the mixture of peoples under Radagaisus was in the last resort caused by expansion by the Huns westwards. These movements of peoples heralded Hunnic expansion, and they affected Moesia Prima which up to that time had generally been spared. Moesia Prima, Dacia Mediterranea and Dardania had often been exposed to raids by the Goths who had·been settled north and south of the Haemus in 376, but these raids concentrated much more on the areas east and south of what had formerly been Upper Moesia, which was only the periphery of the endangered area. Nothing is reported of any emigration worth mentioning at the beginning of the fifth century, and continuity of the church hierarchy can therefore be traced back in some towns of Upper Moesia into the fifth century,[55] in contrast to Pannonia, where except at Sirmium[56] no bishop is attested in the fifth century. The surroundings of Sirmium shared the fate of Moesia from the twenties of the fifth century onwards, and in 437 were ceded to the eastern Roman empire, evidently for the simple reason that they were cut off from the western empire by Hunnic expansion in the area of Pannonia.

The proof that by 420–5 parts of Pannonia had come for practical purposes into Hunnic possession is given by the chronicler Marcellinus *comes*, who under the year 427 mentions the Roman reconquest of Pannonia which had been occupied for the last fifty years by the Huns.[57] It is obvious that Marcellinus counted the round fifty years' rule by the Huns from 379, that is from the date when the peoples of Alatheus and Saphrac, part of whom consisted of a Hunnic group, obtained a footing in Pannonia. The Huns who occupied Pannonia in 427 were certainly not these old Hunnic federates but the Hunnic empire, whose centre at that very period was moved into the Hungarian plain. The same conclusion follows from the fact that two decades later the bishop of Margum was accused by Attila of having robbed Hunnic princes' graves in the approaches of Margum.[58] Hunnic princes were thus already buried in the thirties in the southern parts of the Lowland.

Under such circumstances we must take the ceding by treaty of large areas of

Pannonia [59] in 433 to be the legal recognition of a state of affairs that had already existed in reality for years. Furthermore, the cession of 433 must mean that in 427 Pannonia was only temporarily regained. It is very likely that in 427 it was East not West Rome which regained parts of the province: Marcellinus *comes* composed his chronicle from the view-point of East Rome. When, therefore, in 437 West Rome ceded Sirmium to East Rome [60] this again was probably merely a legal settlement of a *de facto* situation. Moreover, since it was possible to negotiate about Sirmium four years after Pannonia had been ceded to the Huns, the conclusion is clear that the Hunnic conquest, sanctioned in 433, did not relate to the surroundings of Sirmium, but to Savia and the areas north of the Drave. We have already seen that the federate peoples of Alatheus and Saphrac, consisting partly of Huns, were mainly settled in Savia. The East Roman reconquest of Pannonia could be connected with a fifty years' rule in Pannonia by the Huns, for the reason that, in 427, from Sirmium, East Rome undertook an operation in the area between the Drave and the Save, that is in Savia.

By *c.* 437 the largest part of Pannonia was in the possession of the Huns. Four years later, in 441, Attila conquered the East Roman possessions on the Danube: the towns of Ratiaria, Viminacium, Singidunum, Sirmium and Naissus were besieged and taken, and a strip of land five days' journey wide had to be evacuated south of the Danube.[61] In 447 the official trading-post between Huns and East Romans was moved from the Danube to Naissus.[62]

With the stabilization of Hunnic rule in our provinces the history of the Roman empire on the middle Danube properly ends. After the collapse of the empire of the Huns the Danube lands came under the rule of the Goths who in their turn were successively replaced by German, Turkish, Slav—and in the Carpathian basin also by Finno-Ugrian—peoples. Though the Emperor Avitus did appear in Pannonia in 455, this demonstration of force had no results. We do not even know how far Avitus advanced from Italy. For a time south-western Pannonia belonged to Ostrogothic Italy, and the region of Sirmium was retaken several times by East Rome until it was finally lost at the end of the sixth century. From what Procopius says [63] we may presume that between the time of Attila and that of Justinian the Moesian provinces on the Danube suffered much. Only Dardania remained permanently in East Roman possession and southern Moesia gave Byzantium a number of distinguished men of the old Illyrican kind: the Emperors Justin and Justinian came from Thracian country stock in the southern part of the territory of Naissus, not far from the northern border of Dardania. The interlude of Justinian's reorganization, however, lies beyond the scope of this book.

This concise account of the course of events, from the time when the Goths and other federates gained a foothold within the empire after the death of Valentinian down to the period of the Hunnic empire's greatest expansion on the Danube under Attila, remains of course in many respects uncertain. This is due above all to a change in our sources. The rich polemic literature of the end of the fourth century contains exceedingly valuable information on the contemporary situation, and in the occasional poetry of the court poet Claudian fairly accurate statements can also be found. But thereafter everything becomes silent. For later times we have only short, occasional notices or out-of-context traditions written down at a much later period. The situation in Pannonia during the early period of *foederati* is described by several writers with accuracy and feeling; but of the advance of the Huns into Pannonia, hardly a quarter of a century later, we have only chance reports, made by authors whose interests were focused on quite different things. In my opinion, this alteration in the interests of contemporaries is an indication of Rome's decreasing interest in Pannonia as a possession. One has only to think of Sidonius Apollinaris praising the enterprise of his father-in-law Avitus in Pannonia (in 455) as the reconquest of a country that Rome had lost many generations ago.[64] For Procopius Pannonia was already a desert devoid of people,[65] and that, though in no way true, is nevertheless significant for a Byzantine's way of thinking under Justinian. The laments over the loss of Pannonia, repeatedly to be heard from the time of Theodosius, are not only characteristic of the attitude of the Romans towards Pannonia, but of the situation of Pannonia too.

As far as the former Upper Moesia is concerned, similar statements in late Roman literature are neither frequent nor tragically portrayed. A crisis such as Pannonia experienced in the first decade of the fifth century did not befall Upper Moesia until the time of Attila. Whilst federates and passing barbarians plundered Pannonia far and wide, there is no mention of oppression in Upper Moesia in the works of Nicetas, bishop of Remesiana at that time. The Roman civilization of Moesia experienced its time of terror only after the middle of the fifth century, and was therefore spared to a greater degree than that of Pannonia. This difference in phasing provided a firm foundation for the reorganization of Justinian, which in all probability could not have succeeded in Pannonia. The Roman inhabitants of Moesia had stayed on the spot in larger numbers. From Priscus' report of his journey some details can be learnt of the towns of northern Moesia: Naissus at that time was still inhabited, and Ratiaria, according to Priscus, was a large and populous town. Many late antique or Christian inscriptions from Viminacium, Naissus and Ratiaria can be dated to the fifth century,[66] but in Pannonia such a date can be suggested, if at all, only for a few inscriptions from Sirmium. To judge by its fate and its political relationship, Sirmium had belonged to the eastern empire from *c.* 420. Thus continuity of church hierarchy is proved in Pannonia only at Sirmium, whereas in Upper Moesia it is established at several towns: in 424 the bishops of Ratiaria and Viminacium, in 449 a bishop of Remesiana, and in 458 a bishop of Scupi are attested, to say nothing of the bishop of Margum who was involved with the Hunnic princes' graves. Some of these sees were extinguished in the later fifth century, but Justinian endeavoured to re-establish them.[67]

As the fates of Pannonia and Upper Moesia were basically so entirely different from after the end of the fourth century, we shall study each separately.

The development in Pannonia can be divided into periods approximately as follows (Sirmium for reasons given is not taken into consideration): (i) 379–401: settlement of the Goth–Alan–Hun federates of Alatheus and Saphrac, leading to repeated serious crises and devastations; (ii) 401–8: Pannonia becomes the corridor of various Gothic and other peoples: emigration and flight of the Romans begin; (iii) *c.* 409: a short period of quiet, with attempts to organize a Roman occupation; (iv) from *c.* 410–20 onwards: gradual expansion of Hunnic control over Pannonia, ratified by the cession by treaty in 433.

The Upper Moesian provinces, on the other hand, were settled only in part with Gothic federates: in Dacia Ripensis, and probably much more in what was formerly Lower Moesia. The periods of development in Upper Moesia can therefore be put as follows: (i) 376–401: settlements of federates in parts of the country only, causing unrest from time to time, but life on the whole was much more peaceful than in Pannonia; (ii) 401–10/20: northern Moesia becomes a corridor for Goths pressing westwards, but without temporary settlement as happened in Pannonia, nor causing large-scale emigration; (iii) 420–41: Upper Moesia is controlled by East Rome, and in the main is spared by the Huns; (iv) 441: conquest by the Huns and evacuation of northern Moesia.

In the light of these differences, the fate of the Romanized population in Pannonia and in Upper Moesia becomes easily understandable. From 379 onwards the Pannonian provinces had had to suffer so much, that the short period of peace after 408 was of only limited advantage to them. The long coexistence of barbarian federates and Roman provincials, disturbed by repeated crises and pictured by contemporaries as a state of siege, was followed by the heavy ordeals of 401–8 when the country became an invasion corridor both in the north and in the south. It is not surprising that after all these ordeals many Romans fled, thus causing a recession in secular and ecclesiastical life alike. The flight and emigration from Pannonia began after two hard decades of the *obsidio,* and continued without interruption until the invasion of the Avars. The fact that some Romans did not emigrate until the later fifth or even the sixth century does not by any means imply that all Romans held out till the appearance of the Avars; equally unjustifiable would be the inference that the emigration in the first decade of the fifth century means that thereafter no Romans were left in Pannonia. The fifth and sixth centuries were a period of gradual emigration; I am inclined to assume that the largest emigration was the earliest. Nor must we forget that the descendants of the followers of Alatheus and Saphrac stayed in Pannonia, and that the comparative quiet after 408 was enjoyed by a Roman population that was already decimated by emigration.

Both emigration in groups and sporadic individual flight can be deduced from various indications in the sources. We have already mentioned the flight of the Romans of the town of Scarbantia, taking the relics of Quirinus with them to Rome.[68] In date it presumably coincides with the flight of large sections of the population before Radagaisus; an edict of December 408 deals with these refugees.[69] Bishop Amantius had possibly fled even earlier, *c.* 400, to Aquileia.[70] Just as Quirinus found a final resting-place on the via Appia at Rome, further Pannonian martyrs came to Rome and to several other towns of Italy.[71] They were brought by the refugees, and this in effect signified the dissolution of the Christian communities in Pannonia. Continued use of Christian cult-buildings in Pannonia after the fourth century does not prove the survival of the Pannonian cult-communities. Goths and other peoples were Arian Christians, too, and continued to use the dilapidated basilicas, though so far this can only be proved at the so-called Basilica II in the large fortified estate-centre at Fenékpuszta.[72] At the majority of sites it is not possible to prove unequivocally whether the reconstruction of a basilica is to be attributed to barbarians or to Romans staying behind. As far as Fenékpuszta is concerned the recent and older excavations have brought to light a unique group of finds that suggest not only Christians but Romans too. This so-called Keszthely culture is a special archaeological assemblage of the early Avar period which cannot be satisfactorily traced back to the local culture of the Roman period. There are, however, finds such as a gold pin with the name BONOSA (Pl. 45b), proving that some ethnic group of Roman complexion remained at Fenékpuszta.[73] An example of such sporadic Romans who had stayed behind in the fifth century was St Antonius of Lirina, who was born in the province of Valeria towards the middle of the fifth century and as an orphaned child was brought to his uncle Constantius, bishop of Lauriacum. The latter must therefore have already emigrated to Noricum earlier.[74] Even later Leonianus travelled from Savaria to Gaul.[75] St Martin of Bracara in Spain is also considered a Pannonian.[76] Even the Lombards dragged Romans along with them to Italy, if credit can be attached to the statement by

Paulus.[77] The last emigration was that of the Sirmians to Salona, as mentioned there on some inscriptions of the beginning of the seventh century.[78]

In view of the early beginning of barbarization and the gradual emigration of the Romans, it becomes understandable that the place-names of Pannonia were preserved to the present day only in exceptional cases (Fig. 58). The names of rivers, of course, are the same. (Danuvius = Danube, Donau, in Hungarian Duna; Dravus = Drave, in Hungarian and south Slavonic Drava; Savus = German Save, in Hungarian Száva, south Slavonic Sava; Sala = in Hungarian Zala; Arabo = German Raab, in Hungarian Rába; and in this group must be included the Hungarian river-name Marcal which goes back to Mursella. Even the names of rivers beyond the frontier are the old ones: for instance Granua = German Gran, in Hungarian Garam, in Slovakian Hron). Roman place-names, however, which have survived to the present time are represented only in Ptuj = Pettau (Poetovio), Sisak (Siscia), Raab (Arrabona, though in Hungarian this town is called Győr) and possibly Wien, Vienna (Vindobona) although this is very questionable. All these places lie on the edge of Pannonia. Not even Sirmium has been preserved as a town-name, although the area between Danube and Save is named Srijem (in German Syrmien, in Hungarian Szörény, Szerém-ség) from it; the town was later named after its great martyr: in Hungarian Szávaszentdemeter ('St Demeter on the Save'), in south Slavonic Mitrovica (Dimitrovica).

The situation in Upper Moesia is a little different: there are two areas there in which the ancient place-names have been well preserved, and not only those of the largest places. One of the areas is part of Dacia Ripensis, where Arčar goes back to Ratiaria, Vidin to Bononia, Flortin to Florentiana, Rgotina to Argent-ares. The other area is the southern and eastern part of Dardania where Naissus is nowadays named Niš, Scupi Skoplje (in Turkish Üsküb) and a place not far from Ulpianum Lipljan. The place-names of Moesia were all still in use in the sixth century; their disappearance therefore is to be attributed to the upheavals caused by the Slav conquest. The survival of names in the case of large towns, which always had an importance as junctions (Naissus, Scupi, Ulpianum), can be compared with the survival of the names of Poetovio, Siscia and Sirmium. But for the survival of the small group of place-names in Dacia Ripensis there must have been a special reason. In particular we may think of the settlement of Romanized elements in the new Dacia, after the evacuation of the Dacia of Trajan, whereby a Latin-speaking population was created, which by comparison with the other Danubian and Balkan countries was compact, if not massive, and could resist the Thracian language of the native population and the influence of the Germanic, Bulgarian–Turkish and Slav population. Ratiaria, moreover, was

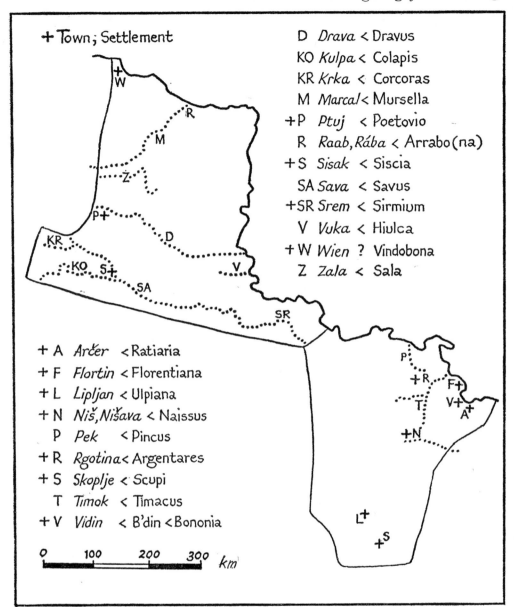

Figure 58 Geographical names of ancient origin

an old military *colonia* where the Latin language always had had stronger roots than in the municipia with their preponderantly local inhabitants.

That these Romans were the precursors of the medieval Vlachs is known; but the subject is beyond the scope of this work. The Vlachs, a Balkan people, who created a neo-Latin language and who are today known as Vlach groups in the

355

Balkan peninsula, and, of course, as Roumanians in Roumania, originated from a people of Latin-speaking mountain-pastoralists; possibly in the case of late Roman Moesia we can identify their origin. When Scupi was afflicted by an earthquake in 518, the inhabitants of the town stayed in the mountains, whither they had fled from the barbarians.[79] In this way the Romanized population of the town could gradually adopt the age-old Dardanian way of life that was based on hill-pastures. The fact that lowland settlements had to be given up in the course of the fifth century, and in later times even more so, and had to be built afresh on mountains and hill-tops for reasons of safety, is proved by the building activity of Justinian.[80] The construction of fortified settlements under him took the new development into account to a large extent, but it was for this very reason that the Justinianic reorganization can only have been an interlude. What had perished in the fifth century could not be reconstructed, and the survivors of the Roman population necessarily started a new development, dictated by the changed situation. Though this enabled Roman elements to survive till modern times in the shape of the Balkan Vlachs and in the Roumanians who derived from them, it could not provide the Justinianic reorganization with a viable basis.

In Pannonia geographical conditions could not have provided the Romanized population with a similar possibility of continued existence by change to upland pastoralism, even if that population had remained as a whole in their homeland. But the emigration of the Romans from Pannonia has deeper reasons which lie in the special development of Pannonia. More than once we have been able to show that the upper class of provincial society in Pannonia, which was the only Romanized group there, enjoyed an existence which was largely tied to the interests of the imperial government. Moreover, the leading class of society owed its position to its possession of a source of income outside the province, whether as members of the army or as representatives of some branch of the administration. Thus, its basis of existence survived only as long as its connections with the central government were not put in jeopardy. As soon as the *foederati* intervened as a new element, and connections with the central government loosened gradually, this social class lost its basis of existence. That was the reason why the Romanized classes of Pannonia took to flight or emigrated gradually. An important detail must be added. We have already pointed out the likelihood that a large part of the great estates of Pannonia were imperial estates. As a result, there was no numerous landed nobility in Pannonia to remain on the spot, like the senatorial aristocracy of Gaul or Spain, to become an equal partner with the barbarian tribal aristocracy. Thus the local Romanized upper class took less and less share in the events of the empire's history and in that of their own country; the active role was taken by the barbarians. This, too, may have been the reason why Rome had to relinquish Pannonia so early.

The situation as just described is presumably applicable to Moesia, too. There, however, the members of the leading class had had no reason at first for emigration, and when the Huns obtained the upper hand they found a new way of preserving their old leading role. This way was also available to the Romans of Pannonia who had stayed behind, down to the time of Attila, and in this respect there are no differences between Pannonian and Moesian Romans. For the Illyrican Romans also this was the last chance to take their share in the direction of events, and thus to preserve their leading role. It is well known that Pannonians and Moesians assembled at the court of Attila in great numbers. The last emperor of West Rome, too, came from a circle which, originating as it did from Sirmium, took sides with Attila.[81] Priscus tells of the trader from Viminacium who gives a charming little account of his reasons for going over to the Hunnic camp.[82] Towards the middle of the fifth century the Romans of Illyricum had no choice but to place themselves at the disposal of the Huns, who were in need of specialists and who probably paid well too. This was brain-stealing which in the last resort deprived Rome of capable minds, but for the Romanized classes under Hunnic rule it was the only possibility of arresting their decline and dissolution in the broad un-Romanized mass of people. It was a short-lived expedient. After the collapse of the Hunnic empire these Romans had to flee. It was probably thus that St Severinus came to Noricum,[83] where he was able to exert his very extensive connections in the interests of the Romans, but had to keep his past in the camp of Attila secret. The disintegration of the Hunnic empire thus contributed to the decline of the Pannonian as well as of the Moesian Romans.

In the fifth century Romanized elements probably concentrated more and more behind the walls of larger towns. Possibly here too the cohesive force of the common language operated. It is an observation of the utmost significance that it is not until the sixth century that barbarian finds can be demonstrated in Pannonian towns.[84] It follows from this, of course, that the remnants of the Roman population could maintain themselves in the towns of Pannonia only during the fifth century. An equally important symptom is the practically complete absence of finds of Roman character of fifth-century date; all finds datable to the fifth century are of distinctly barbarian character. This absence is the result of gradual pauperization and of the decline of industrial production. Production was insignificant in quantity, and the results were twofold: finds are small in quantity and old forms subsist. Owing to the impossibility of distinguishing between Roman finds of the fourth and fifth centuries it is difficult to prove the decrease and pauperization of the Romanized population by archaeological means. Nor do imports exist from the south and west.

Accordingly one might expect that the local native population were able to resume the use of their Celtic, Illyrian and Thracian languages in the fifth century, for the tenacious survival of the local languages of the original population can be proved in Pannonia as well as in Moesia. Jerome was still able to speak the Illyrian dialect of his homeland (*gentilis sermo*).[85] Illyrian nomenclature can be proved well into the Middle Ages,[86] and in the monasteries of the Orient a Thracian liturgy in the *lingua Bessica* was introduced.[87] Even in the later fifth century the inhabitants of the Dardanian countryside still carried Thracian names; the two travel-companions of the future emperor Justin were named Zimarchos and Ditybistos [88]—genuinely Thracian names that were presumably in use at the time of the Slav annexations in the Balkan peninsula. But from the end of the fourth century the rural population became more and more intermixed with barbarians and they lacked the political and social basis for conscious preservation of their native tongue. For the provincial upper class the Latin language formed not only the means and prerequisite of their social position but also the symbol of social relationship and at the same time a force of cohesion. Just as the Sarmatians of the Hungarian Lowlands, who are still identifiable in the second half of the fifth century,[89] later vanished without trace among the Gepidae and Avars, so the Celts, Illyrians and Thracians dissolved in the sea of later conquerors, simply for the reason that during the long period of Roman rule they had lost their native culture and were unable to utilize their language as a means to a political life of their own.

Abbreviations

Acta Ant.	*Acta Antiqua Academiae Scientiarum Hungaricae* (Budapest).
Acta Arch.	*Acta Archaeologica Academiae Scientiarum Hungaricae* (Budapest).
AIJ	V. Hoffiller and B. Saria, *Antike Inschriften aus Jugoslawien* (Zagreb, 1938).
Ann. Ép.	*L'Année Épigraphique* (Paris).
Ant. Tan.	*Antik Tanulmányok—Studia Antiqua* (Budapest).
Archäol. Anzeiger	*Archäologischer Anzeiger*, see *Jahrb. DAI*.
Arch. Ért.	*Archaeologiai Értesitö* (Budapest).
Arch. Iugosl.	*Archaeologia Iugoslavica* (Belgrade).
Arch. Rozhl.	*Archeologické Rozhledy* (Prague).
ARD	*see* Ferri.
Arh. Pregl.	*Arheološki Pregled* (Belgrade).
Arh. Vestn.	*Arheološki Vestnik—Acta Archaeologica* (Ljubljana).
Barb-Festschr.	*Festschrift für A. A. Barb* (Eisenstadt, 1966).
Barkóczi, *Brigetio*	L. Barkóczi, *Brigetio* (*Diss. Pann.* ii, 22, Budapest, 1944–51).
Beiträge	*see* Patsch.
Bevölk.	A. Mócsy, *Die Bevölkerung von Pannonien bis zu den Markomannenkriegen* (Budapest, 1959).
Bp. Müeml.	M. Horler (ed.), *Budapest Müemlékei*, Vol. ii (Budapest, 1962).
Bp. Rég.	*Budapest Régiségei* (Budapest).
Bp. Tört.	A. Alföldi, Gy. László, L. Nagy, T. Nagy, J. Szilágyi and F. Tompa, *Budapest Története*, Vol. i (ed. K. Szendy, Budapest, 1942).
BRGK	*Bericht der Römisch-Germanischen Komission des Deutschen Archäologischen Instituts* (Frankfurt).
Bronzegefässe	*see* Radnóti.

Abbreviations

Burgenl. Heimatb.	*Burgenländische Heimatblätter* (Eisenstadt).
C.	*CIL* iii (Berlin, 1873–1902).
CAH	*The Cambridge Ancient History* (Cambridge).
Carn.-Jb.	*Carnuntum—Jahrbuch* (Graz-Köln).
Carnuntum	*see* Swoboda.
Chron. Min.	*Chronica Minora saec. iv, v, vi, vii*, ed. Th. Mommsen, Berlin, 1892–8, Vols i-iii (*Monumenta Germaniae Historica, Auctorum Antiquissimorum* tom. ix, xi, xiii).
CIL	*Corpus Inscriptionum Latinarum* (Berlin).
Corpus Scr. Eccl. Lat.	*Corpus Scriptorum Ecclesiasticorum Latinorum*
DCU	*see* Dobiáš.
Diss. Pann.	*Dissertationes Pannonicae ex Instituto Numismatico et Archaeologico Universitatis de Petro Pázmány nominatae Budapestinensis provenientes* (Budapest).
Dittenberger, *Syll.*	W. Dittenberger, *Sylloge Inscriptionum Graecarum*, 2nd ed., Berlin.
Dobiáš, *DCU*	J. Dobiáš, *Dějiny československého území před vystoupením Slovanů* (Prague, 1964).
Dobó, *Verwaltung*	Á. Dobó, *Die Verwaltung der römischen Provinz Pannonien von Augustus bis Diocletianus* (Budapest–Amsterdam, 1968).
Egger, *RAFC*	R. Egger, *Römische Antike und frühes Christentum. Ausgewählte Schriften*, Vols i-ii (Klagenfurt, 1962–3).
Ferri, *ARD*	S. Ferri, *Arte Romana sul Danubio* (Milan, 1933).
Fol. Arch.	*Folia Archaeologica* (Budapest).
Garašanin, *Nalazišta*	D. and M. Garašanin, *Arheološka nalazišta u Srbiji* (Belgrade, 1951).
GMKM	*Godišnjak Muzeja Kosova i Metohija* (Pristina).
Grbić, *Plastika*	M. Grbić, *Odabrana grčka i rimska plastika u Narodnom Muzeji* (Belgrade, 1960).
GSND	*Glasnik Skopskog Naučnog Društva* (Skopje).
GZM	*Glasnik Zemaljskog Muzeja, Arheologija* (Sarajevo).
IBAD	*Izvestija na Bălgarskoto Arheologičesko Družestvo—Bulletin de la Société Archéologique Bulgare* (Sofia).
IBAI	*Izvestija na Bălgarskija Arheologičeski Institut—Bulletin de l'Institut Archéologique Bulgare* (Sofia).
ILCV	E. Diehl, *Inscriptiones Christianae Veteres* i-iii (Berlin).
ILJug.	A. and J. Šašel, 'Inscriptiones Latinae quae in Iugoslavia inter annos MCMXL et MCMLX repertae et editae sunt', *Situla* 5 (Ljubljana, 1963).
ILS	H. Dessau, *Inscriptiones Latinae Selectae* i-iii (Berlin).
Inscr. teg.	*see* Szilágyi.
Intercisa	M. R. Alföldi, L. Barkóczi, G. Erdélyi, F. Fülep, A. Radnóti, K. Sági, etc., *Intercisa, Geschichte der Stadt zur Römerzeit* i-ii (Budapest, 1954–7).
Jahrb. DAI	*Jahrbuch des Deutschen Archäologischen Instituts* (Berlin).
JÖAI	*Jahreshefte des Österreichischen Archäologischen Instituts* (Vienna).
JÖAIB	*JÖAI, Beiblatt.*
JRS	*Journal of Roman Studies* (London).

Kanitz, *RS*	F. Kanitz, *Römische Studien in Serbien. Denkschriften der Philos.-hist. Klasse der Akademie* xvii (Vienna, 1892).
Kubitschek, *Römerfunde*	W. Kubitschek, *Römerfunde von Eisenstadt* (Vienna, 1926).
Laur. Aqu.	*Laureae Aquincenses memoriae V. Kuzsinszky dicatae* i-ii (*Diss. Pann.* ii, 10–11) (Budapest 1938–41).
Limeskongress 1969	*Congress of Roman Frontier Studies. Cardiff 1969* (ed. E. Birley, Cardiff, 1974).
LJP	*see* Várady.
LRKN	*Limes Romanus Konferenz Nitra 1957* (Bratislava, 1959).
LuJ	*Limes u Jugoslaviji* (Belgrade).
MFMÉ	*A Móra Ferenc Muzeum Évkönyve* (Szeged).
Mirković, *RGD*	M. Mirković, *Rimski gradovi na Dunavu* (Belgrade, 1968).
Monum. Germ. Hist. Auct. Ant.	*Monumenta Germaniae Historica: Auctores Antiquissimi.*
MPL	*Patrologiae Latinae Cursus Completus* (Paris) (Migne's *Patrologia*).
MS	A. Mócsy, *Gesellschaft und Romanisation in der römischen Provinz Moesia Superior* (Budapest–Amsterdam, 1970).
Nagy, *Mumienbegräbnisse*	L. Nagy, *Mumienbegräbnisse aus Aquincum* (*Diss. Pann.* i, 4, Budapest, 1935).
Nagy, *PS*	L. Nagy, *Pannonia Sacra*, in *Szent István Emlékkönyv*, vol. i (Budapest, 1938).
Nalazišta	*see* Garašanin.
Not. Occ.	*Notitia dignitatum . . . in partibus Occidentis* (ed. O. Seeck).
Not. Or.	*Notitia dignitatum . . . in partibus Orientis* (ed. O. Seeck).
Num. Közl.	*Numizmatikai Közlöny* (Budapest).
OMRTÉ	*Az Országos Magyar Régészeti Társulat Évkönyve—Jahrbuch des Ungarischen Archäologischen Gesellschaft* (Budapest).
Osj. Zb.	*Osječki Zbornik* (Osijek).
Pam. Arch.	*Památky Archeologické* (Prague).
Pannonia	A. Mócsy, *Pannonia*, in *PWRE* Suppl., Vol. ix (Stuttgart, 1962).
Patsch, *Beiträge*	C. Patsch, 'Beiträge zur Völkerkunde von Südosteuropa' in *Sitzungsberichte der Akademie, Phil.-hist. Kl.* (Vienna, 1928–37).
P. Dura	C. B. Welles, R. O. Fink, J. F. Gilliam, *The Excavations at Dura-Europos, Final Report V, Part i: The Parchments and Papyri* (New Haven, 1959).
Plastika	*see* Grbić.
PS	*see* Nagy.
PWRE	*Paulys Realencyclopädie der classischen Altertumswissenschaft*, neu begonnen von G. Wissowa, fortgeführt von H. Kroll, K. Ziegler, etc. (Stuttgart).
Radnóti, *Bronzegefässe*	A. Radnóti, *Die römischen Bronzegefässe von Pannonien* (*Diss. Pann.* ii, 6, Budapest, 1938).
Rad vojv. muz.	*Rad vojvodjanskih muzeja* (Novi Sad).
RAFC	*see* Egger.
Regesten	*see* Seeck.

Abbreviations

*R*GD	*see* Mirković.
*R*IC	H. Mattingly, E. A. Sydenham and others, *Roman Imperial Coinage* (London).
Riv. Ital. di Numismatica	*Rivista Italiana di Numismatica.*
*R*LiÖ	*Der Römische Limes in Österreich* (Vienna).
Römerfunde	*see* Kubitschek.
*R*S	*see* Kanitz.
*R*VP	*see* Thomas.
Saxer, *Vexillationen*	R. Saxer, 'Untersuchungen zu den Vexillationen des römischen Kaiserheeres', *Epigraphische Studien* i (Cologne–Graz, 1967).
Schober	A. Schober, *Die römischen Grabsteine von Noricum und Pannonien* (Vienna, 1923).
SCIV	*Studii si cercetări de istorie veche* (Bucharest).
Seeck, *Regesten*	O. Seeck, *Regesten der Kaiser und Päpste* (Stuttgart, 1919).
SHA	Scriptores Historiae Augustae.
Slov. Arch.	*Slovenska Archeológia* (Bratislava).
Spomenik	*Srpska (kraljevska) Akademija, Spomenik* (Belgrade).
Stud. Mil.	*Studien zu den Militärgrenzen Roms. Vorträge des 6. Internationalen Limes- kongresses in Süddeutschland* (Cologne–Graz, 1967).
Swoboda, *Carnuntum*	E. Swoboda, *Carnuntum. Seine Geschichte und seine Denkmäler* (4th ed., Graz–Cologne, 1964).
Syll.	*see* Dittenberger.
Symposium 1964	*Simpozijum o teritorijalnom i hronološkom razgraničenju ilira u praistorijsko doba—Symposium sur la délimitation territoriale et chronologique des Illyriens à l'époque préhistorique, 15–16 mai, 1964* (ed. A. Benac, Sarajevo, 1964).
Symposium 1966	*Simpozijum o ilirima u antičko doba—Symposium sur les Illyriens à l'époque antique, 10–12 mai, 1966* (ed. A. Benac, Sarajevo, 1967).
Szilágyi, *Inscr. teg.*	J. Szilágyi, *Inscriptiones tegularum Pannonicarum* (*Diss. Pann.* ii, 1, Budapest, 1933).
Thomas, *R*VP	E. B. Thomas, *Römische Villen in Pannonien* (Budapest, 1964).
Várady, *L*JP	L. Várady, *Das letzte Jahrhundert Pannoniens 376–476* (Budapest– Amsterdam, 1969).
Verwaltung	*see* Dobó.
Vexillationen	*see* Saxer.
*V*HAD	*Vjesnik hrvatskoga arheološkoga društva* (Zagreb).
*V*MMK	*A Veszprém Megyei Muzeumok Közleményei* (Veszprém).
*Ž*A	*Živa Antika—Antiquité Vivante* (Skopje).
*Z*FF	*Zbornik Filosofskog Fakulteta Univerziteta u Beogradu* (Belgrade).
*Z*NMB	*Zbornik Radova Narodnog Muzeja* (Belgrade).

Notes

CHAPTER I

1 Homer, *Iliad* xiii, 4–6.

2 Strabo vii, 3, 2–10.

3 S. Borzsák, 'Die Kenntnisse des Altertums über das Karpatenbecken', *Diss. Pann.* 1, 6 (Budapest, 1936).

4 Typical is, e.g., Apollonius Rhodius, *Argonautica* iv, 325–6. Further passages in Borzsák, *op. cit.*, 16–18.

5 See note 80.

6 Hecataeus, fragment 92 (Jacoby). Herodotus V, 9. Apollonius Rhodius, *Argonautica* iv, 320, 324.

7 M. V. Garašanin, *ZFF* vii (i), 1963, 45, traces the name Singidunum back to the Sindoi.

8 J. Harmatta, *Arch. Ért.* xciv, 1967, 133–6.

9 J. Gy. Szilágyi, 'Trouvailles grecques sur le territoire de la Hongrie: Le Rayonnement des civilisations grecque et romaine sur les cultures périphériques', *VIIIe Congr. Internat. d'Archéol. Class. Paris 1963*, 389–90; *Ant. Tan.* xiii, 1966, 102–6. On Greek finds in the northern Balkans see M. Parović-Pešikan, *Arch. Iugosl.* v, 1964, 61–81.

10 iv, 49.

11 Strabo vii, 5, 11.

12 Polyaenus vii, 42, and Theopompus in Athenaeus x, p. 443 A–C, where, however, the name of the Autariatae was wrongly handed down and then incorrectly emended to Ardiaioi. A. Mócsy, *Rivista Storica dell' Antichità* ii, 1972, 13 ff.

13 Strabo vii, 3, 8. Justin xxiv, 4.

14 Details in Chapter 3 (pp. 58 ff., 63 ff.).

15 See in particular the works of J. Untermann, H. Kronasser and R. Katičič listed in bibliography by G. Alföldy, *Die Personennamen in der röm. Provinz Dalmatia* (Heidelberg, 1969), 383–5.

16 M. Suić, *Rad Jugoslovenske Akademije znanosti i umjetnosti* cccvi, 1955, 121–81.

17   G. Alföldy, *Beiträge zur Namenforschung* xv, 1964, 55 ff., but chiefly R. Katičič in several essays and recently *Godišnjak* iii (Centar za Balkanološka Ispitivanja, 1, Sarajevo, 1965), 54–76.

18   Z. Marić in several works; see especially *Symposium 1964*, 191–213.

19   Strabo vii, 3, 13.

20   Strabo vii, 3, 10. In general, I. I. Russu, *Die Sprache der Thrako-Daker* (Bucharest, 1969), 40 f.

21   This theory was put forward by V. Georgiev. See, e.g., his *Trakijskijat ezik* (Sofia, 1957). A map of toponymous suffixes appears in *Arheologija* (Sofia), ii, 1960, Part 2, 13 ff. See also *La Toponymie ancienne de la péninsule Balkanique et la thèse méditerranéenne* (Sofia, 1961). Cf. also note 41 and A. Fol, *Rivista Storica dell'Antichità* i, 1971, 3 ff.

22   Illyrians in Strabo vii, 5, 6.

23   For instance, Dassius, Luccaius, *C.* 15134; Dasmenus, *C.* 10212; Terso, Precio, *C.* 3400.

24   Strabo vii, 3, 2; 5, 2; 5, 10. See also Florus i, 39, Rufus Festus *brev.* 9, Ammianus Marcellinus xxvii, 4, 10.

25   A. Mócsy, *Symposium 1966*, 195–200.

26   Livy v, 34.

27   Justin xxiv, 4.

28   See map of finds in I. Hunyady, 'Die Kelten im Karpatenbecken', *Diss. Pann.* ii, 18 (Budapest, 1944); cf. T. Nagy, *Acta Arch.* ix, 1958, 350 ff.

29   Strabo vii, 3, 8.

30   A. Mócsy, *Acta Ant.* xiv, 1966, 90–2.

31   Theophrastus in Seneca, *quaest. nat.* iii, 11, 2.

32   xxiv, 6, 1.

33   xxii, 9, 1.

34   Justin xxiv, 6; Pausanias x, 19, 7.

35   Justin xxv, 1, 2; Polybius iv, 46, 1.

36   J. Gy. Szilágyi, *op. cit.* (note 9), 389.

37   Mócsy, *op. cit.* (note 30), 93 f.

38   Justin xxxii, 3; Athenaeus vi, p. 234.

39   For the Celtic finds in the Save valley, see Z. Marić, *GZM* xviii, 1963, 83; for those in Scordiscan territory, see J. Todorović, *Kelti u jugoistočnom Evropi* (Belgrade, 1968).

40   Justin xxxii, 3; Diodorus xxii, 9, 3; 18; Athenaeus vi, p. 234; Pausanias x, 23.

41   If there is any justification for dividing the Thracian ethnic group, then, unlike V. Georgiev who suggests splitting it into the Thraco-Getae and the Daco-Mysi (see note 21), I consider a division into the Thraco-Mysi and the Daco-Getae the more likely. In antiquity the Dacians and Getae were taken to be one and the same race in the main, and the native population of Moesia Superior seems to have been the western Thracian element. In later imperial times the native population in Scupi still spoke the *lingua Bessica*. See *MS* 249 f. The personal names of the Upper Moesian natives are Thracian, without any notable local colouring: A. Mócsy, 'Vorarbeiten zu einem Onomasticon von Moesia Superior', *Godišnjak* iii, Centar za Balkanološka Ispitivanja 6 (Sarajevo, 1970), 175.

42   The ancients were already well aware of these and other identical names: Strabo vii, 3, 2; xiii, 1, 21.

43   Justin viii, 6, 3.

44   Livy xxxviii, 16.

45   Livy xl, 57, 6; Justin xxx, 4, 12; Pliny, *Nat. Hist.* iv, 81.

46 For incidents which have been handed down, see A. Mócsy, *Acta Ant.* xiv, 1966, 94–6.

47 Livy xl, 57–8.

48 Livy, *periocha* xliii and xliii, 18.

49 See note 23.

50 Dittenberger, *Sylloge* 710.

51 Livy xlv, 29, 12.

52 Julius Obsequens 16.

53 Livy, *per. Oxyr.* 174–5; Julius Obsequens 22.

54 *Pannonia*, 527–8.

55 Polybius, fragment 64 (122).

56 A. Mócsy, *op. cit.* (note 46), 97–8.

57 vii, 5, 12.

58 *Ibid.*

59 Appian, *Illyric.* 22; cf. *Pannonia*, 528.

60 Strabo vii, 2, 2; cf. 3, 2.

61 vii, 5, 2; cf. iv, 6, 10.

62 See note 55.

63 Strabo vii, 5, 3; Appian, *Illyric.* 14, 22 and *passim*.

64 R. Katičič, *Symposium 1964*, 31 ff.

65 Strabo v, 1, 6.

66 E.g. Tacitus, *Germ.* 28, cf. 42; Strabo vii, 1, 3; Velleius Pat. ii, 109, 5.

67 *Pannonia*, 529; B. Benadík, *Slov. Arch.* x, 1962, 394 ff.; J. Meduna, *Pam. Arch.* lii, 1961, 280 f.; cf. liii, 1962, 135 f. From anthropological angle, M. Stloukál, *Pam. Arch.* liii, 1962, 173.

68 E. Kolniková, *Slov. Arch.* xii, 1964, 402 ff., sees a connection between an *aes grave* found at Zabor in Slovakia and the migration of the Boians from Italy. Whether she is right has still to be investigated. Cf. a similar find from south Pannonia in M. Bahrfeld, *Der Münzfund von Mazin* (Berlin, 1902).

69 Cf. B. Benadík, *Slov. Arch.* xi, 1963, 372 ff.; *Germania* xliii, 1965, 63 ff.

70 Julius Obsequens 48.

71 Livy, *periocha* lxxxiii; Plutarch, *Sulla* 23, 1; Appian, *Mithr.* 55; Granius Licinianus 35; Eutropius v, 7, 1; Ps.-Aurelius Victor, *vir. ill.* 75, 7.

72 Appian, *Illyric.* 5; cf. A. Mócsy, *Acta Ant.* xiv, 1966, 98, n. 109.

73 Plutarch, *Pomp.* 41.

74 Strabo vii, 3, 11; 5, 2.

75 *Bell. Gall.* i, 5, 4.

76 P. Petru, 'Hišaste Žare Latobikov', *Situla* xi (Ljubljana, 1971).

77 *Bell. Gall.* i, 5, but cf. Swoboda, *Carnuntum*, 231.

78 Livy, *periocha* xci; Florus i, 39; Rufus Festus, *brev.* 9; Eutropius vi, 2; Orosius v, 23.

79 Frontinus, *strat.* iv, 1, 43.

80 Sallust, *hist.* ii, 80; Livy, *periocha* xcii, xcv, xcvii; Julius Obsequens 59; Ammianus Marcellinus xxix, 5, 2; Rufus Festus, *brev.* 7; Eutropius vi, 2. 7–8; Eusebius, *chron.* 152k (Helm).

81 Rufus Festus, *brev.* 7; Eutropius vi, 2.

82 Rufus Festus, *brev.* 7; Jordanes, *Rom.* 216; Sallust, *hist.* iv, 18; cf. A. Mócsy, *Acta Ant.* xiv, 1966, 99.

83 Appian, *Mithr.* 102, 109, 110, 119; Plutarch, *Pomp.* 41.

84   Strabo vii, 3, 11; Suetonius, *Caes.* 44; Appian, *bell. civ.* ii, 110.

85   Appian, *bell. civ.* iii, 25; cf. ii, 110, and for Caesar's military plans Strabo vii, 3, 5; Livy, *periocha* cxvii; Suetonius, *Caes.* 44; *Aug.* 8.

86   Dio xxxviii, 10; Cicero, *pro Sestio* 43 (94); Julius Obsequens 61.

87   vii, 3, 11.

88   *Bell. Gall.* vi, 25; cf. *Pannonia*, 532.

89   iii, 8, 3.

90   *Pannonia*, 532–3.

91   vii, 5, 2.

92   Strabo vii, 3, 11; 5, 2. Pliny, *Nat. Hist.* iii, 147. On the problem of this term, see B. Saria, in *Omagiu lui C. Daicoviciu* (Bucharest, 1960), 497, n. 2, and Mócsy, *Pannonia*, 533.

93   On the distribution of Dacian pottery in the Carpathian basin (outside Dacia), see A. Točik, *Arch. Rozhl.* xi, 1959, 843 ff.; M. Lamiová-Schmiedlová, *Slov. Arch.* xvii, 1969, 459 ff.; S. Nagy, *Rad vojv. muz.* ix, 1960, 112 f.; B. Benadik, *Germania* xliii, 1965, 79 f.; 90 f.; R. Rašajski, *Rad vojv. muz.* x, 1961, 23 f.; on the large *oppidum* of Gomolava see *Rad vojv. muz.* xiv, 1965, 109–250; D. Dimitrijevič, *Osj. Zb.* xii, 1969, 89; É. B. Bónis, *Die spätkeltische Siedlung Gellért-hegy-Tabán in Budapest* (Budapest, 1969), 188–91; also *Pannonia*, 533–4, and most recently Zs. Visy, *MFMÉ*, 1970, 5 ff.

94   *CIL* vi, 32542 b, v. 9. 12. c–d, v. 6–7.

95   Pliny, *Nat. Hist.* iv, 80.

96   Strabo vii, 3, 11.

97   Suetonius, *Aug.* 63; Plutarch, *Ant.* 63; Dio li, 22, 8; Frontinus, *strat.* 1, 40, 4, etc.

98   Appian, *bell. civ.* v, 75.

99   Ptolemy iii, 7, 1; cf. 8, 1.

100   Appian, *Illyric.* 22.

101   See note 85.

102   Velleius Pat. ii, 59, 4.

103   See note 97.

104   Horace, *sat.* ii, 6, 53.

105   Appian, *Illyric.* 22 ff.; Dio xlix, 34–8.

106   Dio xlix, 36, 1.

107   Appian, *Illyric.* 23.

108   *Ibid.*, 15.

109   *Ibid.*, 22; Strabo vii, 5, 2, etc.

110   A. Mócsy, *Historia* xv, 1966, 511 ff.

111   *CIL* i², p. 50; cf. Livy, *epit.* 134, 135; Florus ii, 26.

112   Dio li, 23, 2.

113   Horace, *carm.* iii, 8, 18.

114   Mócsy, *op. cit.* (note 110).

115   Dio li, 23–7.

116   *Ibid.*, liv, 20, 3.

117   Deduced by A. v. Premerstein, *JÖAIB* i, 1898, 158 f., from Velleius Pat. ii, 39, 3; Eusebius, *chron.*, 166h (Helm). Cf. *Pannonia*, 540, and J. J. Wilkes, *Dalmatia* (London, 1969), 61.

118·   Dio liv, 31, 3.

119   Cf. *Pannonia*, 535–7; A. Mócsy, *Acta Ant.* xiv, 1966, 108 ff., and now F. Papazoglu, *Srednjobalkanska plemena u predrimskog doba* (Sarajevo, 1969).

120   Ptolemy iii, 10, 4–5; *CIL* v, 1838 = *ILS* 1349.

121   *Athenaeus* x, p. 443 A–C.

122   Strabo vii, 5, 2.

123   Livy v, 34.

124   Polybius iv, 46, viii, 22; Pompeius Trogus, *prol.* 25; cf. G. Mihailov, *Athenaeum*, N.S. xxxix, 1961, 39 f.

125   E. B. Vágó, *Alba Regia* i, 1960, 49 ff.; E. F. Petres, *Fol. Arch.* xvii, 1965, 96 ff.

126   J. Todorović, *Kelti u jugoistočnoj Evropi* (Belgrade, 1968).

127   Strabo vii, 5, 7; *Expositio tot. mundi* 51.

128   Athenaeus vi, p. 272.

129   xiii, 1, 21.

130   Strabo vii, 5, 2–4, 10; Appian, *Illyric.* 14, 22.

131   Cf. Justin xxiv, 4: *ibi domitis Pannoniis*.

132   The basic work is K. Pink, 'Die Münzprägung der Ostkelten und ihrer Nachbarn', *Diss. Pann.* ii, 15 (Budapest, 1939).

133   Ö. Gohl, *Num. Közl.* xxi–xxii, 1922–3, 3 ff.; M. Parović-Pešikan, *Starinar* iv/11, 1961, 41 f.

134   Pink, *op. cit.* (note 132).

135   V. Ondrouch, *Nálezy keltskych, antickych a byzantskych minci na Slovensku* (Bratislava, 1964), 194, and E. Kolniková, *Slov. Arch.* xii, 1964, 402 ff.

136   For finds of dies together with tools for striking coins, see Ö. Gohl, *Num. Közl.* vi, 1907, 47 ff. More recently B. Benadik, *Germania* xliii, 1965, 87.

137   v, 1, 8; *Pannonia*, 681.

138   J. Šašel, 'Contributo alla conoscenza del commercio con gli schiavi norici ed illirici alla fine del periodo repubblicano', in *Atti del iii Congresso Internaz. di Epigrafia greca e latina* (Rome, 1959), 143 ff.

139   *C.* 3776, 3777, 3780, 10721, etc.; cf. J. Šašel, *PWRE* Suppl. xi, 561–2.

140   A. Mócsy, *Num. Közl.* lx–lxi, 1961–2, 25 ff.

141   *Pannonia*, 691.

142   *Ibid.*, 692; B. Mitrea, *Ephemeris Dacoromana* x, 1945, 85 ff.; Parović-Pešikan, *op. cit.* (note 133), 41; *MS* 257; *LuJ* i, 1961, 93.

CHAPTER 2

1   *Illyric.* 15.

2   Strabo v, 1, 8.

3   J. Šašel, in *Corolla memoriae E. Swoboda dedicata* (Cologne–Graz, 1966), 198 ff.

4   Livy xliii, 5.

5   *C.* 3776, 3777; cf. J. Šašel, *PWRE* Suppl. xi, 561–2.

6   *Mon. Ancyr.* 30; Dio liv, 34, 4.

7   Cf. *Pannonia*, 539, 46–60.

8   Dio l, 24, 4.

9   Tacitus, *Ann.* l, 80; vi, 39; Dio lviii, 25, 4.

10 Summary of different views given by J. Szilágyi, *Laur. Aqu.* i, 304 ff. Even Mommsen considered that Augustus advanced only as far as the Drave.

11 Dio liv, 24, 3.

12 *Ibid.*, liv, 28, 1; Velleius Pat. ii, 96; Florus ii, 24.

13 Josephus, *Ant.* xvi, 4, 1; Suetonius, *Tib.* 7; Gardthausen, *Augustus und seine Zeit* ii, 668.

14 Velleius Pat. ii, 96.

15 *Ibid.*; Suetonius, *Tib.* 9; Frontinus, *Strat.* ii, 1, 15; Dio liv, 31; 33, 5; Rufius Festus, *brev.* 7, cf. Velleius Pat. ii, 39; Livy, *epitome* 141; Eutropius vii, 9.

16 Dio liv, 34; 36, 2; lv, 2, 4; Eusebius, *chron.* 167 f., 168b (Helm); Cassiodorus chron., *Chron. Min.* ii, 135.

17 *Mon. Ancyr.* 30; Dio liv, 34, 4.

18 Tacitus, *Ann.* iv, 44; Dio lv, 11a, 2, cf. Tacitus, *Germ.* 41; Suetonius, *Nero* 4.

19 Tacitus, *Germ.* 42; cf. Velleius Pat. ii, 108.

20 Tacitus, *Ann.* ii, 62.

21 Tacitus, *Germ.* 41.

22 *Mon. Ancyr.* 30; Dio liv, 36, 2.

23 *ILS* 8965; cf. *Pannonia*, 543 f.

24 Florus ii, 24; Velleius Pat. ii, 96.

25 Florus ii, 28–9.

26 Strabo vii, 3, 10 and 13; *Consol. ad Liviam* 385; Florus ii, 28–9; Dio lv, 30, 4, etc.

27 Dio lv, 29, 3; Strabo vii, 3, 10; Florus ii, 29.

28 E. Stein, 'Die Legaten von Moesien', *Diss. Pann.* i, 10 (Budapest, 1940), 16.

29 Tacitus, *Ann.* iv, 5: *ripamque Danuvii legionum duae in Pannonia, duae in Moesia attinebant.* Of these four legions only two, XV Apollinaris and V Macedonica, were stationed on the Danube, at Carnuntum and Oescus respectively.

30 Strabo vii, 3, 10.

31 Florus ii, 28–9.

32 They are not mentioned on the fragmentary inscription *ILS* 8965. Jordanes, *Get.* 74, possibly refers to their location at that time, i.e. west of the river Alutus (Olt) and south of the Carpathians.

33 Lucan, *Phars.* iii, 94; Seneca, *quaest. nat.* i, praef., 9.

34 Pliny, *Nat. Hist.* iv, 80.

35 For the earliest Iazygan finds, see A. Mócsy, *Acta Arch.* iv, 1954, 124 ff.

36 Velleius Pat. ii, 110.

37 Tacitus, *Ann.* ii, 46.

38 The main sources for the rebellion are Velleius Pat. ii, 110 ff., and Dio lv, 28 ff.; lvi, 1, 11–17. Cf. also Suetonius, *Tib.* 16; Orosius vi, 21, 27, etc.

39 *MS* 48 n. 13.

40 *Pannonia*, 526–7.

41 *CIL* i², p. 248 = x, 6638, col. ii, v. 3.

42 *CIL* xiv, 3613 = *ILS* 918.

43 This follows from Velleius Pat. ii, 112, 2 and 116, 2.

44 For first time *CIL* x, 6225 = *ILS* 985 and *CIL* xvi, 14.

45 On the cohortes Breucorum, see J. E. Bogaers, *Berichten van de Rijksdienst voor Oudheidkundig Bodemonderzoek* xix, 1969, 27 ff.

46 *Bevölk.* 121.

47 *CIL* xvi, 20; *Acta Arch.* ix, 1958, 407 ff.

48 Suetonius *Tib.* 9: *Pannonico (bello) Breucos et Sarmatas subegit.*

49 Dio lv, 30, mentions Roman allies in the vicinity of Alma mons = Fruška Gora.

50 Dio lv, 30.

51 On the basis of the so-called Drusus inscription found at Aquincum it was unanimously agreed to date the immigration of the Iazyges to between 17 and 20. See *Pannonia*, 549. The new reading, however, shows this to be wrong. See Chapter 4 (p. 80).

52 Velleius Pat. ii, 123, 1; cf. Suetonius, *Tib.* 21.

53 Tacitus, *Ann.* i, 16–29; Dio lvii, 4, 4; Velleius Pat. ii, 125, 4; cf. J. Šašel, *Historia* xix, 1970, 122–4.

54 J. Šašel and I. Weiler, *Carn.-Jb.* 1963–4, 40.

55 It is not known where from. Formerly the deduction in Emona was linked with the evacuation of the legionary fortress there. Cf. B. Saria, *Laur. Aqu.* i, 245 ff. Cogent reasons for doubting that there was a fortress at Emona have now been put forward by J. Šašel, *PWRE* Suppl. xi, 562 ff., who inclines to the view that the rebellion in A.D. 14 took place at Siscia: *Historia* xix, 1970, 123.

56 Tacitus, *Ann.* ii, 44, 46.

57 Tacitus, *Ann.* ii, 62 ff.; iii, 11; Velleius Pat. ii, 129, 3; Suetonius, *Tib.* 37; *CIL* xiv, 244. The situation within the kingdom of Vannius, Pliny *Nat. Hist.* iv, 80–1, is controversial, cf. *Pannonia*, 549; R. Hanslik, *PWRE* viii A, 346 f.; B. Saria, *ibid.*, 338 ff.; I. Bóna, *Acta Arch.* xv, 1963, 303; J. Fitz, *Alba Regia* ii-iii, 1963, 207 f.; iv-v, 1965, 77 ff.; J. Dobiáš, *DCU* 364 ff.

58 Tacitus, *Ann.* xii, 29–30.

59 Tacitus, *Germ.* 43.

60 See Chapter 3.

61 *CIL* xiv, 3608 = *ILS* 986, cf. D. M. Pippidi, *Contributi la istorie veche a Romaniei* (Bucharest, 1967), 287 ff.

62 Tacitus, *Hist.* ii, 85 f.; 96; iii, 1–5, 24; Josephus, *Bell.* iv, 11, 2.

63 See previous note and *Ann. Ép.* 1966, 68.

64 Tacitus, *Hist.* i, 79; iii, 24.

65 *Ibid.*, iii, 5.

66 *Ibid.*, iii, 46; iv, 4; cf. ii, 83.

67 Josephus, *Bell.* vii, 89–91; Tacitus, *Hist.* iv, 54.

68 Josephus, *Bell.* vii, 91–5.

69 *C.* 10878 = *AIJ* 371; *C.* 10879 = *AIJ* 381; *C.* 4060 = 10869 = *AIJ* 260; *AIJ* 262.

70 Tacitus, *Ann.* i, 16–31.

71 J. Šašel, *PWRE* Suppl. xi, 563.

72 Dio lv, 29, 4.

73 B. Gerov, *Acta Ant.* xv, 1967, 85 ff.

74 Florus ii, 30.

75 Tacitus, *Ann.* iv, 5.

76 Appian, *Illyric.* 30.

77 E. Stein, 'Die Legaten von Moesien', *Diss. Pann.* i, 10 (Budapest, 1940), 19.

78 *MS* 68 f.

79 *ŽA* xii, 1962, 365; *C.* 8250; cf. *MS* 95, n. 47.

80 Tacitus, *Ann.* i, 20. The Via Gemina from Aquileia over the Alps was probably built soon after 35 B.C.: J. Šašel, *PWRE* Suppl. xi, 573.

81 The distribution of pre-imperial coins in particular proves that this was a communication route. Cf. note 142 in previous chapter (p. 367), and A. Alföldi jun., *Magyar Muzeum* 1946, 52 ff.

82 G. Alföldy, *Acta Arch.* xvi, 1964, 247 ff.

83 Tacitus, *Ann.* xii, 30.

84 Proof that both legions took part in the building work comes from *C.* 13813 b = *ILJug.* 57 and *C.* 1698 = *ILJug.* 60.

85 A detailed description is given in E. Swoboda's scholarly work, 'Forschungen am obermoesischen Limes', *Schriften der Balkankommission Antiquarische Abteilung* x (Vienna, 1939).

86 *ILJug.* 55; 58: *iter Scorfularum*; cf. the road-staging post *ad Scrofulas* on the Tabula Peutingeriana, east of Viminacium.

87 Strabo vii, 3, 13.

88 *ILJug.* 57; 60.

89 *Ibid.*, pp. 31–9.

90 See *MS* 52 for a provisional view.

91 In this description I depend mainly on E. Swoboda, *op. cit.* (note 85).

92 For example the diploma *CIL* xvi, 4 from the year 60 lists seven cohorts in Pannonia; inscriptions on the other hand attest roughly the same number of *alae* in the province.

93 *Römische Geschichte* v, 21, 178. Mommsen, however, was not so precise in his wording as were his successors.

94 L. Barkóczi and É. B. Bónis, *Acta Arch.* iv, 1954, 129 ff.

95 Only brief preliminary reports: E. B. Vágó, *Arch. Ért.* xci, 1964, 255; *Alba Regia* xi, 1971, 112 ff. (Intercisa) and E. B. Thomas, *Arch. Ért.* xci, 1964, 257 (Matrica).

96 *Bevölk.* 243, No. 158/1–8.

97 *C.* 4269.

98 E.g. *C.* 10514; 14349[8]; 15163.

99 *C.* 3322.

100 *C.* 3271.

101 *C.* 3256 could perhaps refer to Malata (Banoštor). The strange way of counting miles *a Malata Cusum* on *C.* 3700–3 (Nerva) could be regarded as evidence that at an early stage Malata had become a junction.

102 D. Gabler, *Arrabona* ix, 1967, 21.

103 *C.* 4269.

104 G. Juhász, 'Die Sigillaten von Brigetio', *Diss. Pann.* ii, 3 (Budapest, no date), 176 ff.

105 K. Sz. Póczy, *Bp. Rég.* xvi, 1955, 41, Abb. 1.

106 See note 47. The establishment of an imperial frontier on the Danube under Claudius mentioned in Aurelius Victor, *Caes.* 4, 2, will probably have to be taken to refer to the establishment of the provinces of Noricum and Thrace.

107 E.g. A. Betz, *RLiÖ* xviii, 1937, 54, 74 ff.

108 *C.* 4227; 4228; 4244 (cf. D. Gabler, *Arrabona* xi, 1969, 44, No. 39); *Ann. Ép.* 1965, 161.

109 J. Fitz, *Gorsium* (Székesfehérvár, 1970), Fig. 7; cf. on ala Scubulorum B. Gerov, *Acta Ant.* xv, 1967, 95 ff.

110 *ŽA* xii, 1962, 365; see also *C.* 8250.

111   *Listy Filologické* vi/lxxxi, 1958, Eunomia 37.
112   *C.* 8261 = *ILS* 2733; *C.* 14589, cf. *MS* 118, n. 89.
113   Tacitus, *Ann.* xii, 29: *ipsaque e provincia lecta auxilia.*
114   *Bevölk.* 121.
115   *CIL* xvi, 2. 4; *C.* 4372; 4373; 4376; 4377.
116   See note 47 and *CIL* xvi, 31.
117   Tacitus, *Hist.* iii, 12.
118   Dio lv, 29, where, however, it is only stated that there was to be recruitment among the Dalmatians.

CHAPTER 3

1   *Nat. Hist.* iii, 148.
2   *CIL* vi, 2385, 1, 12.
3   On the settlement areas of the Pannonian tribes, see *Bevölk.* 15–80. On the Arabiates, *Acta Arch.* xxi, 1969, 348.
4   Velleius Pat. ii, 96.
5   Dio liv, 31, 5.
6   Typical south Pannonian–Dalmatian names are Bato, Breucus, Dases, Dasmenus, Liccaius, Lucius. Cf. also Germus, Lascus, Madena, Racio, etc. Katičič, *Symposium 1964*, 31 ff.
7   B. Gerov, *Acta Ant.* xv, 1967, 91 ff. and in *Antičnoe Obščestvo* (Moscow, 1967), 23 ff.; *Primus [Iuli lib.] Asalus d[upl. ala] Cap(itoniana).* Cf. *Ann. Ép.* 1912, 187: *Iulius Saturio Iuli lib. domo Haeduus missic. ala Capitoniana.*
8   Velleius Pat. ii, 96.
9   *Ibid.*, ii, 109, 5.
10   For these imitations of *denarii*, see Ö. Gohl, *Num. Közl.* i, 1902, 17 ff.; *Bp. Rég.* viii, 1904, 182; also A. Alföldi, *Karpatenbecken* (see note 27), 29 ff.; A. Mócsy, *Num. Közl.* lx-lxi, 1961-2, 25 ff.; *Pannonia*, 692.
11   See note 54.
12   *C.* 3377.
13   Further literature: I. Bóna, *Acta Arch.* xv, 1963, 304 ff.; S. Foltiny, *Barb-Festschr.* 79 ff.; K. Horedt, *Acta Musei Napocensis* v, 1968, 419 ff.; *Pannonia*, 712, 33–62; G. Reinfuss, *Carn.-Jb.* 1960-1, 65, Nos 6–7 and 91; A. Neumann *PWRE* ixA, 79; B. Saria, *PWRE* viiiA, 344, etc.
14   Ptolemy ii, 14, 2.
15   *Itineraria Ant.* 263, 7.
16   *CIL* vi, 32542.
17   *Germania* i, 1917, 132 ff.; Rossius Vitulus was *praepositus gentis Onsorum.*
18   Tacitus, *Germ.* 43.
19   *C.* 4149, 4224, 5421.
20   *Germ.* 42.
21   *C.* 3598 = 10552.
22   Listed in *Bevölk.*; cf. also *Acta Arch.* xxi, 1969, 342 f.
23   Tacitus, *Germ.* 43.
24   *Germ.* 28.
25   M. Szabó, *Ant. Tan.* x, 1963, 220 ff.

26   A. Mócsy, 'Die Lingua Pannonica', *Symposium 1966*, 195 ff.

27   Primarily Alföldi, 'Zur Geschichte des Karpatenbeckens im 1. Jh. v. Chr.', *Archivum Europae Centro-Orientalis* viii, 1942, and also *Ostmitteleuropäische Bibliothek*, 1942.

28   *CIL* xvi, 4: *Iantumarus Andedunis f. Varcianus* had a Celtic name: contrast *C.* 4372; cf. p. 2280: *Bato Buli f. Colapianus* had an Illyrian–south Pannonian one.

29   West Celtic analogies with Pannonian names have been investigated by M. Szabó, *Arch. Ért.* xci, 1964, 165 ff.

30   L. Nagy, *Laur. Aqu.* ii, 232 ff.; cf. J. M. de Navascués in *Akte des iv. Internat. Kongr. für gr. u. lat. Epigr. Wien 1962* (Vienna, 1964), 281 ff.

31   P. Petru, *Arb. Vestn.* xvii, 1966, 361 ff.; and see ch. 1, n. 76 (p. 365).

32   Caesar, *bell. Gall.* i, 5, 4.

33   Justin xxxii, 3, 12.

34   The most recent study is that of J. Garbsch, *Die norisch-pannonische Frauentracht* (Munich, 1965). Cf. J. Fitz, *Gnomon* xxxvii, 1965, 619 ff.

35   *MS* 139 ff.

36   *Ibid.*, 83 ff.

37   See, e.g., *Spomenik* lxxi, 116, 117, 252, 254, 273, 275, 276, 278, 280; lxxv, 212; lxxvii, 47; xcviii, 272. A pair of brooches on the no longer extant part of a tombstone is mentioned in *Spomenik* lxxi, 520; *JÖAIB* vi, 1903, 32, No. 40. On the other stelae the dress is fastened on the right shoulder by a button or by a round, button-like brooch.

38   *C.* 14507.

39   Cf. F. Papazoglu, *Srednjobalkanska plemena* (Sarajevo, 1969), 173 ff.

40   *CIL* vi, 3205.

41   *C.* 8242 = *Spomenik* lxxi, 254.

42   G. Alföldy, *Acta Ant.* xii, 1964, 109 ff.; but cf. R. Katičič, *Godišnjak* iii (Centar za Balkano-loška Ispitivanja i, Sarajevo, 1965), 63 ff.

43   Strabo vii, 3, 2, 10; 5, 2; cf. *Acta Ant.* xiv, 1966, 97, n. 100.

44   Strabo vii, 3, 10.

45   *ILS* 986.

46   *CIL* xvi, 13.

47   Pliny, *Nat. Hist.* iii, 149; Ptolemy iii, 9, 2.

48   *Ibid.*, iii, 9, 3: their chef-lieu Ratiaria. *ILS* 1349: neighbours of the Triballi.

49   *MS* 25–9.

50   B. Gerov, *loc. cit.* (note 7).

51   Ptolemy iii, 9, 4.

52   *CIL* v, 1838 = *ILS* 1349.

53   *C.* 14387 = *ILS* 9199.

54   *CIL* ix, 5363 = *ILS* 2737.

55   Evidence that the south Pannonian civitates had the status of *civitas stipendiaria* is possibly provided by Aurelius Victor, *epit. Caes.* 1, 7: *Pannonios stipendiarios adiecit.*

56   *C.* 3224.

57   Cf. S. Dušanić, *Arch. Iugosl.* viii, 1967, 67.

58   *Ann. Ép.* 1937, 138.

59   Claudius: *ILS* 212, col. ii. Aelius Arist., *Rom.* 108. Antoninus Pius: *CIL* v, 532.

60   Citizens: *C.* 3546, 15134 (cf. *ŽA* xv, 1965, 89, n. 17, and xvii, 1967, 198 ff.); *Burgenl.*

Heimatb. xiii, 1951, 3, No. 3 = xiv, 1952, 100; *Historia* vi, 1957, 490. For peregrini: *Ann. Ép.* 1937, 138; *C.* 10358.

61 Primarily, of course, the big cemetery at Emona, see Lj. Plesničar-Gec and S. Petru in a forthcoming book on the cemetery. Poetovio: É. B. Bónis, 'Die kaiserzeitliche Keramik von Pannonien i', *Diss. Pann.* ii, 20 (Budapest, 1942), 253 ff. Carnuntum: *RLiÖ* xvi, 1926; xviii, 1937, etc. See also Swoboda, *Carnuntum* 245.

62 E.g. Bónis, *loc. cit.* (note 61), 262–3; A. Barb in A. Radnóti, 'Die römischen Bronzegefässe von Pannonien', *Diss. Pann.* ii, 6 (Budapest, 1938), 177 ff. For *tumuli*, see note 137 in Chapter 5 (p. 381).

63 Velleius Pat. ii, 110, 5.

64 Tacitus, *Ann.* ii, 62.

65 Most recently J. Wielowiejski, *Kontakty Noricum i Pannonii z ludami pólnocnymi* (Wroclaw–Kraków, 1970), 32 ff.

66 Owing to the difficulty and in some cases the impossibility of dating more closely the epigraphic material relating to freedmen in the first century, the reader is referred to the synopsis in *Acta Ant.* iv, 1956, 222 ff.

67 M. Abramić, *JÖAIB* xvii, 1914, 138.

68 *RLiÖ* xvi, 1926, 45 ff.

69 *AIJ* 575.

70 Material listed in Mócsy, *Bevölk.*

71 B. Vikić-Belančić, *Starinar* iv, 13–14, 1965, 89 ff.; D. Gabler, *Arch. Ért.* xci, 1964, 94 ff.; *Arrabona* vi, 1964, 5 ff.; ix, 1967, 21 ff.; *Ant. Tan.* xiv, 1967, 58 ff.; *Acta Arch.* xxiii, 1971, 83 ff.

72 É. B. Bónis, *Die spätkeltische Siedlung Gellérthegy-Tabán in Budapest* (Budapest, 1969).

73 H. Mitscha-Mährheim, *Mitteilungen des Vereins der Freunde Carnuntums* iii, 1950, 2 ff.; cf. Swoboda, *Carnuntum* 270.

74 E. Nowotny and O. Menghin, *Wiener Prähist. Zeitschrift* xiii, 1926, 101 ff.; xiv, 1927, 127 ff., 135 ff.

75 S. Nagy, *Rad vojv. muz.* ix, 1960, 112 ff.; and see ch. 1, n. 93 (p. 366).

76 K. Sz. Póczy, *Bp. Rég.* xvi, 1955, 41 ff.; K. M. Kaba, *ibid.*, 272 ff.

77 *JÖAIB* xii, 1909, 226 ff.

78 *Nat. Hist.* iii, 147.

79 *Ann. Ép.* 1912, 8; 1913, 57.

80 B. Saria, *Pannonia*, viii, 1935, 171 f.

81 Hence I cannot share Swoboda's doubts concerning my interpretation of the *dec. Scarb.* See his *Carnuntum* 242 f. Both in Scarbantia and in Savaria donors of inscriptions, particularly in the early years, saw to it that all the titles and names by which the town was known were transcribed.

82 It is unnecessary to do more than cite J. Šašel's excellent article on Emona (*PWRE* Suppl. xi), which contains all the important bibliographical sources.

83 Pliny, *Nat. Hist.* iii, 147; Velleius Pat. ii, 109, 5.

84 *Bevölk.* No. 88/1 = *C.* 10921; 89/1–2; 90/5 = *C.* 4171.

85 *Ibid.*, 38.

86 *Ibid.*, No. 89/2.

87 A. Alföldi jun., *Arch. Ért.* 1943, 71 ff.

88 *C.* 4156.

89 J. Šašel, *ŽA* x, 1960, 201 ff.
90 E.g. *CIL* v, 943; 1011. Cf. *Bevölk.* 90/30: *P. Opponius Iustus Cl. Savaria Aculeiensis.*
91 *CIL* v, 8336.
92 *C.* 10936 = 4225; cf. *ILS* 8507.
93 *C.* 4009, 4250, 4251, 11259; *Ann. Ép.* 1938, 163.
94 *C.* 4196.
95 E. Türr, *Arch. Ért.* lxxx, 1953, 130. Previous investigations and finds are listed by T. P. Buocz in *Savaria topográfiája* (Szombathely, 1968). A new attempt to reconstruct the street-network in the *colonia* is to be found in *Arch. Ért.* xcviii, 1971, 193 ff., by E. Tóth.
96 Mócsy, *Arch. Ért.* xcii, 1965, 27 ff.
97 *C.* 4224; *Bevölk.* 97/1.
98 *C.* 4212, 10926.
99 *C.* 4199, 4200; *Bevölk.* 38, n. 151.
100 *Ibid.,* 180/1.
101 A. Mócsy, *Arch. Ért.* lxxxi, 1954, 167 ff.

CHAPTER 4

1 Josephus, *bell.* vii, 4, 3.
2 *C.* 11194–7.
3 E. Tóth and G. Vékony, *Acta Arch.* xxii, 1970, 133 ff.
4 Cf. Swoboda, *Carnuntum,* 36.
5 *C.* 4591.
6 *Starinar* iv/18, 1967, 21 ff.
7 K. Wachtel, *Historia* xv, 1966, 247.
8 *C.* 8261 = *ILS* 2733.
9 See note 6.
10 *Spomenik* xcviii, 441.
11 Suetonius, *Domitianus* 6; Eutropius vii, 23, 4; Jordanes, *Get.* 76.
12 Most recently R. Syme, *Arh. Vestn.* xix, 1968, 103.
13 Suetonius, *Domitianus* 6, 5; Jordanes, *Get.* 77.
14 Dio lxvii, 10.
15 The expansion of the auxiliary forces in Pannonia between 80 and 85 can easily be traced from the diplomas: *CIL* xvi, 26, 30, 31.
16 Tacitus, *Ann.* xii, 30; *Hist.* iii, 5; 21.
17 Dio lxvii, 7.
18 *ILS* 9200.
19 Dio lxvii, 5.
20 *Ibid.,* lxvii, 5, 2.
21 Suetonius, *Domitianus* 6; Eutropius vii, 23, 4; Tacitus, *Agr.* 41.
22 Martial vii, 8; viii, 8; ix, 31. His stay at Carnuntum is attested by *C.* 4497.
23 J. Dobiáš, *Corolla memoriae E. Swoboda dedicata* (Cologne–Graz, 1966), 117, n. 10; cf. *Historica* iv, 1962, 27.
24 *CIL* v, 7425 = *ILS* 2720. Pliny, *paneg.* 8, 2; 16, 1.
25 Szilágyi, *Inscr. teg.* 69.

26 *C.* 10224; *VHAD* ix, 1906–7, 103, Nr. 220; see also *C.* 13360.

27 *C.* 3468; *Bp. Rég.* viii, 1904, 162, No. 1; *C.* 143572; cf. G. Alföldy, *Acta Arch.* xi, 1959, 132 ff.

28 *C.* 14349, [2], [4], [9].

29 *C.* 10513, 10514, 14349[8], 15163. T. Nagy, *Bp. Müeml.* ii, 36.

30 A. Mócsy, *Acta Arch.* iv, 1954, 115 ff.

31 *C.* 14359[22]; cf. A. Neumann, *PWRE* ixA, 67 f.

32 *C.* 15196[4] = *Jahrbuch des Vereins für Geschichte der Stadt Wien* xvii–xviii, 1961–2, 8, No. 1.

33 Szilágyi, *Inscr. teg.* 83 f.; L. Barkóczi, *Brigetio* 20.

34 Rufus Festus, *brev.* 8.

35 L. Barkóczi and É. B. Bónis, *Acta Arch.* iv, 1954, 129 ff.

36 Cf. Chapter 2, notes 96–101 (p. 370), and, e.g., *C.* 10246 = 3262, *CIL* xvi, 54, etc.

37 *C.* 1642.

38 A. R. Neumann, *Forschungen in Vindobona 1948 bis 1967*, i, 'Lager und Lagerterritorium', *RLiÖ* xxiii (Vienna, 1967). Barkóczi, *Brigetio.*

39 *CIL* vi, 1548; Tacitus, *Germ.* 42.

40 V. Ondrouch, *Historica Slovaca* ii, 1941, 22 ff., Nos 112–24, 188. T. Kolnik, *LRKN* 40.

41 For the buildings: *Pannonia*, 643 f., with bibliography. Purpose of the buildings: *Acta Arch.* xxi, 1969, 355 f. The discovery of a hitherto unknown building at Pác could not be taken into consideration: see T. Kolnik, *Arch. Rozhl.* xxiv, 1972, 59 ff.

42 Dio lxviii, 10, 3

43 *Oratio* xii, 16–20 (transl. J. W. Cohoon).

44 *CIL* xvi, 39 and 46.

45 Barkóczi, *Brigetio* 20.

46 *C.* 10517. K. M. Kaba, *Bp. Rég.* xvi, 1955, 275; cf. *Acta Arch.* xxi, 1969, 352, n. 111.

47 *C.* 15162. *Bevölk.* 185/33; *Bp. Rég.* xxi, 1964, 247.

48 *CIL* xvi, 47.

49 SHA, *Hadrianus* 3, 9.

50 Doubts about this, expressed by S. Dušanić, *Arch. Iugosl.* viii, 1967, 69, are in my opinion unfounded.

51 J. Šašel, *PWRE* Suppl. xi, 571–6.

52 This can also be deduced from the reliefs of Trajan's Column, which show Roxolanian troopers in coats of mail fighting the Romans.

53 Cf. Jordanes, *Get.* 75.

54 Dio lxviii 10, 3; cf. *Acta Arch.* iv, 1954, 125.

55 SHA, *Hadrianus* 3, 9; Dio lxviii, 10, 3; Eusebius, *chron.* 194b (Helm); Eutropius viii, 3, 1.

56 A. Mócsy, *Acta Arch.* iv, 1954, 124 ff.

57 Ptolemy iii, 7, 1; 8, 1.

58 A. Mócsy, *loc. cit.* (note 56); M. Párducz, *MFMÉ* 1956, 15 ff.; Zs. Visy, *MFMÉ* 1970, 3 ff.

59 D. Protase, *Acta Mus. Napocensis* iv, 1967, 47 ff.

60 H. Nesselhauf, *CIL* xvi, p. 224 ad. n. 164.

61 *MS* 50 ff.

62 *Starinar* iv/15–16, 1965, 173 ff.

63 *CIL* xvi, 111 (A.D. 160).

64 *MS* 50 ff.

65 *C.* 8262, 14575, 14577, 14579; *JÖAIB* xii, 183, No. 52.
66 Patsch, *Beiträge* V(2), 90 ff.; D. Tudor, *Podurile romane de la Dunarea de jos* (Bucharest, 1971), 53 ff.
67 Dio lxviii, 13, 6.
68 *MS* 18–25.
69 Ptolemy, *Geogr.*, *passim*; *Itineraria Ant.* 242–8; Dio lv, 23; *CIL* vi, 3492 = *ILS* 2288, etc.
70 *Arch. Ért.* lxxviii, 1951, 135; *Bp. Rég.* xvii, 1956, 165; *Acta Arch.* xi, 1959, 256; *Bp. Rég.* xx, 1963, 27 f.
71 SHA, *Hadrianus* 6, 6.
72 *Ibid.*, 6, 6; cf. R. Syme, *JRS* lii, 1962, 87.
73 SHA, *Hadrianus* 6, 7; 7, 3.
74 *CIL* v, 32, 33 = *ILS* 852, 853.
75 On the war, see SHA, *Hadrianus* 5, 2; Eusebius, *chron.* 198d (Helm); Orosius vii, 13, 4, and C. and H. Daicoviciu, *Acta of the 5th International Congress of Greek and Latin Epigraphy, Cambridge 1967* (Oxford, 1971), 347.
76 Dio lxix, 15, 2.
77 *Caes.* 13, 3.
78 P. Lakatos, *Ant. Tan.* xii, 1965, 91 ff., and *MFMÉ* 1964/5, 65 ff.
79 Most recently N. Gostar, *SCIV* ii, 1951, 2, 169 f.
80 See note 59.
81 SHA, *Hadrianus* 9, 1; Eutropius viii, 6, 2.
82 L. Barkóczi, *Intercisa* ii, 504.
83 E. Jónás, *Bp. Rég.* xii, 1937, 287 ff.
84 SHA, *Hadrianus* 12, 7.
85 *C.* 6818 = *ILS* 1017, cf. J. Dobiáš in *Omagiu lui C. Daicoviciu* (Bucharest, 1960), 147 ff.; *Historica* iv, 1962, 28 f.; Dobiáš connects him with the wars under Hadrian. This cannot be right as legio XIII Gemina had no legionary legate at that time. To connect him with the wars under Domitian would be a possibility.
86 Listed in *Pannonia*, 554; cf. J. Fitz, *Acta Ant.* xi, 1963, 255, and L. Balla, *Acta Classica Debrecen* i, 1965, 44 ff.
87 SHA, *Aelius* 3, 2; cf. *Hadrianus* 23, 12.
88 *CIL* xi, 5212 = *ILS* 1058.
89 *RIC* iii, 620.
90 SHA, *Marcus* 12, 13; cf. R. Göbl, *Rheinisches Museum* civ, 1961, 70 ff.
91 R. Noll, *Archaeologia Austriaca* xiv, 1954, 43 ff.; cf. E. Swoboda, *Carn.-Jb.* 1956, 5 ff.; *Pannonia*, 555; J. Fitz, *Acta Ant.* xi, 1963, 262, 266; Nagy, *Bp. Müeml.* ii, 41 f.
92 R. Göbl, *Zwei römische Münzhorte, Illmitz und Apetlon* (Eisenstadt, 1967).
93 Tác: J. Fitz, *Acta Arch.* xxiv, 1972, 3 ff. Doboj: *Arh. Pregl.* viii, 1966, 122; ix, 1967, 93. The fort of Doboj in the valley of the Bosna was perhaps a frontier station on the Pannonian–Dalmatian provincial boundary.
94 There are several detailed studies of the history of Pannonian army troop-movement in the second century. See especially L. Barkóczi and A. Radnóti, *Acta Arch.* i, 1951; T. Nagy, *Acta Arch.* vii, 1956 and J. Fitz, *Acta Ant.* vii, 1959.
95 Nagy, *Bp. Müeml.* ii, 519 ff.
96 *Itineraria Ant.* 244, 4.

97 Swoboda, *Carnuntum* 15. T. Nagy's recent excavations of Aquincum have shown that even the legionary fortress there had to be moved back (see Fig. 23).

98 *Pannonia*, 634–8.

99 D. Gabler, *Arrabona* viii, 1966, 67 ff.; *Arch. Ért.* xciv, 1967, 221; xcv, 1968, 130 f.; xcviii, 1971, 269.

100 See Chapter 2, note 95 (p. 370).

101 *Arch. Ért.* lxxxii, 1955, 67 ff. Cf. also Nagy, *Bp. Müeml.* ii, 93, n. 51, on first-century finds under the paving stones.

102 Gabler, *op. cit.* (note 99); cf. *Acta Arch.* xxi, 1969, 354, n. 138.

103 M. v. Groller, *RLiÖ* iv, 1903, 23 ff.; S. Soproni, *Arch. Ért.* lxxxvii, 1960, 234; Barkóczi, *Brigetio* 6.

104 'Forschungen am obermösischen Limes', *Schriften der Balkankommission, Antiquarische Abteilung* x (Vienna, 1939).

105 *C.* 3700; *Burgenl. Heimatb.* i–ii, 1932–3, 77, No. 82; *C.* 4641–53, etc.

106 *Prooim.* 7.

107 *Rom.* 80–4.

108 *ILJug.* 63; Swoboda, 'Forschungen am obermösischen Limes', 78 ff.; H. U. Instinsky, *JÖAIB* xxxv, 1943, 33 ff. For the new Trajanic inscription, see P. Petrović, *Arch. Iugosl.* ix, 1968, 83 ff. On the canal, Procopius, *aedif.* iv, 6, 8–16; cf. *MS* 116, note 71.

109 E.g. Swoboda, *Carnuntum*, 271.

110 *Bp. Tört.* i, 761, Abb. 36.

111 B. Swoboda, *Slov. Arch.* x, 1962, 422 ff.

112 *RLiÖ* i, 1900, 17 ff.; cf. Swoboda, *Carnuntum* 253.

113 *Bp. Tört.* i, 748, Abb. 31; cf. *Pannonia*, 641.

114 L. Nagy, *Az Eskü-téri római eröd, Pest város öse* (Budapest, 1946).

115 T. Nagy, *Acta Ant.* vi, 1958, 429 f.

116 A. Mócsy, *Fol. Arch.* x, 1958, 96 ff.

117 Aquincum (Budapest, District xiii): 89 by 67 m. Lussonium: 100 by 55 m. Lugio: 85 by 59 m.

118 G. Juhász, 'Die Sigillaten von Brigetio', *Diss. Pann.* ii, 3 (Budapest, no date), 196 ff.

119 J. Szilágyi, *Bp. Rég.* xviii, 1958, 53 ff.; *Bp. Müeml.* ii, 352 ff.; R. Egger, 'Das Praetorium als Amtssitz und Quartier römischer Spitzenfunktionäre', *Sitzungsberichte der Österreichischen Akademie* ccl, 1966, 28–36. Should Egger's assumption be correct, the governor's palace at Carnuntum was also on the bank of the Danube. On the date of the palace at Aquincum, see I. Wellner, *Arch. Ért.* xcvii, 1970, 116 ff., who shows two main periods, the first of Trajanic date, later destroyed, and the second dating to the middle of the second century or later.

CHAPTER 5

1 *Nat. Hist.* iii, 148.

2 *CIL* xvi, 18.

3 S. Dušanić, *Arch. Iugosl.* viii, 1967, 78, n. 59, and R. Syme, *Arh. Vestn.* xix, 1968, 103, date the founding of Sirmium to the reign of Domitian, without good grounds in my opinion.

4 *CIL* xvi, 14.

5 Tacitus, *Hist.* iii, 12.

6 *C.* 3971.

7 *CIL* xvi, 15.

8 E.g. *C.* 3896, 3898, 3917, 3922, 3928, 3929, 3932, 3935, 4009, 10802, 10805, 10809, 10812, 10824, 14042 and 14354[22]; *AIJ* 489, 501.

9 Only brief preliminary reports in *Arh. Pregl.* Also Dj. Mano-Zisi, *ZNMB* iv, 1964, 93 ff., and M. Parović-Pešikan, *Starinar* iv/15–16, 1966, 31 ff.; iv/19, 1969, 75 ff.; iv/20, 1970, 265 ff.

10 Dio xlix, 37, 3.

11 A. F. Marsigli, *Danubius Pannonico-Mysicus* ii (Hagae et Amstelodami, 1726), 47 f.; Tab. 20, Fig. x. C. Veith, 'Die Feldzüge des C. Iul. Octavianus in Illyrien', *Schriften der Balkankommission, Antiquarische Abteilung* vii (Vienna, 1904), 51 ff. *AIJ*, p. 237 f.

12 *C.* 10865 = 4001 = *AIJ* 588; *C.* 3685 = 10249; *CIL* xvi, 18, etc.

13 *Ann. Ép.* 1911, 237.

14 *ŽA* iv, 1954, 196, No. 2; *MS* 68 f.

15 *JÖAIB* xiii, 216, No. 29; *Spomenik* lxxi, 650; cf. *MS* 67, nn. 40–1.

16 *Spomenik* xcviii, 441.

17 *MS* 68 ff. For further details, *ibid.*, 62–75.

18 N. Vulić, *Nekoliko pitanja iz antičke istorije naše zemlje* (Belgrade, 1961), 87 ff.; Ć. Truhelka, *GSND* v (2), 1929, 78 ff.

19 *JÖAIB* xiii, 218, No. 31.

20 *MS* 30 n. 9.

21 Listed in *MS* 68.

22 *AIJ* 374, 375; *C.* 4056, 4057.

23 *C.* 4057 = *AIJ* 373.

24 *AIJ* 375.

25 B. Saria, *PWRE* xxi, 1167 ff.; I. Mikl-Curk, *Arh. Vestn.* xv-xvi, 1965, 259 ff.; for the bridge, see *AIJ* 361.

26 *C.* 8087, 14217 = 14500; *JÖAIB* xxxi, 106, No. 11 (cf. *MS* 107, n. 66); *IBAI* xiv, 1940–2, 272.

27 See the tombstones from the oldest cemetery-road at Poetovio: M. Abramić, *JÖAIB* xvii, 1914, 138; cf. A. Mócsy, *Acta Arch.* iii, 1953, 181.

28 *Gromatici veteres* (ed. Lachmann) i, pp. 204 f.

29 *C.* 3279. Stephanus Byz. *s. v.* Mursa.

30 *Noctes Att.* xvi, 13, 5.

31 D. Pinterović, *Osj. Zb.* v, 1956, 55 ff.; xi, 1967, 23 ff.; *LuJ* i, 1961, 35 ff.

32 For details, see *Bevölk.*

33 For the Canii, see *C.* 10901 = 3689; *Bevölk.* 82/1; *C.* 4250; *Ann. Ép.* 1912, 8 = 1913, 57; *C.* 3599; *Fol. Arch.* ix, 1957, 83 ff. For the Caesernii, see J. Šašel, *ŽA* x, 1960, 201 ff.

34 *Bevölk.* 74/1–2; cf. 90/30.

35 *Ibid.*, 73/2; *C.* 10289 = 3308 = *Bevölk.* 216/5.

36 *Bevölk.* 186/11.

37 *C.* 11047.

38 *Bevölk.* 164/8, 24, 25.

39 See Chapter 3, note 71 (p. 373).

40 Barkóczi, *Brigetio* 37; for new evidence, see *Acta Ant.* xiii, 1965, Plate xxiv.

41 T. Szentléleky, *Bp. Rég.* xix, 1959, 197.

42 A. Mócsy, *Fol. Arch.* ix, 1957, 83 ff.; cf. also xvi, 1964, 43 ff.

43 D. Laczkó and Gy. Rhé, *Balácza* (edited by K. Hornig, Veszprém, 1912).

44 *Arch. Ért.* lxxx, 1953, Taf. xvii, 28.

45 *C.* 6480 = 10954 = *Arch. Ért.* xlii, 1928, 207 ff.

46 *CIL* v, 8973.

47 *C.* 4122; cf. E. Ritterling, *PWRE* xii, 1252; *C.* 4123, 4148, 14355[6].

48 *C.* 4153, 4499; cf. *Bevölk.* 164/12, 14.

49 Full length: T. Nagy, *Bp. Rég.* xiii, 1943, 464 f. Eagle: *C.* 14349[8] = Schober No. 57.

50 L. Nagy, *Germania* xvi, 1932, 290 f.

51 *C.* 10548; L. Nagy, *Germania* xv, 1931, 263; xvi, 1932, 288 ff.

52 *Bevölk.* 186/11.

53 *Ibid.,* 186/30: *domo Nemes* = the singular of Nemetes (J. Szilágyi).

54 *Ann. Ép.* 1933, 110.

55 *C.* 10430.

56 *C.* 3505.

57 Barkóczi, *Brigetio* No. 35.

58 Carnuntum: *RLiÖ* vii, 1906, 95 ff.; viii, 1907, 7 ff.; ix, 1908, 43 ff. Aquincum: *Bp. Müeml.* ii, Taf. vi. Vindobona: A. R. Neumann, *Forschungen in Vindobona* i (see Chapter 4, note 38 (p. 375). Brigetio: Barkóczi, *Brigetio* 20.

59 Carnuntum: *RLiÖ* vii, 1906, 83 ff. A similar building at Aquincum (J. Szilágyi, *Bp. Rég.* xiv, 1945, 133; *Bp. Müeml.* ii, 518) was situated, as recent excavations by T. Nagy have shown, within the walls of the legionary fortress.

60 *C.* 4500, 4501, 11301, 13379, 14355[15], 14359[4]; *Ann. Ép.* 1929, 217; 1939, 261, etc.

61 For the time being, see B. Rutkówski, *Acta Rei Cretariae Romanae Fautorum,* v–vi, 47 ff.; cf. *MS* 259 f.

62 *MS* 200, Abb. 45.

63 *MS* 148 ff.

64 *C.* 14505.

65 *Spomenik* xcviii, 9.

66 *CIL* v, 1047 = *ILS* 7526.

67 *MS* 191 f.

68 M. Vasić, *Jahrb. DAI* xx, 1905; *Archäol. Anzeiger* 102 ff.; *MS* 145 f.

69 *MS* 126.

70 *JÖAIB* vi, 23, No. 30.

71 R. Mowat, *Revue de Numismatique* iii (12), 1894, 372 ff.; cf. K. Regling, *PWRE* xv, 1322 ff.

72 *Aedif.* iv, 4, 3.

73 *Itinerarium Ant.* 134, 1; *Itinerarium Burdig.* 565, 1 and *Tab. Peut.*

74 O. Davies, *Roman Mines in Europe* (Oxford, 1935), 209 ff.

75 M. Veličkovič, *ZNMB* 1, 1958, 95 ff.

76 *C.* 6313 = 8333; 14536.

77 *C.* 14606; *Spomenik* lxxi, 251.

78 *Spomenik* lxxi, 217; cf. *MS* 35.

79 The hitherto unpublished results of Emil Čerškov's excavations will provide detailed information on the *metalla Dardanica.* This energetic representative of the younger generation of Serbian archaeologists was unfortunately killed in a road accident in 1969. In 1965 he showed

me some of the results of his highly successful excavations at Ulpianum and Sočanica. See now E. Čerškov, *Municipium DD* (Priština–Belgrade, 1970).

80 *Historia* vi, 1957, 490 ff.

81 *Burgenl. Heimatb.* xiii, 1951, 3, No. 103; xiv, 1952, 100.

82 See *Bevölk.* 118/1, 130/1, 131/1-2, 132/1, 133/2, 136/4, 14, 15.

83 *CIL* ix, 5363 = *ILS* 2737; on the dating, see K. Wachtel, *Historia* xv, 1966, 247.

84 *C.* 3679, 10783; *AIJ* 234.

85 *C.* 3919, 3921, 3925, 10804. *Itinerarium Ant.* 259, 13.

86 *C.* 3679.

87 *C.* 3896, 3917, 3922, 3928, 3929, 3932, 10791, 10802, 10805, 10809, 10811, 14042 and 14354[22].

88 *C.* 4009, 10824, 10866, 11463; *AIJ* 489.

89 *C.* 3925, 10801.

90 *Bevölk.* 24/2, 25/1, 32/1, 33/1, 34/2-3, 35/1-2, 36/1, 37/1, 38/1.

91 *C.* 3546.

92 E.g. Ulpii: *C.* 3375, 3407, 3410, 10334, 10377, 14341[6], 15151. Aelii: *C.* 10355, 10408, 10993, 11043; *Bevölk.* 117/1; *Ann. Ép.* 1953, 14.

93 Athenaeus vi, p. 272D (from Agatharchides).

94 *MS* 86 f.

95 *Spomenik* lxxi, 182, 276; lxxvii, 31; *JÖAIB* vi, 40, No. 46. For further details, see *MS* 83.

96 A certain *Rufinus Dasi pr(inceps?)* is mentioned on an inscription from the Metohija, *Spomenik* lxxi, 278. This is possibly the only epigraphic evidence of a dignitary of the *civitates peregrinae* of Upper Moesia.

97 *CIL* xii, 1122.

98 *C.* 3676; Dio lxix, 9, 6.

99 *Intercisa* i, No. 294.

100 Aquincum: L. Nagy, *Bp. Tört.* i, 372, 467; *Laur. Aqu.* ii, 191; T. Nagy, *Bp. Rég.* xxi, 1964, 9 ff., and especially *Acta Arch.* xxiii, 1971, 59 ff. Carnuntum: Swoboda, *Carnuntum* 154.

101 At present, see A. Mocsy, *Stud. Mil.* 211 ff. and *Acta Ant.* xx, 1974.

102 *C.* 4554; *Ann. Ép.* 1938, 167.

103 *C.* 14359[2].

104 *C.* 8127, 8128, 12659, 13805, 14217[2]. *Spomenik* lxxi, 308, 311.

105 *C.* 1655.

106 *C.* 3347, 10334, 10355, 10377, 14341[6]. *Intercisa* i, No. 294. *Ann. Ép.* 1953, 14; 1965, 12.

107 *C.* 10430, 10525. *Intercisa* i, No. 25. *Ann. Ép.* 1933, 110; 1939, 8-9. *Bevölk.* 185/10; 186/11, 30, 43.

108 *C.* 10418; cf. A. Alföldi, *Arch. Ért.* lii, 1939, 108.

109 *Fol. Arch.* xxi, 1970, 59 ff. J. Fitz holds, though without good grounds, that Gorsium was a municipium: see *Acta Arch.* xxiii, 1971, 47 ff.; xxiv, 1972, 3 ff.

110 *C.* 10408; cf. A. Alföldi, *op. cit.* (note 108).

111 *Alba Regia* xi, 1971, 127, No. 460.

112 *CIL* xvi, 112, 123, 179, and diploma of A.D. 157 to be published by Zs. Visy.

113 *Ann. Ép.* 1953, 14.

114 See note 108.

115   *C.* 10336; *Ann. Ép.* 1953, 11.

116   *C.* 10305.

117   G. Alföldy, *Epigraphica* xxvi, 1965, 95 ff.

118   *CIL* vi, 3297.

119   E.g. the attack on Plautius Silvanus' troops by the Pannonians in A.D. 7, Gallienus' victory over Ingenuus in 258 or 259, Constantine's victory over Licinius in 314 and the battle between Constantius II and Magnentius in 351. For recent excavations at Cibalae, see B. Vikić-Belančić, *Vjesnik arheološkog muzeja u Zagrebu* iv, 1970, 159 ff.

120   S. Dušanić, *Arch. Iugosl.* viii, 1967, 67 ff.

121   See note 117.

122   *C.* 3267.

123   S. Dušanić, *ŽA* xv, 1965, 85 ff.

124   *C.* 10993, 11043.

125   *C.* 10900, 15188^1.

126   *Bevölk.* 53 f.

127   *C.* 4267, 4490.

128   Information kindly supplied by M. Gorenc.

129   Cf. A. Mócsy, *Arch. Ért.* xci, 1964, 16 f.

130   *MS* 75 f.; E. Čerškov, *Rimljani na Kosovu i Metohiji* (Belgrade, 1970).

131   *Spomenik* lxxi, 204; lxxv, 161; xcviii, 222.

132   *MS* 86, Abb. 27.

133   E.g. J. Csalog, *Arch. Ért.* 1943, 41 ff.; Gy. Török, *Arch. Ért.* lxxvii, 1950, 4 ff. Cf. also I. Hunyady, 'Die Kelten im Karpatenbecken', *Diss. Pann.* ii, 18 (Budapest, 1944), 152 f.; M. Párducz, *Acta Arch.* ii, 1952, 143 ff.; iv, 1954, 25 ff. etc.

134   Listed by K. Sági, *Arch. Ért.* 1944–5, 214 ff.; lxxviii, 1951, 75.

135   There is no exhaustive study: see *Pannonia*, 723 f.; also M. Šeper, *Arheološki radovi i rasprave* ii, 1962, 335 ff.; A. Radnóti, *Bayerische Vorgeschichtsblätter* xxviii, 1963, 67 ff.

136   *Fol. Arch.* xiv, 1962, 35 ff.

137   K. Sági, *Arch. Ért.* 1943, 113 ff.; *Pannonia*, 718 f., also H. Kerchler, *Die römerzeitlichen Brandbestattungen unter Hügel* (Vienna, 1967); M. Amand, *Latomus* xxiv, 1965, 614 ff; A. Barb, *Gnomon* xl, 1968, 501; É. B. Bónis, *Fol. Arch.* xiv, 1962, 23 ff.; S. Pahič, *Starinar* vii-viii, 1958, 310; *Arh. Vestn.* xi-xii, 1960–1, 116 ff.; A. Neumann, *Forschungen in Vindobona* ii, 30 ff.

138   Tombstones of the native population are arranged according to place of discovery in A. Mócsy, *Bevölk.* See map, *ibid.*, which is the basis of the map (Fig. 26) in this volume.

139   E.g. *C.* 4224; cf. also A. Barb, *Burgenl. Heimatb.* i, 1932, 78, No. 87; xiii, 1951, 216 ff.; xxii, 1960, 166 ff.

140   *C.* 10895.

141   T. P. Buocz, *Vasi Szemle* 1962, 1, 107.

142   SHA, *Marcus* 21, 7.

143   *MS* 166–78.

144   *C.* 8239; *Spomenik* lxxi, 253, 280.

145   *Spomenik* lxxv, 141 = xcviii, 179.

146   See in particular *C.* 14507; cf. *MS* 172 ff.

147   E.g. *C.* 4371 and 4367; but see also *C.* 4368, which commemorates a Batavian in the ala I Ityraeorum.

148 For instance, *RLiÖ* xii, 321; xviii, 43, No. 8; 61, No. 22; *Ann. Ép.* 1929, 212; *C.* 4473, 14358²¹ᵃ and 4491. Cf. R. Syme, *Arh. Vestn.* xix, 1968, 108.

149 E.g. *Ann. Ép.* 1929, 205, 206, 208, 209, 220; 1934, 174; 1937, 267.

150 E.g. *C.* 3528, 3530, 10497–500.

151 Pannonius also of course occurs as indication of the place of birth; but there is some evidence that a legal implication was involved. Pannonian soldiers discharged from the *exercitus Pannonicus* were indicated by the adjective of the tribe or civitas, whereas those discharged from non-Pannonian armies were indicated by the province. The only exception was a Boian discharged from the *exercitus Raetiae* and mentioned in *CIL* xvi, 55.

152 *C.* 10897: a *Tib. Claudius . . . mari f.*

153 *C.* 3679.

154 *Ann. Ép.* 1909, 235 = 1938, 13.

155 *C.* 14507.

156 Ulcisia Castra, Albertfalva, Campona, Matrica, Vetus Salina, Intercisa, Annamatia, Lussonium, Alta Ripa, Alisca(?).

157 The *origo castris* has no connection with birth in the *canabae*; see *Acta Ant.* xiii, 1965, 425–31.

158 See, e.g., the list in *C.* 14507 or the names in legio II Adiutrix, *Bevölk.* 185/7, 10, 13, 15, 16, 17, 18, 19, 20, 21, 25, etc.

159 See following note.

160 Diplomas found in Pannonia and their find-spots can be divided as follows: from the Claudian and Flavian periods seven, including one or two from fort-sites (*CIL* xvi, 26, 30?) and five elsewhere (*CIL* xvi, 2, 17, 18, 20 and 31). Between A.D. 96 and 167 thirty-one diplomas, including nineteen from fort-sites (*CIL* xvi, 47, 49, 61, 69, 71, 73, 77, 89, 97, 99, 109, 113, 116, 119, 123, 132, 164, 175; *Acta Arch.* ix, 413) and twelve elsewhere (*CIL* xvi, 42, 64, 84, 92, 96, 100, 103, 104, 112, 178, 179, 180).

161 For what follows, see *Bp. Tört.*, *Bp. Müeml.* and the relevant section of the bibliography (p. 425). On the wells of the aqueduct see K. Sz. Póczy, *Arch. Ért.* xcix, 1972, 15 ff.

162 Swoboda, *Carnuntum* is the standard work.

163 Swoboda, *Carnuntum* 83 ff.

164 A. Betz, *Carn.-Jb.* 1960 (1962), 29 ff.; H. G. Kolbe, *ibid.*, 1963–4 (1965), 48 ff.

165 Treated in detail by Swoboda, *Carnuntum* 154 ff.

166 The latest study is that of P. Petru, T. Knez and A. Uršič, *Arh. Vestn.* xvii, 1966, 491 ff.

167 M. Grbić, *Antiquity* x, 1936, 275.

168 K. Sz. Póczy, *Arch. Ért.* xciv, 1967, 137 ff.

169 D. Bošković, *Starinar* iii/4, 1928, 270, Abb. 1.

170 M. Vasić, *Jahrb. DAI* xx, 1905; *Archäol. Anzeiger* 102 ff.

171 Carnuntum: L. Klima and H. Vetters, *RLiÖ* xx, 1953. Aquincum: T. Nagy, *Bp. Rég.* xiii, 1943, 368 ff.; *Bp. Müeml.* ii, 405 ff.

172 Baláca: D. Laczkó and Gy. Rhé, *Balácza* (edited by K. Hornig, Veszprém, 1912). Eisenstadt: W. Kubitschek, *Römerfunde* 21 ff. Gyulafirátót—Poganytelek: Gy. Rhé, *Ös–ès ókori nyomok Veszprém körül* (Budapest, 1905).

173 Parndorf: B. Saria, *Barb-Festschr.* 252 ff.

174 Deutschkreuz: A. Barb, *Geschichte der Altertumsforschung im Burgenland* (Eisenstadt, 1954), 19. Donnerskirchen: Kubitschek, *Römerfunde* 49. Regelsbrunn: *RLiÖ* iii, 1902, 14 ff.

175 Winden am See: B. Saria, *Der römische Gutshof von Winden am See* (Eisenstadt, 1951).

176 Baláca, Parndorf, Donnerskirchen, Regelsbrunn and Šmarje-Grobelce: F. Lorger, *Časopis za zgodovino in narodopisje* xxix, 1934, 147 ff.; xxxi, 1936, 77 ff.

177 St Georgen: A. Barb in Radnóti, *Bronzegefässe, Diss. Pann.* ii, 6 (Budapest, 1938), 194. Fertörákos: D. Gabler, *Arch. Ért.* xcii, 1965, 235; xciii, 1966, 294, and probably also Winden am See (see note 175).

178 É. B. Bónis, *Die spätkeltische Siedlung Gellérthegy-Tabán in Budapest* (Budapest, 1969), 119–36.

179 É. Kocztur, unpublished excavations. See meanwhile J. Fitz, *Gorsium* (Székesfehérvár, 1970), Pl. 43.

180 E.g. J. Peškař, *Pam. Arch.* lii, 1961, 414 ff.; J. Pavúk, *Arch. Rozhl.* xvi, 1964, 323 ff.; M. Lamiová-Schmiedlová, *Slov. Arch.* xvii, 1969, 402 ff.; T. Kolnik, *Slov. Arch.* xix, 1971, 507.

181 A. Mócsy, *Arch. Ért.* lxxxii, 1955, 62 ff.

182 Gy. Török, *Fol. Arch.* xiii, 1961, 63 ff. For the burials from this site, see É. B. Bónis, *Fol. Arch.* xii, 1960, 91 ff. Alternatively, these houses could have been built with turf (a suggestion of R. Müller).

183 See note 181; also T. Nagy, *Bp. Müeml.* ii, 37, 50 f. (Albertfalva); E. B. Vágó, *Arch. Ért.* xci, 1964, 255 (Intercisa), etc.

184 J. Hampel, *Arch. Ért.* v, 1885, 24 ff. The diploma: *CIL* xvi, 96. The bronze vessels: A. Radnóti, *Bronzegefässe*, Taf. xxii, 5; xxv, 1–2.

185 On Tác (= Gorsium), see most recently J. Fitz, *Gorsium* (Székesfehérvár, 1970), and *Acta Arch.* xxiv, 1972, 3 ff.

186 This is attested by tiles stamped TE PR found there and by a votive inscription set up by a sacerdos called L. Virius L. f. Mercator *pro salute templensium*. Virius was possibly a native of Sirmium (cf. *L. Virrius Iustus Sirmiensis* at Aquileia, S. Panciera, *La vita economica di Aquileia*, 1957, 76).

187 L. Barkóczi and É. B. Bónis, *Acta Arch.* iv, 1954, 150 ff.; E. B. Thomas, *Acta Arch.* vi, 1955, 114 ff.

188 É. Bónis, 'Die kaiserzeitliche Keramik von Pannonien i', *Diss. Pann.* ii, 20 (Budapest, 1942); K. Sz. Póczy, *Acta Arch.* xi, 1959, 151 ff.; E. B. Vágó, *Alba Regia* i, 1960, 54 ff., etc.

189 *Pannonia*, 676 f.

190 *Pannonia*, 679 f.; also É. F. Petres, *Fol. Arch.* xvii, 1965, 96 ff.; B. Vikić-Belančić, *Starinar* iv/13–14, 1965, 97 ff.; G. Reinfuss, *Carn.-Jb.* 1960 (1962), 78 f.

191 *Pannonia*, 679, cf. D. Gabler, *Arrabona* vi, 1964, 5 ff.

192 In an area south-east of Sala Roman sites do not yield coins or *sigillata*: see R. Müller, *Régészeti terepbejárások* (Zalaegerszeg, 1971), 79 ff.

193 B. Kuzsinszky, *Bp. Rég.* xi, 1932; K. Kiss, *Laur. Aqu.* i, 212 ff.; K. Sz. Póczy, *Acta Arch.* vii, 1956, 102 ff.

194 A. Alföldi, *Fol. Arch.* i-ii, 1939, 97 ff.

195 L. Nagy, *Bp. Rég.* xiv, 1945, 305 ff.; B. Rutkowski, *Acta Rei Cretariae Romanae Fautorum* x, 1968, 18 ff. A bowl mould found at Gorsium is also attributable to the so-called 'Siscia' pottery, but does not prove that Gorsium was the centre of this industry.

196 E.g. Schober Nos 120, 159, 170, 171 and 267.

197 E.g. *ibid.*, Nos 111, 113, 125, 152, 186, 209, 210 and 246.

198 Provisional reference may be made to *MS* 62–160, Abb. 20, 24, 29, 32, 35, 37, 39, 41 and 42.

199 A. Sz. Burger, *Bp. Rég.* xix, 1959, 9 ff.

200 Scarbantia: C. Praschniker, *JÖAI* xxx, 1937, 111 ff. Savaria: I. Paulovics, *Arch. Ért.* 1940, 34 ff.

201 *Pannonia*, 744; *MS* 243 ff.

202 *C.* 4156, 10908 (Isis), 10913 (Sphinx). From Scarbantia *C.* 4234, cf. V. Wessetzky, *Das Altertum* x, 1964, 154 ff.

203 A. Betz in *Carnuntum 1885–1935* (Vienna, 1935), 28 ff.

204 *C.* 4418; *RLiÖ* xii, 1914, 321 f.

205 *AIJ* pp. 133 ff. At Aquincum the Mithraeum dates to 161–3: *C.* 3479.

206 E.g. *C.* 3416, 4009, 10395.

CHAPTER 6

1 SHA, *Marcus* 12, 13.

2 *ILS* 8977; cf. *CIL* vi, 1497 = *ILS* 1094, etc. On other units, Saxer, *Vexillationen* 33; A. Radnóti, *Germania* xxxix, 1961, 109; T. Nagy, *Bp. Müeml.* ii, 94, n. 108, etc.

3 E. Ritterling, *PWRE* xii, 1544 f.; tile-stamp: *Bp. Rég.* xvii, 1956, 105.

4 The participation of legio VII Claudia is attested, see e.g., *JÖAIB* viii, 19, No. 58.

5 Cf. H. G. Pflaum, *Essai sur les procurateurs équestres* (Paris, 1951), 71 f.

6 SHA, *Marcus* 24, 5; 27, 10; cf. Dio lxxi, 33, 4².

7 F. Lorger, *JÖAI* xix–xx, 1919, 107 ff. For the history of II Italica, see G. Winkler, *Jahrbuch des oberösterreichischen Musealvereins* cxvi, 1971, 86 ff.

8 iii, 12; cf. ii, 17; iv, 12.

9 *De pallio*, ii, 7.

10 Tacitus, *Germ.* 42.

11 SHA, *Marcus* 14, 1; *aliis etiam gentibus, quae pulsae a superioribus barbaris fugerant, nisi reciperentur, bellum inferentibus.*

12 For some archaeological evidence of new peoples, see J. Tejral, *Pam. Arch.* lxi, 1970, 184 ff.

13 Barbarian chieftains wished to be incorporated into the empire: Appian, *preface*, 7.

14 Dio lxxi 3, 1a.

15 Cf. SHA, *Verus* 9, 9.

16 See note 11.

17 SHA, *Marcus* 22, 1: *gentes omnes . . . conspiraverant.*

18 Lucian, *pseudomant.* 48; Ammianus Marcellinus xxviii, 6, 1.

19 SHA, *Marcus* 14, 2; *Verus* 9, 7–10; Galen xiv, 649 f.; xix, 17 f.

20 SHA, *Marcus* 14, 6.

21 *ILS* 8977.

22 SHA, *Verus* 9, 9–10; *Marcus* 14, 3.

23 *CIL* iii, pp. 213 f.

24 This would be so if Claudius Fronto, legate of Moesia, was simultaneously *comes* of Verus. On the situation in Dacia, see L. Balla, *Acta Classica Univ. Scient. de L. Kossuth nominatae Debreceniensis* vii, 1971, 73 ff.

25   *CIL* vi, 1377 = *ILS* 1098.

26   A. Birley, *Septimius Severus* (London, 1971), 113.

27   Dobó, *Verwaltung*, 64 ff.

28   *Spomenik* lxxv, 2.

29   Especially *C.* 6302 = *ILS* 2606; cf. *MS* 16, n. 50.

30   Dio lxxi, 3, 5.

31   Dio lxxi, 3, 1²; Eutropius viii, 13, 1; Eusebius, *chron.* 207a (Helm).

32   *CIL* vi, 8878 = *ILS* 1685; Philostratus, *vit. soph.* ii, 1, 26 ff. A moving document is dated 18 March 175—the will of a young Italian who died at Sirmium, probably of the plague still raging there, after all his servants had died save one slave named Aprilis: *CIL* x, 7457.

33   Dio lxxi, 8–10; Eusebius, *chron.* 206i (Helm); cf. the forged letter of Marcus in R. Merkelbach, *Acta Ant.* xvi, 1968, 339 ff.

34   Dio lxxi, 11, 1–6. Furtius: Dio lxxi, 13, 3.

35   *Ibid.*, lxxi, 12, 1–2.

36   *Ibid.*, lxxi, 13, 3–4.

37   *Ibid.*, lxxi, 15.

38   *Libyca* iii (1955), 145 = *Ann. Ép.* 1956, 124.

39   Cf. the affair of Tarruntenus Paternus, Dio lxxi, 12, 3.

40   Dio lxxi, 12, 3.

41   *Ibid.*, lxxi, 12, 3, on their defeat. *Cives Cotini* from round Mursa and Cibalae served in the praetorian guard under the Severi, see *CIL* vi, 32542.

42   *Ibid.*, lxxi, 21.

43   *Ibid.*, lxxi, 7.

44   *Ibid.*, lxxi, 15.

45   lxxi, 13, 1–2.

46   lxxi, 14, 1.

47   lxxi, 16, 1.

48   lxxi, 16, 2.

49   lxxi, 17.

50   *CIL* vi, 1599 = *ILS* 1326; cf. SHA, *Marcus* 13, 5.

51   Dio lxxi, 33; SHA, *Commodus* 12, 6.

52   Dio lxxi, 18; 19, 2.

53   *Ibid.*, lxxi, 20; *CIL* viii, 619 = *ILS* 2747; *C.* 13439 = *ILS* 9120; *Ann. Ép.* 1956, 124.

54   lxxi, 20.

55   lxxii, 3, 2.

56   Dio lxxii, 15, 3; Eutropius viii, 15, 1; Aurelius Victor, *Caes.* 17, *epit.* 17; SHA, *Commodus* 11, 8; 12, 7.

57   *Ann. Ép.* 1956, 124.

58   Dio lxxii, 2.

59   *Germania* i (1917), 132 ff.

60   For a new summary of the material, see D. Gabler, *Bayerische Vorgeschichtsblätter* xxxiii, 1968, 100 ff.; *Arch. Ért.* xcv (1968), 211 ff. See also J. Tejral, *Arch. Rozhl.* xxii, 1970, 389 ff.

61   Dio lxxi, 20, 2; cf. 33, 4².

62   SHA, *Marcus* 24, 5; 27, 10; Herodian i, 5, 6.

63 Dio lxxii, 2, 4.

64 *Ibid.*, lxxi, 11, 2; 13, 4; 16, 2.

65 *Pannonia*, 562; J. Fitz, *Acta Arch.* xiv, 1962, 76; T. Nagy, *Bp. Müeml.* ii, 44 ff.; A. Neumann, *PWRE* ixA, 78, etc.

66 L. Barkóczi, *Intercisa* ii, 519; A. Mócsy, *Arch. Ért.* lxxxii, 1955, 65 ff.; *Eirene* iv, 1965, 143, n. 147; T. Nagy, *Bp. Müeml.* ii, 101, n. 287; J. Fitz, *Acta Arch.* xiv, 1962, 76.

67 *CIL* viii, 619, 14605, 16553, 25740, 25894, 27512; iii, 3680, 4472, 10419, 10515, 13372; J. Fitz, *Alba Regia* vi–vii, 1966, 208; *Ann. Ép.* 1956, 124.

68 *C.* 3324, 3444, 3542, 3545, 3675; *Arch. Ért.* lxxxii, 1955, 62 f.

69 The extensive information about this unit is given in *Intercisa* i–ii; for the arrival of the cohort, see J. Fitz, *Acta Ant.* xi, 1963, 277.

70 *C.* 14542, 6302 = 8162 = *ILS* 2606; cf. *MS* 16, n. 50.

71 *C.* 14217[6], 14537, 14541, 14545; *JÖAIB* xii, 189, No. 59; *Starinar* iv/5–6, 358.

72 N. Vulić, *Jahrb. DAI* xxvii, 1912; *Archäol. Anzeiger*, 549 ff.; xxviii, 1913, 339 ff.; xxix, 1914, 412.

73 *C.* 14537.

74 SHA, *Marcus* 21, 7.

75 *MS*, 110 ff.; cf. 194 ff.

76 *C.* 8251; *Spomenik* lxxi, 253; cf. *MS* 195.

77 *C.* 3385; *Intercisa* i, Nos 297–307.

78 SHA, *Commodus* 6, 1; cf. also Dio lxxii, 8, 1.

79 Herodian i, 9, 1.

80 SHA, *Commodus* 13, 5. For the disgrace of the governor Cornelius Felix Plotianus, see J. Fitz, *PWRE* Suppl. ix, s.v. Cornelius No. 284.

81 SHA, *Commodus* 12, 8; *CIL* v, 2155 = *ILS* 1574.

82 Herodian ii, 9, 12; SHA, *Severus* 5, 3.

83 An inscription (*Arch. Ért.* lxxviii, 1951, 135) establishes the presence of legio IIII Flavia at Aquincum in the governorship of Q. Caecilius Rufinus Crepereianus (Dobó, *Verwaltung*, 78) which can be dated *c.* 202–6, since legio II Adiutrix was definitely back in Pannonia by 207; M. Mirković, *ZFF* vii (i), 1963, 113 ff. On the absence of this legion under Caracalla, see M. Mirković, *ŽA* xi, 1962, 319 ff.

84 *C.* 3660; *Intercisa* i, No. 326; cf. G. Alföldy, *Arch. Ért.* lxxxviii, 1961, 29 f.

85 *Pannonia*, 587 f.

86 Dio lxxvii, 20, 3–4.

87 Herodian iv, 7, 3–5; 8, 1.

88 Dio lxxvii, 20, 3–4.

89 *C.* 10505; 14349[5]; *Arch. Ért.* 1944–5, 178.

90 See note 41.

91 See, for example, Herodian vi, 7.

92 *P. Dura* 54, col. ii, 3.

93 E. Ritterling, *PWRE* xii, 1686. There may be a connection between the initial reluctance of X Gemina and the troubles in Noricum, cf. *CIL* ii, 4114 = *ILS* 1140.

94 *Ann. Ép.* 1941, 166; cf. A. Betz, in *Corolla memoriae E. Swoboda dedicata* (Cologne–Graz, 1966), 39 ff.

95 Herodian ii, 13; cf. Dio lxxv, 2; cf. A. Passerini, *Le coorti pretorie* (Rome, 1939), 174 ff.;

Pannonia, 646; M. Pavan, *Athenaeum* N.S., xl, 1962, 85 ff.; Á. Dobó, 'Inscriptiones ad res Pannonicas pertinentes', *Diss. Pann.* i, 1 (Budapest, 1940), No. 1–62, 144a–5.

96   i, 9; ii, 8–11, 13–15, etc.

97   Appian, *Illyric.* 6.

98   Dio lxxx, 4–5.

99   *Princ. hist.* 13.

100   Herodian ii, 9; cf. Dio xlix, 36, 2.

101   E.g. Dio lxxv, 2, 6.

102   SHA, *Severus* 10, 3.

103   *CIL* viii, 7062 = *ILS* 1143. The young Caracalla was entrusted to the governor of Pannonia, Fabius Cilo, who was an intimate friend of the Dynasty; cf. Dio lxxvii, 4, 2.

104   Herodian iii, 1, 10; cf. J. Fitz, *Acta Arch.* xi, 1959, 237 ff.; F. Papazoglu, *Bulletin de Correspondance Hellénique* lxxxv, 1961, 162 ff.

105   Cf. G. Alföldy, *Acta Ant.* vi, 1958, 192. The Pannonians were the most loyal of the supporters of Maximinus Thrax, according to Herodian vii, 8; viii, 6.

106   Herodian vi, 7.

107   Both hoards in the hinterland of the *ripa Sarmatica*: S. Soproni, *Acta Arch.* xvii, 1965, 275 ff.

108   Herodian vii, 2, 9; SHA, *Maximinus Thrax* 13, 3.

109   Its importance is already emphasized by Herodian vii, 2, 9.

110   SHA, *Aurelianus* 3, 1; cf. under Carus, SHA, *Carus* 4, 3.

111   E. Jónas, *OMRTÉ* i, 1923–6, 137 ff.; A. Radnóti, *Num. Közl.* xliv-xlv, 1945–6, 6 ff; K. B. Sey, *Num. Közl.* lxiv-lxv, 1965–6, 9 ff.

112   Zosimus 1, 21, 2.

113   For these senior commanders, see J. Fitz, *Stud. Mil.* 113 ff.

114   Zosimus i, 21, 2–3.

115   *CIL* ii, 2220; viii, 1430.

116   Listed by K. B. Sey, *Arch. Ért.* xcviii, 1971, 199; cf. the map, *ibid.*, 196, fig. 6.

117   Eutropius ix, 8, 2; SHA, *trig. tyr.* 10, 1; *Panegyrici Latini* viii (Baehrens), 10, 2, etc.

118   For example, *Starinar* iii/6, 32 ff.; *IBAI* xxii, 360.

119   *MS* 257 f.

120   SHA, *trig. tyr.* 9; *Chron. Min.* i, 521; Aurelius Victor, *Caes.* 33, 2.

121   Zonaras xii, 24; Eutropius ix, 8, 1.

122   SHA, *trig. tyr.* 10; Aurelius Victor, *Caes.* 33, 2; *epit.* 32, 3.

123   R. Göbl, *Der römische Schatzfund von Apetlon* (Eisenstadt, 1954); cf. *Pannonia*, 569.

124   Aurelius Victor, *Caes.* 33, 1; *epit.* 33, 1.

125   *C.* 3228; cf. p. 2382[182].

126   J. Šašel, 'Bellum Serdicense', *Situla* iv (Ljubljana, 1961).

127   It became virtually a commonplace among late Roman historians that the candidate was found to become emperor against his will and only after threats to his life.

128   Most recently J. Fitz in *Mélanges Carcopino* (Paris, 1966), 353 ff. There is no explanation for the absence of some legiones from the series of coin-reverses, VII P(ia) and VII F(idelis).

129   Most recently E. Pašalić, *Stud. Mil.* 127 ff.; cf. *Pannonia*, 662 f., 666.

130   M. Mirković, *RGD*, 66 f.

131   Jordanes, *Get.* xvii; cf. I. Bóna, in *Orosháza Története*, i (Orosháza, 1965), 114.

132 SHA, *trig. tyr.* 9; cf. J. Harmatta, *Studies in the History of the Sarmatians* (Budapest, 1950), 60 f.

133 *AIJ*, pp. 144 ff.

134 Cf. C. Daicoviciu, *La Transylvanie dans l'Antiquité* (Cluj, 1945), 184 f.; cf. C. Daicoviciu, *Dacica* (Cluj, 1970), 378, and D. Tudor, *Historia* xiv, 1965, 369 ff.; *Historica* i (Bucharest, 1970), 67 ff.

135 SHA, *Gallienus* 13; *Claudius* 6, 11; Zosimus i, 43, 45, etc.; cf. A. Alföldi, *CAH* xii, 149 f., 721 ff.

136 Claudius: SHA, *Claudius* 13, 2. Aurelianus: SHA, *Aurelianus* 3, 1–2; Eutropius ix, 13, 1. Heraclianus has a name whose Latin variant appears frequently in Upper Moesia; see A. Mócsy, *Godišnjak* viii (Centar za Balkanološka Ispitivanja vi, 1970), 160 f.; for the cult of Hercules in Moesia Superior, see *MS* 245; and now a votive stone to Omphale: *Starinar* iv/19, 1969, 225.

137 SHA, *Aurelianus* 3, 1.

138 Aurelius Victor, *Caes.* 39, 26; J. Straub, in *Vom Kaiserideal der Spätantike* (Berlin, 1938), 209, explains the remarkable combination of *satis* with a superlative by saying that *optimus*, the normal adjective for good emperors after Trajan, had lost its superlative character. A more probable explanation, however, for which I thank Professor I. Borzsák, is that *satis* appears frequently in later Latin with the meaning of 'as possible'. Cf. also Aurelius Victor, *satis commodans*, 40, 9; *satis clarus*, 32, 2 (the comparable instance occurs in *epit.* 32, 1 with *splendidissimus*). *Satis optimus* comes close to the expression *necessarius magis quam bonus* in SHA, *Aurelianus* 37, 1.

139 SHA, *Aurelianus* 18, 2; 30, 5; Zosimus i, 48–9; Dexippus, frg. 7; cf. A. Alföldi, *Serta Kazaroviana* i (Sofia, 1950), 21 ff.

140 For example the bath-house of the Aquincum legionary fortress was rebuilt (*C.* 3525 = 10492), and the third Mithraeum at Poetovio was built (note 133 above), etc.

141 Ammianus Marcellinus xxxi, 5, 17; cf. SHA, *Aurelianus* 22, 1–2.

142 *Panegyrici Latini* xi (Baehrens), 17; Jordanes, *Get.* 116–20.

143 Eutropius ix, 15, 1.

CHAPTER 7

1 The reforms of Commodus have been studied by J. Fitz in several articles; see especially *Klio* xxxix, 1961, 199 ff.

2 P. Petrović, *Starinar* iv/18, 1967, 57 ff.

3 *CIL* vi, 2386.

4 *Spomenik* lxxi, 248; lxxv, 168.

5 *MS* 35, 43, 90 ff.

6 J. Deininger, *Die Provinziallandtage der römischen Kaiserzeit* (Munich, 1965), 119; cf. *MS* 92.

7 *MS* 42 ff.

8 *C.* 1674–6, 8244, 8249, 12672 = 14561; *Spomenik* lxxvii, 37.

9 *Spomenik* lxxi, 594.

10 *MS* 142 f.; 159.

11 Kanitz, *RS* 69 f.; B. Saria, *BRGK* xvi, 1925–6, 93; D. Piletić, *Arh. Pregl.* iv, 1962, 176 f.

12   *C.* 7591; *CIL* vi, 2388.

13   *JÖAIB* xii, 1909, 23, No. 30.

14   *Starinar* iv/1, 1950, 143 ff.; iv/2, 1951, 115.

15   *C.* 8141.

16   *MS* 144 f.

17   *MS* 134 ff.

18   *C.* 6302 = 8162 = *ILS* 2606.

19   *MS* 134 ff.; 226 ff.

20   *Spomenik* lxxi, 584.

21   *C.* 14217[7]; *JÖAIB* vi, 60, No. 99; xiii, 223, No. 39; xv, 236, No. 39; *ZNMB* iv, 1964, 127; cf. *JÖAIB* xii, 188, No. 58, etc.

22   *C.* 14541[1]; cf. *MS* 140, n. 56.

23   Both were *coloniae Septimiae*: see *C.* 14347, 14359[3]; *RLiÖ* xvi, 1926, 117. On the date 194, see *Pannonia*, 599: *contra* A. Betz, *Carn.-Jb.* 1960 (1962), 31, who believes that the colony at Carnuntum was probably founded in 198. But whereas Aquincum certainly existed as a colony in 198, its *lustrum* years (i.e. the years in which the *quinquennales* held office) fell in the years 4 and 9 of any decade (e.g. A.D. 214, 259) as *C.* 10439, 10440 show us. Thus it cannot have been founded in 198, and the date must be pushed back to 194—the first available date in the reign of Septimius Severus.

24   *MS* 31.

25   *C.* 14509; cf. *MS* 145, n. 2.

26   A. Mócsy, in *Stud. Mil.* 211 ff.

27   Cf. A. Mócsy, *Germania* xliv, 1966, 312 ff.

28   Birth in the *canabae* could be included on inscriptions as a juridically meaningless indication of the birth place, e.g. *CIL* vi, 32783.

29   *C.* 3973, 3976, 4193 and 10836.

30   *C.* 4152; cf. *Pannonia*, 602.

31   See, e.g., *ILS* 6090 = *Fontes iuris Romani anteiustiniani* i[2], No. 92: *iisdem maxime pollicentibus, quod . . . decurionum sufficiens futura sit copia. . . .*

32   The only case is that of the civitas Eraviscorum on an altar of the time of Philip (see A. Alföldi, *Arch. Ért.* lii, 1939, 266). But this *civitas peregrina* had already been in practice dissolved under Hadrian and existed only in name after 124.

33   Cf. A. Mócsy, *Acta Ant.* x, 1962, 367 ff.

34   *C.* 11007.

35   Barkóczi, *Brigetio*, No. 226.

36   *C.* 4557.

37   L. Barkóczi, *Acta Arch.* iii, 1953, 201 f.

38   *Itinerarium Burdig.* 561, 10.

39   Egger, *RAFC* i, 57 ff.

40   *CIL* vi, 2494a; 3241.

41   *C.* 3974.

42   *C.* 10243.

43   *Intercisa* i, No. 124.

44   Cf. *Bevölk.* 74.

45   *MS* 33 f.; M. Mirković, *RGD* 37 ff.

46 Complete list of finds: D. Garašanin, *Godišnjak Muzeja Grada Beograda* i, 1954, 60 ff.; cf. Mirković, *RGD* 48 f.

47 *C.* 8297; cf. *MS* 34 f.; E. Čerškov, *Municipium DD* (Priština–Belgrade, 1970).

48 *C.* 14610.

49 See Chapter 5, note 71 (p. 379).

50 Procopius, *aedif.* iv, 4, 3.

51 *Ibid.*

52 *MS* 38 f., 110 ff.

53 M. Dimić, *Starinar* viii, 1891, 21 ff.; J. Mirković, *Starinar* v, 1888, 72 ff.; *MS* 111.

54 *MS* 158 f.; Mirković, *RGD* 68 f.

55 *Itinerarium Burdig.* 564, 5.

56 *Codex Theod.* i, 32, 5 = *Codex Iust.* xi, 7, 4.

57 E. Čerškov, *op. cit.* (note 47), 64, No. 11.

58 *MS* 141 f.

59 *MS* 117 ff.; Mirković, *RGD* 89 ff.

60 *C.* 14038 (Cibalae); 10197 (Bassiana). In *CIL* vi, 32542b, 9. 17 Cibalae belongs to the pseudo-*tribus* Aurelia, indicating a promotion, but one which I consider improbable under Marcus.

61 Mirković, *RGD* 65 f.

62 *C.* 4335.

63 Cf. A. Mócsy, *Acta Ant.* x, 1962, 367 ff.

64 *C.* 8089, 12646, 8263+14580; cf. *MS* 107.

65 *C.* 10495, 10496; *Ann. Ép.* 1968, 423.

66 *Intercisa* i-ii (Budapest, 1954–7). See also J. Fitz's recent book *Les Syriens à Intercisa* (Brussels, 1972).

67 *C.* 14359².

68 *C.* 10597.

69 *MS* 191 f.; cf. V. Velkov, in *Études Historiques du xiie Congrès International des Sciences Historiques à Vienne, 1965* (Sofia, 1965), 25 ff.

70 For example *C.* 10920, 11076; Barkóczi, *Brigetio,* No. 95; *ŽA* x, 1960, 193; *JÖAIB* xii, 1909, 158, No. 22, etc.

71 *C.* 2006; Z. Kádár and L. Balla, *Ant. Tan.* vi, 1959, 112 ff.; L. Balla, *Acta Arch.* xv, 1963, 225 ff.; I. Tóth, *Arch. Ért.* xcviii, 1971, 80 ff.

72 D. Pinterović, *Osj. Zb.* ix-x, 1965, 72 f.

73 *Intercisa* i, No. 329.

74 S. Scheiber, *Corpus Inscriptionum Hungariae Iudaicarum* (Budapest, 1960).

75 L. Nagy, *Arch. Ért.* lii, 1939, 116 ff.

76 *C.* 4281, 10481, 10533, 10570; Barkóczi, *Brigetio,* Nos 93, 102, 209.

77 E.g. *C.* 3490, 11076, 10982, 15167; cf. *PWRE* Suppl. xi, *s.v.* Toleses; *Bp. Rég.* xiv, 1945, 541.

78 *Intercisa* i, Nos 38, 138; *C.* 4220.

79 E.g. *C.* 3490; Barkóczi, *Brigetio* Nos 93, 113; Kádár and Balla, *op. cit.* (note 71).

80 For example *C.* 11076.

81 For example *C.* 10515, 14349.

82 G. Alföldy, in *Epigraphische Studien* iv (Cologne–Graz, 1967), 26 ff.

83 A. Alföldi, *Zu den Schicksalen Siebenbürgens im Altertum* (Budapest, 1944), 39 ff.
84 Cf. the catalogue in *Bevölk.* and L. Barkóczi, *Acta Arch.* xvi, 1964; also the map in *MS* 200, Abb. 45.
85 *MS* 67.
86 For Pannonia, see D. Gabler, *Arch. Ért.* xciii, 1966, 20 ff.
87 Cf. *MS* 133, 154, 174.
88 *CIL* xiii, 1766; SHA, *Severus* 10, 7; cf. G. Alföldy, *Acta Ant.* viii, 1960, 145 ff.
89 For example *C.* 3349, 3628, 3659, 4146, 4282, 10347, 15188[4], etc.
90 For the evidence of inscriptions, see Dobó, *Verwaltung* 159 ff. For the *beneficiarii* in Upper Moesia, see *MS* 22, Abb. 7 and 24.
91 *C.* 4146, 4268, 4375, 10956, etc.
92 *C.* 6322, 12666–8; *Spomenik* lxxi, 594, 595.
93 *C.* 8258 = 1689; 1684 with *Spomenik* lxxi, 243; *C.* 14555, 14556.
94 *MS* 99 f., 143.
95 M. Vasić, *Jahrb. DAI* xx, 1905; *Archäol. Anzeiger* 102 ff.
96 See Gabler, *op. cit.* (note 86).
97 Cf. the papers by H. Kenner, Á. Kiss and D. Mano-Zisi in *La mosaïque gréco-romaine* (Paris, 1965), 89 ff., 290 ff., 297 ff.
98 L. Nagy, *Römische Mitteilungen* xl, 1925, 51 ff.; *Bp. Rég.* xiii, 1943, 79 ff.
99 I. Wellner, *Acta Arch.* xxi, 1969, 235 ff.
100 See, for example, L. Nagy, *Römische Mitteilungen* xli, 1926, 79 ff.; K. Sz. Póczy, *Bp. Rég.* xvi, 1955, 49 ff.; xviii, 1958, 103 ff.; T. Nagy, *Bp. Rég.* xviii, 1958, 149 ff.; H. Brandenstein, *Carn.-Jb.* 1961–2 (1963), 5 ff., etc.
101 *MS* 152.
102 B. Filow, *IBAD* i, 1910, 8 ff.; cf. *MS* 104.
103 See, for example, Barkóczi, *Brigetio* 40 ff.; G. Erdélyi, *Intercisa* i, 199 f.; *Pannonia*, 720; A. Cermanović-Kuzmanović, *Arch. Iugosl.* vi, 1965, 89 ff.
104 See Chapter 5, note 177 (p. 383).
105 D. Stričević, *Starinar* iv/7–8, 1958, 411 ff.
106 A. Mócsy, *Arch. Ért.* lxxxii, 1955, 62 ff.; G. Alföldy, *Arch. Ért.* xc, 1963, 302; S. Soproni, *Arch. Ért.* xciii, 1966, 296; T. Nagy, *Bp. Müeml.* ii, 37, 50 f.; E. B. Vágó, *Arch. Ért.* xci, 1964, 255.
107 See, e.g., the map in K. Sági and K. Bakay, *Épités és Épitészettudomány* ii, 1971, 406.
108 T. Nagy, *Arch. Ért.* lxxxii, 1955, 97; *Bp. Rég.* xvii, 1956, 9; L. Nagy, *Mumienbegräbnisse* 3 f.; K. Sz. Póczy, *Bp. Rég.* xxii, 1971, 98 ff. A large house, probably a villa, has recently been excavated by A. Sz. Burger at Komló north of Sopianae, dating from Severan times, *Janus Pannonius Muzeum Évkönyve*, 1967, 61 ff. (but see Chapter 9, note 18 (p. 399)).
109 S. Garády, *Arch. Ért.* xlix, 1936, 88 ff.
110 *C.* 10521, 10536.
111 *C.* 11076, 12659, 14217[4]; Barkóczi, *Brigetio*, No. 114.
112 See, e.g., *C.* 4311, 10962, 11076; cf. Barkóczi, *Brigetio*, Nos 91–127. See on the other hand the sarcophagi in Upper Moesia which belonged to the highest social order: *MS* 104, 130, 152.
113 For example *JÖAIB* xiii, 1910, 203, No. 13; cf. *MS* 168.
114 *C.* 8169, 8238 = *JÖAIB* vi, 38, No. 44; 8240 = *Spomenik* lxxi, 326; *Spomenik* xcviii, 450; *JÖAIB* vi, 28, Nos 33–4.

115 Fragment Vat. 220.

116 *MS* 117 ff., and notes 93–4 above.

117 Aurelius Victor, *Caes.* 39, 26; cf. Chapter 6, note 138 (p. 388).

118 See, e.g., the excavation by Barkóczi, *Acta Arch.* xvi, 1964, 273 f., 278; *C.* 3464, 3478, 3515, 3539, 3561, 3571, 3572, 4285, 4286, 4299, 4327, 10389, 11082 and 14349[3]; Barkóczi, *Brigetio*, Nos 141, 208, 216, etc.

119 Á. Kiss, *Acta Arch.* xi, 1959, 159 ff.

120 Aurelius Victor, *Caes.* 29, 1; *epit.* 29, 1; Eutropius ix, 4.

121 SHA, *Probus* 21, 2; Aurelius Victor, *Caes.* 37, 4; cf. *epit.* 37, 1.

122 At present I can only refer to the ample unpublished material, e.g., in the Museum of Sremska Mitrovica.

123 V. Velkov, *Eirene* v, 1966, 174.

124 Philostratus, *vit. soph.* ii, 560.

125 T. Szentléleky, in *Neue Beiträge zur Geschichte der Alten Welt* ii (Berlin, 1965), 381 ff.; *Das Isis-Heiligtum von Szombathely* (Szombathely, 1965); cf. A. Mócsy, *Acta Arch.* xxi, 1969, 370, n. 331.

126 *C.* 3925, 4495; *CIL* vii, 341; xvi, 179–80.

127 Dio lxxviii, 13, 4. He was descended from a Celtic family (*Triccus > Triccianus*) whose citizenship went back to Hadrian or to Antoninus Pius (*Aelius*).

128 Ti. Claudius Marinus Pacatianus, Q. Messius Traianus Decius, M. Aemilius Aemilianus, P. C(. . .) Regalianus, L. Domitius Aurelianus, C. Valerius Diocletianus, etc. On the other hand Claudius Gothicus, Probus, Maximianus and others were *Marci Aurelii*.

129 D. Gabler, *Arch. Ért.* xci, 1964, 94 ff.; *Bayerische Vorgeschichtsblätter* xxxi, 1966, 123 ff.

130 E. T. Szőnyi, *Acta Arch.* xxv, 1973, 87 ff.

131 L. Barkóczi, *Fol. Arch.* xviii, 1966, 67 ff.

132 I. Sellye, 'Les bronzes émaillés de la Pannonie', *Diss. Pann.* ii, 8 (Budapest, 1939).

133 N. Láng, *Bp. Rég.* xii, 1937, 5 ff.

134 *Ibid.*, x, 1923, 3 ff.; Z. Kádár, *Fol. Arch.* xiv, 1962, 41 ff.

135 Radnóti, *Bronzegefässe*, 82 ff., 105 ff., 137 ff.; *Intercisa* ii, 190 f., 212 f.

136 E.g. B. Kuzsinszky, *Bp. Rég.* x, 1923, 60 f.; E. Swoboda, *Carn.-Jb.* 1955 (1956), 15 ff. There are now two further examples of the glass cameo with the inscription MNHMONEYE (Swoboda, *op. cit.*), namely D. Tudor, *Romula* (Bucharest, 1968), Abb. 16, and an unpublished example found at Brigetio, now in the Hungarian National Museum.

137 For example E. B. Thomas in *Studien zur Geschichte und Philosophie des Altertums*, edited by J. Harmatta (Budapest, 1968), 337 ff.; L. Nagy, *Mumienbegräbnisse*, 18 f., etc.

138 B. Filow, *IBAD* i, 1910, 8 ff.; Ferri, *ARD* 366 ff.; *Archeologija* (Sofia), vi (1964), i, 24 ff.; *MS* 104; *Spomenik* xcviii, 317.

139 See note 130.

140 For example I. Kovrig, 'Die Haupttypen der kaiserzeitlichen Fibeln in Pannonien', *Diss. Pann.*, ii, 4 (Budapest, 1937), Taf. xxvi, 4; xxvii, 1.

141 *Pannonia*, 688; J. Fromols, *Jahrbuch des Römisch-Germanischen Zentralmuseums* v, 1958, 259 ff.; J. M. C. Toynbee, *Art in Britain under the Romans* (Oxford, 1964), 167 f.; cf. also *Acta Arch.* xxi, 1969, 358 f.

142 Pliny, *Nat. Hist.* xvi, 7, 43–4; Lucan, *Pharsalia* vi, 220; Martial, xiii, 69; Nemesianus, *cynegetica* 126; *CIL* vi, 10005 = *ILS* 5285; cf. also *Acta Arch.* xxi, 1969, 359.

143 *Expositio tot. mundi*, 51.

144 E.g. *Arch. Ért.* lxxxii, 1955, 60, and K. Sz. Póczy, *Intercisa* ii, 95.

145 Except of course temporary deliveries, such as the corn which the Marcomanni had to produce under the peace-treaty of 180 (Dio lxxii, 2, 2). This corn probably went to the army in Pannonia.

146 See note 89.

147 *CIL* vi, 32542.

148 Compiled by A. Sz. Burger, 'Áldozati jelenet Pannonia köemlékein', *Régészeti Füzetek*, ii, 5 (Budapest, 1959); cf. *Pannonia*, 725. The scene appears sporadically also in Upper Moesia, e.g. *Starinar* iv/1, 1950, 205, Fig. 53.

149 *C.* 14507; cf. *MS* 172 ff.

150 *CIL* vi, 3205; cf. *MS* 249 f.

151 SHA, *Severus* 10, 7.

152 K. Wigand, *JÖAIB* xviii, 1915, 189 ff.; *AIJ*, pp. 154 ff.; *Pannonia*, 744.

153 *Pannonia*, 741–3.

154 Cf. G. E. F. Chilver, *Cisalpine Gaul* (Oxford, 1941), 193.

155 I. Szántó, *Arch. Ért.* lxxviii, 1951, 43.

156 Ferri, *ARD* 165, Fig. 169; L. Balla, Z. Kádár, A. Mócsy and others, *Die römischen Steindenkmäler von Savaria* (Budapest, 1971), Nos 81–4.

157 A. Dobrovits, *Fol. Arch.* iii-iv, 1942, 304.

158 For example *C.* 4410.

159 A. D. Nock, *CAH* xii, 415, n.1.

160 Magla, Messor (if not referring to harvest) and Vidasus.

161 *AIJ*, 230 ff.; cf. M. Abramić, *Festschrift R. Egger* (Klagenfurt, 1952), 317 ff.

162 E.g. *Bp. Rég.* iii, 1891, 69, 71; v, 1897, 126; ix, 1906, 47 ff.; xi, 1932, 381; xii, 1932, 139; Kubitschek, *Römerfunde*, 104, etc.; cf. H. Kenner, *JÖAI* xliii, 1956-8, 91.

163 *C.* 3498.

164 *C.* 3491.

165 *C.* 4425–32, 13469–73.

166 E. M. Štaerman, *Vjestnik drevnej istorii*, 1946, 3, 264.

167 *C.* 13368.

168 See the map in Zs. Bánki, *Alba Regia* vi-vii, 1967, 165 ff.

169 Strabo vii, 5, 10; Pliny, *Nat. Hist.* iii, 148; Appian, *Illyric.* 14; Aurelius Victor, *Caes.* 40, 9; *Gromatici Veteres* (ed. Lachmann), p. 205.

170 E.g. *C.* 10414; *Intercisa* i, No. 369; *JÖAI* xlii, 1955, 109 f.

171 *C.* 3617.

172 *C.* 3474.

173 *Pannonia*, 740, 745.

174 *MS* 243 ff.

175 E.g. *C.* 1677, 1678, 8171, 14563, 14565; *Spomenik* lxxi, 251; xcviii, 222; cf. *MS* 244 f.

176 *Spomenik* lxxi, 210 f., 325, xcviii, 223.

177 *JÖAIB* vi, 41, No. 48; xii, 172, No. 35; *Spomenik* lxxi, 191; xcviii, 171-2.

178 *C.* 6303, 8148; *Spomenik* lxxi, 3, 111, 322; lxxvii, 36; xcviii, 10, 173. *Godišnjak Muzeja Grada Beograda* ii, 36; *JÖAIB* xv, 235, No. 38; 236, No. 40. A god with the name Atta or Atto may be connected with Paternus: see P. Petrović, *Vranjski Glasnik* v, 1969, 368.

179 *C.* 8184.

180 *JÖAIB* vi, 40, No. 46; *Spomenik* lxxi, 182, 278, 323; lxxv, 167.

181 D. Tudor, *Corpus monumentorum equitum Danuvinorum* (Leiden, 1969).

182 See, e.g., note 141.

183 See Chapter 5, note 205 (p. 384); also *Spomenik* lxxvii, 28–30 and 64; lxxv, 155.

184 M. Vermaseren, *Corpus Inscriptionum et Monumentorum Religionis Mithraicae*, ii (Hagae, 1960), 170 ff.; cf. also T. Nagy, *Arch. Ért.* lxxxix, 1962, 281 ff.

185 Z. Kádár, *Die kleinasiatischen und syrischen Kulte in Ungarn* (Leiden, 1962), 2 f., shows that the oriental cults were not widely distributed among the garrisons of Syrian units.

186 N. Láng, *Arch. Ért.* 1943, 64 ff.; cf. Z. Kádár, *op. cit.* (note 185), 39 f.

187 *C.* 3343; cf. R. Egger, in *Omagiu lui C. Daicoviciu* (Bucharest, 1960), 167 ff.; for a different view, see *Pannonia*, 595.

188 J. Dell, *Archäologisch-Epigraphische Mittheilungen aus Österreich-Ungarn* xvi, 1893, 176 ff.; N. Láng, *Laur. Aqu.* ii, 165 ff.

189 D. Vučković-Todorović, *Starinar* iv/15–16, 1966, 173 ff.

190 *Arh. Pregl.* v, 1963, 116 ff.

191 L. Vidman, *Sylloge Inscriptionum religionis Isiacae et Serapiacae*, *Religionsgeschichtliche Versuche und Vorarbeiten* xxviii (Berlin, 1969), Nos 652–74, 700–700a.

192 *Ibid.*, Nos 669, 670.

193 Cf. note 185.

194 For example, *CIL* xiii, 6646.

195 *AIJ* pp. 144 ff.

196 See *Pannonia*, 770 f.; *MS* 222 ff., 234 ff.; a list of the Greek inscriptions found in Austria is given by A. Betz, *Wiener Studien* lxxix, 1966, 593 ff.

197 *Itinerarium Burdig.* 566, 7.

198 *Spomenik* xcviii, 448, 449.

199 E.g. *Starinar* iv/7–8, 293.

200 SHA, *Aurelianus* 24, 3.

201 E.g. *CIL* vi, 2552, 2662, 32783.

202 G. Erdélyi, *Arch. Ért.* lxxvii, 1950, 72 ff.

203 *C.* 8098 = 6298, 12659; *Spomenik* lxxi, 311.

204 For bibliography, see *Pannonia*, 768 f.; recently G. Erdélyi, *Acta Arch.* xiii, 1961, 89 ff.; *Acta Ant.* xiv, 1966, 211 ff.; E. Diez, *Carn.-Jb.* 1963–4 (1965), 43 ff.; *Actes du viii*e *Congrès Internat. d'Archéologie Classique Paris 1963* (Paris, 1965), 209; Z. Kádár, *ibid.*, 381 ff.; A. Sz. Burger, *Fol. Arch.* xiii, 1961, 51 ff.

205 E.g. *Intercisa* i, Nos 150–2, 165.

206 Ferri, *ARD* 226, Fig. 265.

207 *C.* 15166; cf. 4490; *Pannonia*, 768; *Acta Arch.* xxi, 1969, 372. For a youth *omnibus studiis liberalibus eruditus* in Brigetio see *Komárom Megyei Múzeumok Közleményei* i, 1968, 81.

208 Gy. Diósdi, *Ant. Tan.* viii, 1961, 99 ff.; *MS* 228 and 252.

209 *C.* 10717, 10864.

210 On these, see *Pannonia*, 769 f.; but cf. J. Gy. Szilágyi, *Arch. Ért.* xc, 1963, 189 ff.; *MS* 232.

211 See the summary of the Vergilian passages in R. P. Hoogma, *Der Einfluss Vergils auf die Carmina Latina Epigraphica* (Amsterdam, 1959). It is evident from this that Vergilian echoes on the inscriptions of our provinces have been taken almost without exception from books i and

vi of the Aeneid. On Vergil's influence on teaching, see, e.g., B. A. Gwynn, *Roman Education from Cicero to Quintilian* (Oxford, 1926), 154 ff.

212 Fragment Vat. 220.

213 *MS* 198–236.

214 E.g. *Arch. Ért.* 1944–5, Taf. xci; *AIJ* 573; Schober, No. 206, etc.

215 E.g. the house with the Hercules-mosaic at Aquincum, I. Wellner, *Acta Arch.* xxi, 1969, 271.

216 Aurelius Victor, *Caes.* 37, 4; *epit.* 37, 4; SHA, *Probus* 18, 8; 21, 2; Eutropius ix, 17, 2; Eusebius, *chron.* 224a (Helm).

CHAPTER 8

1 And for a short time to Florian also: *Ann. Ép.* 1970, 495.

2 Aurelius Victor, *Caes.* 40, 7.

3 *C.* 10275 = *VHAD* ix, 1906–7, 112, No. 233.

4 Aurelius Victor, *Caes.* 39, 11; SHA, *Carinus* 18, 2; *Itinerarium Hieros.* 564, 9.

5 SHA, *Probus* 18, 2; cf. *Pannonia*, 567; T. Nagy, *Bp. Müeml.* ii, 56.

6 SHA, *Carinus* 8, 1; 9, 4; Eutropius ix, 18, 1.

7 On the medallion of Numerianus, see G. Elmer, *Der Münzsammler* viii, 1935, 17 f.; cf. *Pannonia*, 567.

8 Cf. Aurelius Victor, *Caes.* 39, 43: Diocletian as conqueror of the Marcomanni.

9 Cf. Aurelius Victor, *Caes.* 40, 9–10.

10 Eutropius, ix, 25, 2, speaks of *ingentes captivorum copiae*.

11 *Itinerarium Burdig.* 565, 8.

12 *Not. Or.* xxviii, 26; *Occ.* xl, 54.

13 *Gentiles Sarmatae: Not. Occ.* xlii, 46–70.

14 Th. Mommsen, *Gesammelte Schriften* ii, 267 ff.

15 Aurelius Victor, *Caes.* 39, 30 (Diocletian in Pannonia); W. Ensslin, *PWRE* viiA, 2438 f.

16 A. Alföldi, *Arch. Ért.* 1941, 51 ff.; R. Egger, *JÖAI*, xxxv, 1943, 27.

17 *Panegyrici Latini* viii (Baehrens), 5, 1.

18 *Codex Iust.* ix, 20, 10–11.

19 And perhaps in 303 on the journey from Rome to Nicomedia: Lactantius, *mort. pers.* 17, 4.

20 *Chron. Min.* i, p. 230.

21 Mostly with PROVIDENTIA or VIRTVS, but at the mint of Rome also with VICTORIAE SARMATICAE, *RIC* vi, p. 354, No. 43.

22 *Not. Occ.* xxxiii, 41; xxxiii, 48.

23 *C.* 10605a; cf. A. Alföldi, *Arch. Ért.* 1941, 49; R. Egger, *JÖAI*, xxxv, 1943, 21 ff.; W. Ensslin, *PWRE* viiA, 2440. On the find-spot of the inscription, see A. Mócsy in Limeskongress 1969.

24 A. Mócsy, *Fol. Arch.* x, 1958, 89 ff.; *Pannonia*, 642; *Osj. Zb.* xii, 1969, 71 ff.; Limeskongress 1969.

25 V. Balás, *Acta Arch.* xv, 1961, 310 ff.; S. Soproni, *Arch. Ért.* xcvi, 1969, 43 ff.

26 The Hungarian name for the earthwork is Ördögárok (Devil's Dyke) or Csörszarok (Ditch of Csörsz). Csörsz is a loan word from the Slav language and also means 'devil'. A date can be derived from this, since, when the Slavs arrived in the Hungarian Lowlands, the origin

of the earthwork was no longer known. This gives a *terminus ante quem* of the sixth or seventh century.

27 S. Soproni, *op. cit.* (note 25).
28 P. Patay, *Arch. Ért.* xcvii, 1970, 312.
29 *Chron. Min.* i, 230; see note 7.
30 Aurelius Victor, *Caes.* 40, 9–10.
31 Ammianus Marcellinus xxviii, 1, 5; cf. Eutropius ix, 25, 2.
32 Aurelius Victor, *Caes.* 40, 10; Ammianus Marcellinus xix, 11, 4.

33 From Aurelius Victor, *Caes.* 40, 10, one might conclude that the drainage was begun only shortly before the death of Galerius, and as the naming of the province Valeria is connected with it, the partition of Pannonia Inferior had to be dated *c.* 310. But the partition must be placed earlier than 303 according to *C.* 10981; cf. T. Nagy, *Bp. Müeml.*, ii, 106, n. 367.

34 E.g. the omission of Savia in *Itinerarium Burdig.*, and the mention of *tres Pannoniae* by Optatus of Milevis, *Corpus Scr. Eccl. Lat.*, xxvi, p. 33. These pieces of evidence suggest that Savia was not founded before the reign of Constantine the Great.

35 *Not. Occ.* xxxii, 28, 40, 45; T. Nagy, *Akten des iv Internat. Kongresses für griechische und lateinische Epigraphik, Wien, 1962* (Vienna, 1964), 274 ff.

36 *Not. Occ.* i, 51, 83, 87; cf. *Pannonia*, 611. The *praeses Valeriae* is missing in the *Notitia*; cf. A. H. M. Jones, *The Later Roman Empire* iii (Oxford, 1964), 351.

37 B. Saria, *PWRE* xxi, 1174.
38 *CIL* vi, 32937.
39 *Not. Or.* i, 77, 121, 122, 124; cf. *MS* 41 ff.
40 Eutropius ix, 25, 1.
41 *Excerpta Valesiana* 3.
42 For the date, see J. Moreau, *Carn.-Jb.* 1960 (1962), 7 ff.
43 *C.* 4413.
44 Aurelius Victor, *Caes.* 40, 8.
45 *Epit.* 41, 5; Eutropius x, 5; *Excerpta Valesiana*, 16–17; Zosimus ii, 19, 1.

46 Z. Kádár, *Fol. Arch.* xii, 1960, 132 ff.; *Arheologija* (Sofia) iii, 1961, 1, 47 ff. I prefer this context for some further hoards, e.g. the gold medallions and jewelry from Petrijanec: see J. Kovrig, 'Die Haupttypen der Kaiserzeitlichen Fibeln', *Diss. Pann.* ii, 4 (Budapest, 1937), Pl. xxxviii.

47 Seeck, *Regesten*, 162 ff.
48 Zosimus ii, 21; Optatianus Porphyr., *carm.* vi.
49 *Codex Theod.* i, 1, 1.
50 An edict from Bononia, dated probably to 6 July; *Codex Theod.* xi, 27, 2.
51 *Chron. Min.* i, 234; Eusebius, *chron.* 233c (Helm).
52 *Excerpta Valesiana* 32.
53 *Excerpta Valesiana* 32; Eusebius, *chron.* 233 f. (Helm); Eusebius, *vita Const.* 4, 6; *Chron. Min.* i, 234.
54 See note 51.
55 See note 53.
56 Ammianus Marcellinus xvii, 12, 19.
57 Seeck, *Regesten*, 182 f.
58 A. Alföldi, *Riv. Ital. di Numismatica* xxiv, 1921, 113 ff.

59 Ammianus Marcellinus xvii, 13, 18–19.
60 Jordanes, *Get.* 115, 161.
61 J. Korek and M. Párducz, *Arch. Ért.* 1946–8, 298 ff.; I. Bóna, in *Orosháza Története* i (Orosháza, 1965), 114 ff.
62 Ammianus Marcellinus xvii, 12, 19.
63 D. Tudor, *Oltenia Romana* (third ed., Bucharest, 1968), 251 ff.
64 Tokod: *Acta Arch.* xxi, 1969, 351, Abb. 3. Esztergom-Hideglelöskereszt: S. Soproni, *LRKN* 137, Abb. 4. Pilimarót: *ibid.* 136, Abb. 3. Visegrád-Sibrik: *ibid.* 134, Abb. 2. Slankamen: D. Dimitrijević, *Starinar* iv/7–8, 1956–7, 304; *Rad. vojv. muz.* v, 1956, 150 ff.
65 For example Budapest-Eskü tér: L. Nagy, *Az Eskü téri római erőd, Pest város öse* (Budapest, 1946). For Visegrád-Sibrik, see note 64.
66 Lébény-Barátföldpustzta: D. Gabler, *Arrabona* viii, 1966, 67 ff.; *Arch. Ért.* xcv, 1968, 131. Almásfüzitő: F. Fülep, *Arch. Ért.* lxxxvii, 1960, 236. Szentendre: T. Nagy, *Arch. Ért.* 1942, 262, 270. Nagytétény: F. Fülep and É. Cserey, *Nagytétény müemlékei* (Budapest, 1957). Intercisa: L. Barkóczi, *Intercisa* i, 32 f. Iža opposite Brigetio: B. Swoboda, *Slov. Arch.* x, 1962, 422 ff.
67 *C.* 15172; *Not. Occ.* xxxiii, 34.
68 Nagytétény (Campona), Intercisa and perhaps Szentendre (Ulcisia Castra = Castra Constantia).
69 See, for example, Tokod, Nagytétény, Intercisa.
70 E. Nowotny, *RLiÖ* xii, 1914.
71 Proved best at Tokod (unpublished), where all the buildings are against the fort-walls.
72 See, for example, Tokod, Pilismarót.
73 Tokod, Budapest-Eskü tér, Visegrád, Szentendre.
74 Čortanovci, see M. Manojlović, *Rad vojv. muz.* xi, 1962, 123 ff.
75 Stamps with names of places and stamps of legiones VII Cl. and IIII Fl. are limited to the *limes* of Moesia Prima, stamps with DRP to Dacia Ripensis.
76 D. Tudor, *SCIV* xi, 1960, 335 ff.
77 A. Alföldi, *Arch. Ért.* 1941, 53.
78 *ILS* 724.
79 Seeck, *Regesten*, 187 ff.
80 *Ibid.*, 186 f.
81 *Ibid.*, 188 f., 191, 195.
82 Aurelius Victor, *Caes.* 41, 26; Eutropius x, 10, 2; Eusebius, *chron.* 237c (Helm).
83 Zosimus ii, 44, 3; Julian, *or.* ii, 76d; Eusebius, *chron.* 238c (Helm), etc.; cf. Seeck, *Geschichte des Untergangs der antiken Welt* iv, 429 f.
84 This villa (Ammianus Marcellinus xxvi, 5, 1) has been partially excavated, see Chapter 9 (p. 302).
85 *Chron. Min.* i, 238; Eusebius, *chron.* 238e (Helm).
86 Cf. A. Mócsy, *Ant. Tan.* xiii, 1966, 242 ff.
87 Zosimus ii, 46–50.
88 Ammianus Marcellinus xvi, 10, 20; Zosimus iii, 1, 1.
89 A coin-hoard in the north-east of Pannonia may also be referred to it: K. B. Sey, *Fol. Arch.* xvi, 1964, 63 ff.
90 Ammianus Marcellinus xvii, 12, 16.
91 Zosimus iii, 2, 2; Julian, *ep. ad Athen.* 279d.

92 Seeck, *Regesten*, 204 ff.

93 xvi, 10; xvii, 12–13; xix, 11.

94 xvii, 13, 1.

95 Seeck, *Regesten*, 206 f.

96 Ammianus Marcellinus xxi, 9–10; Zosimus iii, 10–11.

97 Ammianus Marcellinus xxi, 12, 21–3; xxii, 2, 2.

98 For example *C.* 10648b = *ILCV*, 11.

99 Ammianus Marcellinus xxx, 5. On the non-existence of a Pannonian party under Valentinian, see A. Demandt, *Historia* xviii, 1969, 618.

100 *Epit.* 44, 1.

101 Ammianus Marcellinus xxx, 7, 2; Eusebius, *chron.* 243e (Helm), etc.

102 Ammianus Marcellinus xxvi, 5, 1.

103 Seeck, *Regesten*, 215 ff.

104 Ammianus Marcellinus xxvi, 4, 5.

105 *Mosella*, 8–9; cf. *Not. Occ.* xlii, 65–70.

106 See, for example, Ammianus Marcellinus xxvi, 4–5.

107 *Codex Theod.* xv, 1, 3.

108 *Pannonia*, 629 ff.; *Eirene* iv, 1965, 143 f.

109 S. Soproni, *Stud. Mil.* 138 ff.

110 *C.* 3653; 10596; Soproni, *Stud. Mil.* 138.

111 See the works of S. Soproni, especially *Stud. Mil.* 138 ff., and *LRKN* 131 ff.

112 *C.* 14358[11].

113 For *burgi*, see *Pannonia*, 638 ff.; S. Soproni, *op. cit.* (note 111), *Studia Comitatensia* i, 1972, 39 ff., and D. Dimitrijević, *Osj. Zb.* xii, 1969, 88 ff. A large number of *burgi* excavated by S. Soproni have not been published—see so far *Eirene* iv, 1965, 141 f.; *Acta Arch.* xxi, 1969, 354 f.

114 xxix, 6.

115 S. Soproni, *1969 Congress of Roman Frontier Studies Report* (Cardiff, 1974).

116 *Not. Occ.* xxxii, 41; xxxiii, 48.

117 xxxix, 6, 1.

118 *In ipsis Quadorum terris, quasi Romano iuri iam vindicatis.*

119 M. R. Alföldi, *Antiquitas Hungarica* iii, 1949, 86 ff.; A. Sz. Burger, *Num. Közl.* lxvi–lxvii, 1967–8, 3 ff.; cf. M. R. Alföldi, *Jahrbuch für Numismatik und Geldgeschichte* xiii, 1963, 93 f., and in addition A. Mócsy, *Eirene* iv, 1965, 141 f.

120 Cf. V. Lányi, *Acta Arch.* xxi, 1969, 33 ff.; K. B. Sey, *Fol. Arch.* xx, 1969, 63 ff. They may belong to the barbarian raids of 379.

121 Ammianus Marcellinus xxx, 5–6.

122 See A. Mócsy, in *Adriatica, Mélanges offerts à Grga Novak* (Zagreb, 1970), 583 ff.

123 Ammianus Marcellinus xxx, 10; *Chron. Min.* i, 242; Socrates iv, 31, 6–7.

CHAPTER 9

1 See, for example, R. Noll, *Barb-Festschr.* 160 f., where the late Roman inscriptions of Austria are listed.

2 Solinus xxi, 2; *Expositio tot. mundi*, 57; Avienus, *descr. orbis*, 456–7; Isidore, *orig.* xiv, 4, 16.

3 *Itinerarium Burdig.* 563, 10; cf. *Tabula Imperii Romani* L 34 (Budapest, 1968), 59.

4   See Chapter 7, note 216 (p. 395).

5   *C.* 10275 = *VHAD* ix, 1906–7, 112, No. 233.

6   *Itinerarium Ant.* 243; *Not. Occ.* xxxii, 45.

7   *Pannonia*, 669, and I. Vincze, *Acta Ethnographica Acad. Sc. Hungaricae* vii, 1958.

8   B. Saria, *Der römische Gutshof von Winden am See* (Eisenstadt, 1951), 16 ff.; *Bp. Rég.* xv, 1950, 313; *Arch. Ért.* lxxviii, 1951, 128 f.

9   xlix, 36. 2.

10 Herodian viii, 2.

11 Ambrose, *ep.* xviii, 21; cf. S. Panciera, *La vita economica di Aquileia* (Venice, 1957), 110 f.; L. Ruggini, *L'Italia Annonaria* (Milan, 1961), 113 ff., 534.

12 See note 2.

13 Thomas, *RVP* 128 ff.

14 Cf. A. Mócsy, *Eirene* iv, 1965, 149 f.; *Acta Arch.* xxi, 1969, 365.

15 See especially the bath-buildings north of Lake Pelso in Thomas, *RVP*: Balatonfökajár: 21 f.; Balatonfüred: 23 f.; Balatongyörök: 25 f.; Rezi: 111, and Abb. 5 on page 21. Nearly all were uncovered in old excavations and not dated.

16 See previous note; also Örvényes: T. Szentléleky, *VMMK* iv, 1965, 103 ff.; Kékkut: K. Sági, *Arch. Ért.* xciii, 1966, 295; Sümeg: Thomas, *RVP* 112; also part of Szentkirályszabadja: Thomas, *RVP* 118 ff.; etc. South of Scarbantia: D. Gabler, *Archäologische Forschungen, Mitteilungen des Archäologischen Institutes der Ungarischen Akademie* ii, 1971, 57 ff.

17 Pécs-Mecsekalja: Thomas, *RVP* 288; Hosszuhetény: Thomas, *RVP* 274; Kékkut: K. Sági, *op. cit.* (note 16); cf. Thomas, *RVP* 55; Sümeg: Thomas, *RVP* 112; Tác I: J. Fitz, *Gorsium* (Szekesfehérvár, 1970).

18 Gyulafirátót III: Thomas, *RVP* 44. Sümeg: Thomas, *RVP* 112. Szentkirályszabadja: Thomas, *RVP* 118 f. Hosszuhetény: Thomas, *RVP* 274 f. Komló: see Chapter 7, note 108 (p. 391); samian and early Roman finds are not known at this villa.

19 Sümeg, Pécs-Mecsekalja, etc.

20 For example, Tüskevár: Á. Kiss, *VMMK* vi, 1967, 37 ff. Purbach: Thomas, *RVP* 192 ff. Egregy: Thomas, *RVP* 33, etc.; see note 15.

21 Csucshegy: L. Nagy, *Bp. Rég.* xii, 1937, 25 ff. and J. Szilágyi, *Arch. Ért.* lxxvi, 1949, 67 ff.; Budakalász: Thomas, *RVP* 214 f., etc.; domestic buildings in the environs of Vindobona also: A. R. Neumann, *Barb-Festschrift* 115 ff.

22 *ILS* 8987; for dating, see *Acta Arch.* xxiii, 1971, 357.

23 D. Sabovljević, *Starinar* v, 1888, 66 ff.

24 M. Dimić, *Starinar* viii, 1891, 21 ff.

25 M. Veličković, *ZNMB* i, 1958, 110 ff.

26 M. Vasić, *Revue Archéologique* 1903, 19 ff.; Garašanin, *Nalazišta*, Taf. xxivb, etc.

27 M. Grbić, *Plastika*, Taf. lxviii.

28 See Chapter 8, note 46 (p. 396).

29 *Epit.* 40, 10.

30 *Epit.* 40, 16; Procopius, *aedif.* iv, 4, 3.

31 Ammianus Marcellinus xxvi, 5, 1.

32 A. Oršić-Slavetić, *Starinar* iii/8–9, 1933, 307 ff.; iii/13, 1938, 199 ff.; *Spomenik* xcviii, 233–48; cf. *MS* 91 f.; A. Nenadović, *LuJ* i, 1961, 169.

33 B. Saria, *Barb-Festschrift* 252 ff.

34   xiv, 11, 20; S. Jenny, *Mittheilungen der Zentral-Commission* xxii, 1896, 1 ff.

35   xxx, 5.

36   D. Mano-Zisi, *ZNMB* ii, 1960, 100 ff.; *Arch. Iugosl.* ii, 1956, 72 ff.; *La mosaïque gréco-romaine* (Paris, 1965), etc.

37   *Pannonia*, 700 f.; K. Sági, *Acta Arch.* ix, 1961, 397 ff.

38   Most recently L. Barkóczi, *Acta Arch.* xx, 1968, 275 ff.; K. Sági, *Acta Ant.* xviii, 1970, 147.

39   K. Sági, *Acta Arch.* xii, 1960, 190 ff.

40   *MS* 111 ff.

41   S. Soproni, *Arch. Ért.* xcvii, 1970, 310; xcviii, 1971, 271.

42   A. Radnóti, *PWRE* viiA, 82 f.

43   A. Sz. Burger, *Acta Arch.* xviii, 1966, 99 ff.

44   A. Radnóti, *Laur. Aqu.* ii, 91 ff.

45   E. Biró, *Arch. Ért.* lxxxvi, 1959, 173.

46   Mursella: *Codex Theod.* xvi, 8, 1; cf. i, 8, 6 and 9, 2 (A.D. 339). Tricciana: *Codex Theod.* xi, 36, 26 (A.D. 379), doubtful, see Seeck, *Regesten* 109.

47   *C.* 4219.

48   *Not. Occ.* xii, 21.

49   L. Barkóczi, *Acta Arch.* xx, 1968, 275 ff.; Á. Kiss, *Arch. Ért.* xcv, 1968, 93 ff.

50   *MS* 67, 72.

51   For example *C.* 3522, 8151; cf. p. 1022.

52   *C.* 4180.

53   For example *C.* 3370, 3576, 4039, 4219, 10527, 10981, 13810, 14594, 15172; *Intercisa* i, No. 32; *Spomenik* xcviii, 229, etc.

54   On the discontinuation of the practice of setting up inscriptions, see *MS* 203. The suggestions made there are not fully valid for Pannonia; cf. also A. Mócsy in G. Alföldy, *Gesellschaft und Bevölkerung der römischen Provinz Dalmatien* (Budapest, 1965), 214.

55   *Panegyrici Latini* ii (Baehrens), 37.

56   *Codex Theod.* i, 32, 5.

57   L. Nagy, *Bp. Rég.* xii, 1937, 25 ff.; cf. K. Sz. Póczy, *Bp. Rég.* xxii, 1971, 98 ff.

58   L. Nagy, *Mumienbegräbnisse*; *Bp. Rég.* xiv, 1945, 535 ff.; Gy. Parragi, *Bp. Rég.* xx, 1963, 311 ff.; K. Sz. Póczy, *Arch. Ért.* xci, 1964, 175, etc.

59   K. Sz. Póczy, *Bp. Rég.* xxi, 1964, 62 ff.; T. Nagy, *Bp. Müeml.* ii, 59 ff.

60   Swoboda, *Carnuntum* 140 f.

61   xxx, 5–6.

62   See, e.g., Chapter 7, note 214 (p. 395); in addition *JÖAIB* xxxi, 101 ff.; cf. *MS* 105 f., 109, 121.

63   See note 51; also *C.* 1661, 4039, 4413, 10981, 12657, etc.

64   *MS* 105, 109, 122; M. Veličković, *ZNMB* iii, 1962, 99 ff.

65   K. Sz. Póczy, *Arch. Ért.* xciv, 1967, 137 ff.

66   *Antiquity* x (1936), 475. In the case of Bassiana a later date for the town-wall cannot be excluded, see Chapter 10 (p. 348).

67   D. Bošković, *Starinar* iii/4, 1928, 270, Abb. 1; B. Saria, *BRGK* xvi, 1925–6, 93; D. Piletić, *Arh. Pregl.* iv, 1962, 176 ff.; D. Mano-Zisi and Lj. Popović, *Starinar* iv/9–10, 1959, 381 ff.

68   For example *AIJ*, p. 269 (Aquae Balisae); J. Korda, *LuJ* i, 1961, 59 ff. (Cibalae) etc.; cf.

Pannonia, 696. The problem of Velike Malence, the large mountain fortress near Neviodunum (Fig. 50), remains unsolved. It may have served as a refuge for the town. For the most recent plan see P. Petru, *Osj. Zb.* xii, 1969, Fig. 14.

69 *C.* 10107 = *ILS* 3458.

70 See the works cited in Chapter 5, note 9 (p. 378).

71 See note 70.

72 A. Hytrek, *Ephemeris Salonitana* (Zara, 1894), 5 ff.; V. Hoffiller, *Bericht über den vi Internat. Kongress für Archäologie* (Berlin, 1939), 522 f.

73 Grbić, *Plastika*, 127 ff.

74 *Starinar* v, 1888, Taf. x; iii/8–9, 1933, 304 ff.; iv/5–6, 1956, 53 ff.; *Arch. Iugosl.* ii, 1956, 85 ff.; *Starinar* iv/9–10, 1959, 382 ff.; iv/11, 1961, 246 ff.; *LuJ* i, 1961, 171 ff.; *Jahrbuch des Römisch-Germanischen Zentralmuseums* x, 1963, 118 ff., etc.

75 L. Mirković, *Arch. Iugosl.* ii, 1956, 85 ff.

76 L. Nagy, *Römische Mitteilungen* xli, 1926, 123 ff.; Fr Gerke, in *Neue Beiträge zur Kunstgeschichte des i. Jahrtausends* (Baden, 1954), 147 ff.

77 Gy. Gosztonyi, *A pécsi ókeresztény temető* (Pécs, no date); F. Fülep, *Acta Arch.* xi, 1959, 401 ff.; *Arch. Ért.* lxxxix, 1962, 23 ff.; xcvi, 1969, 3 ff.

78 I. Paulovics, *Szombathely Szent Márton egyházának rómaikori eredete* (Szombathely, 1944).

79 The publication of the excavations by I. Paulovics, *Savaria-Szombathely topográfiája* (Szombathely, 1942), is in some respects not satisfactory. A new compilation of the evidence from the excavation, based on photographs and notes, has been made by my student E. Tóth, who has solved the problem of the *aula*.

80 *Not. Or.* ix, 44. For the cemetery, see *GMKM* i, 1956, 320 ff.; ii, 1957, Abb. 3; iv–v, 1960, 371 ff.; D. Mano-Zisi, *La mosaïque gréco-romaine* (Paris, 1965), Fig. 18.

81 *Starinar* iv/7–8, 1957, 289 ff.; *ŽA* vi, 1956, 292; xiii–xiv, 1964, 148 ff., etc.

82 *Moravski arheološki Glasnik* iii, 1936, 43; see also *MS* 92 f.

83 M. Valtrović, *Starinar* i, 1884, 62 ff., 90 ff.; M. Vasić, *Jahrb. DAI* xx, 1905; *Archäol. Anzeiger*, 108; N. Vulić, *ibid.* xxvii, 1912, 548.

84 The best-known is the discovery of the sarcophagus at Szekszárd: L. Nagy, *PS* 48 ff.

85 *Not. Or.* ix, 40, 43, 44.

86 On Tác = Gorsium, see most recently J. Fitz, *Acta Arch.* xxiv, 1972, 3 ff. Matrica: A. Mócsy, *Arch. Ért.* lxxxii, 1955, 62 ff.

87 Cf. *Eirene* iv, 1965, 147 f.; *Acta Arch.* xxi, 1969, 365.

88 E. Nowotny, *RLiÖ* xii (1914).

89 Unpublished.

90 See notes 26, 27, 84, 141, 144, and 145.

91 For a comprehensive study of the late Roman cemeteries of Pannonia, see V. Lányi, *Acta Arch.* xxiv, 1972, 53 ff.

92 See note 91.

93 *Pannonia*, 765 with bibliography; see also Z. Kádár, *Arch. Ért.* xcv, 1968, 90 ff.; *Fol. Arch.* xv, 1963, 69 ff.; B. Jeličić, *ZNMB* iii, 1963, 109 ff.; and D. Gáspár, *Spätrömische Kästchenbeschläge in Pannonien* (Szeged, 1971).

94 L. Nagy, *Arch. Ért.* xliv, 1930, 111 ff.; *Bp. Rég.* xii, 1937, 189 ff.; L. Barkóczi, *Fol. Arch.* xii, 1960, 121 ff.

95 M. R. Alföldi, *Intercisa* ii, 404 f.

96   There is no study of coin-circulation for Pannonia and Upper Moesia; see *Pannonia*, 691 ff.; *MS* 257 ff.

97   E.g. L. Nagy, *Mumienbegräbnisse* 18 f.; K. Sz. Póczy, *Arch. Ért.* xci, 1964, 176 ff.; L. Barkóczi, *Acta Ant.* xiii, 1965, 251.

98   This is proved by finds of hoards, see, for example, *Num. Közl.* xxviii-xxix, 1929–30, 30 ff.; *Arch. Ért.* xciii, 1966, 296. On captives see e.g. Optatianus Porphyrius, *carm.* vi.

99   *C.* 3653.

100 I. Paulovics, *Arch. Ért.* xlvii, 1934, 163.

101 E.g. the martyrs of Sirmium: Demetrius, Basilla, Anastasia, Synerotas, Hermagoras, etc.

102 The earliest-known bishops are Eusebius, Irenaeus, Quirinus, Domnus and Victorinus, but for the latter see note 103.

103 Jerome, *vir. ill.* 74.

104 For Syrians in Sirmium in post-Severan times, see *C.* 2006 and 6443.

105 The Christian tombstones are difficult to date. Those from Sirmium are not completely listed in *CIL* iii; for these, see *VHAD* x, 1908–9, Nos. 374–466 (pp. 231–63).

106 *C.* 4413; see also *C.* 4039 (Poetovio).

107 See, for example, the objects from the Mithraeum at Sárkeszi: T. Nagy, *Bp. Rég.* xv, 1950, 50.

108 S. Soproni, *Arch. Ért.* lxxxi, 1954, 50; *Spomenik* lxxv, 171 = *Starinar* iii/8–9, 1933, 310, No. 11.

109 M. Abramić, *Führer durch Poetovio* (Vienna, 1925), 68; B. Saria, *Starinar* iii/3, 1924–5, 163 = *BRGK* xvi, 1925–6, 92.

110 M. K. Kubinyi, *Arch. Ért.* 1946–8, 276 ff.

111 Fr Kenner, *Mitteilungen der Zentral-Commission* xii, 1867, 127 ff.

112 For example, Soproni, *Szemle* x, 1956, 346.

113 T. Nagy, *Bp. Tört.* i, 446 n. 215; C. Praschniker, *JÖAI* xxx, 1937, 120 f.

114 *Pannonia*, 749 f.

115 A. Alföldi, *Arch. Ért.* 1940, 214 ff. on *C.* 3343; but cf. R. Egger in *Omagiu lui C. Daicoviciu* (Bucharest, 1960), 167 f., and *Pannonia*, 595.

116 L. Barkóczi, *Acta Ant.* xiii, 1965, 238 ff.

117 E.g. *Passio Irenaei* 3–4.

118 *Acta Sanctorum, Aprilis*, iii, p. 571.

119 *Corpus Scr. Eccl. Lat.* xlix.

120 Cf. *Passio Irenaei* 5.

121 See most recently D. Simonyi, *Acta Ant.* viii, 1960, 165 ff.

122 For the following, see J. Zeiller, *Les origines chrétiennes dans les provinces Danubiennes* (Paris, 1918); T. Nagy, 'A pannoniai kereszténység története', *Diss. Pann.*, ii, 12 (Budapest, 1939); R. Egger, *Der Heilige Hermagoras* (Klagenfurt, 1948).

123 For example *C.* 10232, 10233.

124 Priscus, fragment 2.

125 Egger, *RAFC* i, 57 ff.

126 For example R. Bratanić, *Arh. Vestn.* iv, 1953, 282 ff.; also *VHAD* viii, 1905, No. 187; ix, 1906–7, Nos 190–1; *JÖAIB*, iii, 134, No. 37; vi, 21, Nos 28–9; xii, 158, No. 23; xiii, 204, Nos 15–16.

127 See note 79.

128   I. Paulovics, *Szombathely Szent Márton egyházának rómaikori eredete* (Szombathely, 1944), 31, Abb. 10.

129   See note 123.

130   L. Nagy, *Az óbudai ókeresztény cella trichora a Raktár-utcában* (Budapest, 1931); V. Hoffiller, *Bericht über den vi Internat. Kongress für Archäologie* (Berlin, 1939), 522 f.; F. Fülep, *Acta Arch.* xi, 1959, 401 ff.

131   See *Pannonia*, 727 f.; F. Fülep, *Arch. Ért.* lxxxix, 1962, 23 ff.; xcvi, 1969, 3 ff.

132   A. Radnóti, *Arch. Ért.* lii, 1939, 152 ff.; E. Čerškov, *GMKM* iv-v, 1960, 374, Abb. 3.

133   *Starinar* viii, 1891, 130 ff.; iii/8–9, 1933, 75, Abb. 3.

134   L. Nagy, *PS* 48 ff.

135   See L. Mirković, *Arch. Iugosl.* ii, 1956, 85 ff.

136   F. Fülep, *Arch. Ért.* xcvi, 1969, 33.

137   *Ibid.*, 3 ff.

138   Gy. Gosztonyi, *Arch. Ért.* 1940, 56 ff.

139   Ulpianum: *GMKM* i, 1956, 322; vii-viii, 1964, 352 ff. Remesiana: *Starinar* iv/9–10, 1959, 381 ff. Aquincum: *Arch. Ért.* 1940, 250 ff. Fenékpuszta: *Acta Ant.* ix, 1961, 397 ff. Kékkut north of Lake Pelso: *Arch. Ért.* xlv, 1931, 32 ff.; but cf. K. Sági, *Arch. Ért.* xciii, 1966, 295; see also *Pannonia*, 755 f., and *Acta Arch.* xxi, 1969, 370.

140   C. Praschniker in Kubitschek, *Römerfunde* 52 ff.; cf. A. Barb, *Mullus, Festschrift für Th. Klauser* (Münster, 1964), 17 ff.; E. B. Thomas, *Acta Ant.* iii, 1955, 261 ff.; Örvényes: unpublished, in the Lapidarium of the museum at Tihany.

141   A. Hekler, *Arch. Ért.* xxxiii, 1913, 210 ff., cf. G. Erdélyi, *Ant. Tan.* viii, 1961, 137 f.

142   Grbić, *Plastika*, Taf. lxviii.

143   A. Alföldi, *Acta Arch.* (Copenhagen) i, 1934, 99 ff.; M. Manojlović-Marijanski, *Kasnorimski šljemovi iz Berkasova* (Novi Sad, 1964).

144   See Chapter 8, note 46 (p. 396); also M. Lenkei, *Fol. Arch.* vii, 1955, 97 ff. and note 26 above.

145   G. Erdélyi, *Arch. Ért.* xlv, 1931, 1 ff.

146   See note 93.

CHAPTER 10

1   Jerome, *ep.* 60, 16, 2 = *MPL* xxii, 600 = ed. Labourt, p. 106 f.; cf. *ep.* 123, 16 = *MPL* xxii, 1058 = ed. Labourt, 92 f.

2   *Chron. Min.* ii, 76.

3   Seeck, *Regesten*, 250 f.

4   The main source for these events is Ammianus Marcellinus xxxi, 4–16. See now T. Nagy, *Acta Ant.* xix, 1971, 299 ff.

5   *Chron. Min.* i, 243, 297; ii, 60.

6   In July he was still in Scupi: *Codex Theod.* vi, 39, 2; two edicts dated to August are from places as yet unidentified: *Codex Theod.* vi, 30, 3; xii, 13, 4.

7   Sidonius Apollinaris v, 107 ff.; for his rank, see Várady, *LJP* 38.

8   *Chron. Min.* i, 243.

9   Jordanes, *Get.* 139–42; Zosimus iv, 31.

10   Vitalianus: Ammianus Marcellinus xxi, 10, 9; Zosimus iv, 34. For destruction by the

people of Alatheus and Saphrac in Pannonia, see *Panegyrici Latini* ii (Baehrens) 32, 3–4; *Collectio Avellana* 39, 4 (*Corpus Scr. Eccl. Lat.* xxxv, 89); Jerome, *vir. ill.* 65; *Comm. in Soph.* i, 676 (*MPL* xxv, 1340 f.); cf. Chapter 8, note 120 (p. 398).

11 Zosimus iv, 32–3.

12 *Panegyrici Latini* ii (Baehrens) 32, 3–4; Jordanes, *Get.* 139–42; for the emperors' meeting: Seeck, *Regesten*, 254 f.; cf. Várady, *LJP* 378. For the Gothic treaty of 382, see, e.g., *Chron. Min.* i, 243; ii, 61; *Panegyrici Latini* ii (Baehrens) 22, 3; Themistius, *or.* xvi, 208.

13 *CIL* v, 1623; cf. Egger, *RAFC* i, 57 ff.

14 Cassiodorus, *varia* v, 14, 6; cf. Várady, *LJP* 522.

15 xxxi, 4, 2.

16 *Num. Közl.* xxviii–xxix, 1929–30, 30 ff.; *Arch. Ért.* xciii, 1966, 296.

17 Ambrose gives a concise summary of the unrest among the barbarians: *in Luc.* x, 10 (= *MPL* xv, 1898 f. = *Corpus Scr. Eccl. Lat.* xxxii, 458 f.).

18 Symmachus, *Relatio* 47.

19 See Chapter 9, note 11 (p. 399).

20 *ILS* 8987; for the date, see *Acta Arch.* xxiii, 1971, 357.

21 Zosimus iv, 42, 5.

22 Zosimus iv, 45, 3; 48–50.

23 *Panegyrici Latini* ii (Baehrens) 34; 37; Seeck, *Regesten* 275.

24 A. Alföldi, *Der Untergang der Römerherrschaft in Pannonien* i (Berlin–Leipzig, 1924), 13.

25 *Ibid.* 15.

26 V. Lányi, *Acta Arch.* xxi, 1969, 33 ff.

27 A. Alföldi, *Egyetemes Philologiai Közlöny* liv, 1930, 2 ff.; L. Barkóczi, *Fol. Arch.* xiii, 1961, 111 f.; S. Soproni, *Fol. Arch.* xx, 1969, 69 ff.

28 Zosimus iv, 48–9.

29 Ambrose, *de obitu Valent.* 2, 4, 22 (*MPL* xvi, 1427 ff.).

30 *Vir. ill.* 65.

31 *Comm. in Soph.* i, 676 (*MPL* xxv, 1340 f.).

32 *Ep.* 66, 14 (*MPL* xxii, 647 = ed. Labourt, 180).

33 For example, Jerome, *ep.* 60, 16, 2; *Comm. in Soph.* 1, 676; Ambrose, *de off. ministr.* i, 15, 70 (*MPL* xvi, 129); *in Luc.* x, 10; *de fide*, ii, 140 (*MPL* xvi, 613); Ps.-Augustine, *quaest. vet. et nov. test.* 115, 46 (*Corpus Scr. Eccl. Lat.* i, 334); *Collectio Avellana*, 38, 1 (*Corpus Scr. Eccl. Lat.* xxxv, 85); see also Claudian, *bell. Goth.* 632 ff.; *in Ruf.* ii, 26–53; Orosius vii, 43, 4. Várady, *LJP* 123 ff., wants to reject these and other points as literary platitudes.

34 Sidonius Apollinaris, vii, 589 f.; Ennodius, *v. Ant.* 12–14.

35 *Cons. iii Stil.* ii, 191 ff.

36 L. Várady describes the part played in the political and military history of this period by the *foederati* settled in Illyricum (*LJP*). I cannot accept his comments on conditions in Pannonia at this time; cf. *Acta Arch.* xxiii, 1971, 347 ff., and T. Nagy, *Acta Ant.* xix, 1971, 299 ff.

37 The evidence for the time and place of the barbarian raids is given by Claudian, *in Ruf.* ii, 124; see also *ibid.* 1, 301–22; ii, 26 ff.

38 E. Polaschek, *Numismatische Zeitschrift* lviii, 1925, 127 ff.

39 Paulinus, *vita Ambrosii*, 36 (*MPL* xiv, 42).

40 *Not. Occ.* xxxiv, 24.

41 Jerome, *ep.* 60, 16, 2; 123, 15; cf. already Ammianus Marcellinus xxxi, 4, 2.

42 E.g. *in Ruf.* i, 310; *epith. Pall.* 88; *fesc.* iv, 15.

43 *Cons. Stil.* iii, 13.

44 ii, 191–207.

45 Cf. Claudian, *bell. Goth.* 279, 363–5.

46 Claudian, *vi cons. Hon.* 227 f.; Zosimus v, 26, 29; Jordanes, *Get.* 147.

47 Zosimus v, 37.

48 There is no evidence for the route of this march and this, in my opinion, shows that the Pannonian territory through which the people of Radagaisus travelled had passed completely outside the sphere of interest of contemporaries. So whether the march followed the Danube valley or was partly outside Roman territory is irrelevant.

49 *Ep.* 123, 15 (*MPL* xxii, 1057).

50 Appendix, *passio Quirini* (Ruinart, *Acta martyrum sincera* 524); cf. T. Nagy, *Regnum* vi, 1947, 244 ff.; Egger, *Der Heilige Hermagoras* (Klagenfurt, 1948), 51 f. T. Nagy, *op. cit.* (note 36), does not exclude the possibility that the body of Amantius came through *translatio* to Aquileia.

51 *Codex Theod.* x, 10, 25; cf. v, 7, 2. A young lady of senatorial rank went to Salona: *C.* 9515, cf. *JRS* lvii, 1967, 299, and J. Wilkes, *Phoenix* 26, 1972, 377 ff.

52 A. Chastagnol, *Epigraphica* xxix, 1967, 105 ff.; *Codex Theod.* xi, 17, 4; xv, 1, 49.

53 Zosimus v, 46, 2.

54 The omission of two frontier forts, Ad Mures and Statuas, between Arrabona and Brigetio is equally notable. They were either the most easterly forts of Pannonia Prima or the most westerly of Valeria, but they occur neither in *Occ.* xxxiii nor in *Occ.* xxxiv. From this it may be deduced that these forts no longer had garrisons when the chapter on Pannonia Prima was brought up to date for the last time. If they had belonged to Valeria they would have been included in the relevant chapter xxxiii, especially as this chapter describes an earlier, or even several earlier situations.

55 J. Zeiller, *Les origines chrétiennes dans les provinces Danubiennes* (Paris, 1918), 148 ff.

56 *Innocentii papae ep.* 42.

57 *Chron. Min.* ii, 76.

58 Priscus, fragment 2.

59 *Ibid.*, fragment 7.

60 Cassiodorus, *var.* xi, 1, 9; cf. Várady, *LJP* 308.

61 *Chron. Min.* ii, 80.

62 Priscus, fragment 7.

63 Procopius, *aedif.* iv, 5.

64 Sidonius Apollinaris vii, 589 ff.

65 *Aedif.* iv, 5, 9.

66 *Spomenik* lxxvii, 38–9; *JÖAI* xxxi, 117, No. 24; cf. also *JÖAI* vi, 57, No. 90; viii, 6, No. 15, etc.

67 See note 55.

68 See note 50.

69 See note 51.

70 Admittedly this has not been recorded as a flight, but as bishop of a Pannonian community he could not otherwise have spent his last years in his Italian home. See also note 50.

71 On the cult of Pannonian martyrs outside Pannonia, see T. Nagy, *A pannoniai kereszténység története* 68 ff.

72 K. Sági, *Acta Ant.* ix, 1961, 397 f.

73 L. Barkóczi, *Acta Arch.* xx, 1968, 275 ff. On Romans transferred by the Avars from the Balkans and from Italy to Pannonia, see I. Bóna, *Arch. Ért.* xcvii, 1970, 258.

74 Ennodius (*Monum. Germ. Hist. Auct. Ant.* vii), *vita Antoni* 12–13.

75 On Leonianus and other refugees, see A. Alföldi, 'Tracce del cristianesimo nell'epoca delle grandi migrazioni in Ungheria', *Quaderni dell'Impero: Roma e le province* (Rome, 1938).

76 Gregory of Tours, *Hist. Francorum* v, 37; Isidore, *vir. illustr.* xxxv; cf. J. Vives, *Inscripciones cristianas de la España Romana y Visigoda* (Barcelona, 1942), 82 f., No. 275.

77 Paulus Diaconus ii, 26.

78 *C.* 9551 = *ILCV* 1653; *C.* 9576 = *ILVC* 4455; *ILCV* 118.

79 *Chron. Min.* ii, 100.

80 Procopius, *aedif.* iv, 5.

81 The father of Romulus, Orestes: Priscus, fragment 7; *Excerpta Valesiana* ii, 38.

82 Priscus, fragment 7; and cf. *ibid.*, to mention only the following: Rusticius, Constantiolus, the builder of the baths of Onegesius, etc.

83 The very plausible hypothesis of I. Bóna, *Ant. Tan.* xvi, 1969, 285 ff.

84 I. Bóna, 'Die pannonischen Grundlagen der langobardischen Kultur', *Problemi della civiltà ed economia langobarda* (Milan, 1964). For stone buildings of post-Roman date at Intercisa, see E. B. Vágó, *Alba Regia* xi, 1971, 112.

85 *Comm. in Isaiam* vii, 19 (292).

86 D. Rendić-Miočević, *Starohrvatska Prosvjeta* iii/1, 1949, 9 ff.

87 V. Beševliev, *Études Balkaniques* i, 1964, 147 ff.

88 Procopius, *anecd.* 6, 1.

89 For example, Jordanes, *Get.* 265, 282.

Select Bibliography

GENERAL WORKS AND WORKS OF REFERENCE

A. ALFÖLDI, *Magyarország népei és a római birodalom* (Budapest, 1934).

A. ALFÖLDI, 'Pannonia rómaiságának kialakulása és történeti kerete', *Századok* lxx, 1936, 1. ff., 129 ff.

A. ALFÖLDI, 'Studi ungheresi sulla romanizzazione della Pannonia', *Gli Studi Romani nel mondo* ii, 1935, 265 ff.

A. BETZ, *Aus Österreichs römischer Vergangenheit* (Vienna, 1956).

S. BORZSÁK, 'Die Kenntnisse des Altertums über das Karpatenbecken', *Diss. Pann.* i, 6 (Budapest, 1936).

Bp. Tört.

M. FLUSS, 'Moesia', *PWRE* xv, 1932, 2354 ff.

B. GEROV, 'Romanizmăt meždu Dunava i Balkana' i-iii, *Godišnik na Sofijskija Universitet, Filologičeski Fakultet* xlv, 1950; xlvii, 1951; xlviii, 1955.

T. KOLNIK, 'Prehl'ad a stav badania v dobe rimskej a stahováne národov', *Slov. Arch.* xix, 1971, 499 ff.

M. MIRKOVIĆ, *Rimski gradovi na Dunavu* (Belgrade, 1968).

A. MÓCSY, *Gesellschaft und Romanisation in der römischen Provinz Moesia Superior* (Budapest–Amsterdam, 1970).

A. MÓCSY, *Pannonia*.

A. MÓCSY, 'Pannonia-Forschung 1961–1963', *Eirene* iv, 1965, 133 ff.

A. MÓCSY, 'Pannonia-Forschung 1964–1968', *Acta Arch.* xxi, 1969, 340 ff.

A. MÓCSY, 'Pannonia-Forschung 1969–1972', *Acta Arch.* xxv, 1973, 375 ff.

T. NAGY, *Bp. Müeml.*

E. NOWOTNY, 'Römische Forschung in Österreich', *BRGK* xv, 1925, 121 ff.; xvi, 1926, 164 ff.

Select Bibliography

P. OLIVA, *Pannonia and the Onset of Crisis in the Roman Empire* (Prague, 1962).

M. PAVAN, 'La Provincia Romana di Pannonia Superior', *Atti dell' Accademia Nazionale dei Lincei* ccclii, 1955; *Memorie* viii, 6, 5 (Rome).

B. SARIA, 'Römische Forschung in Jugoslawien', *BRGK* xvi, 1926, 90 ff.

B. SARIA, 'Noricum und Pannonien', *Historia* i, 1950, 436 ff.

W. SCHMID, 'Römische Forschung in Österreich', *BRGK* xv, 1925, 178 ff.

A. SCHOBER, *Die Römerzeit in Österreich* (2nd ed., Vienna, 1955).

E. SWOBODA, *Carnuntum.*

V. VELKOV, *Gradăt v Trakija i Dakija prez kăsnata antičnost* (Sofia, 1959).

H. VETTERS, 'Dacia Ripensis', *Schriften der Balkankommission, Antiquarische Abteilung* xi, 1 (Vienna, 1951).

T. D. ZLATKOVSKAYA, *Mezija v i-ii. vekah našej ery* (Moscow, 1951).

INSCRIPTIONS

A. BARB, 'Die römischen Inschriften des Burgenlandes', *Mitteilungen des Burgenländischen Heimat- und Naturschutzvereins* v, 1931; *Burgenländische Heimatblätter* i, 1932 ff.

L. BARKÓCZI, 'Brigetio', *Diss. Pann.* ii, 22 (Budapest, 1944–51).

V. BEŠEVLIEV, *Epigrafski prinosi* (Sofia, 1952).

A. BETZ, 'Die griechischen Inschriften Österreichs', *Wiener Studien* lxxix, 1966, 597 ff.

J. ČEŠKA and R. HOŠEK, *Inscriptiones Pannoniae Superioris in Slovacia Transdanubiana asservatae* (Brno, 1967).

A. DOBÓ, 'Inscriptiones ad res Pannonicas et Dacicas pertinentes extra fines earundem pro-vinciarum repertae', 2nd ed., *Diss. Pann.* i, 1 (Budapest, 1940), third ed. forthcoming.

S. DUŠANIĆ, 'Bassianae and its Territory', *Arch. Iugosl.* viii, 1967, 67 ff.

J. FITZ, 'Römische Inschriften im Komitat Fejér', *Alba Regia* viii-ix, 1968, 197 ff.

B. GEROV, *Romanizmăt*, see GENERAL WORKS.

V. HOFFILLER and B. SARIA, *AIJ.*

Intercisa

A. NEUMANN, 'Inschriften aus Vindobona,' *Jahrbuch des Vereins für Geschichte der Stadt Wien* xvii-xviii, 1962, 7 ff.; xix-xx, 1964, 7 ff.

A. VON PREMERSTEIN, N. VULIĆ and others, *JÖAIB* iii, 1900–xv, 1912.

B. SARIA, 'Die römischen Inschriften des Burgenlandes', *Burgenl. Heimatb.* xiii, 1951, 1 ff.

A. and J. ŠAŠEL, *ILJug.*

S. SOPRONI, 'Römische Meilensteine aus Százhalombatta', *Fol. Arch.* xxi, 1970, 90 ff.

J. SZILÁGYI, *Inscr. teg.*

E. B. VÁGÓ, 'Neue Inschriften aus Intercisa und Umgebung', *Alba Regia* xi, 1971, 120 ff.

E. VORBECK, *Militärinschriften aus Carnuntum* (Vienna, 1954).

N. VULIĆ, 'Antički spomenici naše zemlje', *Spomenik* lxx, lxxv, lxxvii, cxviii, 1931–48.

E. WEBER, 'Die römischen Meilensteine aus dem österreichischen Pannonien', *JÖAI* xlix, 1968–71, 121 ff.

Forthcoming: *Die Römischen Inschriften Ungarns* i (Budapest, 1972), ii (1974) etc.

See also SCULPTURE IN STONE.

HISTORY: PRE-ROMAN PERIOD

A. ALFÖLDI, 'Zur Geschichte des Karpatenbeckens im i. Jahrhundert v. Chr.', *Archivum Europae Centro-Orientalis* viii, *Ostmitteleuropäische Bibliothek* xxxvii, 1942.

G. ALFÖLDY, 'Des territoires occupés par les Scordisques', *Acta Ant.* xii, 1964, 107 ff.

G. ALFÖLDY, 'Taurisci und Norici', *Historia* xv, 1966, 224 ff.

B. BENADIK, 'Obraz doby latenskej na Slovensku', *Slov. Arch.* xix, 1971, 465 ff.

É. B. BÓNIS, *Die spätkeltische Siedlung Gellérthegy-Tabán in Budapest* (Budapest, 1968).

É. B. BÓNIS, 'Die Siedlungsverhältnisse der pannonischen Urbevölkerung—einige Fragen ihres Weiterlebens', *Acta Arch.* xxiii, 1971, 33 ff.

J. FILIP, *Keltové ve střední Evropě* (Prague, 1956).

O. H. FREY, 'Zur latènezeitlichen Besiedlung Unterkrains', *Festschrift Dehn* 1969, 7 ff.

M. GARAŠANIN, 'K problematici kasnog latena u donjem Podunavlji', *Naučni Zbornik Matice Srpske, Odelenje društvenih nauke* xviii, 1958, 92 ff.

M. GARAŠANIN, 'Contribution à l'archéologie et à l'histoire des Scordisques', *A Pedro Bosch-Gimpera* (Mexico, 1963).

M. GARAŠANIN, 'Ka konfrontacija pisanih i arheoloških izvora o keltima u našoj zemlji', Materijali, iii, *Simpozijum praistorijske i srednjevekovne sekcije arheološkog društva Jugoslavije Novi Sad 1965* (Belgrade, 1966), 17 ff.

B. GAVELA, 'O etničkim problemima latenske kulture Židovara', *Starinar* xx, 1970, 119 ff.

I. HUNYADY, 'Die Kelten im Karpatenbecken', *Diss. Pann.* ii, 18 (Budapest, 1942–4).

'Keltische Oppida, Symposium 1970', *Arch. Rozhl.* xxiii, 1971, 261–4, with contributions from B. BENADIK, É. B. BÓNIS, J. FILIP, J. REITINGER, J. TODOROVIĆ, V. ZIRRA and others.

F. LOCHNER-HÜTTENBACH, 'Illyrier und Illyrisch', *Das Altertum* xvi, 1970, 216 ff.

M. MACREA, 'Burębista şi Celti de la Dunarea de mijloc', *SCIV* vii, 1956, 119 ff.; cp. *Dacia* ii, 1958, 143 ff.

Z. MARIĆ, 'Problème des limites septentrionales du territoire illyrien', *Symposium 1964*, 177 ff.

A. MÓCSY, 'Die Vorgeschichte Obermösiens im hellenistisch-römischen Zeitalter', *Acta Ant.* xiv, 1966, 87 ff.

A. MÓCSY, 'Der vertuschte Dakerkrieg des M. Licinius Crassus', *Historia* xv, 1966, 511 ff.

F. PAPAZOGLU, *Srednjobalkanska plemena u predrimsko doba* (Sarajevo, 1969).

P. PETRU, 'Vzhodnoalpski Taurisci', *Arh. Vestn.* xix, 1968, 339 ff.

P. PETRU, 'Die Hausurnen der Latobiker', *Situla* xi (Ljubljana, 1971).

M. SZABÓ, *The Celtic Heritage in Hungary* (Budapest, 1971).

J. GY. SZILÁGYI, 'Trouvailles grecques sur le territoire de la Hongrie', *Actes du viiie Congrès International d'Archéologie Classique Paris 1963* (Paris, 1965), 386 ff.

J. TEJRÁL, 'Zur Frage der Stellung Mährens um die Zeitrechnungswende', *Pam. Arch.* lix, 1968, 488 ff.

A. TOČIK, 'K otázce osidlenia juhozapadého Slovenska na zlome letopočtu', *Arch. Rozhl.* xi, 1959, 841 ff.

J. TODOROVIČ, *Kelti u jugoistočnom Evropi* (Belgrade, 1968).

H. VETTERS, 'Zur ältesten Geschichte der Ostalpenländer', *JÖAI* xlvi, 1961–3, 201 ff.

ZS. VISY, 'Die Daker im Gebiet von Ungarn', *MFMÉ* 1970, 5 ff.

V. ZIRRA, 'Beiträge zur Kenntnis des keltischen La Tène in Rumänien', *Dacia* xv, 1971, 171 ff.

Select Bibliography

Pre-Roman coinage

K. CASTELIN, *Germania* xxxviii, 1960, 32 ff.

E. GOHL, *Numismatische Zeitschrift* xxxv, 1903, 145 ff.

E. GOHL, *Num. Közl.* i, 1902, 17 ff.; vi, 1907, 47 ff.

A. KERÉNYI, *Fol. Arch.* xi, 1959, 47 ff.

E. KOLNIKOVÁ, *Slov. Arch.* xii, 1964, 402 ff.

A. MÓCSY, *Num. Közl.* lx-lxi, 1961–2, 25 ff.

V. ONDROUCH, *Keltské mince typu Biatec* (Bratislava, 1959).

K. PINK, 'Die Münzprägung der Ostkelten und ihrer Nachbarn', *Diss. Pann.* ii, 15 (Budapest, 1939).

HISTORY: ROMAN PERIOD

A. ALFÖLDI, 'The Central Danubian Provinces', *CAH* xi, 1936.

A. ALFÖLDI, 'Epigraphica IV', *Arch. Ért.* 1941, 40 ff.

A. ALFÖLDI, 'Über die Juthungeneinfälle unter Aurelian, Serta Kazaroviana i', *IBAI*, xvi, 1950, 21 ff.

A. ALFÖLDI, 'The Moral Barrier on Rhine and Danube', *The Congress of Roman Frontier Studies 1949* (Durham, 1952), 1 ff.

A. ALFÖLDI, 'Wo lag das Regnum Vannianum?' *Südostforschungen* xv, 1956, 48 ff.

G. ALFÖLDY, 'Serapis-oltár Nyergesujfaluból', *Arch. Ért.* lxxxviii, 1961, 26 ff.

G. ALFÖLDY, 'Der Friedensschluss des Kaisers Commodus mit den Germanen', *Historia* xx, 1971, 84 ff.

L. BALLA, 'Guerre iazyge aux frontières de la Dacie', *Acta Classica Universitatis de Ludovico Kossuth nominatae Debreceniensis* v, 1969, 111 ff.

L. BARKÓCZI, 'Dák tolmács Brigetioban', *Arch. Ért.* 1944–5, 178 ff.

L. BARKÓCZI, 'Die Naristen zur Zeit der Markomannenkriege', *Fol. Arch.* ix, 1957, 91 ff.

L. BARKÓCZI, *Intercisa* ii, 497 ff.

G. BARTA, 'Lucius Verus and the Marcomannic Wars', *Acta Classica Universitatis de Ludovico Kossuth nominatae Debreceniensis* vii, 1971, 67 ff.

A. R. BIRLEY, 'The invasion of Italy in the Reign of Marcus Aurelius', *Provincialia, Festschrift Laur-Belart* (Basel, 1966), 214 ff.

C. DAICOVICIU, *Dacica* (Cluj, 1970).

DOBIÁŠ, *DCU.*

J. DOBIÁŠ, 'Rom und die Völker jenseits der mittleren Donau', *Corolla E. Swoboda dicata* (Graz, 1966), 115 ff.

R. EGGER, 'Ein zweimal beschriebener Weihestein', *RAFC* i, 312 ff.

J. FITZ, 'Der Besuch des Septimius Severus in Pannonien im Jahre 202', *Acta Arch.* xi, 1959, 237 ff.

J. FITZ, 'Il soggiorno di Caracalla in Pannonia nel 214', *Accademia d'Ungheria, Quaderni di Documentazione* ii, 2 (Rome, 1961).

J. FITZ, 'Die Vereinigung der Donauprovinzen in der Mitte des 3. Jahrhunderts', *Stud. Mil.* 113 ff.

J. FITZ, 'Pannonien und die Klientelstaaten an der Donau', *Alba Regia* iv-v, 1965, 73 ff.

J. FITZ, 'Der markomannisch-quadische Angriff gegen Aquileia und Opitergium', *Historia* xv, 1966, 336 ff.

J. FITZ, *Ingenuus et Régalien* (Brussels, 1966).

J. FITZ, 'Zur Geschichte der Praetentura Italiae et Alpium im Laufe der Markomannenkriege', *Arh. Vestn.* xix, 1968, 43 ff.

R. GÖBL, 'Rex Quadis datus', *Rheinisches Museum* civ, 1961, 70 ff.

F. HAMPL, 'Kaiser Marc Aurel und die Völker jenseits der Donaugrenze', *Festschrift R. Heuberger* (Innsbruck, 1960), 33 ff.

J. HARMATTA, *Studies in the History and Language of the Sarmatians* (Szeged, 1972).

H. J. KELLNER, 'Raetien und die Markomannenkriege', *Bayerische Vorgeschichtsblätter* xxx, 1965, 173 ff.

J. KLOSE, *Roms Klientelrandstaaten am Rhein und an der Donau* (Breslau, 1934).

J. K. KOLOSOVSKAYA, 'Zavoevanija Pannonii Rimom', *Vestnik drevnei istorii*, 1961, 1, 60 ff.

E. KÖSTERMANN, 'Der pannonisch-dalmatische Krieg 6-9 n. Chr.', *Hermes* lxxxi, 1953, 345 ff.

A. MÓCSY, 'Die Expansionsfrage im i. und ii. Jahrh.', *Annales Universitatis Scientiarum de R. Eötvös nominatae, Sectio Historica* v, 1963, 3 ff.

A. MÓCSY, 'Tampius Flavianus Pannoniában', *Arch. Ért.* xciii, 1966, 203 ff.

A. MÓCSY, 'Das Gerücht von neuen Donauprovinzen unter Marcus Aurelius', *Acta Classica Universitatis de Ludovico Kossuth nominatae Debreceniensis* vii, 1971, 63 ff.

R. NOLL, 'Zwei unscheinbare Kleinfunde aus Emona', *Arh. Vestn.* xix, 1968, 79 ff.

C. PATSCH, *Beiträge*.

H. G. PFLAUM, 'Deux carrières équestres de Lambèse et de Zana', *Libyca* iii, 1955, 135 ff.

A. VON PREMERSTEIN, 'Die Anfänge der Provinz Mösien', *JÖAIB* i, 1898, 146 ff.

A. VON PREMERSTEIN, 'Der Daker- und Germanensieger M. Vinicius', *JÖAI* xxviii, 1933, 140 ff.; xxix, 1934, 60 ff.

E. RITTERLING, 'Die Osi in einer afrikanischen Inschrift', *Germania* i, 1917, 132 ff.

J. ŠAŠEL, 'Bellum Serdicense', *Situla* iv (Ljubljana, 1962).

J. ŠAŠEL, 'Drusus Ti. f. in Emona', *Historia* xix, 1970, 122 ff.

J. ŠAŠEL, 'Über Umfang und Dauer der Militärzone Praetentura Italiae et Alpium zur Zeit Marc Aurels', *Acta of the 5th International Congress of Greek and Latin Epigraphy, Cambridge 1967* (Oxford, 1971), 317.

E. SWOBODA, 'Der pannonische Limes und sein Vorland', *Carn.-Jb.* 1959 (1961), 17 ff.

E. SWOBODA, 'Traian und der pannonische Limes', *Carn.-Jb.* 1963-4 (1965), 9 ff.

R. SYME, 'Lentulus and the Origin of Moesia', *JRS* xxiv, 1934, 133 ff.

I. WEILER, 'Huic Severo Pannoniae et Italiae urbes et Africae contigerunt', *Historia* xiii, 1964, 373 ff.

I. WEILER, 'Orbis Romanus und Barbaricum', *Carn.-Jb.* 1963-4 (1965), 94 ff.

HISTORY: THE END OF THE ROMAN PERIOD

A. ALFÖLDI, *Der Untergang der Römerherrschaft in Pannonien* i-ii (Berlin and Leipzig, 1924-6).

L. BARKÓCZI, 'A 6th Century Cemetery from Keszthely-Fenékpuszta', *Acta Arch.* xx, 1968, 275 ff.

Select Bibliography

I. BÓNA, 'Ein Vierteljahrhundert Völkerwanderungszeitforschung in Ungarn', *Acta Arch.* xxiii, 1971, 265 ff.

R. EGGER, 'Die Zerstörung Pettaus durch die Goten', *RAFC* i, 36 ff.

R. EGGER, 'Civitas Noricum', *RAFC* i, 116 ff.

R. EGGER, 'Von den letzten Romanen Vindobonas', *RAFC* ii, 226 ff.

R. EGGER, 'Historisch-epigraphische Studien in Venetien', *RAFC* i, 45 ff.

Á. KISS, 'Pannonia lakossága népvándorláskori helybenmaradásának kérdéséhez', *Janus Pannonius Muzeum Évkönyve*, 1964 (1965), 81 ff.

H. LADENBAUER-OREL, 'Archäologische Stadtkernforschung in Wien', *Jahrbuch des Vereins für Geschichte der Stadt Wien* xxi–xxii, 1965-6, 18 ff.

A. MÓCSY (review of VÁRADY, *LJP*), *Acta Arch.* xxiii, 1971, 347 ff.

T. NAGY, 'Reoccupation of Pannonia from the Huns in 427', *Acta Ant.* xv, 1967, 159 ff.

T. NAGY (review of VÁRADY, *LJP*), *Acta Ant.* xix, 1971, 299 ff.

GY. SZÉKELY, 'Le sort des agglomérations pannoniennes au début du moyen âge et les origines de l'urbanisme en Hongrie', *Annales Universitatis Scientiarum Budapestinensis de Rolando Eötvös nominatae, Sectio Historica* iii, 1961, 59 ff.

L. VÁRADY, *LJP*.

H. VETTERS, 'Zur Spätzeit des Lagers Carnuntum', *Österreichische Zeitschrift für Kunst und Denkmalpflege* xvii, 1963, 157 ff.

ADMINISTRATION

L. BALLA, 'Die Inschrift eines Senators aus Savaria', *Epigraphische Studien* iv (Cologne-Graz, 1967), 61 f.

A. BETZ, 'Zum Sicherheitsdienst in den Provinzen', *JÖAIB* xxxv, 1943, 137 ff.

A. R. BIRLEY, *Acta Antiqua Philippopolitana* 1963, 107 ff.

Á. DOBÓ, 'Publicum portorium Illyrici', *Diss. Pann.* ii, 16 (Budapest, 1940).

Á. DOBÓ, *Verwaltung*.

R. EGGER, 'Das Praetorium als Amtssitz und Quartier römischer Spitzenfunktionäre', *Sitzungsberichte der Österreichischen Akademie, Philosophisch-historische Klasse* ccl/4 (Vienna, 1966).

J. FITZ, 'The Governors of Pannonia Inferior', *Alba Regia* xi, 1971, 145 ff.

J. FITZ, 'Le iscrizioni del Capitolium di Gorsium', *Rivista Storica dell'Antichità* 1, 1971, 145 ff.

J. FITZ, 'A pannoniai bányák igazgatása', *Alba Regia* xi, 1971, 154 ff.

T. NAGY, 'Zu den Militär- und Verwaltungsreformen Diocletians im pannonischen Raum', *Akte des iv. Internationalen Kongresses für griechische und lateinische Epigraphik, Wien, 1962* (Vienna, 1964), 274 ff.

A. STEIN, 'Die Legaten von Moesien', *Diss. Pann.* i, 11 (Budapest, 1940).

R. SYME, 'Governors of Pannonia Inferior', *Historia* xiv, 1965, 342 ff.

R. SYME, 'Hadrian and Moesia', *Arh. Vestn.* xix, 1968, 101 ff.

I. WEILER, 'Beiträge zur Verwaltung Pannoniens zur Zeit der Tetrarchie', *Situla* viii (Ljubljana, 1965), 141 ff.

MUNICIPALITIES

G. ALFÖLDY, 'Augustalen- und Sevirkörperschaften in Pannonien', *Acta Ant.* vi, 1958, 433 ff.

G. ALFÖLDY, 'Eine Strassenbauinschrift aus Salona', *Acta Arch.* xvi, 1964, 247 ff.

G. ALFÖLDY, 'Municipium Iasorum', *Epigraphica* xxvi, 1965, 95 ff.

S. DUŠANIĆ, 'Bassianae and its Territory', *Arch. Iugosl.* viii, 1967, 67 ff.

J. FITZ, 'Angaben zu den Gebietsveränderungen der Civitas Eraviscorum', *Acta Arch.* xxiii, 1971, 47 ff.

M. MIRKOVIĆ, *RGD*.

A. MÓCSY, 'Zur Geschichte der peregrinen Gemeinden in Pannonien', *Historia* vi, 1957, 488 ff

A. MÓCSY, 'Scribák a pannoniai kisvárosokban', *Arch. Ért.* xci, 1964, 16.

A. MÓCSY, 'Decurio Eraviscus', *Fol. Arch.* xxi, 1970, 59 ff.

V. VELKOV, 'Ratiaria', *Eirene* v, 1966, 155 ff.

MILITARY TERRITORIES

L. BARKÓCZI, 'Beiträge zum Rang der Lagerstadt am Ende des ii. und am Anfang des iii Jhs.', *Acta Arch.* iii, 1953, 201 ff.

R. EGGER, 'Bemerkungen zum Territorium pannonischer Festungen', *RAFC* ii, 135 ff.

A. MÓCSY, 'Das Territorium Legionis und die Canabae in Pannonien' *Acta Arch.* iii, 1953, 179 ff.

A. MÓCSY, 'Il problema delle condizioni del suolo attribuito alle unità militari nelle province danubiane', *Atti del Convegno Internazionale sui diritti locali nelle province dell'Impero Romano, Roma, 1971* (Rome, forthcoming).

F. VITTINGHOFF, 'Die Bedeutung der Legionslager für die Entstehung der römischen Städte an der Donau und in Dakien', *Studien zur europäischen Vor- und Frühgeschichte* (Neumünster, 1968), 132 ff.

THE ARMY

G. ALFÖLDY, 'Die Truppenverteilung der Donaulegionen am Ende des i. Jhs.', *Acta Arch.* xi, 1959, 122 ff.

G. ALFÖLDY, 'Thrakische und illyrische Soldaten in den rheinischen Legionen', *Epigraphische Studien* iv, 1967, 26 ff.

L. BARKÓCZI, 'A new military diploma from Brigetio', *Acta Arch.* ix, 1959, 413 ff.

A. BETZ, 'Zur Geschichte der Legio x. Gemina', *Corolla E. Swoboda dedicata* (Graz, 1966), 39 ff.

S. DUŠANIĆ, 'Rimska vojska na istočnom Sremu', *ZFF* x, 1968, 107 ff.

J. FITZ, 'Legati legionum Pannoniae Superioris', *Acta Ant.* xi, 1961, 159 ff.

J. FITZ, 'A Military History of Pannonia from the Marcomannic Wars to the Death of Alexander Severus', *Acta Arch.* xiv, 1962, 25 ff.

J. FITZ, 'Massnahmen zur militärischen Sicherheit von Pannonia Inferior unter Commodus', *Klio* xxxix, 1961, 199 ff.

J. FITZ, 'Réorganisation militaire au début des guerres marcomannes', *Hommages à M. Renard* ii (Brussels, 1969), 262 ff.

B. GEROV, 'Epigraphische Beiträge zur Geschichte des mösischen Limes in vorclaudischer Zeit', *Acta Ant.* xv, 1967, 91 ff.

O. V. KUDRIAVCEV, 'Dunaiskie legioni i ih značenie v istorii rimskoi imperii', *Issledovanija po istorii balkansko-dunaiskih oblasti* (Moscow, 1957), 147 ff.

M. MIRKOVIĆ, 'Cohors i. Cantabrorum i posada kastela Aquae', *ZFF* viii/1, 1964, 87 ff.

M. MIRKOVIĆ, 'Die Auxiliareinheiten in Mösien unter den Flaviern', *Epigraphische Studien* v, 1968, 177 ff.

A. MÓCSY, 'Zur frühesten Besatzungsperiode in Pannonien', *Acta Arch.* xxiii, 1971, 41 ff.

T. NAGY, 'The Military Diploma of Albertfalva', *Acta Arch.* vii, 1956, 17 ff.

T. NAGY, 'Commanders of Legions in the Age of Gallienus', *Acta Arch.* xvii, 1965, 295 ff.

H. NESSELHAUF, 'Zwei Inschriften aus Belgrad', *ŽA* x, 1960, 191 ff.

M. PAVAN, 'Iscrizioni latine ad Albano Laziale', *Athenaeum* xl, 1962, 85 ff.

D. PROTASE, 'La légion IV Flavia au nord du Danube et la première organisation de la Dacie romaine', *Acta of the 5th International Congress of Greek and Latin Epigraphy, Cambridge 1967* (Oxford, 1971), 337 ff.; cp. I. GLODARIU, *ibid.*, 327 ff.

A. RADNÓTI and L. BARKÓCZI, 'The Distribution of Troops in Pannonia Inferior during the 2nd Century A.D.', *Acta Arch.* i, 1951, 191 ff.

A. RADNÓTI, 'Zur Dislokation der Auxiliartruppen in den Donauprovinzen', *Limes-Studien* (Basel, 1957), 142 ff.

S. SOPRONI, 'Two Inscribed Relics of the Cohors XIIX Vol. c. R.', *Fol. Arch.* xvi, 1964, 33 ff.

S. SOPRONI, 'Der Stempel der Legio XIV Gemina in Brigetio', *Fol. Arch.* xvii, 1965, 119 ff.

R. SYME, 'The First Garrison of Trajan's Dacia', *Laur. Aqu.* i, 267 ff.

E. TÓTH and G. VÉKONY, 'Beiträge zu Pannoniens Geschichte im Zeitalter des Vespasianus', *Acta Arch.* xxii, 1970, 133 ff.

D. VUČKOVIĆ-TODOROVIĆ, 'Vojnička diploma iz kastruma Taliatae', *Starinar* iv/18, 1967, 21 ff.

K. WACHTEL, 'Kritisches und Ergänzendes zu neuen Inschriften aus Mainz', *Historia* xv, 1966, 247 ff.

ETHNIC COMPOSITION, SOCIAL AND ECONOMIC HISTORY

M. R. ALFÖLDI, 'Der Geldverkehr in Intercisa', *Intercisa* i, 142 ff.

G. ALFÖLDY, 'Municipális középbirtok Aquincum környékén', *Ant. Tan.* vi, 1959, 19 ff.

G. ALFÖLDY, 'Die Valerii in Poetovio', *Arh. Vestn.* xv-xvi, 1964-5, 137 ff.

L. BALLA, 'Östliche ethnische Elemente in Savaria', *Acta Arch.* xv, 1963, 225 ff.

L. BALLA, 'Gesellschaft und Geschichte von Savaria', *Die römischen Steindenkmäler von Savaria* (Budapest, 1971), 19 ff.

L. BALLA and I. TÓTH, 'A propos des rapports entre la Pannonie et la Dacie', *Acta Classica Universitatis Scientiarum de Ludovico Kossuth nominatae Debreceniensis* iv, 1968, 69 ff.

L. BARKÓCZI, 'The Population of Pannonia from Marcus Aurelius to Diocletian', *Acta Arch.* xvi, 1964, 257 ff.

O. DAVIES, *Roman Mines in Europe* (Oxford, 1935).

S. DUŠANIĆ, 'Novi antinojev natpisi Metalla Municipii Dardanorum', *ŽA* xxi, 1971, 291 ff.

G. ELMER, 'Der römische Geldverkehr in Carnuntum', *Numismatische Zeitschrift* lxvi, 1933, 55 ff.

J. FITZ, 'Die domus Heraclitiana in Intercisa', *Klio* i, 1968, 159 ff.

J. FITZ, *Les syriens à Intercisa* (Brussels, 1972).

S. FOLTINY, 'Eine dakische Henkelschale aus Müllendorf', *Barb-Festschr.* 79 ff.

F. FREMERSDORF, 'Rheinischer Export nach dem Donauraum', *Laur. Aqu.* i, 168 ff.

J. FROMOLS, 'Découverte d'une plaque danubienne à Port sur Saône', *Jahrbuch des Römisch-Germanischen Zentralmuseums* v, 1958, 259 ff.

D. GABLER, 'Munera Pannonica', *Arch. Ért.* xciii, 1966, 20 ff.

J. GARBSCH, *Die norisch-pannonische Frauentracht* (München, 1965).

E. GREN, *Der Münzfund von Viminacium* (Uppsala–Leipzig, 1934).

E. GREN, *Kleinasien und der Ostbalkan in der wirtschaftlichen Entwicklung der römischen Kaiserzeit* (Uppsala, 1941).

R. KATIČIĆ, 'Die neuesten Forschungen über die einheimische Sprachschicht in den illyrischen Provinzen', *Symposium 1964*, 31 ff.

R. KATIČIĆ, 'Zur Frage der keltischen und pannonischen Namengebiete im römischen Dalmatien', *Godišnjak* iii, Centar za Balkanološka Ispitivanja i (Sarajevo, 1965), 53 ff.

R. KATIČIĆ, 'Keltska osobna imena u antičko Sloveniji', *Arh. Vestn.* xvii, 1966, 145 ff.

H. G. KOLBE, 'Die Laufbahn des Faustinianus aus Carnuntum', *Carn.-Jb.* 1963-4 (1965), 48 ff.

J. K. KOLOSOVSKAYA, 'Veteranskoe zemlevladenie v Pannonii', *Vestnik drevnei istorii* 1963, 96 ff.

I. MIKL-CURK, 'Gospodarstvo na ozemlju današnje Slovenije v zgodnji antiki', *Arh. Vestn.* xix, 1968, 307 ff.

A. MÓCSY, *Bevölk.*

A. MÓCSY, 'Dekentioi,' *Ant. Tan.* xiii, 1966, 242 ff.

A. MÓCSY, 'Vorarbeiten zu einem Onomasticon von Moesia Superior', *Godišnjak* iii, Centar za Balkanološka Ispitivanja 6 (Sarajevo, 1970), 139 ff.

A. MÓCSY, 'Die lingua Pannonica', *Symposium 1967*, 195 ff.

GY. NOVÁKI, 'Überreste des Eisenhüttenwesen in Westungarn', *Barb-Festschr.* 163 ff.

V. ONDROUCH, *Nálezy keltskych, antickych a byzantskych minci na Slovensku* (Bratislava, 1964).

S. PANCIERA, *La vita economica di Aquileia* (Venice, 1957).

C. PATSCH, 'Die Saveschiffahrt in der Kaiserzeit', *JÖAI*, viii, 1905, 139 ff.

C. PATSCH, 'Zur Geschichte von Sirmium', *Strena Buliciana* (Split, 1924), 229 ff.

T. PEKÁRY, 'Aquincum pénzforgalma', *Arch. Ért.* lxxx, 1953, 106 ff.

K. PINK, 'Der Geldverkehr am österreichischen Donaulimes', *Jahrbuch für Landeskunde von Niederösterreich* xxv, 1932, 49 ff.

M. I. ROSTOWZEW, 'Ein Speculator auf der Reise. Ein Geschäftsmann bei der Abrechnung', *Römische Mitteilungen* xxvi, 1911, 278 ff.

J. ŠAŠEL, 'Caesernii', *ŽA* x, 1960, 201 ff.

J. ŠAŠEL, 'Barbii', *Eirene* v, 1966, 117 ff.

415

Select Bibliography

J. ŠAŠEL, 'Keltisches Portorium in den Ostalpen', *Corolla E. Swoboda dedicata* (Graz, 1966), 198 ff.

S. SOPRONI, 'Über den Münzumlauf in Pannonien zu Ende des 4. Jhs.', *Fol. Arch.* xx, 1969, 69 ff.

E. M. ŠTAERMAN, 'Etničeski i socialni sostav rimskogo voiska na Dunae', *Vestnik drevnei istorii* 1946, 3, 256 ff.

M. SZABÓ, 'Néhány nyelvészeti szempont a pannoniai kelta személynévanyag vizsgálatában', *Ant. Tan.* x, 1963, 220 ff.

M. SZABÓ, 'A pannoniai kelta személynévanyag vizsgálata', *Arch. Ért.* xci, 1964, 165 ff.

J. SZILÁGYI, 'Belgae im Vorort von Pannonia Inferior', *Hommages à M. Renard* ii (Brussels, 1969), 708 ff.

I. TÓTH, 'Eine Tempelbauinschrift eines ritterlichen Dekurios aus Brigetio', *Acta Classica Universitatis Scientiarum de Ludovico Kossuth nominatae Debreceniensis* vii, 1971, 91 ff.

I. TÓTH, 'A savariai Iuppiter Dolichenus-szentély feliratos emlékeiről', *Arch. Ért.* xcviii, 1971, 80 ff.

GY. ÜRÖGDI, 'A bankélet nyomai Aquincumban', *Bp. Rég.* xxi, 1964, 239 ff.

V. VELKOV, 'Kleinasiaten und Syrier in den Balkangebieten', *Études Historiques de XIIᵉ Congrès International des Sciences Historiques à Vienne 1965* (Sofia, 1965), 25 ff.

P. VEYNE, 'Epigraphica', *Latomus* xxiii, 1964, 30 ff.

J. WIELOWIEJSKI, *Kontakty Noricum i Pannonii z ludami północnymi* (Warsaw, 1970).

MINTS

Viminacium

S. DUŠANIĆ, *Starinar* iv/12, 1961, 141 ff.

J. FITZ, *Num. Közl.* lxii-lxiii, 1963–4, 19 ff.

Siscia

A. ALFÖLDI, 'Vorarbeiten zu einem Corpus der in Siscia geprägten Römermünzen', *Num. Közl.* xxvi, 1927–xxxix, 1940.

V. LÁNYI, 'The Coinage of Valentinian I in Siscia', *Acta Arch.* xxi, 1969, 33 ff.

RELIGION

Paganism

G. ALFÖLDY, 'Geschichte des religiösen Lebens in Aquincum', *Acta Arch.* xiii, 1961, 103 ff.

G. ALFÖLDY, 'Ein Denkmal des Isis-Sarapis-Kultes in Pannonien', *Alba Regia* iv-v, 1965, 87 ff.

L. BALLA, 'Deorum Prosperitati', *Acta Classica Universitatis Scientiarum de Ludovico Kossuth nominatae Debreceniensis* ii, 1966, 96 ff.

L. BALLA, 'Religion und religiöses Leben', *Die römischen Steindenkmäler von Savaria* (Budapest, 1971), 37 ff.

ZS. BÁNKI, 'A táci Iuppiter—Silvanus Domesticus-oltár', *Alba Regia* vi-vii, 1967, 165 ff.

A. CERMANOVIĆ-KUZMANOVIĆ, 'Die Denkmäler des thrakischen Heros in Iugoslavien', *Arch. Iugosl.* iv, 1963, 31 ff.

A. CERMANOVIĆ-KUZMANOVIĆ, 'Nekoliko spomenika tračkog konjanika', *Starinar* iv/13–1, 1965, 114 ff.

A. DOBROVITS, 'Az egyiptomi kultuszok emlékei Aquincumban', *Bp. Rég.* xiii, 1943, 45 ff.

F. FÜLEP, 'New Remarks on the Synagoga at Intercisa', *Acta Arch.* xviii, 1966, 93 ff.

Z. KÁDÁR, *Die kleinasiatisch-syrischen Kulte zur Römerzeit in Ungarn, Études Préliminaires* ii (Leiden, 1962).

G. I. KAZAROW, 'Denkmäler des Dolichenus-Kultes', *JÖAI* xxvii, 1932, 168 ff.

V. KOLŠEK, 'Pregled antičkih kultov na slovenskem ozemlju', *Arh. Vestn.* xix, 1968, 273 ff.

N. LÁNG, 'Das Dolichenum von Brigetio', *Laur. Aqu.* ii, 165 ff.

N. LÁNG, 'A savariai Dolichenus-csoportozat', *Arch. Ért.* 1943, 64 ff.

N. LÁNG, 'Die Dolichenus-Votivhand des Budapester Nationalmuseums', *Arch. Ért.* 1946–8, 183 ff.

R. MARIČ, *Antički kultovi u našoj zemlji* (Belgrade, 1933).

T. NAGY, 'Vallási élet Aquincumban', *Bp. Tört.* 386 ff.

T. NAGY, 'A sárkeszi Mithraeum', *Bp. Rég.* xv, 1950, 46 ff.

T. NAGY, 'Das Mithras-Relief von Paks', *Acta Ant.* vi, 1958, 407 ff.

D. PINTEROVIĆ, 'Da li je u rimsko koloniji Mursi postojalo sinagoga?', *Osj. Zb.* ix-x, 1965, 72 ff.

G. SEURE, 'Votivni reljefi u beogradskom muzeji', *Starinar* iii/1, 1922–3, 238 ff.

R. M. SWOBODA, 'Denkmäler des Magna-Mater-Kultes in Slowenien und Istrien', *Bonner Jahrbücher* clxix, 1969, 195 ff.

T. SZENTLÉLEKY, 'Das Iseum von Szombathely', *Neue Beiträge zur Geschichte der Alten Welt* ii (Berlin, 1965), 381 ff.

T. SZENTLÉLEKY, *Das Isis-Heiligtum von Szombathely* (Szombathely, 1965).

M. SZÖKE, 'Building Inscription of a Silvanus-Sanctuary from Cirpi', *Acta Arch.* xxiii, 1971, 221 ff.

I. TÓTH, 'Megjegyzések Mithras pannoniai kultuszának történetéhez', *Ant. Tan.* xii, 1965, 86 ff.

V. WESSETZKY, *Die ägyptischen Kulte zur Römerzeit in Ungarn, Études Préliminaires* i (Leiden, 1961).

V. WESSETZKY, 'Der Isis-Altar von Sopron', *Das Altertum* x, 1964, 154 ff.

K. WIGAND, 'Die Nutrices Augustae von Poetovio', *JÖAIB* xviii, 1915, 189 ff.

E. WILL, 'Les fidèles de Mithra à Poetovio', *Adriatica, Miscellanea Grga Novak* (Zagreb, 1970), 633 ff.

L. ZOTOVIĆ, 'Kult Jupitera Depulzora', *Starinar* iv/17, 1966, 43 ff.

L. ZOTOVIĆ, *Les cultes orientaux sur le territoire de la Mésie Supérieure, Études Préliminaires* xi (Leiden, 1966).

Christianity

A. ALFÖLDI, *Tracce del cristianesimo nell'epoca delle grandi migrazioni in Ungheria* (Rome, 1938).

R. EGGER, *Der heilige Hermagoras* (Klagenfurt, 1948).

Select Bibliography

L. NAGY, *Pannonia Sacra, Szent István Emlékkönyv* i (Budapest, 1938).
T. NAGY, 'A pannoniai kereszténység története', *Diss. Pann.* ii, 12 (Budapest, 1939).
R. NOLL, *Frühes Christentum in Österreich* (Vienna, 1954).
H. VETTERS, 'Drei Silberlöffel aus Carnuntum', *BRGK* xlix, 1968, 149 ff.
J. ZEILLER, *Les origines chrétiennes dans les provinces danubiennes de l'Empire Romain* (Paris, 1918).
(see also GRAVES AND CEMETERIES: INHUMATION)

CULTURE, EDUCATION, LATIN

S. DUŠANIĆ, 'Cuillula, an Epigraphic Hapax Legomenon', *Epigraphische Studien* v, 1968,
   158 ff.
R. EGGER, 'Epikureische Nachklänge aus Aquincum', *RAFC* ii, 153 ff.
A. HEKLER, 'Kunst und Kultur Pannoniens in ihren Hauptströmungen', *Strena Buliciana*
   (Split, 1924), 107 ff.
J. HERMAN, 'Posit = posuit et questions connexes dans les inscriptions pannoniennes', *Acta*
   Ant. ix, 1961, 321 ff.
J. HERMAN, 'Latinitas Pannonica', *Filológiai Közlöny* 1968, 364 ff.
H. MIHĂESCU, *Limba latina în provinciile dunarene ale imperiului Roman* (Bucharest, 1960).
L. NAGY, *Az aquincumi orgona* (Budapest, 1934).
J. GY. SZILÁGYI, 'Megjegyzések az uj szentendrei verses feliratról', *Arch. Ért.* xc, 1963, 189 ff.
W. WALCKER-MAYER, *Die römische Orgel von Aquincum* (Stuttgart, 1970).

THE ARCHAEOLOGICAL MATERIAL

Coarse pottery

É. BÓNIS, 'Die kaiserzeitliche Keramik von Pannonien', i, *Diss. Pann.* ii, 20 (Budapest, 1942).
É. BÓNIS, *Fol. Arch.* xxi, 1970, 71 ff.
P. PETRU, *Razprave, Slovenska Akademija* vi (Ljubljana, 1969), 197 ff.
K. SZ. PÓCZY, *Intercisa* ii, 29 ff.
K. SZ. PÓCZY, *Arch. Ért.* lxxxii, 1955, 56 ff.
K. SZ. PÓCZY, *Acta Arch.* vii, 1956, 73 ff.
A. SALAMON, *Fol. Arch.* xx, 1969, 53 ff.
A. SCHÖRGENDORFER, *Die römerzeitliche Keramik in den Ostalpenländern* (Vienna, 1942).
U. TRINKS, *Carn.-Jb.* 1957, 51 ff.; 1959, 70 ff.
B. VIKIĆ-BELANČIĆ, *Starinar* iv/13–14, 1965, 89 ff.
I. WELLNER, *Arch. Ért.* xcii, 1965, 42 ff.

Lamps

D. GAJ-POPOVIĆ, *ZNMB* iii, 1963, 129 ff.
D. IVÁNYI, 'Die pannonischen Lampen', *Diss. Pann.* ii, 2 (Budapest, 1935).
V. KONDIĆ and J. TODOROVIĆ, *Godišnjak Muzeja Grada Beograda* iii, 1956, 63 ff.

A. NEUMANN, 'Lampen und andere Beleuchtungsgeräte aus Vindobona', *RLiÖ* xxii (Vienna, 1967).

T. SZENTLÉLEKY, *Bp. Rég.* xix, 1959, 167 ff.

Kilns, potteries

L. BARKÓCZI, *Fol. Arch.* viii, 1956, 81 ff.

B. KUZSINSZKY, *Bp. Rég.* xi, 1932.

J. MEDUNA, *Germania* xlviii, 1970, 44.

L. NAGY, *Arch. Ért.* 1942, 162 ff.

GY. PARRAGI, *Arch. Ért.* xcviii, 1971, 60 ff.

Z. ŠUBIČ, *Arh. Vestn.*, xix, 1968, 455 ff.

Sigillata (samian ware)

F. EICHLER, *Germania* xxv, 1941, 30 ff.

D. GABLER, *Arrabona* vi, 1964, 5 ff.; ix, 1967, 21 ff.

D. GABLER, *Arch. Ért.* xci, 1964, 94 ff.; xcv, 1968, 211 ff.

D. GABLER, *Acta Ant.* xvi, 1968, 297 ff.

D. GABLER, *Acta Arch.* xxiii, 1971, 83 ff.

D. GABLER, *Bayerische Vorgeschichtsblätter* xxxi, 1966, 123 ff.; xxxiii, 1968, 100 ff.

G. JUHÁSZ, 'Die Sigillaten von Brigetio', *Diss. Pann.* ii, 3 (Budapest, 1935).

G. JUHÁSZ, *Arch. Ért.* xlix, 1936, 33 ff.

F. KŘIŽEK, *Slov. Arch.* ix, 1961, 301 ff.; xiv, 1966, 97 ff.

I. MIKL-CURK, *Terra sigillata in sorodne vrste keramike iz Poetovija* (Belgrade-Ljubljana, 1969).

L. NAGY, *Bp. Rég.* xiv, 1945, 303 ff.

B. RUTKOWSKI, *Wiadomosci Archeologiczne* xxx, 1964, 75 ff.

B. RUTKOWSKI, *Archeologia* xviii (Warsaw, 1967), 55 ff.

V. SAKAŘ, *Arch. Rozhl.* xxi, 1969, 202 ff.

J. TEJRAL, *Arch. Rozhl.* xxii, 1970, 389 ff.

B. VIKIČ-BELANČIĆ, *Arh. Vestn.* xix, 1968, 509 ff.

ZS. VISY, *Alba Regia* x, 1969, 87 ff.

The potter Pacatus

A. ALFÖLDI, *Fol. Arch.* i-ii, 1939, 97 ff.

K. KISS, *Laur. Aqu.* i, 188 ff.

Terracotta

A. ALFÖLDI, *Laur. Aqu.* i, 312 ff.

L. CASTIGLIONE, *Actes du viiie Congrès International d'Archéologie Classique, Paris, 1963* (Paris, 1965), 361 ff.

A. CERMANOVIĆ-KUZMANOVIĆ, *ZNMB* iv, 1964, 141 ff.

M. VELIĆKOVIĆ, *Katalog grčkih i rimskih terakota. Narodni Muzej, Antika* iii (Belgrade, 1957).

Select Bibliography

Bronze

zs. BÁNKI, 'Objets romains figurés en bronze, argent et plomb', *Collection du Musée Roi Saint Étienne* (Székesfehérvár, 1972).

D. GÁSPÁR, 'Spätrömische Kästchenbeschläge in Pannonien', *Acta Antiqua et Archaeologica Universitatis de Attila József nominatae* xv/1–2 (Szeged, 1971).

I. KOVRIG, 'Die Haupttypen der kaiserzeitlichen Fibeln in Pannonien', *Diss. Pann.* ii, 4 (Budapest, 1937).

A. MOTYKOVÁ-SNEIDROVÁ, 'Norisch-pannonische Gürtelbeschläge und ihre Nachbildungen in Böhmen', *Pam. Arch.* lv, 1964, 350 ff.

E. PATEK, 'Verbreitung und Herkunft der römischen Fibeltypen von Pannonien', *Diss. Pann.* ii, 19 (Budapest, 1942).

Lj. POPOVIC, DJ. MANO-ZISI, M. VELIČKOVIĆ and others, *Greek, Roman, and Early Christian Bronzes in Yugoslavia* (Belgrade, 1969).

A. RADNÓTI, *Bronzegefässe*.

A. RADNÓTI, *Intercisa* ii, 173–364.

I. SELLYE, 'Les bronzes émaillés de la Pannonie', *Diss. Pann.* ii, 8 (Budapest, 1939).

I. SELLYE, 'A pannoniai attört fémmunkák áttekintése', *Arch. Ért.* 1940, 236 ff.; 1941, 62 ff.

I. SELLYE, 'Recueil des bronzes ajourés en Pannonie', *Hommages à M. Renard* iii (Brussels 1969), 516 ff.

I. SELLYE, 'Adatok az arrabonai fémmüvességhez', *Arrabona* xii, 1970, 69 ff.

Sculpture in bronze

I. BÓNA, *Acta Arch.* xxiii, 1971, 225 ff.

R. FLEISCHER, *Die römischen Bronzen aus Österreich* (Mainz, 1967).

G. GAMER, *Germania* xlvi, 1968, 53 ff.

Z. KÁDÁR, *Fol. Arch.* xvii, 1965, 111 ff.

I. PAULOVICS, *Pannonia* i, 1935, 21 ff.

D. PINTEROVIĆ, *Osj. Zb.* viii, 1962, 117 ff.; ix-x, 1965, 92 ff.

T. SZENTLÉLEKY, *VMMK* iv, 1965, 103 ff.

Glass

L. BARKÓCZI, *Fol. Arch.* xviii, 1966, 67 ff.; xix, 1968, 59 ff.; xx, 1969, 47 ff.; xxii, 1971, 71 ff.

L. BARKÓCZI and Á. SALAMON, *Arch. Ért.* xcv, 1968, 29 ff.

A. BENKÖ, 'Üvegcorpus', *Régészeti Füzetek* ii, 11 (Budapest, 1962).

É. BÓNIS, *Bp. Rég.* xiv, 1945, 561 ff.

F. FÜLEP, *Acta Ant.* xvi, 1968, 401 ff.

K. M. KABA, *Bp. Rég.* xviii, 1958, 425 ff.

S. PETRU, *Razprave, Slovenska Akademija* vi (Ljubljana, 1969), 163 ff.

A. RADNÓTI, *Intercisa* ii, 141 ff.

R. SUNKOWSKY, *Antike Gläser in Carnuntum und Wien* (Vienna, 1956).

Sculpture in stone

L. BALLA, T. P. BUOCZ, Z. KÁDÁR, A. MÓCSY and T. SZENTLÉLEKY, *Die römischen Steindenk-mäler von Savaria* (Budapest, 1971).

J. BRUNŠMID, 'Kameni spomenici Hrvatskoga Narodnoga Muzeja u Zagrebu', *VHAD* vii, 1903–xi, 1911.

A. SZ. BURGER, 'Áldozati jelenet Pannonia köemlékein', *Régészeti Füzetek* ii, 5 (Budapest, 1959).

A. SZ. BURGER, 'Collegiumi köfaragómühelyek Aquincumban', *Bp. Rég.* xix, 1959, 9 ff.

A. CERMANOVIĆ-KUZMANOVIĆ, 'Die dekorierten Sarkophage in den römischen Provinzen von Jugoslawien', *Arch. Iugosl.* vi, 1965, 89 ff.

E. DIEZ, 'Die Aschenkisten von Poetovio', *JÖAI* xxxvii, 1948, 151 ff.

E. DIEZ, 'Der Giebel des carnuntinischen Fahnenheiligtums', *Corolla E. Swoboda dedicata* (Graz, 1966), 105 ff.

D. P. DIMITROV, *Nadgrobnite ploči ot rimsko vreme v severna Bălgarija* (Sofia, 1942).

G. ERDÉLYI, 'Steindenkmäler', *Intercisa* i, 169 ff.

G. ERDÉLYI, 'Adatok a pannoniai siraediculákhoz', *Arch. Ért.* lxxxviii, 1961, 184 ff.

D. GABLER, 'Arrabona környékének köplasztikai emlékei', *Arrabona* x, 1968, 51 ff.

D. GABLER, 'Scarbantia környékének köplasztikai emlékei', *Arrabona*, xi, 1969, 5 ff.

M. GORENC, 'Klesarna kiparska manufaktura u našim krajima', *Arh. Vestn.* xix, 1968, 195 ff.

G. KRĂSTEVA-NOŽAROVA, *Ukazatel na predmeti izobrazeni vărhu anticni pametnici ot Bălgarija* (Sofia, 1958).

M. L. KRÜGER, 'Die Rundskulpturen des Stadtgebietes von Carnuntum', *Corpus der Skulpturen der römischen Welt, Österreich* i, 2 (Vienna, 1967).

M. L. KRÜGER, 'Die Reliefs des Stadtgebietes von Carnuntum, i, Die figürlichen Reliefs', *Corpus der Skulpturen der römischen Welt, Österreich* i, 3 (Vienna, 1970); 'ii, Die Dekatíven Reliefs', *ibid.*, i. 4 (Vienna, 1972).

A. MÓCSY, *Fol. Arch.* ix, 1957, 83 ff.; xvi, 1964, 43 ff.

L. NAGY, *Laur. Aqu.* ii, 232 ff.

T. NAGY, 'Köfaragás és szobrászat Aquincumban', *Bp. Rég.* xxii, 1971, 103 ff.

A. NEUMANN, 'Die Skulpturen des Stadtgebietes von Vindobona', *Corpus der Skulpturen der römischen Welt, Österreich* i, 1 (Vienna, 1970).

D. PINTEROVIĆ, *Osj. Zb.* xi, 1967.

A. SCHOBER, *Die römischen Grabsteine von Noricum und Pannonien* (Vienna, 1923).

(See also INSCRIPTIONS)

Mythological scenes

E. DIEZ, *Carn.-Jb.* 1963–4 (1965), 43 ff.

G. ERDÉLYI, *Arch. Ért.* lxxvii, 1950, 72 ff.

G. ERDÉLYI, *Acta Ant.* xiv, 1966, 211 ff.

G. ERDÉLYI, *Acta Arch.* xiii, 1961, 89 ff.

Z. KÁDÁR, *Actes du viiie Congrès International d'Archéologie Classique, Paris 1963* (Paris, 1965), 381 ff.

Select Bibliography

Wall-paintings, mosaics, stuccos

K. M. KABA, *Bp. Rég.* xvi, 1955, 255 ff.

H. KENNER, *La Mosaïque gréco-romain* (Paris, 1965), 89 ff.

Á. KISS, *ibid.*, 297 ff.

DJ. MANO-ZISI, *ibid.*, 290 ff.

L. NAGY, *Arch. Ért.* xli, 1927, 114 ff.

L. NAGY, *Römische Mitteilungen* xl, 1925, 51 ff.; xli, 1926, 79 ff.

I. WELLNER, *Acta Arch.* xxi, 1969, 235 ff.

I. WELLNER, 'A magyarországi rómaikori épületek belső diszitő müvészete', *Épités és Épité-szettudomány* ii, 1971, 327 ff.

Art in general: various finds and treasures

É. BÓNIS, 'Emaillierte Palästrageräte aus Brigetio', *Fol. Arch.* xix, 1968, 25 ff.

S. FERRI, *ARD.*

M. GRBIĆ, *Plastika.*

Z. KÁDÁR, 'Adatok a Duna vidéki késöantik ezüstedények problematikájához', *Fol. Arch.* xii, 1960, 133 ff.

M. MANOJLOVIĆ-MARIJANSKI, *Kasnorimski šljemovi iz Berkasova* (Novi Sad, 1964).

DJ. MANO-ZISI, *Antika u Narodnom Muzeju u Beogradu* (Belgrade, 1956).

DJ. MANO-ZISI, *Nalaz iz Tekije* (Belgrade, 1957).

R. NOLL, *Kunst der Römerzeit in Österreich* (Salzburg, 1949).

V. ONDROUCH, *Boháté hroby z doby rimskej-na Slovensku* (Bratislava, 1957).

D. PINTEROVIĆ, 'Geme s terena Murse', *Osj. Zb.* ix-x, 1965, 52 ff.

K. SZ. PÓCZY, 'Gemaltes Männerporträt aus einem Mumiengrab in Aquincum', *Studien zur Geschichte und Philosophie des Altertums* (editor J. HARMATTA) (Budapest, 1968), 331 ff.

K. B. SEY, M. KÁROLYI and T. SZENTLÉLEKY, 'A balozsameggyesi római ékszer- és éremlelet', *Arch. Ért.* xcviii, 1971, 190 ff.

E. B. THOMAS, 'Severisches Goldgewebe aus Viminacium', *Studien zur Geschichte und Philosophie des Altertums* (editor J. HARMATTA) (Budapest, 1968), 337 ff.

E. B. THOMAS, *Helme, Schilde, Dolche* (Budapest, 1971).

Graves and cemeteries

Cremation

É. BÓNIS, *Fol. Arch.* xii, 1960, 91 ff.

A. SZ. BURGER, *Arch. Ért.* xciii, 1966, 254 ff.

D. DŽONOVA, *Arheologija* iv (Sofia, 1962), 3, 32 ff.

F. FÜLEP, *Acta Arch.* ix, 1958, 373 ff.

M. V. GARAŠANIN, *Godišnjak*, vi, *Akademija Nauka i Umjetnosti Bosne i Hercegovine*, Centar za Balkanološka Ispitivanja, 4 (Sarejevo, 1968), 5 ff.

T. KNEZ, *Arh. Vestn.* xix, 1968, 221 ff.

T. KNEZ, P. PETRU and S. PETRU, *Razprave, Slovenska Akademija* vi (Ljubljana, 1969), 7 ff.,
    85 ff., 109 ff.
A. MÓCSY, *Arch. Ért.* lxxxi, 1954, 167 ff.
E. F. PETRES, *Fol. Arch.* xvii, 1965, 96 ff.
A. RADNÓTI, *Barb-Festschr.* 199 ff.
A. SCHOBER, *JÖAIB* xvii, 1914, 222 ff.
D. SREJOVIĆ, *Starinar* iv/13–14, 1965, 52 ff.
I. WEILER, *Carn.-Jb.* 1961–2 (1963), 61 ff.

Tumuli and cart-graves

A. ALFÖLDI and A. RADNÓTI, 'Zügelringe und Zierbeschläge von römischen Jochen und
    Kummeten aus Pannonien', *Serta Hoffilleriana* (Zagreb, 1940), 309 ff.
É. BÓNIS, *Fol. Arch.* ix, 1957, 76 ff.; xiv, 1962, 23 ff.
H. KERCHLER, 'Die römerzeitlichen Brandbestattungen unter Hügeln in Niederösterreich',
    Archaeologia Austriaca, Beiheft viii (Vienna, 1967).
R. MÜLLER, *Arch. Ért.* xcviii, 1971, 3 ff.
S. PAHIĆ, *Arh. Vestn.* xi-xii, 1960–1, 116 ff.; xix, 1968, 321 ff.
K. SÁGI, *Arch. Ért.* 1943, 113 ff.; lxxviii, 1951, 73 ff.
M. ŠEPER, *Arheološki Radovi i Rasprave* ii, 1962, 335 ff.

Inhumation

L. BARKÓCZI, *Acta Ant.* xiii, 1965, 215 ff.
L. BARKÓCZI, *Komárom Megyei Muzeumok Közleményei* i, 1968, 75 ff.
A. SZ. BURGER, *Acta Arch.* xviii, 1966, 99 ff.
A. SZ. BURGER, *Fol. Arch.* xix, 1968, 87 ff.
F. FÜLEP, *Arch. Ért.* lxxxix, 1962, 23 ff.; xcvi, 1969, 3 ff.
V. LÁNYI, *Acta Arch.* xxiv, 1972, 53 ff.
L. NAGY, 'Mumienbegräbnisse aus Aquincum', *Diss. Pann.* i, 4 (Budapest, 1935).
R. NOLL, *Barb-Festschr.* 149 ff.
K. SZ. PÓCZY, *Arch. Ért.* xci, 1964, 176 ff.
E. SPAJIĆ, *Osj. Zb.* xi, 1967, 101 ff.
E. B. VÁGÓ, *Alba Regia* i, 1960, 46 ff.
M. VALTROVIĆ, *Starinar* ii, 1885, 33 ff.; ii/1, 1906, 128 ff.
L. ZOTOVIĆ, *LuJ* i, 1961, 171 ff.

ARCHAEOLOGICAL RESEARCH: BOOKS AND PERIODICALS GEOGRAPHICALLY ARRANGED

PANNONIA

A. GRAF, 'Übersicht der antiken Geographie von Pannonien', *Diss. Pann.* i, 5 (Budapest, 1936).
Tabula Imperii Romani, Sheet L 34, Aquincum—Sarmizegethusa—Sirmium (Budapest, 1968).
E. B. THOMAS, *RVP*.

Select Bibliography

Austria

Burgenl. Heimatb.
Fundberichte aus Österreich
Jahrbuch für Altertumskunde
JÖAI
Pro Austria Romana
RLiÖ

L. FRANZ and A. NEUMANN, *Lexikon Ur- und Frühgeschichtlicher Fundstätten Österreichs* (Vienna, 1965).

W. KUBITSCHEK, *Römerfunde.*

W. KUBITSCHEK, 'Ältere Berichte über den römischen Limes in Pannonien, i', *Sitzungsberichte der Österreichischen Akademie, Philosophisch-historische Klasse* ccix/1 (Vienna, 1929).

G. PASCHER, 'Römische Siedlungen und Strassen im Limesgebiet zwischen Enns und Leitha', *RLiÖ* xix (Vienna, 1949).

B. SARIA, 'Der römische Gutshof von Winden am See', *Burgenländische Forschungen*, xiii (Eisenstadt, 1951).

H. VETTERS and H. MITSCHA-MÄHRHEIM, *Die Römerzeit in Niederösterreich, Atlas von Niederösterreich* (Vienna, 1958).

Vindobona

A. NEUMANN, *Vindobona. Die römische Vergangenheit Wiens, Geschichte–Erforschung–Funde* (Vienna–Cologne–Graz, 1972).

Carnuntum

Bericht des Vereins Carnuntum
Carn.-Jb.
E. SWOBODA, *Carnuntum.*

Hungary

Acta Ant.
Acta Arch.
Alba Regia
Ant. Tan.
Arch. Ért.
Arrabona
Fol. Arch.
VMMK

K. BAKAY, N. KALICZ and K. SÁGI, *Veszprém megye régészeti topográfiája, A keszthelyi járás* (Budapest, 1966). *A devecseri és sümegi járás* (Budapest, 1970).

J. BANNER and I. JAKABFFY, *Archäologische Bibliographie des Mittel-Donau-Beckens* (Budapest, 1954), *1954–1959* (Budapest, 1961), *1960–1966* (Budapest, 1966).

I. ÉRI, M. KELEMEN, P. NÉMETH and I. TORMA, *Veszprém megye régészeti topográfiája, A vesz-prémi járás* (Budapest, 1969).

J. FITZ, *A római kor Fejér megyében* (Székesfehérvár, 1970).

F. FÜLEP and É. CSEREY, *Nagytétény müemlékei* (Budapest, 1959).

Intercisa i-ii.

É. KOCZTUR, 'Somogy megye régészeti leletkatasztere', *Régészeti Füzetek* ii, 13 (Budapest, 1964).

B. KUZSINSZKY, *A Balaton környékének archaeologiája* (Budapest, 1920).

R. MÜLLER, *Régészeti terepbejárások a göcseji 'szegek' vidékén* (Zalaegerszeg, 1971).

M. WOSINSKY, *Tolna vármegye története az öskortól a honfoglalásig* i-ii (Budapest, 1896).

Savaria

Savaria, Bulletin der Museen des Komitates Vas
Vasi Szemle
T. P. BUOCZ, *Savaria—Szombathely topográfiája* (Szombathely, 1968).

Scarbantia

Soproni Szemle
K. SZ. PÓCZY, *Sopron rómaikori emlékei* (Budapest, 1965).

Brigetio

BARKÓCZI, *Brigetio*.

Aquincum

Bp. Müeml.
Bp. Rég.
L. NAGY, *Az óbudai ókeresztény cella trichora a Raktár utcában* (Budapest, 1934).
L. NAGY, *Az Eskü téri eröd, Pest város öse* (Budapest, 1946).
J. SZILÁGYI, *Aquincum* (Budapest–Berlin, 1956).

Gorsium

Alba Regia
J. FITZ, *Gorsium* (Székesfehérvár, 1970).

Sopianae

Janus Pannonius Muzeum Évkönyve
F. FÜLEP, *Pécs rómaikori emlékei* (Budapest, 1964).

425

Select Bibliography

Yugoslavia

Arch. Iugosl.
Arh. Pregl.
LuJ
ŽA

Slovenia

Arh. Vestn.
Glasnik Muzejskega Društva za Slovenijo
Situla
Varstvo Spomenikov
J. KLEMENC and B. SARIA, *Archäologische Karte von Jugoslawien, Blatt Ptuj* (Zagreb, 1936);
    Blatt Rogatec (Zagreb, 1939).

Emona

J. ŠAŠEL, *Vodnik po Emoni* (Ljubljana, 1955).
M. DETONI and T. KURENT, *Modularna rekonstrukcija Emone* (Ljubljana, 1963).

Poetovio

M. ABRAMIĆ, *Führer durch Poetovio* (Vienna, 1924).
J. KLEMENC, *Ptujski grad v kasni antiki* (Ljubljana, 1950).

Neviodunum
T. KNEZ, P. PETRU and S. ŠKALER, *Neviodunum* (Dolenjske Založbe, 1960).

Croatia

Arheološki Radovi i Rasprave
Osj. Zb.
VHAD
Vjesnik arheološkog muzeja u Zagrebu
J. KLEMENC and B. SARIA, *Archäologische Kater von Jugoslawien, Blatt Zagreb* (Zagreb, 1938).
B. VIKIĆ and M. GORENC, *Prilog istraživanja antičkih naselja i putova u sjeverno-zapadnoj Hrvatski*
    (Zagreb, 1969).

Serbia and Vojvodina

Rad vojv. muz.
Starinar
ZNMB
D. and M. GARAŠANIN, *Nalazišta.*

426

MOESIA SUPERIOR

Yugoslavia (see above)

Serbia (see also above)

Arheološki nalazišta i spomenici u Srbiji, ii, *Centraljna Srbija, Gradja x. Arheološkog Instituta* (Belgrade, 1956).
Stare Kulture u Djerdapu—Anciennes Cultures du Djerdap (Belgrade, 1969).

Kosovo and Metohija

Godišnjak Muzeja Kosova i Metohije
E. ČERŠKOV, *Rimljani na Kosovu i Metohiji* (Belgrade, 1970).
E. ČERŠKOV, *Municipium DD, Sočanica* (Pristina—Belgrade, 1970).

Macedonia

ŽA
Zbornik, Recueil des Travaux, Publications du Musée Archéologique de Skoplje
N. VULIĆ, *Nekoliko pitanja iz antičke istorije naše zemlje* (Belgrade, 1961).

Bulgaria

Arheologija
IBAI
S. GEORGIJEVA and V. VELKOV, *Bibliografija na bălgarskata arheologija 1879-1955* (Sofia, 1957).

BARBARIAN LANDS BEYOND THE FRONTIER

Czechoslovakia

Arch. Rozhl.
Pam. Arch.
Slov. Arch.

Hungary (see also above)

Debreceni Déri Muzeum Évkönyve
Egri Muzeum Évkönyve
MFMÉ
Miskolci Herman Ottó Muzeum Évkönyve
M. PÁRDUCZ, *Denkmäler der Sarmatenzeit in Ungarn* i-iii (Budapest, 1941-51).

228; foreign policy, 193; rebellion against, 200; wars, 191 f., 197

Constans, emperor, 285, 291, 306, 308

Constantia, daughter of Constantine, 285

Constantine I, emperor, 27, 267, 285, 291, 302, 307, 311, 313 f.; Christianity under, 323 f., 328, 332; estates, 306; *limes* under, 282, 285; prosperity under, 312; provincial administration, 273 ff.; wars, 277 f., 288 f.

Constantius I, emperor, 267
  bishop of Lauriacum, 353
  II, emperor, 286 ff., 293 f., 329 f.

Cornacates, civitas of, 51, 66, 137; organized by Romans, 53 f.; Pannonian tribe, 14

Cornelius Felix Plotianus, governor of Lower Pannonia, 197
  Fuscus, procurator of Pannonia, 41; praetorian prefect, 82 f.
  Scipio Asiagenus, L., 15, 18 f., 32

Cosmius, customs official, 228

Cotini (Kytnoi), incorporated into empire, 57 f., 188, 190, 192, 199, 209, 222, 248, 272; in pre-Roman period, 19, 35; language of, 59 f.; pay tribute to Vannius, 40

Cotiso, Dacian king, 23 f.

Crassus, *see* Licinius

Critasirus, king of Taurisci and Boii, 19, 27

Dacians, 27, 47; civitas, 66, 68; early Roman campaign against, 21 ff., 32, 35 ff., 42; invasion of Moesia, 41 ff.; language of, 4; recruitment of praetorians from, 200; settlement of free Dacians, 191; wars against Celts, 17 ff., 25, 35, 37, 55, 61; wars against Rome, 82 ff., 91, 94 f., 100, 129

Daesidiates, rebellion of, 37 ff.

Daizo, Thracian name, 65

Dalmatians, 4, 52, 117, 217

Dardanians, 34, 240; civitas, 68 f., 145, 223; exports of, 246; in pre-Roman period, 5, 9 ff., 25 ff., 153; names, 65, 254; pacification, 23 f.; recruitment from, 236; wars of, 15 ff., 21

Dardanoi, 9

Dardanos, 274

Dases, Pannonian name, 59

Dasmenus, Pannonian leader, 38
  Pannonian name, 59

Das(s)ius, Illyrian name, 65

Dea Syria, 258

Decebalus, king of Dacia, 82 ff., 86, 95

Decius, Traianus, emperor, 203 ff., 244 f.

Decoratus, *cognomen*, 248

Deianira, on mosaic, 237

Demetrius, king of Macedonia, 9
  (Demetrios, Dmitrij) deacon of Sirmium, 327

Dentheletae, 23 f., 34

Diana, goddess, 231, 251 ff.

Diegis, 84

Dimenses, in Pliny, 68

Dio Cassius, *see* Cassius
  Chrysostom, 91

Diocletian, emperor, 312; abdication of, 276; formation of Tetrarchy, 267; frontier policy, 269 ff., 280, 285; palace of, 306; persecution of Christians, 259, 326; provincial reorganization, 223, 273 ff.; wars, 268 f.

Diodorus, 7

Dioscuri, 254

Ditybistos, 358

Diurpaneus, king of Dacians, 82

Dolens, Thracian name, 65

DOMISA, legend on coin, 56

Domitian, emperor, colony of, 116; death, 91; frontier under, 47, 89; provincial garrison under, 81 ff., 85 f., 120, 124; trade under, 120, 122, 124, 129; wars, 82 ff., 102, 129

Domitius Ahenobarbus, governor of Illyricum, 35
  Zmaragdus, C., decurion of Carnuntum, 141, 227

Domnus, bishop of Sirmium, 330

Donatus, priest, 327

Drigissa, Thracian name, 65

Drusus, 39 f., 44

Dubius, *scriba*, 144

DVTEVTA, on coin, 56

Ennodius, 344

Epona, worship of, 253 f.

Eppii, from Gaul, 136

Eravisci, 152; civitas, 66, 137, 141 f., 151, 155 ff.; coins, 56; names, 59 ff.; Pannonian tribe, 5, 54 f.

Eusebius, bishop of Cibalae, 325

Eutherius, bishop of Sirmium, 330

Farnobius, Gothic leader, 339 f.

Firmidii, at Neviodunum, 136

Flavii, native, 135 f., 138

Flavius Biturix, T., 135
  Bonio, T., 135
  Cobromarus, T., 135
  Lupus, senator, 348
  Proculus, T., tombstone of, 134 f.
  Samio, T., 135

Florus, Christian martyr, 326
  historian, 36, 39, 185
  princeps of Eravisci, 137

Fonteius Agrippa, governor of Moesia, 42, 80, 83

Fortunatus, Christian martyr, 327, 333

Fragiledus, 288

Frigeridus, campaign in Illyricum, 339 f.; tile-stamps of, 291, 293

Fritigern, Gothic leader, 339 f., 342

Fritigil, queen of Marcomanni, 345

Fronto, 201

Fufius Germinus, army commander, 22

Funisulanus Vettonianus, L., governor of Moesia Superior, 82

Furii, estate of, 138, 241; freedman of, 263

Index of Deities, Peoples and Persons

Pacatianus, 206
Pacatus, potter, 177 f.
Paionians, 65
Palpellius Hister, governor of Pannonia, 41
Pan, on pottery, 177
Parthians, 201 f., 228
Paul, St, in painting, 334
Paulinus, 345
Paulus, 354
Pelagonians, 65
Peregrinus, *sutor caligarius*, 129
Perseus, king of Macedonia, 10
Persians, 202
Pertinax, emperor, 188
Pescennius Niger, 200 f., 228
Peter, St, in painting, 334
Petronii, Aquileian family, 77
Petronius L. lib. Licco, L., 124
Philip II, king of Macedonia, 9
    V, king of Macedonia, 9 f.
Philippus, emperor, 203 ff.
Photeinos, bishop of Sirmium, 330
Phrygians, 9
Picenses, Picensii, Moesian tribe, 68
    part of Limigantes, 279
Pinnes, leader of Pannonian rebellion, 39
Pipa, daughter of Attalus, 206
Plautius Silvanus, crushes Pannonian rebellion,
    38 f.
    Aelianus, governor of Moesia, 41, 66 ff.
Pliny the elder, 19, 37, 53, 59 f., 66 ff., 74 ff., 112
Pollio, *lector* at Cibalae, 325, 327
Polybius, 12 f., 28
Pompeius Sextus, governor of Macedonia, 38
Pompeius Trogus, 5, 7
Pontii, estate of, 138, 241
Poppaeus Sabinus, governor of Moesia, 33, 44
Posidonius, 1 f., 13 ff., 17 f., 28
Priam, on grave reliefs, 262
Priscus, 351, 357
Probus, emperor, economic policy, 265 ff., 272,
    298 f.; estates of, 244
    praetorian prefect of Illyricum, 294 f., 310
Procopius, 131 ff., 350 f.
Pseudo-Augustine, 344
Ptolemy, 18, 21, 53, 57 ff., 66 ff., 110

Quadi, 35, 37, 60, 89, 342, 345; fort in territory of,
    293 f.; under Vannius, 40 f., 57 f.; wars,
    83 f., 101, 103, 186 ff., 199, 203, 268, 286 ff.,
    291, 294 ff.
Quadriviae, *see* Silvanae
Quartus Adnamati f., 151
Quintilius Condianus, army commander, 188
    Maximus, army commander, 188
Quintillus, 210
Quirinus, bishop of Siscia, 326 ff., 348, 353;
    basilica of, 313, 332

Radagaisus, barbarian leader, 347 ff., 353
Rausimodus, Sarmatian leader, 277 f.

RAVIZ, on coins, 56
Regalianus, usurper, 206 ff., 264
Resatus, potter, 177
Rhea Sylvia, on grave reliefs, 262
Rhinelanders, 125, 130 f.
Richomeres, army commander, 339
Rider-gods, Danubian, 246, 254, 324
Romula, mother of Galerius, 302
Romulus, Christian martyr, 327
Roumanians, 356
Roxolani, 94 f., 191, 193; migration of, 209; wars,
    41 f., 100
Rubrius Gallus, governor of Moesia, 42, 80
Runo, 288
Rustii, at Neviodunum, 136

Sabinii, customs-farmers, 226
Saco, Celtic name, 59
Saphrac, Alannic leader, 340 ff., 345, 349 f., 352 f.
Sarapis, cult of, 258
Sarmatians, 47, 59 f., 89, 93, 186, 293, 296, 322,
    345, 358; trade with, 245; treaties, 89, 91,
    271, 288 f.; wars, 36 ff., 42, 84, 95, 99, 103,
    187, 192 f., 197, 202 f., 205 f., 209 ff., 264,
    267 ff., 276 ff., 286 f., 290 f., 294 f., 310, 342;
    see also Iazyges
Savarina, freedwoman, 77
Savus, god, 136, 182
Scerviaedus, Illyrian name, 65
    Sitaes (Sitae filius), 65
Scipio, *see* Cornelius
Scordisci (Scordistae), 25 ff., 30, 53; allies of
    Rome, 34, 39; campaign of Tiberius, 23 f.,
    34; Celtic tribe, 5 ff.; civitas, 66, 137; de-
    cline, 15 ff.; hegemony, 9 ff., 202; names of,
    60, 65 f.
Scorilo Ressati lib., 129
Scribonius Curio, C., Macedonian campaign of,
    2, 17 f., 32
Scythians, 68
Secconius Paternus, C., 125
Sedatus, god, 253
Septimius Severus, emperor, and Danubian army,
    197 f., 200 f., 204, 217; immigration into
    towns under, 228; municipal reorganization
    by, 140, 214 f., 218 ff., 225 f.; religion under,
    232 ff., 256, 258
    House of, 110, 126, 148, 193, 198, 248, 251, 256,
    263, 265, 307; client-kingdoms under, 198 f.,
    209; Danubian army of, 200 f.; immigration
    under, 227 ff.; mining under, 134, 216;
    prosperity under, 230 ff.; provincial ad-
    ministration, 119, 217 ff., 240; trade, 247
Serapilli, Pannonian tribe, 54; civitas, 66, 145, 155
Serretes, Pannonian tribe, 54; civitas, 66
Severi, *see* Septimius Severus, House of
Severinus, St, 357
Severus Alexander, emperor, 202, 216
Sido, prince of Quadi, 41, 83 f.
Sidonius Apollinaris, 344, 351
Sigovesus, Celtic prince, 5, 25

434

2 Index of Places

Achaea, 44
Actium, battle of, 21, 23, 33
Acumincum (Slankamen), 279
Ad Flexum (Magyaróvár), fort, 88
Ad Militare (Kiskőszeg-Batina), fort, 88
Adony, see Vetus Salina
Adrianople, 340, 344
Aegean, the, 212
Aelianum, 225
Africa, 44, 194 f., 338
Albertfalva, fort, 88, 105; civil settlement, 238
Alessio, see Lissus
Alexandria, 109, 245
Alisca (Őcsény), fort, 106
Alma Mons (Fruška Gora), 38, 246, 298
Alutus (Olt), R., 96, valley, 97
Amida, 287
Andautonia, municipium, 115, 136, 152, 155, 213
Angros, R., 2
Annamatia, fort, 105
Antioch, 340; immigrants from, 81, 141, 227
Apollonia, coins of, 30
Apulum, 218
Aquae, Chora of, 133, 224, 303; see at, 329
Aquae, fort, see Prahovo
Aquae Balizae (Aquae Balissae), (Daruvar), 143
Aquae Iasae, 144
Aquileia, 22, 33 f., 124, 181, 187, 228, 250, 299, 330, 341, 348, 353; coins from, 321; immigrants from, 76 ff., 120, 130; mosaics, 302, 313, 337; road to Danube, 44, 71; trading centre, 28, 31 f., 123, 246
Aquincum, 88, 175, 182, 196, 230, 262, 269, 277, 290, 292 f., 295, 337; amphitheatre (civil), 162; (military), 169; Christianity, 333, 336; citizenship grants at, 57 f., 145; civil settlements, 126, 140 ff., 226; cults at, 256 ff.;

decline of, 310 f.; excavation, 159 ff.; foreign settlers, 120 ff., 124 ff., 228; fort (auxiliary), 50, 73, 80 f., 110, 143; fortress (legionary), 85 f., 92, 99, 101 f., 105, 110 f., 152, 183, 198; granted rank of colonia, 218; headquarters of governor, 94, 111, 237; houses, 161, 164 ff., 237, 239, 300, 310, 317; industry, 122, 161 f., 177 f., 181, 237; municipium, 139 ff., 157; roads, 118, 122 f., 137, 144, 152, 162, 244; slave-trade, 129; walls, 161, 168; and see Budapest
Arabo (Rába, Raab), R., 54, 121, 144, 198, 354
Arčar, 354; and see Ratiaria
Arelate (Arles), mint, 343
Argentares, 354
Arrabona (Raab, Győr), 144, 354; fort, 49 f., 69, 81, 88, 135, 186
Asia Minor, 1, 7, 9, 25, 27, 210, 238, 328; cults from, 258; immigrants from, 117, 228; legionaries from, 154 f.; trade with, 245, 322
Astibus (Bregalnica), R., 13
Ászár, hoard, 175
Athens, 210
Atrans, 286
Augst, silver-hoard, 337
Aureliana (= municipium Aurelianum?), 131, 133, 224 f., 273, 276, 310
Aureus Mons, mountain, 298
Aureus Mons (Seona), municipium, 224 f., 298
Axius (Vardar), R., 9, 44; as route, 322; site of Scupi on, 115 ff.; valley, 12, 214, 340

Bakony, mountains, 144
Baláca, villa, 124, 171, 243, 307
Balaton, lake, see Pelso
Banat, the, 19, 97, 101
Banatska Palanka, see Lederata

437

Bardovce, 116
Bassiana, 240; municipium, 143, 152, 159, 166 ff.; promoted to colony, 225 f.; walls of, 312
Bathinus (Bosna), R., 56
Batuša, 224
Bela Palanka, *see* Remesiana
Belgrade (Beograd), 63, 68, 223; *and see* Singidunum
Berkasovo, 337
Beroia, 117
Berytos, 117
Berzovia, 101
Black Sea, 1, 82, 322, 342
Bohemia (Boiohaemum), 14, 71
Bononia, 269, 278, 326, 354
Bor, 131
Bosnia, 12, 30
Botivo, *see* Jovia
Boževac, 224
Brač, 312
Bracara, 353
Braunsberg, 73
Brazda lui Novac, *see* Devil's Dyke
Bregalnica, R., *see* Astibus
Brestovik, burial vault, 238
Brigetio (Szőny), 106, 144, 186, 203, 230, 288, 292, 294 f., 302, 325; civil settlements, 140; cults at, 256 ff.; foreign settlers, 120 ff., 124, 126, 228 f., 260; fort (auxiliary), 49 f., 88; fortress (legionary), 85, 88 f., 110; garrison, 92, 99; joins Pannonia Inferior, 198; municipium, 221; promotion to colony, 225 f.; roads, 122, 162, 305; tombstone reliefs, 122
Britain, 107
Brongos, R., *see* Morava
Brundisium, 23, 98
Brza Palanka, *see* Egeta
Budalia, 244
Budapest, native coins minted at, 56; native *oppidum* at, 18, 72, 173; *and see* Aquincum
Bulgaria, 55, 277
Byzantium, 350; road to Singidunum, 196, 213 f., 246, 260; *and see* Constantinople

Caesariana, 244, 307
Campania, 72
Campona (Budapest-Nagytétény), coin-hoard, 279; fort, 101, 105, 277, 282 ff.
Cannabiaca (Klosterneuburg), 88
Caričin Grad, 215
Carnuntum, 45, 56, 73, 104 f., 110, 188, 200 f., 206, 262, 293, 295; amphitheatre (civil), 164, 227; (military), 169; *auxilia* at, 50; Boian aristocracy at, 151 f., 238, 243; civil settlements, 126 ff., 140; conference of emperors, 276, 302, 324; cults at, 181, 256, 324; decay of, 310; entrepôt at, 120; excavations, 159, 162 ff., 237, 310; foreign settlers, 120, 124, 227; fortress (legionary), 48, 69, 71, 80 f., 85, 284, 317; garrison, 40, 43, 92, 99; headquarters of governor, 94, 186; market place,

126, 168; municipium, 139 f.; promotion to colony, 218; roads, 44, 50, 71, 76 f., 115, 152; settlement of veterans, 77; slave-trade, 129
Castra Constantia, *see* Ulcisia Castra
Castra Martis, 340
Castra Regina (Regensburg), 184
Celeia (Celje = Zilli), 77
Celamantia, *see* Iža-Leányvár
Čerević, 114
Čezava, *see* Novae
Ciabrus (Cibrica), R., 68
Cibalae (Vinkovci), 203, 267, 277, 286, 290, 315; Christianity at, 325 ff., 329; Cotini settled in territory, 199, 248, 272; municipium, 143, 152; promoted to colony, 225 f.; road, 244
Cibrica, R., *see* Ciabrus
Colapis (Kulpa), R., 12, 22, 54, 114; valley, 51
Cologne, 125, 130
Constantinople, 277, 287, 290, 340; *and see* Byzantium
Cornacum (Sotin), 53, 143
Cremona, 83
Cserszegtomaj, cemetery, 124
Csopak, 336
Cuppae (Golubac), 47 f., 96
Cynoscephalae, battle of, 9

Dacia, 4, 18, 21, 84, 86, 91, 183, 265, 279, 340, 342; barbarian attacks on, 203, 205, 208 f.; clearing of barbarians from, 187 f.; colonization of, 130, 227; conquest by Rome, 89; establishment of province, 94 f., 97 f., 100, 130; evacuation, 101, 209 ff., 267 f., 280, 311, 354; flight from, 60; garrison, 88, 92, 95 ff., 198, 203; reliefs of Rider-god in, 254; roads, 97 f., 100 f., 118, 138, 214; Sarmatians granted access, 191, 193, 197, 209; settlement of barbarians, 189, 191 f.; trade, 130
Aurelian's province, 267, 272 ff., 311, 354
Dacia Mediterranea, establishment, 274 ff.; Gothic raids on, 349; peace in, 296; persecution of Christians, 326; sees in, 329; settlement of barbarians in, 341
Dacia Ripensis, 302 f., 349, 354; establishment, 274; failure to export, 321; invasion of, 340, 342; *limes*, 280, 285, 296; peace in, 296; persecution of Christians, 326; sculpture, 311; sees in, 329; settlement of Goths in, 339, 343, 352; Vandal migration from, 347
Dalj, *see* Teutoburgium
Dalmatia, 34, 38 f., 65, 82, 139, 223, 348; mines in, 131, 208; native population, 4, 13 f., 27, 55; provincial administration, 134 f., 140; recruitment in, 154, 242; settlers from, 113, 134
Damascus, 98
Danube, *passim*
Dard (. . . .), municipium, 133; *and see* Sočanica
Dardania, 354; emperor from, 249; establishment of province, 273 ff.; exports, 299, 322; in pre-Roman period, 9 ff., 15; legionary fortress in, 69; part of East Roman empire,

I Noricorum equitata, 81
I Ulpia Pannoniorum, 216
I Raetorum, 81
I Thracum Syriaca, 81, 97, 195
XXXII Voluntariorum c.R., 81
Cohortes Aureliae Dardanorum, 195
 Aureliae novae, 195, 216
 Latobicorum, 39
 Maurorum, 195
 Pannoniorum, 39
 Varcianorum, 39
cohorts, 42, 52, 155; on frontier, 81, 120; in
 Moesia Superior, 91, 97
coin-dies, moulds for, 28
 -hoards, 56, 101 ff., 194, 198, 202 f., 205 f., 264,
 279, 294, 321, 345
coins, coinage, Celtic, 28 ff.; from Apollonia and
 Dyrrhachium, 30; *denarii*, 30, 33, 56, 104,
 206; from eastern mints, 321; of Eravisci, 56;
 legionary, of Gallienus, 208; mine-, 131 ff.;
 minted in Pannonia, 208; offering of to
 water, 324; the Philippus, 28; rarity in late
 fourth century, 345; of Regalianus and Dryantilla, 206; representing fort-gates, 269, 285;
 Roman, 30, 45, 49, 103, 107, 294, 322; tetradrachms, 28, 30; *see also* money
collegia, 125 f., 141, 164, 181
Collegium centonariorum, 161
 fabrum et centonariorum, 125
 funeraticium, 125
 negotiantium, 125
coloni, 243; lease mines, 134
 argentariorum, shrine erected by, 134
Colonia (Aelia) Mursa, 222; *see also* Index 2,
 Mursa
 Claudia Savaria, 151, 308; *see also* Index 2,
 Savaria
 Flavia Felix Domitiana, 136; *see also* Index 2,
 Scupi
 Julia Emona, 40; *see also* Index 2, Emona
 Septimia Siscia, 219; *see also* Index 2, Siscia
 Ulpia Traiana Dacica (Sarmizegethusa), 136
 Ulpia (Poetovio), 151; *see also* Index 2, Poetovio
colonies, 70 f., 145, 159, 226, 241, 355; *colonia*
 (*nova*) at Siscia, 244; foundation of, 40, 50,
 74 ff., 98, 112 ff., 118 f., 130, 135, 137, 178;
 promotion of towns to rank of, 218, 225 f.;
 recruitment from, 94, 154, 157 f., 236; territories of, 76 f., 79, 114 f., 135, 158
colonnades, 161, 164, 168
columns, depicted on tombstones, 186; of Marcus
 Aurelius, 175, 188; of Trajan, 107
commanders, supreme, of Danubian forces, 204 ff.
Concilium Provinciae, 175, 214
Conference of emperors at Carnuntum, 268, 276,
 302, 308, 324
Constitutio Antoniniana, 221
consulares, as provincial governors, 273, 276
Consularia Constantinopolitana, 269
Conventus civium Romanorum, at Margum, 216

copper, mining, 133; trade, 246
corn, trade, 299, 321 f., 342
corrector, of Savia, 273
corridors, in houses, 164, 171, 173
costume, late Roman, 338; local, 58, 63, 147, 150,
 152 f., 176, 261, 319
country-houses, *see* villas
courtyards, in buildings, 91, 161, 164 ff., 171 ff.,
 300; of gates, 303
craftsmen, 71, 122, 124, 126, 162, 178, 181; *see also*
 potters; stonemasons
cremation, 238; vessels, 17
crypt, of cathedral, 335
cult-buildings, Christian, 313, 333, 335, 353
cults, *see* religion
culture, native, 147 ff.; abandonment of, 234, 247,
 250, 259, 319, 358
curatores civium Romanorum, 126, 131, 142, 216
curiales, in mines-administration, 224, 308 f.; of
 towns, 310 f., 317, 323
Customs (*portorium*), 200; officials of, 181, 228,
 231, 255 f.
custor cymiteri, 333

Dacianization, 19, 57, 60, 199, 248
damnatio memoriae, of Domitian, 116
deacons, Christian, 327
decoration, on bone bracelets, 319; on pottery,
 176 f.
decurions, 159, 175; of *canabae*, 142, 162, 168,
 218; of *cives Romani*, 126; of towns, 74, 77,
 116 ff., 136, 141 ff., 158, 162, 175, 214, 221 ff.,
 225 ff., 237 ff., 244, 249, 263, 308
deductio, *see* colonies, foundation of
deductus (*deducticius*), 116 f.
denarii, *see* coins
diplomas, military, 39, 51, 60, 66, 68, 81, 85, 91,
 104, 112, 142, 155, 157, 175
dishes, 19, 176 f., 277, 300, 337
ditches, of forts, 49, 106, 284; of towns, 74, 310; of
 villas, 171
diversorium (inn), 161
dogs, in representations of Silvanus, 252; pots,
 177
Dolichena, *see* temples
drainage, 162, 266, 272 f., 298
drains, at Aquincum, 310
dress, *see* costume
dromos, in *tumuli*, 151
duces, of Moesia Prima, 294; of Noricum Ripense
 and Pannonia Prima, 348; of Valeria, 291 f.,
 294, 339
duoviri, of Aquincum, 142, 182
dux et praeses, of Valeria, 340
dykes, 114

eagles, on tombstones, 125
earthquake, at Scupi, 356
earthworks, 271 f., 280 ff.; near forts, 106
economic policy, in late third century, 265 ff., 272,
 297 f.

paganism, in late Roman period, 290, 323 ff., 332
paintings, in tombs, 315; *see also* wall-paintings
palaces, imperial, 202, 295, 300 ff., 312 ff., 332; of governor, 111, 237; *see also* villas
palaestra equipment, found in graves, 148, 150, 175 f.
paving, of streets, 78
peregrini, 58 f., 61, 63, 79, 137, 148, 217
peristyles, of market, 161; in villas, 300
Philippus, the, *see* coins
piers, of aqueduct, 159; of bridge, 215
pig's head, on gravestone, 261
pilasters, on tombstones, 180
piles, of bridge, 110
pin, gold, 353
pitcher, on tombstones, 148
pit-dwellings, 175, 317
plague, 187, 210
plaque, dedication, 74
plates, on tombstones, 148, 261
plinth, of statue, 139
porphyry, quarry, 326; statues, 303, 337
portico, 110, 126
portorium, see customs
portraits, gilt-bronze, 337; on tombstones, 180
posterns, of Emona, 74
post-holes, of forts, 106
pots, on tombstones, 180
potters, 176 ff.; 'first master', 177; *see also* craftsmen
pottery, Aco-beakers, 72; barbarian production of, 247; Dacian, 19; found at late Roman forts, 282; industry, 176 ff., 245 ff., 319 f.; Iron Age, 176; -quarter at Aquincum, 162; red and white, 72; Roman, 71, 176; samian, 49 f., 86, 114, 123, 177, 193, 245 f., 320; *sigillata*, 72, 79, 88, 111, 120, 122, 129, 161, 177 f., 320; workshops, 161, 178
praefecti, of alae, 50, 114 f., 135; auxiliary, 69, 189; of cohorts, 69; legionary, 86, 264; military of civitates, 69, 134 f.; native of civitates, 134 f.; praetorian, 50, 82 f., 188, 191, 197, 210, 294 f., 310; *ripae Danuvii et civitatium duarum Boiorum et Azaliorum,* 69; *vehiculorum,* 101
praefecturae, military of civitates, 69 f., 134 f.
praepositi, native in Dalmatia, 135
praesides, 273, 276, 310
praesidia, on Danube bank, 196 f.
Praetentura Italiae et Alpium, 187
praetoria, see road-stations
Praetorian guard, praetorians, 117, 214, 222; Cotini in, 248; Illyriciani in, 200 f., 204
prata legionis, 140
priests, Christian, 326 f., 332; of Dolichenus, 256, 325
pr(inceps) c(ivitatis) B(oiorum), 135
praef(ectus) Scord(iscorum), 134 f.
principes, native, 70, 134 ff., 141 f., 159, 175
primipili, 69
primipilus leg. V. Mac. praefectus civitatium Moesiae et Treballiae, 69

procurators, of mines, 131, 133, 308; provincial, 41
protectores, Illyriciani as, 210
publicum portorium Illyrici, 200, 255
punches, 177

quattuorvirate, of municipium, 144

ramparts, at forts, 91, 106 f., 284
reading, teaching of, 262 f.
receptio, of barbarians into empire, 186, 189, 192, 209, 268 f., 278 ff., 288, 290 f., 296, 339, 342, 345
recruitment, *see* auxiliary troops; legions; praetorian guard
Regnum Noricum, 76
Thraciae, 68
relics, of martyrs, 327 f., 348, 353
reliefs, bronze, 148, 337; on column of Marcus, 175, 188; on column of Trajan, 107; cult, 253, 336; in Mithraeum, 256; of Rider-gods, 254, 324; of Silvanus, 250, 252; *and see* tombstones
religion, cults, 142, 153, 181 f., 232 ff., 250 ff., 308, 322 ff., 325, 336, 338; *see also* Christianity; paganism
res privata, administrator, 306
rings, 320
roads, amber road, 71 f., 74, 76 f., 115, 117 f., 120, 122, 152; army units stationed on, 50 f.; building, 43 ff., 100 f., 109, 280 ff.; Danube road, 45 ff., 96 ff., 105 ff., 109, 119, 153, 225, 284; development of towns and settlements along, 72, 74, 98, 115, 117 ff., 124, 137, 143 f., 196, 213 ff., 225, 241; trade along, 71, 120 ff., 137, 152, 246
road-stations (*praetoria*), 100, 106, 215, 298
robbers, *see latrones*
Roman buildings beyond frontier, 89 ff.
Romanization, 70 f., 79, 113, 117, 124, 130, 147 ff., 176, 181 f., 234 ff., 247 ff., 253, 256, 259, 263, 319 f., 336, 338
roofs, of houses, 173
rosettes, on tombstones, 180

sailors, settlement of, 113 f
Sancti Quattuor Coronati, 326 f.
sanctuaries, *see* temples
sarcophagi, 237 ff., 246, 264, 310, 325, 334
schools, 262
scribae, municipal, 144 f., 153, 222, 263
Scrofulae, rapids on Danube, 45; *see also iter Scorfularum*
sculptors, foreign, 181, 337
sculpture, 209, 244, 336; ivory, 245; marble, 303; porphyry, 303; stone, 100, 124 f., 179 ff., 234, 246, 297, 311
sea, products of, 28
sees, of bishops, 214, 222, 303, 310, 312, 328 ff., 336, 352
serfs, of Dardanians, 27

PLATE 1a (above)   Greek bronze vessel from the Celtic cemetery at Szob (p. 7)

PLATE 1b (below)   Late Celtic vessel with Latin inscription DA BIBERE from the Romano-Celtic cemetery at Cserszegtomaj (p. 124)

PLATE 2   Some types of barbarian coins (obverse) (1/1) (pp. 28–30, 56)

PLATE 3   Reverses of coins on Plate 2

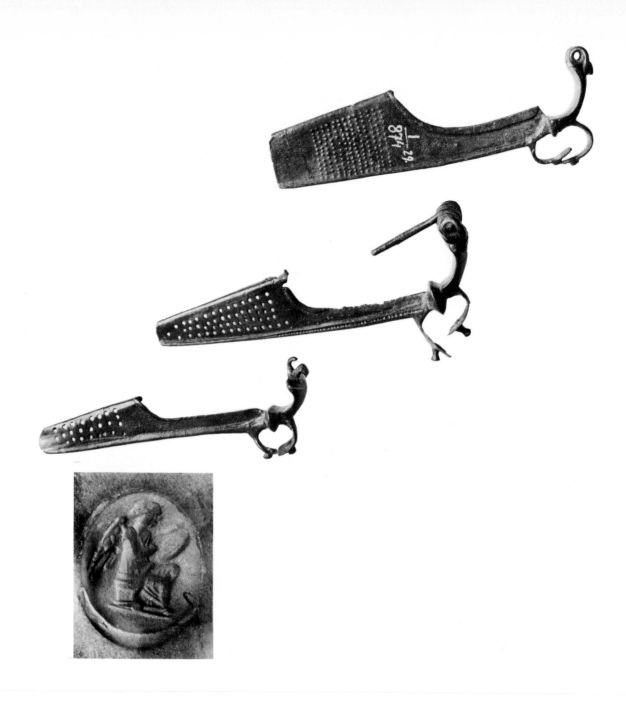

PLATE 4a (above)   Norican-Pannonian fibulae (p. 63)

PLATE 4b (below)   The impression of an intaglio showing Victoria on a late Celtic red-painted vessel from Aquincum (4/1) (p. 72)

PLATE 5a (right)   Tombstone of Scerviaedus, son of Sita, slain by robbers (pp. 65, 180)

PLATE 5b (centre)   Tombstone of Atta, son of Bataio, a trader from Savaria (pp. 78f., 152)

PLATE 5c (left)   Tombstone of Tiberius Satto of Cambodunum (Kempten), veteran of legio X Gemina from Aquincum (p. 162)

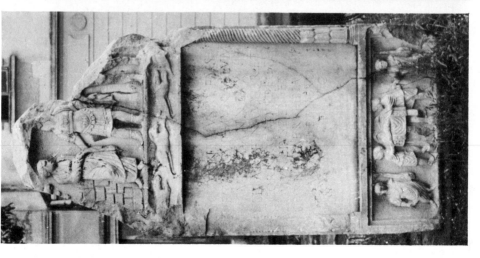

PLATE 6a (right)   Tombstone of Verondacus, son of Veruicus from Törökbálint

PLATE 6b (centre) Tombstone of Cornelius Zosimus from Viminacium (p.130)

PLATE 6c (left)   Tombstone of Cornelius Rufus from Viminacium (pp. 180f., 261)

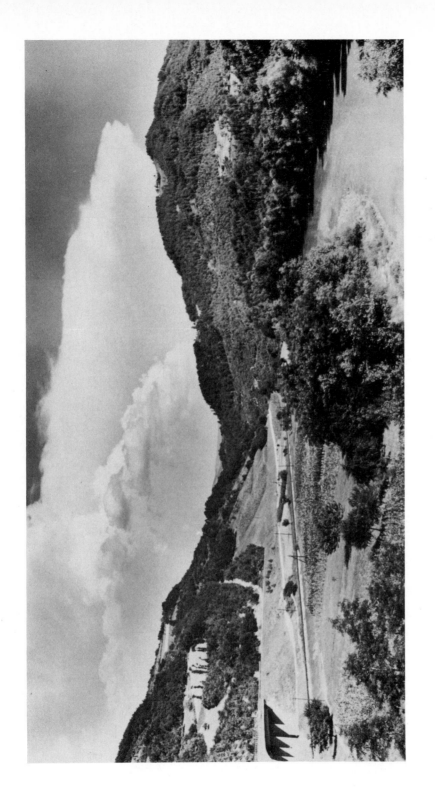

PLATE 7   The Morava valley

PLATE 8a (above) The late Roman fort at Boljetin in the Djerdap (p. 107)
PLATE 8b (below) View of the Djerdap (p. 45)

PLATE 9a (above) The rock-cut Roman road in the Djerdap (pp. 43ff.)
PLATE 9b (below) Trajan's road-inscription (the *Tabula Traiani*) in the Djerdap (p. 47)

PLATE 10a (right) Tombstone of L. Licinius Lepidus from Dozmat

PLATE 10b (left) Tombstone of Reginus son of Troucetissa of Trier, from Aquincum (pp. 124f.)

PLATE 11a (right)   Tombstone of C. Sextilius Senecio, decurion of Scarbantia

PLATE 11b (centre)   Tombstone of Sempronius Marcellinus and his family from Savaria (pp. 179f., 261)

PLATE 11c (left)   Tombstone of Aurelius Rufinus from Intercisa

PLATE 12a
Tombstone of
Aelius Munatius
from Intercisa

PLATE 12b (left)   Tombstone of Aelius Septimus from Brigetio
PLATE 12c (right)   Tombstone of Bozi daughter of Vellasa from Ercsi

PLATE 13a (left)   Stone slab from an *aedicula* at Bölcske (p. 261)

PLATE 13b (right)   Coins of Trajan referring to the mines (pp. 131–3)
(i) left, *RIC* 706, *metalli Ulpiani Pan* (2/1)
(ii) right, *RIC* 704, *Dardanici* (2/1)

PLATE 14a (above)   The mountains of the *metalla Ulpiana* near Ulpiana (Gračanica)
(p. 131)
PLATE 14b (below)   The Danube at Visegrád (p. 293)

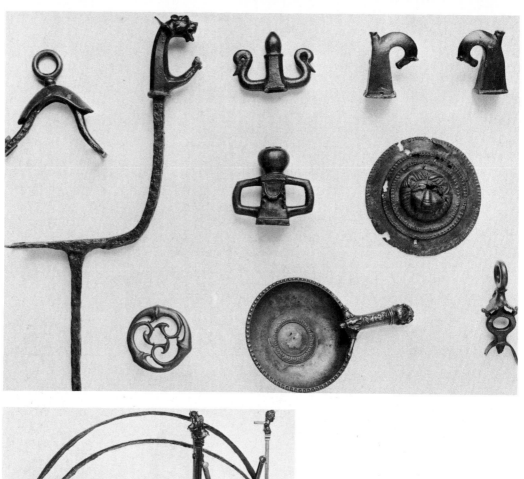

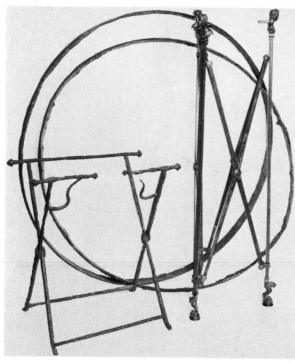

PLATE 15a, b Finds
from cart-graves
(p. 148)

PLATE 16a (above)   Aquincum municipium: the *macellum* (p. 161)

PLATE 16b (below)   Aquincum municipium: the 'big house' (pp. 161, 166)

PLATE 17a, b (opposite)   Aquincum municipium: the so-called 'large baths' (p. 161)

PLATE 18a (above)   Carnuntum municipium: House VI (pp. 162–4)

PLATE 18b (below)   Carnuntum municipium: House IV

PLATE 19a (above)   Carnuntum: the civil amphitheatre (p. 164)
PLATE 19b (below)   Carnuntum: the four-way arch (Heidentor)

PLATE 20a (above)   Excavations at the municipium Dardanorum: the *horreum* (p. 223)

PLATE 20b (below)   The Roman city wall of Scarbantia as reconstructed in the Middle Ages (pp. 166, 312)

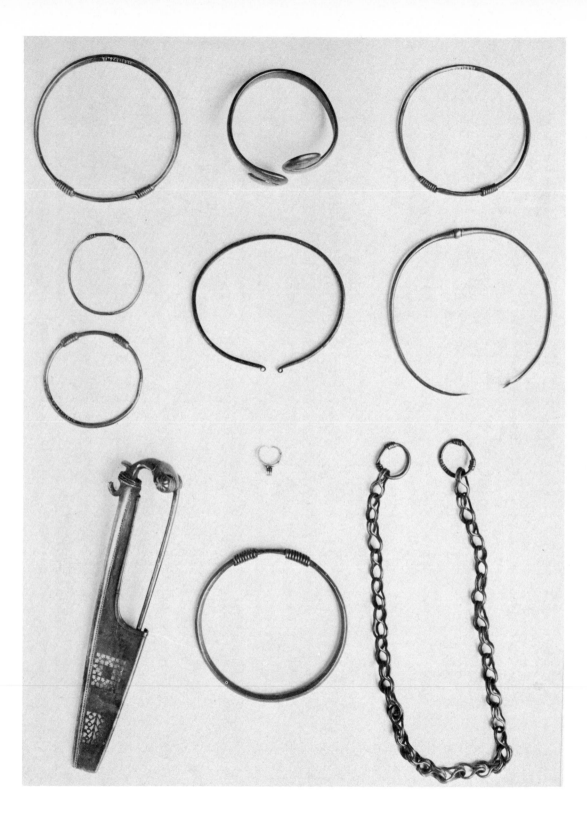

PLATE 21   The Ászár hoard (p. 175)
(a) (opposite) Jewelry   (b) (above) Bronze Vessel 1   (c) (below) Bronze Vessel 3

PLATE 22 The military diploma from the Ászár hoard (pp. 155, 175)

PLATE 23a   The Nymphaeum of
the Sanctuary at Gorsium (p. 175)

PLATE 23b Sarcophagus of *cline*
type from Viminacium

PLATE 24a (above)   Some types of Pannonian stamped pottery (pp. 176f.)

PLATE 24b (below)   Samian ware from the 'Siscia' pottery, found at Viminacium
(p. 178)

PLATE 25a A mould of the potter Pacatus from Aquincum (p. 176)

PLATE 25b The Emperor Marcus Aurelius: stamp for a potter's mould from Aquincum

PLATE 26a (left)   Tombstone of Trebia Lucia from Timacum minus (p. 180)

PLATE 26b (right)   Tombstone of Aelia Clementilla from Ulpiana

PLATE 27a (above)   Scarbantia: the Capitoline Triad, Juppiter, Juno and Minerva
(pp. 181, 244, 324)
PLATE 27b (below)   Savaria: relief from the Iseum (p. 181)

PLATE 28a, b (above)   Mosaic pavements from Aquincum (p. 237)

PLATE 29a (opposite above)   The sarcophagus of Pia Celerina from Aquincum (p. 237)

PLATE 29b (opposite centre)   The sarcophagus of the scholasticus L. Septimius Fuscus from Aquincum (p. 262)

PLATE 29c (opposite below)   The sarcophagus of the *Interpres Dacorum* from Brigetio (p. 199)

PLATE 30a   Statuette of Negro boy
from Aquincum

PLATE 30b Burial-vault from
Brestovik, east of Singidunum
(p. 238)

PLATE 31a Stone slab with relief of the god Silvanus from Aquincum (p. 251)

PLATE 31b Relief of Diana from Balatonvilágos (pp. 251ff.)

PLATE 32a   Stone slab with mythological scene: Bellerophon and the Chimaera.
From Intercisa (pp. 261f.)

PLATE 32b   Stone slab with mythological scene: Aeneas escaping from Troy.
From Intercisa (pp. 261f.)

PLATE 33a (above)   Stone slab with mythological scene: Priam and Achilles. From Aquincum (pp. 261ff.)

PLATE 33b (below)   Stone slab with mythological scene: Achilles and Hector (pp. 261ff.)

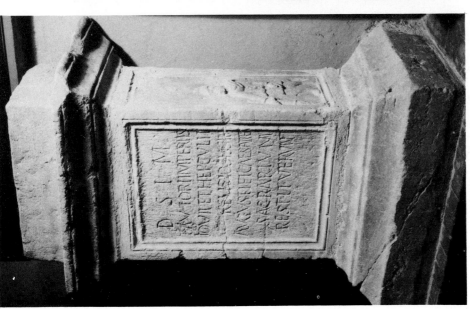

PLATE 34a (right) Lead tablet showing the Danubian Rider-gods (p. 254)

PLATE 34b (left) Altar erected at Carnuntum to Mithras by the Tetrarchs in 308 (pp. 276, 324)

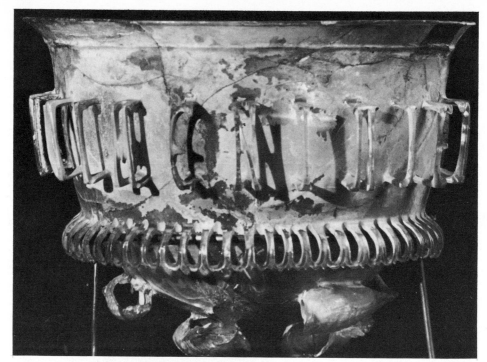

PLATE 35a Glass *vas diatretum* from Szekszárd

PLATE 35b Silver vessel of Licinius from Esztergom (pp. 277, 300)

PLATE 36a Late Roman fort at Tokod (p. 317)

PLATE 36b Valentinianic *burgus* at Visegrád (p. 292)

PLATE 37a, b Gamzigrad: late Roman imperial palace (p. 303)

PLATE 38a (above)   The early Christian basilica at Fenékpuszta (pp. 303, 336)

PLATE 38b (below)   The early Christian basilica at Ulpianum (pp. 332, 334)

PLATE 39a (above) Sirmium: the late Roman baths (p. 312)
PLATE 39b (below) Sirmium: part of the late Roman palace (p. 312)

PLATE 40a Sopianae (Pécs): painted tomb No. 2 (pp. 313, 335)

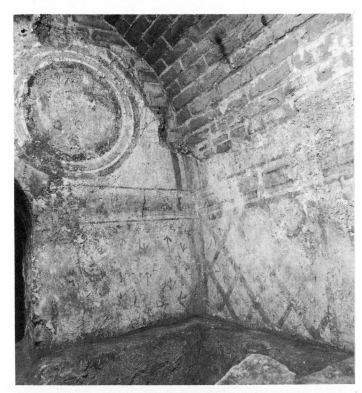

PLATE 40b Sopianae (Pécs): tomb with painting of SS. Peter and Paul (pp. 313, 334)

PLATE 41a (above)   Sirmium: fragments of a *cancellum* (screen) from a Christian basilica, built into a modern wall

PLATE 41b (below)   Painted tomb in the Christian cemetery at Jagodin Mahala

PLATE 42a, b   Details of painted figures in the Christian cemetery of Jagodin Mahala

PLATE 43a (right)   Painted figure from a tomb at Beška
PLATE 43b (above)   Mosaic from the Constantinian villa of
Mediana (p. 302)

PLATE 44a (above)   The augur's staff from Brigetio (p. 325)

PLATE 44b (centre)   Bronze plate from late Roman box: Kisárpás (pp. 320, 337)

PLATE 44c (below)   Bronze plates from late Roman boxes: Ságvár (pp. 320, 337)

PLATE 45a (left)   Silver tripod from Polgárdi (p. 337)

PLATE 45b (right)   Late Roman inscribed gold pin from Fenékpuszta (p. 353)